P L Green

INTRODUCTION TO
PERTURBATION TECHNIQUES

Introduction to Perturbation Techniques

ALI HASAN NAYFEH

University Distinguished Professor
Virginia Polytechnic Institute and State University
Blacksburg, Virginia
and
Yarmouk University, Irbid, Jordan

Wiley Classics Library Edition Published 1993

A Wiley-Interscience Publication
JOHN WILEY & SONS, INC.
New York • Chichester • Brisbane • Toronto • Singapore

This text is printed on acid-free paper.

Copyright © 1981 by John Wiley & Sons, Inc.

Wiley Classics Library Edition Published 1993.

All rights reserved. Published simultaneously in Canada.

Reproduction or translation of any part of this work beyond that permitted by Sections 107 or 108 of the 1976 United States Copyright Act without the permission of the copyright owner is unlawful. Requests for permission or further information should be addressed to the Permissions Department, John Wiley & Sons, Inc.

Library of Congress Cataloging in Publication Data:

Nayfeh, Ali Hasan, 1933–

   Introduction to perturbation techniques.

   "A Wiley-Interscience publication."
   Bibliography: p.
   Includes index.
   1. Differential equations—Numerical solutions.
2. Equations—Numerical solutions. 3. Perturbation (Mathematics) I. Title.

QA371.N32   515.3′5   80-15233
ISBN 0-471-31013-1

Printed in the United States of America

10 9 8 7 6 5 4 3 2 1

To my parents Hasan and Khadrah
my wife Samirah
and my children Mahir, Tariq, Samir, and Nader

Preface

Many of the problems facing physicists, engineers, and applied mathematicians involve such difficulties as nonlinear governing equations, variable coefficients, and nonlinear boundary conditions at complex known or unknown boundaries that preclude solving them exactly. Consequently, solutions are approximated using numerical techniques, analytic techniques, and combinations of both. Foremost among the analytic techniques are the systematic methods of perturbations (asymptotic expansions) in terms of a small or a large parameter or coordinate. This book is concerned only with these perturbation techniques.

The author's book *Perturbation Methods* presents in a unified way an account of most of the perturbation techniques, pointing out their similarities, differences, and advantages, as well as their limitations. Although the techniques are described by means of examples that start with simple ordinary equations that can be solved exactly and progress toward complex partial-differential equations, the material is concise and advanced and therefore is intended for researchers and advanced graduate students only. The purpose of this book, however, is to present the material in an elementary way that makes it easily accessible to advanced undergraduates and first-year graduate students in a wide variety of scientific and engineering fields. As a result of teaching perturbation methods for eight years to first-year and advanced graduate students at Virginia Polytechnic Institute and State University, I have selected a limited number of techniques and amplified their description considerably. Also I have attempted to answer the questions most frequently raised by my students. The techniques are described by means of simple examples that consist mainly of algebraic and ordinary-differential equations.

The material in Chapters 3 and 15 and Appendices A and B cannot be found in *Perturbation Methods*. Chapter 3 discusses asymptotic expansions of integrals. Chapter 15 is devoted to the determination of the adjoints of homogeneous linear equations (algebraic, ordinary-differential, partial-differential, and integral equations) and the solvability conditions of linear inhomogeneous problems. Appendix A summarizes trigonometric identities, and Appendix B summarizes the properties of linear ordinary-differential equations and describes the symbolic method of determining the solutions of homogeneous and inhomogeneous ordinary-differential equations with constant coefficients.

The reader should have a background in calculus and elementary ordinary-differential equations.

Each chapter contains a number of exercises. For more exercises, the reader is referred to *Perturbation Methods* by Nayfeh and *Nonlinear Oscillations* by Nayfeh and Mook. Since this book is elementary, only a list of the pertinent books is included in the bibliography without any attempt of citing them in the text.

I am indebted to K. R. Asfar and D. T. Mook for reading the whole manuscript and to L. Watson, M. Williams, C. Prather, S. A. Ragab, I. Wickman, A. Yen, Y. Liu, H. Reed, J. Dederer, Y. Ma, and W. S. Saric for reading parts of the manuscript. Many of the figures were drawn by T. H. Nayfeh, K. R. Asfar, I. Wickman, T. Dunyak, and T. McCawly; and I wish to express my appreciation to them. Finally, I wish to thank Patty Belcher, Janet Bryant, and Sharon Larkins for typing the manuscript.

ALI HASAN NAYFEH

Blacksburg, Virginia
April 1980

Contents

1 **INTRODUCTION**   1

    1.1 Dimensional Analysis, 1
    1.2 Expansions, 10
    1.3 Gauge Functions, 12
    1.4 Order Symbols, 17
    1.5 Asymptotic Series, 18
    1.6 Asymptotic Expansions and Sequences, 22
    1.7 Convergent Versus Asymptotic Series, 23
    1.8 Elementary Operations on Asymptotic Expansions, 24
        Exercises, 24

2 **ALGEBRAIC EQUATIONS**   28

    2.1 Quadratic Equations, 28
    2.2 Cubic Equations, 39
    2.3 Higher-Order Equations, 43
    2.4 Transcendental Equations, 45
        Exercises, 48

3 **INTEGRALS**   51

    3.1 Expansion of Integrands, 52
    3.2 Integration by Parts, 56
    3.3 Laplace's Method, 65
    3.4 The Method of Stationary Phase, 79

3.5 The Method of Steepest Descent, 88
Exercises, 101

4 THE DUFFING EQUATION 107

4.1 The Straightforward Expansion, 109
4.2 Exact Solution, 113
4.3 The Lindstedt-Poincaré Technique, 118
4.4 The Method of Renormalization, 121
4.5 The Method of Multiple Scales, 122
4.6 Variation of Parameters, 127
4.7 The Method of Averaging, 129
Exercises, 131

5 THE LINEAR DAMPED OSCILLATOR 134

5.1 The Straightforward Expansion, 135
5.2 Exact Solution, 136
5.3 The Lindstedt-Poincaré Technique, 139
5.4 The Method of Multiple Scales, 142
5.5 The Method of Averaging, 144
Exercises, 146

6 SELF-EXCITED OSCILLATORS 147

6.1 The Straightforward Expansion, 148
6.2 The Method of Renormalization, 151
6.3 The Method of Multiple Scales, 152
6.4 The Method of Averaging, 155
Exercises, 157

7 SYSTEMS WITH QUADRATIC AND CUBIC NONLINEARITIES 159

7.1 The Straightforward Expansion, 160

7.2 The Method of Renormalization, 162
7.3 The Lindstedt-Poincaré Technique, 164
7.4 The Method of Multiple Scales, 166
7.5 The Method of Averaging, 168
7.6 The Generalized Method of Averaging, 169
7.7 The Krylov-Bogoliubov-Mitropolsky Technique, 173
Exercises, 175

8 GENERAL WEAKLY NONLINEAR SYSTEMS   177

8.1 The Straightforward Expansion, 177
8.2 The Method of Renormalization, 179
8.3 The Method of Multiple Scales, 181
8.4 The Method of Averaging, 182
8.5 Applications, 184
Exercises, 188

9 FORCED OSCILLATIONS OF THE DUFFING EQUATION   190

9.1 The Straightforward Expansion, 191
9.2 The Method of Multiple Scales, 193
    9.2.1 Secondary Resonances, 193
    9.2.2 Primary Resonance, 205
9.3 The Method of Averaging, 209
    9.3.1 Secondary Resonances, 209
    9.3.2 Primary Resonance, 212
Exercises, 213

10 MULTIFREQUENCY EXCITATIONS   216

10.1 The Straightforward Expansion, 216
10.2 The Method of Multiple Scales, 219
    10.2.1 The Case $\omega_2 + \omega_1 \approx 1$, 220

10.2.2 The Case $\omega_2 - \omega_1 \approx 1$ and $\omega_1 \approx 2$, 222

10.3 The Method of Averaging, 226

10.3.1 The Case $\omega_1 + \omega_2 \approx 1$, 230

10.3.2 The Case $\omega_2 - \omega_1 \approx 1$ and $\omega_1 \approx 2$, 230

Exercises, 230

11 THE MATHIEU EQUATION 234

11.1 The Straightforward Expansion, 235

11.2 The Floquet Theory, 236

11.3 The Method of Strained Parameters, 243

11.4 Whittaker's Method, 247

11.5 The Method of Multiple Scales, 249

11.6 The Method of Averaging, 253

Exercises, 254

12 BOUNDARY-LAYER PROBLEMS 257

12.1 A Simple Example, 257

12.2 The Method of Multiple Scales, 268

12.3 The Method of Matched Asymptotic Expansions, 270

12.4 Higher Approximations, 279

12.5 Equations with Variable Coefficients, 284

12.6 Problems with Two Boundary Layers, 296

12.7 Multiple Decks, 304

12.8 Nonlinear Problems, 307

Exercises, 320

13 LINEAR EQUATIONS WITH VARIABLE COEFFICIENTS 325

13.1 First-Order Scalar Equations, 326

13.2 Second-Order Equations, 329

13.3 Solutions Near Regular Singular Points, 331

13.4 Singularity at Infinity, 342
13.5 Solutions Near an Irregular Singular Point, 344
Exercises, 355

14 DIFFERENTIAL EQUATIONS WITH A LARGE PARAMETER 360

14.1 The WKB Approximation, 361
14.2 The Liouville-Green Transformation, 364
14.3 Eigenvalue Problems, 366
14.4 Equations with Slowly Varying Coefficients, 369
14.5 Turning-Point Problems, 370
14.6 The Langer Transformation, 375
14.7 Eigenvalue Problems with Turning Points, 379
Exercises, 383

15 SOLVABILITY CONDITIONS 388

15.1 Algebraic Equations, 389
15.2 Nonlinear Vibrations of Two-Degree-of-Freedom Gyroscopic Systems, 394
15.3 Parametrically Excited Gyroscopic Systems, 397
15.4 Second-Order Differential Systems, 401
15.5 General Boundary Conditions, 406
15.6 A Simple Eigenvalue Problem, 412
15.7 A Degenerate Eigenvalue Problem, 414
15.8 Acoustic Waves in a Duct with Sinusoidal Walls, 418
15.9 Vibrations of Nearly Circular Membranes, 426
15.10 A Fourth-Order Differential System, 432
15.11 General Fourth-Order Differential Systems, 438
15.12 A Fourth-Order Eigenvalue Problem, 441
15.13 A Differential System of Equations, 445
15.14 General Differential Systems of First-Order Equations, 447
15.15 Differential Systems with Interfacial Boundary Conditions, 452

15.16 Integral Equations, 454
15.17 Partial-Differential Equations, 458
       Exercises, 462

APPENDIX A   TRIGONOMETRIC IDENTITIES   472

APPENDIX B   LINEAR ORDINARY-DIFFERENTIAL EQUATIONS   480

BIBLIOGRAPHY   501

INDEX   507

CHAPTER 1

Introduction

1.1. Dimensional Analysis

Exact solutions are rare in many branches of fluid mechanics, solid mechanics, motion, and physics because of nonlinearities, inhomogeneities, and general boundary conditions. Hence, engineers, physicists, and applied mathematicians are forced to determine approximate solutions of the problems they are facing. These approximations may be purely numerical, purely analytical, or a combination of numerical and analytical techniques. In this book, we concentrate on the purely analytical techniques, which, when combined with a numerical technique such as a finite-difference or a finite-element technique, yield very powerful and versatile techniques.

The key to solving modern problems is mathematical modeling. This process involves keeping certain elements, neglecting some, and approximating yet others. To accomplish this important step, one needs to decide the order of magnitude (i.e., smallness or largeness) of the different elements of the system by comparing them with each other as well as with the basic elements of the system. This process is called *nondimensionalization* or making the variables *dimensionless*. Consequently, one should always introduce dimensionless variables before attempting to make any approximations. For example, if an element has a length of one centimeter, would this element be large or small? One cannot answer this question without knowing the problem being considered. If the problem involves the motion of a satellite in an orbit around the earth, then one centimeter is very very small. On the other hand, if the problem involves intermolecular distances, then one centimeter is very very large. As a second example, is one gram small or large? Again one gram is very very small compared with the mass of a satellite but it is very very large compared with the mass of an electron. Therefore, expressing the equations in dimensionless form brings out the important dimensionless parameters that govern the behavior of the system. Even if one is not interested in approximations, it is recommended that one perform this important step before analyzing the system or presenting

2 INTRODUCTION

experimental data. Next, we give a few examples illustrating the process of nondimensionalization.

EXAMPLE 1

We consider the motion of a particle of mass m restrained by a linear spring having the constant k and a viscous damper having the coefficient μ, as shown in Figure 1-1. Using Newton's second law of motion, we have

$$m\frac{d^2u}{dt^2} + \mu\frac{du}{dt} + ku = 0 \qquad (1.1)$$

where u is the displacement of the particle and t is time. Let us assume that the particle was released from rest from the position u_0 so that the initial conditions are

$$u(0) = u_0 \qquad \frac{du}{dt}(0) = 0 \qquad (1.2)$$

In this case, u is the dependent variable and t is the independent variable. They need to be made dimensionless by using a characteristic distance and a characteristic time of the system. The displacement u can be made dimensionless by using the initial displacement u_0 as a characteristic distance, whereas the time t can be made dimensionless by using the inverse of the system's natural frequency $\omega_0 = \sqrt{k/m}$. Thus, we put

$$u^* = \frac{u}{u_0} \qquad t^* = \omega_0 t$$

where the asterisk denotes dimensionless quantities. Then,

$$\frac{du}{dt} = \frac{d(u_0 u^*)}{dt^*} \cdot \frac{dt^*}{dt} = \omega_0 u_0 \frac{du^*}{dt^*}$$

$$\frac{d^2u}{dt^2} = \omega_0^2 u_0 \frac{d^2u^*}{dt^{*2}}$$

so that (1.1) becomes

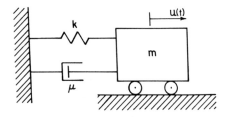

Figure 1-1. A mass restrained by a spring and a viscous damper.

$$m\omega_0^2 u_0 \frac{d^2 u^*}{dt^{*2}} + \mu\omega_0 u_0 \frac{du^*}{dt^*} + k u_0 u^* = 0$$

Hence,

$$\frac{d^2 u^*}{dt^{*2}} + \mu^* \frac{du^*}{dt^*} + \frac{k}{m\omega_0^2} u^* = 0$$

or

$$\frac{d^2 u^*}{dt^{*2}} + \mu^* \frac{du^*}{dt^*} + u^* = 0 \tag{1.3}$$

where

$$\mu^* = \frac{\mu}{m\omega_0} \tag{1.4}$$

In terms of the above dimensionless quantities, (1.2) becomes

$$u^*(0) = 1 \quad \text{and} \quad \frac{du^*}{dt^*}(0) = 0 \tag{1.5}$$

Thus, the solution to the present problem depends only on the single parameter μ^*, which represents the ratio of the damping force to the inertia force or the restoring force of the spring. If this ratio is small, then one can use the dimensionless quantity μ^* as the small parameter in obtaining an approximate solution of the problem, and we speak of a lightly damped system. We should note that the system cannot be considered lightly damped just because μ is small; $\mu^* = \mu/m\omega_0 = \mu/\sqrt{km}$ must be small.

EXAMPLE 2
Let us assume that the spring force is a nonlinear function of u according to

$$f_{\text{spring}} = ku + k_2 u^2 \tag{1.6}$$

where k and k_2 are constants. Then, (1.1) becomes

$$m\frac{d^2 u}{dt^2} + \mu \frac{du}{dt} + ku + k_2 u^2 = 0 \tag{1.7}$$

Again, using the same dimensionless quantities as in the preceding example, we have

$$mu_0 \omega_0^2 \frac{d^2 u^*}{dt^{*2}} + \mu u_0 \omega_0 \frac{du^*}{dt^*} + k u_0 u^* + k_2 u_0^2 u^{*2} = 0$$

or

4 INTRODUCTION

$$\frac{d^2 u^*}{dt^{*2}} + \mu^* \frac{du^*}{dt^*} + u^* + \epsilon u^{*2} = 0 \tag{1.8}$$

where

$$\mu^* = \frac{\mu}{m\omega_0} \quad \text{and} \quad \epsilon = \frac{k_2 u_0}{k} \tag{1.9}$$

The initial conditions transform as in (1.5). Thus, the present problem is a function of the two dimensionless parameters μ^* and ϵ. As before, μ^* represents the ratio of the damping force to the inertia force or the linear restoring force. The parameter ϵ represents the ratio of the nonlinear and linear restoring forces of the spring.

When we speak of a weakly nonlinear system, we mean that $k_2 u_0 /k$ is small. Even if k_2 is small compared with k, the nonlinearity will not be small if u_0 is large compared with k/k_2. Thus, ϵ is the parameter that characterizes the nonlinearity.

EXAMPLE 3

As a third example, we consider the motion of a spaceship of mass m that is moving in the gravitational field of two fixed mass-centers whose masses m_1 and m_2 are much much bigger than m. With respect to the Cartesian coordinate system shown in Figure 1-2, the equations of motion are

$$m \frac{d^2 x}{dt^2} = -\frac{mm_1 Gx}{(x^2 + y^2)^{3/2}} - \frac{mm_2 G(x-L)}{[(x-L)^2 + y^2]^{3/2}} \tag{1.10}$$

$$m \frac{d^2 y}{dt^2} = -\frac{mm_1 Gy}{(x^2 + y^2)^{3/2}} - \frac{mm_2 Gy}{[(x-L)^2 + y^2]^{3/2}} \tag{1.11}$$

where t is the time, G is the gravitational constant, and L is the distance between m_1 and m_2.

In this case, the dependent variables are x and y and the independent variable is t. Clearly, a characteristic length of the problem is L, the distance between the two mass centers. A characteristic time of the problem is not as obvious. Since

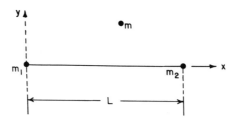

Figure 1-2. A satellite in the gravitational field of two fixed mass centers.

DIMENSIONAL ANALYSIS

the motions of the masses m_1 and m_2 are assumed to be independent of that of the spaceship, m_1 and m_2 move about their center of mass in ellipses. The period of oscillation is

$$T = \frac{2\pi L^{3/2}}{\sqrt{G(m_1 + m_2)}}$$

so that the frequency of oscillation is

$$\omega_0 = L^{-3/2}\sqrt{G(m_1 + m_2)} \quad (1.12)$$

Thus, we use the inverse of ω_0 as a characteristic time. Then, we introduce dimensionless quantities defined by

$$x^* = \frac{x}{L} \qquad y^* = \frac{y}{L} \qquad t^* = \omega_0 t \quad (1.13)$$

so that

$$\frac{dx}{dt} = \frac{d(x^*L)}{dt^*}\frac{dt^*}{dt} = L\omega_0\frac{dx^*}{dt^*} \qquad \frac{d^2x}{dt^2} = L\omega_0^2\frac{d^2x^*}{dt^{*2}}$$

$$\frac{dy}{dt} = \frac{d(y^*L)}{dt^*}\frac{dt^*}{dt} = L\omega_0\frac{dy^*}{dt^*} \qquad \frac{d^2y}{dt^2} = L\omega_0^2\frac{d^2y^*}{dt^{*2}}$$

Hence, (1.10) and (1.11) become

$$mL\omega_0^2\frac{d^2x^*}{dt^{*2}} = -\frac{mm_1 GLx^*}{[L^2(x^{*2} + y^{*2})]^{3/2}} - \frac{mm_2 GL(x^* - 1)}{[L^2(x^* - 1)^2 + L^2 y^{*2}]^{3/2}}$$

$$mL\omega_0^2\frac{d^2y^*}{dt^{*2}} = -\frac{mm_1 GLy^*}{[L^2(x^{*2} + y^{*2})]^{3/2}} - \frac{mm_2 GLy^*}{[L^2(x^* - 1)^2 + L^2 y^{*2}]^{3/2}}$$

or

$$\frac{d^2x^*}{dt^{*2}} = -\frac{m_1 G}{L^3\omega_0^2}\frac{x^*}{(x^{*2} + y^{*2})^{3/2}} - \frac{m_2 G}{L^3\omega_0^2}\frac{(x^* - 1)}{[(x^* - 1)^2 + y^{*2}]^{3/2}} \quad (1.14)$$

$$\frac{d^2y^*}{dt^{*2}} = -\frac{m_1 G}{L^3\omega_0^2}\frac{y^*}{(x^{*2} + y^{*2})^{3/2}} - \frac{m_2 G}{L^3\omega_0^2}\frac{y^*}{[(x^* - 1)^2 + y^{*2}]^{3/2}} \quad (1.15)$$

Using (1.12), we have

$$\frac{m_1 G}{L^3\omega_0^2} = \frac{m_1}{m_1 + m_2} \qquad \frac{m_2 G}{L^3\omega_0^2} = \frac{m_2}{m_1 + m_2}$$

Hence, if we put

$$\frac{m_2}{m_1 + m_2} = \epsilon \quad \text{then} \quad \frac{m_1}{m_1 + m_2} = 1 - \epsilon \quad (1.16)$$

and (1.14) and (1.15) become

$$\frac{d^2x^*}{dt^{*2}} = -\frac{(1-\epsilon)x^*}{(x^{*2}+y^{*2})^{3/2}} - \frac{\epsilon(x^*-1)}{[(x^*-1)^2+y^{*2}]^{3/2}} \tag{1.17}$$

$$\frac{d^2y^*}{dt^{*2}} = -\frac{(1-\epsilon)y^*}{(x^{*2}+y^{*2})^{3/2}} - \frac{\epsilon y^*}{[(x^*-1)^2+y^{*2}]^{3/2}} \tag{1.18}$$

Therefore, the problem depends only on the parameter ϵ, which is usually called the reduced mass. If m_1 represents the mass of the earth and m_2 the mass of the moon, then

$$\epsilon \approx \frac{\frac{1}{80}}{1+\frac{1}{80}} = \frac{1}{81}$$

which is small and can be used as a perturbation parameter in determining an approximate solution to the motion of a spacecraft in the gravitational field of the earth and the moon.

EXAMPLE 4

As a fourth example, we consider the vibration of a clamped circular plate of radius a under the influence of a uniform radial load. If w is the transverse displacement of the plate, then the linear vibrations of the plate are governed by

$$D\nabla^4 w - P\nabla^2 w - \rho\frac{\partial^2 w}{\partial t^2} = 0 \tag{1.19}$$

where t is the time, D is the plate rigidity, P is the uniform radial load, and ρ is the plate density per unit area. The boundary conditions are

$$\begin{aligned} w = 0 \quad & \frac{\partial w}{\partial r} = 0 \quad \text{at } r = a \\ w < \infty \quad & \quad \text{at } r = 0 \end{aligned} \tag{1.20}$$

In this case, w is the dependent variable and t and r are the independent variables. Clearly, a is a characteristic length of the problem. The characteristic time is assumed to be T and it is specified below. Then, we define dimensionless variables according to

$$w^* = \frac{w}{a} \quad r^* = \frac{r}{a} \quad t^* = \frac{t}{T}$$

Hence,

$$\frac{\partial w}{\partial r} = \frac{\partial(aw^*)}{\partial r^*}\frac{dr^*}{dr} = \frac{\partial w^*}{\partial r^*}$$

$$\frac{\partial w}{\partial \theta} = \frac{\partial(aw^*)}{\partial \theta} = a\frac{\partial w^*}{\partial \theta}$$

$$\frac{\partial w}{\partial t} = \frac{\partial(aw^*)}{\partial t^*}\frac{dt^*}{dt} = \frac{a}{T}\frac{\partial w^*}{\partial t^*}$$

Since

$$\nabla^2 = \frac{\partial^2}{\partial r^2} + \frac{1}{r}\frac{\partial}{\partial r} + \frac{1}{r^2}\frac{\partial^2}{\partial \theta^2}$$

(1.19) becomes

$$\frac{D}{a^3}\left(\frac{\partial^2}{\partial r^{*2}} + \frac{1}{r^*}\frac{\partial}{\partial r^*} + \frac{1}{r^{*2}}\frac{\partial^2}{\partial \theta^2}\right)^2 w^* - \frac{P}{a}\left(\frac{\partial^2}{\partial r^{*2}} + \frac{1}{r^*}\frac{\partial}{\partial r^*} + \frac{1}{r^{*2}}\frac{\partial^2}{\partial \theta^2}\right)w^*$$

$$-\frac{\rho a}{T^2}\frac{\partial^2 w^*}{\partial t^{*2}} = 0$$

or

$$\frac{D}{a^2 P}\nabla^{*4}w^* - \nabla^{*2}w^* - \frac{\rho a^2}{PT^2}\frac{\partial^2 w^*}{\partial t^{*2}} = 0 \qquad (1.21)$$

We can choose T to make the coefficient of $\partial^2 w^*/\partial t^{*2}$ equal to 1, that is, $T = a\sqrt{\rho/P}$. Then, (1.21) becomes

$$\epsilon \nabla^{*4}w^* - \nabla^{*2}w^* - \frac{\partial^2 w^*}{\partial t^{*2}} = 0 \qquad (1.22)$$

where

$$\epsilon = \frac{D}{a^2 P} \qquad (1.23)$$

In terms of dimensionless quantities, the boundary conditions (1.20) become

$$w^* = \frac{\partial w^*}{\partial r^*} = 0 \quad \text{at} \quad r^* = 1$$

$$w^* < \infty \quad \text{at} \quad r^* = 0 \qquad (1.24)$$

Therefore, the problem depends on the single dimensionless parameter ϵ. If the radial load is large compared with D/a^2, then ϵ is small and can be used as a perturbation parameter.

8 INTRODUCTION

EXAMPLE 5

As a final example, we consider steady incompressible flow past a flat plate. The problem is governed by

$$\frac{\partial u}{\partial x} + \frac{\partial v}{\partial y} = 0 \tag{1.25}$$

$$\rho \left(u \frac{\partial u}{\partial x} + v \frac{\partial u}{\partial y} \right) = -\frac{\partial p}{\partial x} + \mu \left(\frac{\partial^2 u}{\partial x^2} + \frac{\partial^2 u}{\partial y^2} \right) \tag{1.26}$$

$$\rho \left(u \frac{\partial v}{\partial x} + v \frac{\partial v}{\partial y} \right) = -\frac{\partial p}{\partial y} + \mu \left(\frac{\partial^2 v}{\partial x^2} + \frac{\partial^2 v}{\partial y^2} \right) \tag{1.27}$$

$$\begin{aligned} u = v = 0 \quad &\text{at} \quad y = 0 \\ u \to U_\infty, v \to 0 \quad &\text{as} \quad x \to -\infty \end{aligned} \tag{1.28}$$

where u and v are the velocity components in the x and y directions, respectively, p is the pressure, ρ is the density, and μ is the coefficient of viscosity.

In this case, u, v, and p are the dependent variables and x and y are the independent variables. To make the equations dimensionless, we use L as a characteristic length, where L is the distance from the leading edge to a specified point on the plate as shown in Figure 1-3, and use U_∞ as a characteristic velocity. We take ρU_∞^2 as a characteristic pressure. Thus, we define dimensionless quantities according to

$$u^* = \frac{u}{U_\infty} \quad v^* = \frac{v}{U_\infty} \quad p^* = \frac{p}{\rho U_\infty^2} \quad x^* = \frac{x}{L} \quad y^* = \frac{y}{L}$$

Then,

$$\frac{\partial u}{\partial x} = \frac{\partial (U_\infty u^*)}{\partial x^*} \frac{dx^*}{dx} = \frac{U_\infty}{L} \frac{\partial u^*}{\partial x^*} \quad \frac{\partial u}{\partial y} = \frac{U_\infty}{L} \frac{\partial u^*}{\partial y^*} \quad \frac{\partial^2 u}{\partial x^2} = \frac{U_\infty}{L^2} \frac{\partial^2 u^*}{\partial x^{*2}}$$

$$\frac{\partial^2 u}{\partial y^2} = \frac{U_\infty}{L^2} \frac{\partial^2 u^*}{\partial y^{*2}}$$

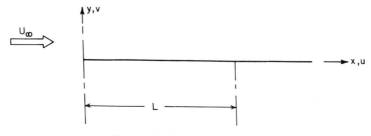

Figure 1-3. Flow past a flat plate.

$$\frac{\partial v}{\partial x} = \frac{U_\infty}{L} \frac{\partial v^*}{\partial x^*} \qquad \frac{\partial v}{\partial y} = \frac{U_\infty}{L} \frac{\partial v^*}{\partial y^*} \qquad \frac{\partial^2 v}{\partial x^2} = \frac{U_\infty}{L^2} \frac{\partial^2 v^*}{\partial x^{*2}} \qquad \frac{\partial^2 v}{\partial y^2} = \frac{U_\infty}{L^2} \frac{\partial^2 v^*}{\partial y^{*2}}$$

$$\frac{\partial p}{\partial x} = \frac{\rho U_\infty^2}{L} \frac{\partial p^*}{\partial x^*} \qquad \frac{\partial p}{\partial y} = \frac{\rho U_\infty^2}{L} \frac{\partial p^*}{\partial y^*}$$

Hence, (1.25) through (1.28) become

$$\frac{U_\infty}{L} \frac{\partial u^*}{\partial x^*} + \frac{U_\infty}{L} \frac{\partial v^*}{\partial y^*} = 0 \tag{1.29}$$

$$\frac{\rho U_\infty^2}{L} u^* \frac{\partial u^*}{\partial x^*} + \frac{\rho U_\infty^2}{L} v^* \frac{\partial u^*}{\partial y^*} = -\frac{\rho U_\infty^2}{L} \frac{\partial p^*}{\partial x^*} + \frac{\mu U_\infty}{L^2} \left(\frac{\partial^2 u^*}{\partial x^{*2}} + \frac{\partial^2 u^*}{\partial y^{*2}} \right) \tag{1.30}$$

$$\frac{\rho U_\infty^2}{L} u^* \frac{\partial v^*}{\partial x^*} + \frac{\rho U_\infty^2}{L} v^* \frac{\partial v^*}{\partial y^*} = -\frac{\rho U_\infty^2}{L} \frac{\partial p^*}{\partial y^*} + \frac{\mu U_\infty}{L^2} \left(\frac{\partial^2 v^*}{\partial x^{*2}} + \frac{\partial^2 v^*}{\partial y^{*2}} \right) \tag{1.31}$$

$$u^* = v^* = 0 \qquad \text{at } y^* = 0$$
$$U_\infty u^* \to U_\infty, v^* \to 0 \qquad \text{as } x^* \to -\infty \tag{1.32}$$

Equations (1.29) through (1.32) can be rewritten as

$$\frac{\partial u^*}{\partial x^*} + \frac{\partial v^*}{\partial y^*} = 0 \tag{1.33}$$

$$u^* \frac{\partial u^*}{\partial x^*} + v^* \frac{\partial u^*}{\partial y^*} = -\frac{\partial p^*}{\partial x^*} + \frac{1}{R} \left(\frac{\partial^2 u^*}{\partial x^{*2}} + \frac{\partial^2 u^*}{\partial y^{*2}} \right) \tag{1.34}$$

$$u^* \frac{\partial v^*}{\partial x^*} + v^* \frac{\partial v^*}{\partial y^*} = -\frac{\partial p^*}{\partial y^*} + \frac{1}{R} \left(\frac{\partial^2 v^*}{\partial x^{*2}} + \frac{\partial^2 v^*}{\partial y^{*2}} \right) \tag{1.35}$$

$$u^* = v^* = 0 \qquad \text{at} \qquad y^* = 0 \tag{1.36}$$

$$u^* \to 1 \qquad v^* \to 0 \qquad \text{as} \qquad x^* \to -\infty \tag{1.37}$$

where

$$R = \frac{\rho U_\infty L}{\mu} \tag{1.38}$$

is called the Reynolds number.

Equations (1.33) through (1.37) show that the problem depends only on the dimensionless parameter R. For the case of small viscosity, namely μ small compared with $\rho U_\infty L$, R is large and its inverse can be used as a perturbation parameter to determine an approximate solution of the present problem. This process leads to the widely used boundary-layer equations of fluid mechanics. When the flow is slow, namely $\rho U_\infty L$ is small compared with μ, R is small and it can be

10 INTRODUCTION

used as a perturbation parameter to construct an approximate solution of the present problem. This process leads to the Stokes-Oseen flow.

1.2 Expansions

In determining approximate solutions of algebraic, differential, and integral equations or evaluating integrals, we need to expand quantities in power series of a parameter or a variable. These power series expansions are usually obtained either as binomial expansions or Taylor series. These are explained next.

BINOMIAL THEOREM

Using straight multiplication, we have

$$(a + b)^2 = a^2 + 2ab + b^2$$
$$(a + b)^3 = a^3 + 3a^2 b + 3ab^2 + b^3$$
$$(a + b)^4 = a^4 + 4a^3 b + 6a^2 b^2 + 4ab^3 + b^4$$

The process can be generalized for general n as

$$(a + b)^n = a^n + na^{n-1}b + \frac{n(n-1)}{2!} a^{n-2}b^2 + \frac{n(n-1)(n-2)}{3!} a^{n-3}b^3 + \cdots \qquad (1.39a)$$

which can be rewritten as

$$(a + b)^n = \sum_{m=0}^{n} \frac{n!}{m!(n-m)!} a^{n-m} b^m \qquad (1.39b)$$

or

$$(a + b)^n = \sum_{m=0}^{n} {}^nC_m a^{n-m} b^m \quad \text{where } {}^nC_m = \frac{n!}{m!(n-m)!} \qquad (1.39c)$$

It turns out that (1.39a) terminates and hence it is valid when n is a positive integer. If it does not terminate, it is valid for any positive or negative number n provided that $|b/a|$ is less than 1; otherwise, the series diverges because

$$\lim_{m \to \infty} \frac{m\text{th term}}{(m-1)\text{th term}} = \lim_{m \to \infty} \frac{(m-1)!n(n-1)(n-2)\cdots(n-m+1)a^{n-m}b^m}{m!n(n-1)(n-2)\cdots(n-m+2)a^{n-m+1}b^{m-1}}$$

$$= \lim_{m \to \infty} \frac{(n-m+1)b}{ma} = -\frac{b}{a}$$

For example,

$$(a+b)^5 = \sum_{m=0}^{5} \frac{5!}{m!(5-m)!} a^{5-m} b^m$$

$$= a^5 + 5a^4 b + 10a^3 b^2 + 10a^2 b^3 + 5ab^4 + b^5$$

$$(a+b)^6 = \sum_{m=0}^{6} \frac{6!}{m!(6-m)!} a^{6-m} b^m$$

$$= a^6 + 6a^5 b + 15a^4 b^2 + 20a^3 b^3 + 15a^2 b^4 + 6ab^5 + b^6$$

$$(a+b)^{1/2} = a^{1/2} + \tfrac{1}{2} a^{-1/2} b + \frac{(\tfrac{1}{2})(-\tfrac{1}{2})}{2!} a^{-3/2} b^2 + \frac{(\tfrac{1}{2})(-\tfrac{1}{2})(-\tfrac{3}{2})}{3!} a^{-5/2} b^3 + \cdots$$

$$= a^{1/2} + \tfrac{1}{2} a^{-1/2} b - \tfrac{1}{8} a^{-3/2} b^2 + \tfrac{1}{16} a^{-5/2} b^3 + \cdots$$

$$(a+b)^{-1} = a^{-1} - a^{-2} b + \frac{(-1)(-2)}{2!} a^{-3} b^2 + \frac{(-1)(-2)(-3)}{3!} a^{-4} b^3 + \cdots$$

$$= a^{-1} - a^{-2} b + a^{-3} b^2 - a^{-4} b^3 + \cdots$$

We note that the first two series corresponding to $n = 5$ and 6 terminate. The last two series corresponding to $n = \tfrac{1}{2}$ and -1 do not terminate, and hence, they are valid only when $|b| < |a|$.

TAYLOR SERIES EXPANSIONS

If a function $f(x)$ is infinitely differentiable at $x = x_0$, we express it in a power series of $(x - x_0)$ as

$$f(x) = a_0 + a_1 (x - x_0) + a_2 (x - x_0)^2 + a_3 (x - x_0)^3 + \cdots$$

$$= \sum_{n=0}^{\infty} a_n (x - x_0)^n \qquad (1.41)$$

where the a_n are constants related to f and its derivatives at $x = x_0$. Putting $x = x_0$ in (1.41), we find that $a_0 = f(x_0)$. Differentiating (1.41) with respect to x, we have

$$f'(x) = a_1 + 2a_2 (x - x_0) + 3a_3 (x - x_0)^2 + 4a_4 (x - x_0)^3 + \cdots \qquad (1.42)$$

which, upon putting $x = x_0$, yields $a_1 = f'(x_0)$. Differentiating (1.42) with respect to x, we have

$$f''(x) = 2!a_2 + 3!a_3 (x - x_0) + 4 \cdot 3a_4 (x - x_0)^2 + \cdots \qquad (1.43)$$

which, upon putting $x = x_0$, yields $a_2 = (1/2!) f''(x_0)$. Differentiating (1.43) with respect to x gives

$$f'''(x) = 3!a_3 + 4!a_4 (x - x_0) + \cdots \qquad (1.44)$$

which, upon putting $x = x_0$, yields $a_3 = (1/3!) f'''(x_0)$. Continuing the process, we obtain

$$a_n = \frac{1}{n!} f^{(n)}(x_0) \qquad f^{(n)} = \frac{d^n f}{dx^n} \qquad (1.45)$$

and $f^{(0)} = f(x_0)$. Therefore, (1.41) can be rewritten as

$$f(x) = \sum_{n=0}^{\infty} \frac{f^{(n)}(x_0)}{n!} (x - x_0)^n \qquad (1.46)$$

which is called the *Taylor series expansion* of $f(x)$ about $x = x_0$.
Since

$$\frac{d}{dx}(\sin x) = \cos x \quad \text{and} \quad \frac{d}{dx}(\cos x) = -\sin x$$

we have

$$\sin x = x - \frac{x^3}{3!} + \frac{x^5}{5!} - \frac{x^7}{7!} + \cdots = \sum_{n=0}^{\infty} \frac{(-1)^n x^{2n+1}}{(2n+1)!} \qquad (1.47)$$

$$\cos x = 1 - \frac{x^2}{2!} + \frac{x^4}{4!} - \frac{x^6}{6!} + \cdots = \sum_{n=0}^{\infty} \frac{(-1)^n x^{2n}}{(2n)!} \qquad (1.48)$$

Since

$$\frac{d}{dx}(e^x) = e^x$$

$$e^x = 1 + \frac{x}{1!} + \frac{x^2}{2!} + \frac{x^3}{3!} + \cdots = \sum_{n=0}^{\infty} \frac{x^n}{n!} \qquad (1.49)$$

Since

$$\frac{d}{dx}[\ln(1+x)] = (1+x)^{-1} \qquad \frac{d}{dx}(1+x)^{-n} = -n(1+x)^{-n-1}$$

$$\ln(1+x) = x - \frac{x^2}{2} + \frac{x^3}{3} - \frac{x^4}{4} + \cdots = \sum_{n=1}^{\infty} \frac{(-1)^{n+1} x^n}{n} \qquad (1.50)$$

The above Taylor series expansions are frequently used in subsequent chapters.

1.3. Gauge Functions

In this book, we are interested in the limit of functions such as $f(\epsilon)$ as ϵ tends to zero, denoted by $\epsilon \to 0$. This limit might depend on whether ϵ tends to zero from below, denoted by $\epsilon\uparrow^{0}$, or from above, denoted by $\epsilon\downarrow_{0}$. For example,

$$\lim_{\epsilon\downarrow_{0}} e^{-1/\epsilon} = 0 \qquad \lim_{\epsilon\uparrow^{0}} e^{-1/\epsilon} = \infty$$

In what follows, we assume that the parameters have been normalized so that $\epsilon \geq 0$. If the limit of $f(\epsilon)$ exists (i.e., it does not have an essential singularity at $\epsilon = 0$ such as $\sin \epsilon^{-1}$), then there are three possibilities

$$\left.\begin{array}{l} f(\epsilon) \to 0 \\ f(\epsilon) \to A \\ f(\epsilon) \to \infty \end{array}\right\} \quad \text{as} \quad \epsilon \to 0, \quad 0 < A < \infty \qquad (1.51)$$

Most often, the above classification is not very useful because there are infinitely many functions that tend to zero as $\epsilon \to 0$. For example,

$$\lim_{\epsilon \to 0} \sin \epsilon = 0 \qquad \lim_{\epsilon \to 0} (1 - \cos \epsilon) = 0$$

$$\lim_{\epsilon \to 0} (\epsilon - \sin \epsilon) = 0 \qquad \lim_{\epsilon \to 0} [\ln(1 + \epsilon)]^4 = 0 \qquad (1.52)$$

$$\lim_{\epsilon \to 0} e^{-1/\epsilon} = 0$$

Also, there are infinitely many functions that tend to ∞ as $\epsilon \to 0$. For example,

$$\lim_{\epsilon \to 0} \frac{1}{\sin \epsilon} = \infty \qquad \lim_{\epsilon \to 0} \frac{1}{1 - \frac{1}{2}\epsilon^2 - \cos \epsilon} = \infty$$

$$\lim_{\epsilon \to 0} e^{1/\epsilon} \qquad \lim_{\epsilon \to 0} \ln \frac{1}{\epsilon} = \infty \qquad (1.53)$$

Therefore, to narrow down the above classification, we subdivide each class according to the rate at which they tend to zero or infinity. To accomplish this, we compare the rate at which these functions tend to zero and infinity with the rate at which known functions tend to zero and infinity. These comparison functions are called *gauge functions*. The simplest and most useful of these are the powers of ϵ

$$1, \epsilon, \epsilon^2, \epsilon^3, \cdots$$

and the inverse powers of ϵ

$$\epsilon^{-1}, \epsilon^{-2}, \epsilon^{-3}, \epsilon^{-4}, \cdots$$

For small ϵ, we know that

$$1 > \epsilon > \epsilon^2 > \epsilon^3 > \epsilon^4 > \cdots$$

and

$$\epsilon^{-1} < \epsilon^{-2} < \epsilon^{-3} < \epsilon^{-4} < \cdots$$

14 INTRODUCTION

Let us determine the rate at which the preceding functions tend to zero or infinity. Using the Taylor series expansion (1.47), we have

$$\sin \epsilon = \epsilon - \frac{\epsilon^3}{3!} + \frac{\epsilon^5}{5!} - \frac{\epsilon^7}{7!} + \cdots$$

so that $\sin \epsilon \to 0$ as $\epsilon \to 0$ because

$$\lim_{\epsilon \to 0} \frac{\sin \epsilon}{\epsilon} = \lim_{\epsilon \to 0} \left(1 - \frac{\epsilon^2}{3!} + \frac{\epsilon^4}{4!} + \cdots \right) = 1$$

Using (1.48), we have

$$1 - \cos \epsilon = \frac{\epsilon^2}{2!} - \frac{\epsilon^4}{4!} + \cdots$$

so that $1 - \cos \epsilon \to 0$ as $\epsilon^2 \to 0$ because

$$\lim_{\epsilon \to 0} \frac{1 - \cos \epsilon}{\epsilon^2} = \lim_{\epsilon \to 0} \left(\frac{1}{2!} - \frac{\epsilon^2}{4!} + \cdots \right) = \frac{1}{2!}$$

Using (1.47), we have

$$\epsilon - \sin \epsilon = \frac{\epsilon^3}{3!} - \frac{\epsilon^5}{5!} + \cdots$$

so that $\epsilon - \sin \epsilon \to 0$ as $\epsilon^3 \to 0$ because

$$\lim_{\epsilon \to 0} \frac{\epsilon - \sin \epsilon}{\epsilon^3} = \lim_{\epsilon \to 0} \left(\frac{1}{3!} - \frac{\epsilon^2}{5!} + \cdots \right) = \frac{1}{3!}$$

Using (1.50), we have

$$[\ln(1 + \epsilon)]^4 = \left(\epsilon - \frac{\epsilon^2}{2} + \frac{\epsilon^3}{3} + \cdots \right)^4$$

so that $[\ln(1 + \epsilon)]^4 \to 0$ as $\epsilon^4 \to 0$ because

$$\lim_{\epsilon \to 0} \frac{[\ln(1 + \epsilon)]^4}{\epsilon^4} = \lim_{\epsilon \to 0} \left(1 - \frac{\epsilon}{2} + \frac{\epsilon^2}{3} + \cdots \right)^4 = 1$$

To determine the rate at which $\exp(-1/\epsilon) \to 0$ as $\epsilon \to 0$, we attempt to expand it in a Taylor series for small ϵ. To accomplish this, we need the derivatives of ϵ at $\epsilon = 0$. But

$$f'(\epsilon) = \frac{d(e^{-1/\epsilon})}{d\epsilon} = \frac{1}{\epsilon^2} e^{-1/\epsilon} \qquad (1.54a)$$

which, at $\epsilon = 0$, gives 0 over 0. Hence, we need to use l'Hospital's rule to determine its limit as $\epsilon \to 0$. Thus,

$$\lim_{\epsilon \to 0} f'(\epsilon) = \lim_{\epsilon \to 0} \frac{e^{-1/\epsilon}}{\epsilon^2} = \lim_{x \to \infty} x^2 e^{-x} = \lim_{x \to \infty} \frac{x^2}{e^x}$$

which, upon differentiating the numerator and denominator twice with respect to x, gives

$$\lim_{\epsilon \to 0} f'(\epsilon) = \lim_{x \to \infty} \frac{2}{e^x} = 0$$

Hence,

$$f'(0) = 0 \qquad (1.54b)$$

Differentiating (1.54a) with respect to ϵ, we have

$$f''(\epsilon) = \left(\frac{1}{\epsilon^4} - \frac{2}{\epsilon^3}\right) e^{-1/\epsilon} \qquad (1.55a)$$

Hence,

$$f''(0) = 0$$

because

$$\lim_{\epsilon \to 0} \left(\frac{1}{\epsilon^4} - \frac{2}{\epsilon^3}\right) e^{-1/\epsilon} = \lim_{x \to \infty} \frac{x^4 - 2x^3}{e^x} = \lim_{x \to \infty} \frac{4!}{e^x} = 0$$

according to l'Hospital's rule. Differentiating (1.55a) with respect to ϵ gives

$$f'''(\epsilon) = \left(\frac{1}{\epsilon^6} - \frac{6}{\epsilon^5} + \frac{6}{\epsilon^4}\right) e^{-1/\epsilon} \qquad (1.56a)$$

Hence,

$$f'''(0) = 0 \qquad (1.56b)$$

because

$$\lim_{\epsilon \to 0} \left(\frac{1}{\epsilon^6} - \frac{6}{\epsilon^5} + \frac{6}{\epsilon^4}\right) e^{-1/\epsilon} = \lim_{x \to \infty} \frac{x^6 - 6x^5 + 6x^4}{e^x}$$

$$= \lim_{x \to \infty} \frac{6!}{e^x} = 0$$

according to l'Hospital's rule. Continuing the process, we find that

$$f^{(n)}(0) = 0 \qquad (1.57)$$

for all n. Therefore, it follows from (1.46) that

$$e^{-1/\epsilon} = 0 + 0 + 0 + 0 + \cdots$$

16 INTRODUCTION

which is certainly not true. The function $\exp(-1/\epsilon)$ cannot be represented by a power series in ϵ. In fact, it tends to zero faster than any power of ϵ because

$$\lim_{\epsilon \to 0} \frac{e^{-1/\epsilon}}{\epsilon^n} = \lim_{x \to \infty} \frac{x^n}{e^x} = \lim_{x \to \infty} \frac{n!}{e^x} = 0$$

according to l'Hospital's rule. Therefore, the powers of ϵ are not complete and must be supplemented by $\exp(-1/\epsilon)$.

Next, we consider the rates at which the functions in (1.53) tend to ∞. Using (1.47), we have

$$\frac{1}{\sin \epsilon} = \frac{1}{\epsilon - \dfrac{\epsilon^3}{3!} + \cdots} = \frac{1}{\epsilon\left(1 - \dfrac{\epsilon^2}{3!} + \cdots\right)}$$

so that $(\sin \epsilon)^{-1} \to \infty$ as ϵ^{-1} because

$$\lim_{\epsilon \to 0} \frac{1/\sin \epsilon}{1/\epsilon} = \lim_{\epsilon \to 0} \frac{\epsilon}{\sin \epsilon} = \lim_{\epsilon \to 0} \frac{1}{1 - \dfrac{\epsilon^2}{3!} + \cdots} = 1$$

Using (1.48), we have

$$\frac{1}{1 - \tfrac{1}{2}\epsilon^2 - \cos \epsilon} = \frac{1}{-\dfrac{\epsilon^4}{4!} + \dfrac{\epsilon^6}{6!} + \cdots}$$

so that $(1 - \tfrac{1}{2}\epsilon^2 - \cos \epsilon)^{-1} \to -\infty$ as $-\epsilon^{-4}$ because

$$\lim_{\epsilon \to 0}\left(\frac{1}{1 - \tfrac{1}{2}\epsilon^2 - \cos \epsilon} \div \frac{1}{\epsilon^4}\right) = \lim_{\epsilon \to 0} \frac{\epsilon^4}{-\dfrac{\epsilon^4}{4!} + \dfrac{\epsilon^6}{6!} + \cdots}$$

$$\lim_{\epsilon \to 0} \frac{-4!}{1 - \dfrac{\epsilon^2}{30} + \cdots} = -4!$$

Since $\exp(-1/\epsilon)$ tends to zero faster than any power of ϵ, $\exp(1/\epsilon)$ tends to infinity faster than any inverse power of ϵ because

$$\lim_{\epsilon \to 0} \frac{e^{1/\epsilon}}{1/\epsilon^n} = \lim_{x \to \infty} \frac{e^x}{x^n} = \lim_{x \to \infty} \frac{e^x}{n!} = \infty$$

according to l'Hospital's rule. Therefore, we need to supplement the gauge functions by $\exp(1/\epsilon)$. The function $\ln(1/\epsilon)$ tends to infinity as $\epsilon \to 0$ more slowly than any power of ϵ^{-1}, that is, $\epsilon^{-\alpha}$, no matter how small α is because

$$\lim_{\epsilon \to 0} \frac{\ln(1/\epsilon)}{\epsilon^{-\alpha}} = \lim_{x \to \infty} \frac{\ln x}{x^\alpha} = \lim_{x \to \infty} \frac{1}{x \cdot \alpha x^{\alpha-1}} = \frac{1}{\alpha} \lim_{x \to \infty} \frac{1}{x^\alpha} = 0$$

Therefore, we need to supplement the gauge functions by $\ln(1/\epsilon)$.

Similarly, we need to supplement the gauge functions by $[\ln(1/\epsilon)]^{-1}$ to represent the functions that tend to zero more slowly than any power of ϵ, that is ϵ^α, no matter how small α is as long as it is positive. The above discussion shows that, to obtain a complete set of gauge functions, the powers of ϵ must be supplemented by logarithms, exponentials and

$$e^{e^{1/\epsilon}}, e^{-e^{1/\epsilon}}, \ln\ln\left(\frac{1}{\epsilon}\right), \ln\ln\ln\left(\frac{1}{\epsilon}\right), \left(\ln\frac{1}{\epsilon}\right)^n, \left(\ln\frac{1}{\epsilon}\right)^{-n}, \text{etc.}$$

1.4. Order Symbols

Instead of saying that $\sin \epsilon$ tends to zero at the same rate that ϵ tends to zero, we say $\sin \epsilon$ is order ϵ as $\epsilon \to 0$ or $\sin \epsilon$ is big "oh" of ϵ as $\epsilon \to 0$ and write it as

$$\sin \epsilon = O(\epsilon) \quad \text{as} \quad \epsilon \to 0$$

In general, we put

$$f(\epsilon) = O[g(\epsilon)] \quad \text{as} \quad \epsilon \to 0 \quad (1.58)$$

if

$$\lim_{\epsilon \to 0} \frac{f(\epsilon)}{g(\epsilon)} = A \quad 0 < |A| < \infty \quad (1.59)$$

Thus, as $\epsilon \to 0$,

$$\cos \epsilon = O(1) \qquad \cos \epsilon - 1 = O(\epsilon^2)$$
$$\sinh \epsilon = O(\epsilon) \qquad \tan \epsilon = O(\epsilon)$$
$$\text{cosec } \epsilon = O(\epsilon^{-1}) \qquad \sec \epsilon = O(1)$$
$$\cot \epsilon = O(\epsilon^{-1}) \qquad \frac{\epsilon^{3/2}}{\sin \epsilon} = O(\epsilon^{1/2})$$
$$\sinh \frac{1}{\epsilon} = O(e^{1/\epsilon}) \qquad \text{sech } \frac{1}{\epsilon} = O(e^{-1/\epsilon})$$

It should be noted that the above mathematical order expressed by the symbol O is formally distinct from the physical order of magnitude because no account is taken of the numerical value of A, that is, the constant of proportionality. Thus, $A\epsilon = O(\epsilon)$ even if A is a hundred thousand. However, one has the mystical hope that they are somehow related. In other words, one has the hope that the

18 INTRODUCTION

constants of proportionality are $O(1)$ so that the numerical value is not very much different from that given by the order symbol.

In many instances, the information available about a given function may be incomplete to determine the rate at which it tends to its limit but sufficient to determine whether the rate is faster or slower than that of a given gauge function. In such instances, we use the order symbol o (little oh) defined as follows

$$f(\epsilon) = o[g(\epsilon)] \quad \text{as} \quad \epsilon \to 0 \tag{1.60}$$

if

$$\lim_{\epsilon \to 0} \frac{f(\epsilon)}{g(\epsilon)} = 0$$

Thus, as $\epsilon \to 0$,

$$\sin \epsilon = o(1) \qquad \sin \epsilon = o(\epsilon^{1/2})$$

$$\cos \epsilon = o(\epsilon^{-1}) \qquad \cos \epsilon = o(\epsilon^{-1/3})$$

$$e^{-1/\epsilon} = o(\epsilon^{-10^{-8}}) \qquad \epsilon^{-10^{10}} = o(e^{1/\epsilon})$$

$$\ln \frac{1}{\epsilon} = o(\epsilon^{-0.00001}) \qquad \ln\ln \frac{1}{\epsilon} = o\left(\ln \frac{1}{\epsilon}\right)$$

$$e^{2/\epsilon} = o(e^{e^{1/\epsilon}}) \qquad \left(\ln \frac{1}{\epsilon}\right)^2 = o(\epsilon^{-0.00001})$$

1.5. Asymptotic Series

We consider the value of the integral

$$f(\omega) = \int_0^\infty \frac{\omega e^{-x}}{\omega + x} dx \tag{1.62}$$

for large positive ω. One method of determining an approximation to $f(\omega)$ is Laplace's method, which is discussed in Section 3.3. It consists of expanding the coefficient of $\exp(-x)$ in powers of x and then integrating the resulting series term by term. Using the binomial theorem, we have

$$\frac{\omega}{\omega + x} = \frac{1}{1 + \omega^{-1}x} = 1 - \frac{x}{\omega} + \frac{x^2}{\omega^2} - \frac{x^3}{\omega^3} + \cdots = \sum_{n=0}^{\infty} \frac{(-1)^n x^n}{\omega^n} \tag{1.63}$$

which converges for $x < \omega$. The basic idea underlying Laplace's method is that $\exp(-x)$ tends to zero faster than any power of x tends to infinity for large x.

ASYMPTOTIC SERIES

Hence, only the immediate neighborhood of the origin contributes to the integral when ω is large. Substituting (1.63) into (1.62), we have

$$f(\omega) = \int_0^\infty \sum_{n=0}^\infty \frac{(-1)^n x^n e^{-x}}{\omega^n} \, dx = \sum_{n=0}^\infty \frac{(-1)^n}{\omega^n} \int_0^\infty x^n e^{-x} \, dx$$

But,

$$\int_0^\infty x^n e^{-x} \, dx = n!$$

when n is an integer. Therefore,

$$f(\omega) = \sum_{n=0}^\infty \frac{(-1)^n n!}{\omega^n} \tag{1.64}$$

Using the ratio test in (1.64), we have

$$\lim_{n \to \infty} \frac{n\text{th term}}{(n-1)\text{th term}} = \lim_{n \to \infty} \frac{(-1)^n n! \, \omega^{n-1}}{\omega^n (-1)^{n-1}(n-1)!} = \lim_{n \to \infty} \frac{-n}{\omega} = -\infty$$

Hence, the series (1.64) diverges for all values of ω. To investigate whether (1.64) is of any value for computing $f(\omega)$, we determine the remainder if we truncate the series after N terms. To do this, we note that

$$\sum_{n=0}^N \frac{(-1)^n x^n}{\omega^n}$$

is a geometric series whose sum is

$$\frac{1 - \left(\frac{-x}{\omega}\right)^{N+1}}{1 + \frac{x}{\omega}}$$

Hence,

$$\frac{\omega}{\omega + x} = \sum_{n=0}^N \frac{(-1)^n x^n}{\omega^n} + \hat{R}_N(x, \omega)$$

where

$$\hat{R}_N = \frac{\omega}{\omega + x} - \frac{1 - \left(\frac{-x}{\omega}\right)^{N+1}}{1 + \frac{x}{\omega}} = \frac{\left(\frac{-x}{\omega}\right)^{N+1}}{1 + \frac{x}{\omega}} = \frac{(-x)^{N+1}}{\omega^N(\omega + x)}$$

Therefore,

$$\frac{\omega}{\omega+x} = \sum_{n=0}^{N} \frac{(-1)^n x^n}{\omega^n} + \frac{(-x)^{N+1}}{\omega^N(\omega+x)} \qquad (1.65)$$

Multiplying (1.65) with exp $(-x)$ and integrating the result from $x = 0$ to $x = \infty$, we obtain

$$f(\omega) = \int_0^\infty \frac{\omega e^{-x}}{\omega+x} dx = \sum_{n=0}^{N} \frac{(-1)^n}{\omega^n} \int_0^\infty x^n e^{-x} dx + R_N(\omega)$$

or

$$f(\omega) = \sum_{n=0}^{N} \frac{(-1)^n n!}{\omega^n} + R_N(\omega) \qquad (1.66)$$

where

$$R_N(\omega) = \frac{(-1)^{N+1}}{\omega^N} \int_0^\infty \frac{x^{N+1} e^{-x}}{\omega+x} dx \qquad (1.67)$$

We note that the remainder is a function of both N and ω.

Instead of using the ratio test, one can check the behavior of $R_N(\omega)$ for fixed ω as $N \to \infty$. In order for the series to converge, $\lim_{N \to \infty} R_N$ must be zero. This is not true in our example; in fact, $R_N \to \infty$ as $N \to \infty$ so that the series diverges for all values of ω in agreement with the result of the ratio test. Thus, if the series (1.66) is to be useful, N must be fixed. Hence, let us investigate the behavior of $R_N(\omega)$ for fixed N. To accomplish this, we need to estimate $R_N(\omega)$.

Since ω and x are positive

$$\frac{1}{\omega+x} < \frac{1}{x}$$

and then

$$|R_N(\omega)| = \frac{1}{\omega^N} \int_0^\infty \frac{x^{N+1} e^{-x}}{\omega+x} dx < \frac{1}{\omega^N} \int_0^\infty x^N e^{-x} dx = \frac{N!}{\omega^N} \qquad (1.68)$$

Hence, the error committed in truncating the series after N terms is numerically less than the first neglected term, namely the $(N + 1)$th term. Moreover, as $\omega \to \infty$ with N fixed, $R_N \to 0$. Therefore, although the series (1.64) diverges, for a fixed N the first N terms in the series can represent $f(\omega)$ with an error that can be made arbitrarily small by taking ω sufficiently large. Such a series is called an *asymptotic series of the Poincaré type* and is denoted by

$$f(\omega) \sim \sum_{n=0}^{\infty} \frac{(-1)^n n!}{\omega^n} \qquad (1.69)$$

In general, given a series $\sum_{n=0}^{\infty} (a_n/\omega^n)$, where a_n is independent of ω, we say that the series is an *asymptotic series* and write

$$f(\omega) \sim \sum_{n=0}^{\infty} \frac{a_n}{\omega^n} \quad \text{as} \quad \omega \to \infty \qquad (1.70)$$

if and only if

$$f(\omega) = \sum_{n=0}^{N} \frac{a_n}{\omega^n} + o\left(\frac{1}{\omega^N}\right) \quad \text{as} \quad \omega \to \infty \qquad (1.71)$$

It follows from (1.71) that

$$f(\omega) = \sum_{n=0}^{N-1} \frac{a_n}{\omega^n} + \frac{a_N}{\omega^N} + o\left(\frac{1}{\omega^N}\right) \quad \text{as} \quad \omega \to \infty$$

Hence the condition (1.71) can be rewritten as

$$f(\omega) = \sum_{n=0}^{N-1} \frac{a_n}{\omega^n} + O\left(\frac{1}{\omega^N}\right) \qquad (1.72)$$

We should note that the utility of an asymptotic series lies in the fact that the error committed in truncating the series is by definition the order of the first neglected term, and hence, it tends rapidly to zero as $\omega \to \infty$. In applications, one usually fixes ω at a large value and attempts to reduce the error by adding more terms. But if the series is divergent, a point is reached beyond which adding terms increases rather than decreases the error, as illustrated in Figure 1-4. Therefore, for a given ω there is an optimum value of N, which yields the

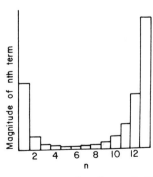

Figure 1-4. Behavior of terms in a divergent asymptotic series.

22 INTRODUCTION

smallest error. In practice, one rarely determines more than one or two terms in the expansion, and hence, one need not worry about the divergence of the series. In the cases in which many terms can be calculated, they are usually obtained by performing the algebraic manipulations on a digital computer. The series is then examined to reveal its analytic structure, and then, it is transformed appropriately using rational fractions, natural coordinates, or the Euler transformation. Improvement of divergent series is not discussed in this book and we refer the reader to van Dyke (1975) and his references.

1.6. Asymptotic Expansions and Sequences

As shown in Section 1.3, there are many functions whose behavior cannot be represented in terms of a power series of the small parameter. Moreover, we found that the powers ϵ must be supplemented by its logarithm, exponential, logarithm of its logarithm, and so on. Thus, to determine an asymptotic representation of a given function, we may be unable to use powers of ϵ alone. Instead, we can use a general sequence of functions $\delta_n(\epsilon)$ as long as

$$\delta_n(\epsilon) = o\left[\delta_{n-1}(\epsilon)\right] \quad \text{as} \quad \epsilon \to 0 \tag{1.73}$$

Such a sequence is called an *asymptotic sequence*. Examples of such asymptotic sequences are

$$\epsilon^n \qquad \epsilon^{n/3} \qquad (\ln \epsilon)^{-n} \qquad (\sin \epsilon)^n \qquad (\cot \epsilon)^{-n} \tag{1.74}$$

In terms of asymptotic sequences, we can define asymptotic expansions as follows. Given $\sum_{n=0}^{\infty} a_n \delta_n(\epsilon)$, where the a_n are independent of ϵ and $\delta_n(\epsilon)$ is an asymptotic sequence, we say that this expansion is an *asymptotic expansion* and write

$$f(\epsilon) \sim \sum_{n=0}^{\infty} a_n \delta_n(\epsilon) \quad \text{as} \quad \epsilon \to 0 \tag{1.76}$$

Clearly, an asymptotic series is a special case of an asymptotic expansion.

We note that an asymptotic representation of a given $f(\epsilon)$ is not unique. In fact, $f(\epsilon)$ can be represented by an infinite number of asymptotic expansions because there exists an infinite number of asymptotic sequences that can be used in the representation. However, given an asymptotic sequence $\delta_n(\epsilon)$, the representation of $f(\epsilon)$ in terms of this sequence is unique as shown below. Let us put

$$f(\epsilon) \sim a_0 \delta_0(\epsilon) + a_1 \delta_1(\epsilon) + a_2 \delta_2(\epsilon) + \cdots \tag{1.77}$$

Dividing (1.77) by $\delta_0(\epsilon)$, we have

$$\frac{f(\epsilon)}{\delta_0(\epsilon)} \sim a_0 + a_1 \frac{\delta_1(\epsilon)}{\delta_0(\epsilon)} + a_2 \frac{\delta_2(\epsilon)}{\delta_0(\epsilon)} + \cdots$$

which, upon letting $\epsilon \to 0$, yields

$$a_0 = \lim_{\epsilon \to 0} \frac{f(\epsilon)}{\delta_0(\epsilon)} \quad \text{because} \quad \lim_{\epsilon \to 0} \frac{\delta_n(\epsilon)}{\delta_0(\epsilon)} = 0 \quad \text{for} \quad n \geq 1$$

Moving $a_0 \delta_0(\epsilon)$ to the left and dividing the resulting equation by $\delta_1(\epsilon)$, we have

$$\frac{f(\epsilon) - a_0 \delta_0(\epsilon)}{\delta_1(\epsilon)} \sim a_1 + a_2 \frac{\delta_2(\epsilon)}{\delta_1(\epsilon)} + \cdots$$

which, upon letting $\epsilon \to 0$, yields

$$a_1 = \lim_{\epsilon \to 0} \frac{f(\epsilon) - a_0 \delta_0(\epsilon)}{\delta_1(\epsilon)}$$

Continuing the process, we find that

$$a_n = \lim_{\epsilon \to 0} \frac{f(\epsilon) - \sum_{m=0}^{n-1} a_m \delta_m(\epsilon)}{\delta_n(\epsilon)} \qquad (1.78)$$

1.7. Convergent Versus Asymptotic Series

In Chapter 13, we determine the two different representations

$$J_0(x) = 1 - \frac{x^2}{2^2} + \frac{x^4}{2^2 \cdot 4^2} - \frac{x^6}{2^2 \cdot 4^2 \cdot 6^2} + \cdots \qquad (1.80)$$

$$J_0(x) \sim \sqrt{\frac{2}{\pi x}} \left[u \cos(x - \tfrac{1}{4}\pi) + v \sin(x - \tfrac{1}{4}\pi) \right] \quad \text{as} \quad x \to \infty \qquad (1.81)$$

where

$$u(x) = 1 - \frac{1^2 \cdot 3^2}{4^2 \cdot 2^2 \cdot 2! x^2} + \frac{1^2 \cdot 3^2 \cdot 5^2 \cdot 7^2}{4^4 \cdot 2^4 \cdot 4! x^4} + \cdots \qquad (1.82)$$

$$v(x) = \frac{1}{4 \cdot 2x} - \frac{1^2 \cdot 3^2 \cdot 5^2}{4^3 \cdot 2^3 \cdot 3! x^3} + \cdots \qquad (1.83)$$

of Bessel's function of order zero. The series (1.80) is uniformly and absolutely convergent for all values of x, whereas the series $u(x)$ and $v(x)$, and hence, (1.81) diverge for all values of x. However, the representation (1.81) is asymptotic because the error committed by truncating the series is the order of the first neglected term.

For small x, the first few terms in (1.80) give fairly accurate results. In fact, the first nine terms give a value of $J_0(2)$ correct to 11 significant figures. However, as x increases, the number of terms needed to yield the same accuracy increases rapidly. At $x = 4$, eight terms are needed to give an accuracy of three significant figures, whereas the first term of the asymptotic expansion (1.81) yields the same accuracy. As x increases further, an accurate result is obtained with far less labor by using the asymptotic divergent series (1.81). In fact, for very large values of x, the convergent series is useless from the computational point of view, owing to the finite word length of modern computers. Thus, any attempt to evaluate $J_0(x)$ for large x using the convergent series (1.80) fails beyond a given value of x; to be sure, this value depends on the skill of the programmer.

1.8. Elementary Operations on Asymptotic Expansions

To determine approximate solutions of algebraic, differential, and integral equations and to evaluate integrals, we assume expansions, substitute them into the equations, and perform on them elementary operations such as addition, subtraction, multiplication, exponentiation, integration, and differentiation. Some of these operations are not justified. When they are not justified, they lead to *singularities* and *nonuniformities*. For example,

$$\sqrt{x+\epsilon} = \sqrt{x}\left(1+\frac{\epsilon}{x}\right)^{1/2} = \sqrt{x}\left[1+\frac{1}{2}\frac{\epsilon}{x}+\frac{(\frac{1}{2})(-\frac{1}{2})}{2}\left(\frac{\epsilon}{x}\right)^2+\cdots\right]$$

$$= \sqrt{x}\left(1+\frac{\epsilon}{2x}-\frac{\epsilon^2}{8x^2}+\cdots\right) \quad \text{as} \quad \epsilon \to 0 \quad (1.84)$$

is not justified when $\epsilon/x = O(1)$ because the second term becomes the order of the first term and the third term becomes the order of the first term. Thus, the error committed by truncating the series after N terms is not $O(\epsilon^N)$, the order of the first neglected term if $x = O(\epsilon)$, and we speak of a *nonuniform expansion*. Similarly,

$$\frac{1}{1+\epsilon x} = 1 - \epsilon x + \epsilon^2 x^2 - \epsilon^3 x^3 + \cdots \quad (1.85)$$

is not justified when $\epsilon x = O(1)$, because the error committed in truncating the series after N terms is not $O(\epsilon^N)$, the order of the first neglected term. Consequently, one needs always to check whether the obtained expansions are uniform or not. This is the major objective of perturbation methods.

Exercises

1.1. For small ϵ, determine three terms in the expansions of
  (a) $(1 - \frac{3}{8}a^2\epsilon + \frac{51}{256}a^4\epsilon^2)^{-1}$
  (b) $\cos(\sqrt{1-\epsilon}\,t)$
  (c) $\sqrt{1 - \frac{1}{2}\epsilon + 2\epsilon^2}$

(d) $\sin(1 + \epsilon - \epsilon^2)$

1.2. Expand each of the following expressions for small ϵ and keep three terms:

(a) $\sqrt{1 - \frac{1}{2}\epsilon^2 t - \frac{1}{8}\epsilon^4 t}$
(b) $(1 + \epsilon \cos f)^{-1}$
(c) $(1 + \epsilon\omega_1 + \epsilon^2\omega_2)^{-2}$
(d) $\sin(s + \epsilon\omega_1 s + \epsilon^2\omega_2 s)$
(e) $\sin^{-1}\left(\dfrac{\epsilon}{\sqrt{1 + \epsilon}}\right)$
(f) $\ln\dfrac{1 + 2\epsilon - \epsilon^2}{\sqrt[3]{1 + 2\epsilon}}$

1.3. Let $\mu = \mu_0 + \epsilon\mu_1 + \epsilon^2\mu_2$ in $h = \frac{3}{2}[1 - \sqrt{1 - 3\mu(1 - \mu)}]$, expand for small e, and keep three terms.

1.4. For small ϵ, determine the order of the functions

$$\sinh\left(\frac{1}{\epsilon}\right) \quad \ln(1 + \sin \epsilon) \quad \ln(2 + \sin \epsilon) \quad e^{\ln(1-\epsilon)}$$

1.5. Determine the order of the following expressions as $\epsilon \to 0$:

$$\sqrt{\epsilon(1 - \epsilon)} \quad 4\pi^2 \epsilon \quad 1000\epsilon^{1/2} \quad \ln(1 + \epsilon) \quad \dfrac{1 - \cos \epsilon}{1 + \cos \epsilon} \quad \dfrac{\epsilon^{3/2}}{1 - \cos \epsilon}$$

$$\operatorname{sech}^{-1} \epsilon \quad e^{\tan \epsilon} \quad \ln\left[1 + \dfrac{\ln(1 + 2\epsilon)}{\epsilon(1 - 2\epsilon)}\right] \quad \ln\left[1 + \dfrac{\ln\dfrac{1 + 2\epsilon}{\epsilon}}{1 - 2\epsilon}\right]$$

$$e^{-\cosh(1/\epsilon)} \quad \int_0^\epsilon e^{-s^2}\, ds$$

1.6. Determine the order of the following expressions as $\epsilon \to 0$:

$$\ln(1 + 5\epsilon) \quad \sin^{-1}\dfrac{\epsilon}{\sqrt{1 + \epsilon}} \quad \dfrac{\sqrt{\epsilon}}{\sin \epsilon} \quad 1 - \tfrac{1}{2}\epsilon^2 - \cos \epsilon$$

$$\ln\left(\sinh\dfrac{1}{\epsilon}\right)$$

1.7. Determine the order of the following as $\epsilon \to 0$:

$$\ln(\cot \epsilon) \quad \sinh\left(\dfrac{1}{\epsilon}\right) \quad \coth\left(\dfrac{1}{\epsilon}\right) \quad \dfrac{\epsilon^{3/4}}{1 - \cos \epsilon} \quad \ln\left[1 + \ln\dfrac{1 + 2\epsilon}{\epsilon}\right]$$

1.8. Arrange the following in descending order for small ϵ:

$$\epsilon^2 \quad \epsilon^{1/2} \quad \ln(\ln \epsilon^{-1}) \quad 1 \quad \epsilon^{1/2}\ln\epsilon^{-1} \quad \epsilon\ln\epsilon^{-1} \quad e^{-1/\epsilon} \quad \ln \epsilon^{-1}$$

$$\epsilon^{3/2} \quad \epsilon \quad \epsilon^2 \ln \epsilon^{-1}$$

1.9. Arrange the following in descending order for small ϵ:

$$\exp\left(-\frac{1}{\epsilon}\right) \quad \ln\left(\frac{1}{\epsilon}\right) \quad \epsilon^{-1/100} \quad \cot(\epsilon) \quad \sinh\left(\frac{1}{\epsilon}\right)$$

1.10. Arrange the following in descending order for small ϵ:

$$\ln(1+\epsilon) \quad \cot(\epsilon) \quad \tanh\left(\frac{1}{\epsilon}\right) \quad \frac{\sin(\epsilon)}{\epsilon^{3/4}}$$

$$\epsilon \ln(\epsilon) \quad \exp\left(-\frac{1}{\epsilon}\right) \quad \sinh\left(\frac{1}{\epsilon}\right) \quad \frac{1}{\ln\left(\frac{1}{\epsilon}\right)}$$

1.11. Arrange the following terms in descending order for small ϵ:

$$e^{-1/\epsilon} \quad \ln\left(\frac{1}{\epsilon}\right) \quad \frac{1}{\epsilon} \quad \epsilon^{1/2} \quad \left(\ln\frac{1}{\epsilon}\right)^2 \quad \epsilon^{1/2}\ln\left(\frac{1}{\epsilon}\right)$$

$$\epsilon^{1/\epsilon} \quad \frac{1}{\epsilon^{3/2}} \quad 1 \quad \epsilon \quad \epsilon^{3/2} \quad \frac{1}{\epsilon^{1/2}} \quad e^{e^{1/\epsilon}} \quad e^{-e^{1/\epsilon}} \quad \ln\left(\ln\frac{1}{\epsilon}\right)$$

$$\epsilon^{0.0001} \quad \epsilon^{-0.0001} \quad \epsilon^{0.0001}\ln\left(\frac{1}{\epsilon}\right) \quad 5^{1/\epsilon} \quad 5^{-1/\epsilon}$$

1.12. Arrange the following terms in descending order for small ϵ:

$$\epsilon^{\nu} \quad \epsilon^{-\nu} \quad \epsilon^{\mu} \quad \epsilon^{-\mu} \quad 1 \quad \epsilon^{3/2} \quad \epsilon^{-3/2} \quad \exp\left(\frac{1}{\epsilon}\right) \quad \ln\left(\frac{1}{\epsilon}\right) \quad \ln\left[\ln\left(\frac{1}{\epsilon}\right)\right]$$

$$\left[\ln\left(\frac{1}{\epsilon}\right)\right]^2 \quad \epsilon\ln\left(\frac{1}{\epsilon}\right) \quad \epsilon/\ln\left(\frac{1}{\epsilon}\right) \quad \exp\left(-\frac{1}{\epsilon}\right) \quad \epsilon$$

where $\nu = 10^{-100}$ and $\mu = 10^{100}$.

1.13. Arrange the following in descending order for small ϵ:

$$\ln(1+\epsilon) \quad \operatorname{sech}^{-1}(\epsilon) \quad \frac{1-\cos\epsilon}{1+\cos\epsilon} \quad \sqrt{\epsilon(1-\epsilon)} \quad e^{-\cosh(1/\epsilon)}$$

$$\ln\left[1+\frac{\ln\frac{1+2\epsilon}{1-2\epsilon}}{\epsilon}\right] \quad \ln\left[1+\frac{\ln(1+2\epsilon)}{\epsilon(1-2\epsilon)}\right] \quad \frac{\epsilon^{1/2}}{1-\cos\epsilon}$$

1.14. Which of the following expansions is nonuniformly valid and what are its regions of nonuniformity?

(a) $\sqrt{2}\,t = \frac{2}{3}x^{3/2} + \epsilon\left(\frac{2}{3}x^{3/2} + \sqrt{x} - \frac{1}{2}\ln\frac{1+\sqrt{x}}{1-\sqrt{x}}\right) + O\left(\frac{\epsilon^2}{1-x}\right)$

(b) $\eta = \epsilon\cos x + \frac{1}{2}\epsilon^2\,\frac{1+\gamma^2}{1-2\gamma^2}\cos 2x + \frac{3}{16}\epsilon^3\,\frac{2\gamma^4+7\gamma^2+2}{(1-2\gamma^2)(1-3\gamma^2)}\cos 3x$
$\quad + O(\epsilon^4)$

(c) $\sigma = \sqrt{k^2 - 1} - \dfrac{\epsilon^2 k^2}{\sqrt{k^2 - 1}} + O(\epsilon^3)$

(d) $f = 1 - \epsilon x + \epsilon^2 x^2 - \epsilon^3 x^3 + O(\epsilon^4)$

(e) $u = a\cos(1 + \tfrac{3}{8}\epsilon a^2)t + \dfrac{\epsilon a^3}{32}\cos 3(1 + \tfrac{3}{8}\epsilon a^2)t + O(\epsilon^2)$

(f) $u = a\cos t + \dfrac{\epsilon a^3}{8}(\tfrac{1}{4}\cos 3t - 3t\sin t) + O(\epsilon^2)$

(g) $c = 1 + \dfrac{2\epsilon}{\tau - 1} + \dfrac{3\epsilon^2}{(\tau - 1)^2} + O(\epsilon^3)$

(h) $y \sim \dfrac{1}{\sqrt[4]{x^2(1-x)}} \cos\left[\lambda \displaystyle\int_0^x \sqrt{x(1-x)}\,dx\right]$ as $\lambda \to \infty$

(i) $f = \sin x + \epsilon \cos x - \tfrac{1}{2}\epsilon^2 \sin x - \tfrac{1}{6}\epsilon^3 \cos x + O(\epsilon^4)$

CHAPTER 2

Algebraic Equations

In this chapter, we discuss approximate solutions of algebraic equations that depend on a small parameter. The solution is represented as an asymptotic expansion in terms of the small parameter. Such expansions are called *parameter perturbations*. To describe the method, we begin by applying it to quadratic equations because their exact solutions are easily obtained for comparison. We consider cubic equations in Section 2.2, higher-order equations in Section 2.3, and transcendental equations in Section 2.4.

2.1. Quadratic Equations

We begin with quadratic equations because their exact solutions are available for comparison. We consider three examples.

EXAMPLE 1
As a first example, we determine the roots of

$$x^2 - (3 + 2\epsilon)x + 2 + \epsilon = 0 \tag{2.1}$$

for small ϵ. When $\epsilon = 0$, (2.1) reduces to

$$x^2 - 3x + 2 = (x - 2)(x - 1) = 0 \tag{2.2}$$

whose roots are $x = 1$ and 2. Equation (2.1) is called the *perturbed equation*, whereas (2.2) is called the *unperturbed* or *reduced equation*. When ϵ is small but finite, we expect the roots to deviate slightly from 1 and 2. The first step in determining an approximate solution is to assume the form of the expansion. In this case, we assume that the roots have expansions in the form

$$x = x_0 + \epsilon x_1 + \epsilon^2 x_2 + \cdots \tag{2.3}$$

where the ellipses stand for all terms with powers of ϵ^n for which $n \geqslant 3$. In most applications, one calculates only one or two terms in the expansion because the algebra needed to calculate the higher-order terms is so involved. The algebra is relegated to a digital computer when possible. We should note that in many physical problems, especially nonlinear problems, determination of higher-order terms is not straightforward even if the routine algebra is relegated to a digital computer. In this book, we are concerned only with the first few terms in the expansion. Usually, one refers to the first term x_0 as the zeroth-order term, the second term ϵx_1 as the first-order term, and the third term $\epsilon^2 x_2$ as the second-order term. In other words, the order of a term is decided by the gauge function rather than its numerical order.

The second step involves substituting the assumed expansion (2.3) into the governing equation (2.1). The result is

$$(x_0 + \epsilon x_1 + \epsilon^2 x_2 + \cdots)^2 - (3 + 2\epsilon)(x_0 + \epsilon x_1 + \epsilon^2 x_2 + \cdots) + 2 + \epsilon = 0 \quad (2.4)$$

The third step involves carrying out elementary operations such as addition, subtraction, multiplication, exponentiation, and so on, and then, collecting coefficients of like powers of ϵ. Using the binomial theorem to expand the first term, we have

$$(x_0 + \epsilon x_1 + \epsilon^2 x_2 + \cdots)^2 = x_0^2 + 2x_0(\epsilon x_1 + \epsilon^2 x_2 + \cdots) + (\epsilon x_1 + \epsilon^2 x_2 + \cdots)^2$$

$$= x_0^2 + 2\epsilon x_0 x_1 + 2\epsilon^2 x_0 x_2 + \epsilon^2 x_1^2 + 2\epsilon^3 x_1 x_2 + \epsilon^4 x_2^2 + \cdots$$

$$= x_0^2 + 2\epsilon x_0 x_1 + \epsilon^2(2x_0 x_2 + x_1^2) + \cdots \quad (2.5)$$

where only terms up to $O(\epsilon^2)$ have been retained, consistent with the assumed expansion (2.3). Had we sought an expansion of $O(\epsilon^n)$, where $n \geq 3$, we would have included terms up to $O(\epsilon^n)$ in (2.5). Multiplying the factors in the second term in (2.4), we have

$$(3 + 2\epsilon)(x_0 + \epsilon x_1 + \epsilon^2 x_2 + \cdots) = 3x_0 + 3\epsilon x_1 + 3\epsilon^2 x_2 + 2\epsilon x_0 + 2\epsilon^2 x_1$$

$$+ 2\epsilon^3 x_2 = 3x_0 + \epsilon(3x_1 + 2x_0) + \epsilon^2(3x_2 + 2x_1) + \cdots \quad (2.6)$$

Here again, only terms up to $O(\epsilon^2)$ have been retained, consistent with the assumed expansion. Substituting (2.5) and (2.6) into (2.4), we have

$$x_0^2 + 2\epsilon x_0 x_1 + \epsilon^2(2x_0 x_2 + x_1^2) - 3x_0 - \epsilon(3x_1 + 2x_0)$$

$$- \epsilon^2(3x_2 + 2x_1) + 2 + \epsilon + \cdots = 0$$

Collecting coefficients of like powers of ϵ yields

$$(x_0^2 - 3x_0 + 2) + \epsilon(2x_0 x_1 - 3x_1 - 2x_0 + 1) + \epsilon^2(2x_0 x_2$$

$$+ x_1^2 - 3x_2 - 2x_1) + \cdots = 0 \quad (2.7)$$

The fourth step involves equating the coefficient of each power of ϵ to zero. To justify this step, we let $\epsilon \to 0$ in (2.7). The result is

$$x_0^2 - 3x_0 + 2 = 0 \qquad (2.8)$$

and (2.7) becomes

$$\epsilon(2x_0 x_1 - 3x_1 - 2x_0 + 1) + \epsilon^2(2x_0 x_2 + x_1^2 - 3x_2 - 2x_1) + \cdots = 0$$

Dividing by ϵ gives

$$2x_0 x_1 - 3x_1 - 2x_0 + 1 + \epsilon(2x_0 x_2 + x_1^2 - 3x_2 - 2x_1) + \cdots = 0 \qquad (2.9)$$

which, upon letting $\epsilon \to 0$, yields

$$2x_0 x_1 - 3x_1 - 2x_0 + 1 = 0 \qquad (2.10)$$

Then, (2.9) becomes

$$\epsilon(2x_0 x_2 + x_1^2 - 3x_2 - 2x_1) + \cdots = 0$$

which, when divided by ϵ, yields

$$2x_0 x_2 + x_1^2 - 3x_2 - 2x_1 + O(\epsilon) = 0 \qquad (2.11)$$

Letting $\epsilon \to 0$ in (2.11), we have

$$2x_0 x_2 + x_1^2 - 3x_2 - 2x_1 = 0 \qquad (2.12)$$

We note that (2.8), (2.10), and (2.12) can be obtained directly from (2.7) by equating the coefficient of each power of ϵ to zero.

The fifth step involves solving the simplified equations (2.8), (2.10), and (2.12) in succession. Equation (2.8) is the same as the reduced equation (2.2), and hence, its solutions are

$$x_0 = 1, 2$$

With x_0 known, we can solve (2.10) for x_1. We note that (2.10) is linear in x_1. In most problems, all perturbation equations are linear, except perhaps the first. When $x_0 = 1$, (2.10) becomes

$$x_1 + 1 = 0 \quad \text{or} \quad x_1 = -1$$

With x_0 and x_1 known, we can solve (2.12) for x_2. When $x_0 = 1$, $x_1 = -1$ and (2.12) becomes

$$x_2 - 3 = 0 \quad \text{or} \quad x_2 = 3$$

When $x_0 = 2$, (2.10) becomes

$$x_1 - 3 = 0 \quad \text{or} \quad x_1 = 3$$

Then, (2.12) becomes

$$x_2 + 3 = 0 \quad \text{or} \quad x_2 = -3$$

The last step involves substituting the values obtained for x_0, x_1, and x_2 into the assumed expansion (2.3). When $x_0 = 1$, $x_1 = -1$ and $x_2 = 3$; therefore, it follows from (2.3) that

$$x = 1 - \epsilon + 3\epsilon^2 + \cdots \qquad (2.13)$$

When $x_0 = 2$, $x_1 = 3$ and $x_2 = -3$; therefore, it follows from (2.3) that

$$x = 2 + 3\epsilon - 3\epsilon^2 + \cdots \qquad (2.14)$$

Equations (2.13) and (2.14) provide approximations for the two roots of (2.1). To determine how good these approximations are, we compare them with the exact solution

$$x = \tfrac{1}{2}[3 + 2\epsilon \mp \sqrt{(3 + 2\epsilon)^2 - 4(2 + \epsilon)}]$$

or

$$x = \tfrac{1}{2}[3 + 2\epsilon \mp \sqrt{1 + 8\epsilon + 4\epsilon^2}] \qquad (2.15)$$

Using the binomial theorem, we have

$$(1 + 8\epsilon + 4\epsilon^2)^{1/2} = 1 + \tfrac{1}{2}(8\epsilon + 4\epsilon^2) + \frac{(\tfrac{1}{2})(-\tfrac{1}{2})}{2!}(8\epsilon + 4\epsilon^2)^2 + \cdots$$

$$= 1 + 4\epsilon + 2\epsilon^2 - \tfrac{1}{8}(64\epsilon^2 + \cdots) = 1 + 4\epsilon - 6\epsilon^2 + \cdots$$

which, when substituted into (2.15), gives

$$x = \begin{cases} \tfrac{1}{2}(3 + 2\epsilon + 1 + 4\epsilon - 6\epsilon^2 + \cdots) \\ \tfrac{1}{2}(3 + 2\epsilon - 1 - 4\epsilon + 6\epsilon^2 + \cdots) \end{cases}$$

or

$$x = \begin{cases} 2 + 3\epsilon - 3\epsilon^2 + \cdots \\ 1 - \epsilon + 3\epsilon^2 + \cdots \end{cases} \qquad (2.16)$$

in agreement with (2.13) and (2.14).

EXAMPLE 2

As a second example, we consider an equation whose roots may involve fractional rather than integral powers of ϵ. Specifically, we consider

$$(x - 1)(x - \tau) = -\epsilon x \qquad (2.17)$$

When $\epsilon = 0$, (2.17) reduces to

$$(x - 1)(x - \tau) = 0$$

whose roots are $x = 1$ and τ. This suggests that we seek approximations to the roots of (2.17) in the form

32 ALGEBRAIC EQUATIONS

$$x = x_0 + \epsilon x_1 + \epsilon^2 x_2 + \cdots \tag{2.18}$$

Here again, we are stopping at $O(\epsilon^2)$. Hence, the resulting expansion is called a second-order expansion. Substituting (2.18) into (2.17), we have

$$(x_0 - 1 + \epsilon x_1 + \epsilon^2 x_2 + \cdots)(x_0 - \tau + \epsilon x_1 + \epsilon^2 x_2 + \cdots)$$
$$= -\epsilon(x_0 + \epsilon x_1 + \epsilon^2 x_2 + \cdots)$$

which, upon expanding, yields

$$(x_0 - 1)(x_0 - \tau) + \epsilon(x_0 - 1)x_1 + \epsilon^2(x_0 - 1)x_2 + \epsilon(x_0 - \tau)x_1$$
$$+ \epsilon^2 x_1^2 + \epsilon^2(x_0 - \tau)x_2 + \epsilon x_0 + \epsilon^2 x_1 + \cdots = 0$$

Collecting coefficients of like powers of ϵ gives

$$(x_0 - 1)(x_0 - \tau) + \epsilon[(2x_0 - 1 - \tau)x_1 + x_0]$$
$$+ \epsilon^2[(2x_0 - 1 - \tau)x_2 + x_1^2 + x_1] + \cdots = 0 \tag{2.19}$$

As before, only terms up to $O(\epsilon^2)$ have been retained, consistent with the assumed expansion. Equating the coefficient of each power of ϵ in (2.19) to zero, we obtain

$$(x_0 - 1)(x_0 - \tau) = 0 \tag{2.20}$$
$$(2x_0 - 1 - \tau)x_1 + x_0 = 0 \tag{2.21}$$
$$(2x_0 - 1 - \tau)x_2 + x_1^2 + x_1 = 0 \tag{2.22}$$

which can be solved in succession for x_0, x_1, and x_2.

The solutions of (2.20) are

$$x_0 = 1 \quad \text{or} \quad \tau$$

When $x_0 = 1$, (2.21) becomes

$$(1 - \tau)x_1 + 1 = 0 \quad \text{so that} \quad x_1 = -\frac{1}{1 - \tau}$$

Then, (2.22) becomes

$$(1 - \tau)x_2 = -\frac{1}{(1 - \tau)^2} + \frac{1}{1 - \tau} = -\frac{\tau}{(1 - \tau)^2}$$

so that

$$x_2 = -\frac{\tau}{(1 - \tau)^3}$$

Hence, one of the roots is

$$x = 1 - \frac{\epsilon}{1-\tau} - \frac{\epsilon^2 \tau}{(1-\tau)^3} + \cdots \qquad (2.23)$$

When $x_0 = \tau$, (2.21) becomes

$$(\tau - 1)x_1 + \tau = 0 \qquad \text{so that} \qquad x_1 = \frac{\tau}{1-\tau}$$

Then, (2.22) becomes

$$(\tau - 1)x_2 = -\frac{\tau}{1-\tau} - \frac{\tau^2}{(1-\tau)^2} = -\frac{\tau}{(1-\tau)^2}$$

so that

$$x_2 = \frac{\tau}{(1-\tau)^3}$$

Hence, the second root is

$$x = \tau + \frac{\epsilon \tau}{1-\tau} + \frac{\epsilon^2 \tau}{(1-\tau)^3} + \cdots \qquad (2.24)$$

Equations (2.23) and (2.24) show that the above expansions *break down* (i.e., are *nonuniform*) as $\tau \to 1$ because the "corrections" to the solution of the reduced equation tend to ∞. In fact, τ need not be exactly equal to 1 for the above expansions to break down. The expansions break down whenever the first-order term, second-order term, and so on are the order of the zeroth-order term, because the corrections to the zeroth-order term will not be small, contrary to the assumption underlying the method. To determine the order of $\tau - 1$ for which the above expansions break down (i.e., *region of nonuniformity*), we determine the conditions under which successive terms are the same order. It follows from (2.23) that the zeroth- and first-order terms are the same order when

$$\frac{\epsilon}{1-\tau} = O(1) \qquad \text{or} \qquad 1 - \tau = O(\epsilon)$$

whereas the first- and second-order terms are the same order when

$$\frac{\epsilon}{1-\tau} = O\left[\frac{\epsilon^2}{(1-\tau)^3}\right] \qquad \text{or} \qquad (1-\tau)^2 = O(\epsilon)$$

or

$$1 - \tau = O(\epsilon^{1/2})$$

Since for small ϵ, $\epsilon^{1/2}$ is bigger than ϵ, the region of nonuniformity is $1 - \tau = O(\epsilon^{1/2})$, the larger of the above two regions.

34 ALGEBRAIC EQUATIONS

As discussed in Chapter 1, nonuniformities in the expansions arise whenever an elementary operation used in obtaining the expansions is not justified. To determine this operation, we investigate the exact solution. To this end, we rewrite (2.17) as

$$x^2 - x - \tau x + \tau + \epsilon x = 0$$

or

$$x^2 - (1 + \tau - \epsilon) x + \tau = 0$$

whose roots are given by

$$x = \tfrac{1}{2} [1 + \tau - \epsilon \mp \sqrt{(1 + \tau - \epsilon)^2 - 4\tau}]$$

or

$$x = \tfrac{1}{2} [1 + \tau - \epsilon \mp \sqrt{(1 - \tau)^2 - 2\epsilon(1 + \tau) + \epsilon^2}] \quad (2.25)$$

Next, we expand (2.25) for small ϵ and compare the result with (2.23) and (2.24). Using the binomial theorem, we have

$$[(1 - \tau)^2 - 2\epsilon(1 + \tau) + \epsilon^2]^{1/2} = (1 - \tau)\left[1 - \frac{2\epsilon(1 + \tau) - \epsilon^2}{(1 - \tau)^2}\right]^{1/2}$$

$$= (1 - \tau)\left\{1 - \frac{1}{2}\frac{2\epsilon(1 + \tau) - \epsilon^2}{(1 - \tau)^2} + \frac{(\tfrac{1}{2})(-\tfrac{1}{2})}{2!}\frac{[2\epsilon(1 + \tau) - \epsilon^2]^2}{(1 - \tau)^4} + \cdots\right\}$$

$$= (1 - \tau)\left[1 - \frac{\epsilon(1 + \tau)}{(1 - \tau)^2} + \frac{\epsilon^2}{2(1 - \tau)^2} - \frac{1}{8}\frac{4\epsilon^2(1 + \tau)^2}{(1 - \tau)^4} + \cdots\right]$$

$$= (1 - \tau)\left[1 - \frac{\epsilon(1 + \tau)}{(1 - \tau)^2} - \frac{2\epsilon^2 \tau}{(1 - \tau)^4} + \cdots\right] \quad (2.26)$$

where again only terms up to $O(\epsilon^2)$ have been retained, consistent with the assumed expansion. Putting (2.26) into (2.25) with the positive sign gives one of the roots as

$$x = \tfrac{1}{2}\left[1 + \tau - \epsilon + 1 - \tau - \frac{\epsilon(1 + \tau)}{1 - \tau} - \frac{2\epsilon^2 \tau}{(1 - \tau)^3} + \cdots\right] \quad (2.27)$$

or

$$x = 1 - \frac{\epsilon}{1 - \tau} - \frac{\epsilon^2 \tau}{(1 - \tau)^3} + \cdots$$

in agreement with (2.23). Putting (2.26) into (2.25) with the negative sign gives the second root as

$$x = \tfrac{1}{2}\left[1 + \tau - \epsilon - 1 + \tau + \frac{\epsilon(1 + \tau)}{1 - \tau} + \frac{2\epsilon^2 \tau}{(1 - \tau)^3} + \cdots\right]$$

or

$$x = \tau + \frac{\epsilon\tau}{1-\tau} + \frac{\epsilon^2\tau}{(1-\tau)^3} + \cdots \qquad (2.28)$$

in agreement with (2.24).

In arriving at (2.27) and (2.28) from the exact solution, we performed only the exponentiation operation in (2.26) and the addition and subtraction operations in (2.27) and (2.28). The subtraction and addition operations are usually justified, and hence, the exponentiation operation is the suspect operation. In approximating

$$(1-u)^{1/2} \quad \text{by} \quad 1 - \tfrac{1}{2}u + \frac{(\tfrac{1}{2})(-\tfrac{1}{2})}{2!}u^2 + \cdots$$

we made the implicit assumption that $|u| < 1$. In the present example,

$$u = \frac{2\epsilon(1+\tau) - \epsilon^2}{(1-\tau)^2} \qquad (2.29)$$

whose magnitude is small compared with 1 only when τ is away from 1. In fact, at $\tau = 1$, $u = \infty$, irrespective of how small ϵ is as long as it is different from zero. It follows from (2.29) that the binomial expansion is not justified when $u = O(1)$ or $(1-\tau)^2 = O(\epsilon)$ or $1 - \tau = O(\epsilon^{1/2})$.

Therefore, to obtain a uniform expansion when $1 - \tau = O(\epsilon^{1/2})$, we need to modify the above procedure by taking this fact into account. This can be formalized by introducing a so-called detuning parameter σ defined by

$$1 - \tau = \epsilon^{1/2}\sigma \qquad (2.30)$$

where σ is independent of ϵ. Putting (2.30) in (2.17) gives

$$(x - 1)(x - 1 + \epsilon^{1/2}\sigma) = -\epsilon x \qquad (2.31)$$

When $\epsilon = 0$, (2.31) reduces to

$$(x - 1)^2 = 0$$

which yields the double root $x = 1$. This fact and the presence of $\epsilon^{1/2}$ in (2.31) suggest trying an expansion in the form

$$x = 1 + \epsilon^{1/2} x_1 + \cdots \qquad (2.32)$$

We stop at $O(\epsilon^{1/2})$ because obtaining the higher-order terms is straightforward. Substituting the first two terms in (2.32) into (2.31) gives

$$(\epsilon^{1/2} x_1 + \cdots)(\epsilon^{1/2} x_1 + \epsilon^{1/2}\sigma + \cdots) = -\epsilon(1 + \epsilon^{1/2} x_1 + \cdots)$$

or

36 ALGEBRAIC EQUATIONS

$$\epsilon x_1^2 + \epsilon \sigma x_1 + \epsilon + \cdots = 0$$

Hence,

$$x_1^2 + \sigma x_1 + 1 = 0$$

whose roots are

$$x_1 = \tfrac{1}{2}(-\sigma \mp \sqrt{\sigma^2 - 4})$$

Therefore, the roots of (2.17) in this case are given by

$$\begin{aligned} x &= 1 - \tfrac{1}{2}\epsilon^{1/2}(\sigma + \sqrt{\sigma^2 - 4}) + \cdots \\ x &= 1 - \tfrac{1}{2}\epsilon^{1/2}(\sigma - \sqrt{\sigma^2 - 4}) + \cdots \end{aligned} \qquad (2.33)$$

which are regular at $\sigma = 0$ or $\tau = 1$.

EXAMPLE 3

As a third example, we consider

$$\epsilon x^2 + x + 1 = 0 \qquad (2.34)$$

in which the small parameter multiplies the highest power of x. Since (2.34) is a quadratic equation, it has two roots. However, as $\epsilon \to 0$, (2.34) reduces to

$$x + 1 = 0 \qquad (2.35)$$

which is of first order, and hence, it has only one solution. Thus, x is discontinuous at $\epsilon = 0$. Such perturbation problems are called *singular perturbation problems*.

Equation (2.35) suggests trying the expansion

$$x = x_0 + \epsilon x_1 + \cdots \qquad (2.36)$$

for one of the roots. To minimize the algebra, we determine only first-order expansions in the remainder of this chapter. Putting (2.36) in (2.34), we have

$$\epsilon(x_0 + \epsilon x_1 + \cdots)^2 + x_0 + \epsilon x_1 + \cdots + 1 = 0$$

or

$$\epsilon(x_0^2 + 2\epsilon x_0 x_1) + x_0 + \epsilon x_1 + 1 + \cdots = 0$$

or

$$x_0 + 1 + \epsilon(x_1 + x_0^2) + \cdots = 0$$

Equating coefficients of like powers of ϵ gives

$$x_0 + 1 = 0$$
$$x_1 + x_0^2 = 0$$

which can be solved successively for x_0 and x_1. Hence, $x_0 = -1$ and $x_1 = -x_0^2 = -1$ so that one of the roots is

$$x = -1 - \epsilon + \cdots \tag{2.37}$$

Thus, as expected, the above procedure yielded only one root. To devise a modified procedure for determining the other root, we investigate the exact solution, that is,

$$x = \frac{1}{2\epsilon}(-1 \mp \sqrt{1 - 4\epsilon}) \tag{2.38}$$

Using the binomial theorem, we have

$$\sqrt{1 - 4\epsilon} = 1 - 2\epsilon + \frac{(\frac{1}{2})(-\frac{1}{2})}{2!}(-4\epsilon)^2 + \cdots$$

$$= 1 - 2\epsilon - 2\epsilon^2 + \cdots \tag{2.39}$$

Substituting (2.39) into (2.38) with the positive sign yields one of the roots as

$$x = \frac{-1 + 1 - 2\epsilon - 2\epsilon^2 + \cdots}{2\epsilon} = -1 - \epsilon + \cdots \tag{2.40}$$

in agreement with (2.37). Substituting (2.39) into (2.38) with the negative sign yields the other root as

$$x = \frac{-1 - 1 + 2\epsilon + 2\epsilon^2 + \cdots}{2\epsilon} = -\frac{1}{\epsilon} + 1 + \epsilon + \cdots \tag{2.41}$$

Therefore, both of the roots go in powers of ϵ but one starts with ϵ^{-1}. Hence, it is not surprising that the assumed form (2.36) of the expansion failed to produce the root (2.41). Consequently, one cannot determine the second root by a perturbation technique unless its form is known. However, for a general problem whose exact solution is not known, the form of the roots is not known a priori and must be determined as part of the solution. In those cases, we recognize that, if the order of the equation is not to be reduced, the other roots tend to ∞ as $\epsilon \to 0$, and hence, assume that the leading term has the form

$$x = \frac{y}{\epsilon^\nu} + \cdots \tag{2.42}$$

where ν must be greater than zero and needs to be determined in the course of analysis. Substituting (2.42) into (2.34), we have

$$\epsilon^{1 - 2\nu} y^2 + \epsilon^{-\nu} y + 1 + \cdots = 0 \tag{2.43}$$

Next, we extract the dominant terms in (2.43). To recover the second root, we must keep the first term $\epsilon^{1 - 2\nu} y^2$; otherwise, we will end up where we started.

Since $\nu > 0$, the second term is much bigger than 1. Hence, the dominant part of (2.43) is

$$\epsilon^{1-2\nu} y^2 + \epsilon^{-\nu} y = 0 \tag{2.44}$$

which demands the powers of ϵ be the same. That is,

$$1 - 2\nu = -\nu \quad \text{or} \quad \nu = 1$$

for y to be different from zero. Then, it follows from (2.44) that

$$y = 0 \quad \text{or} \quad -1$$

The first value $y = 0$ corresponds to the first root (2.37) since in the region $O(\epsilon^{-1})$ it appears to be zero, whereas $y = -1$ corresponds to the second root. Thus, it follows from (2.42) that to the first approximation the second root is given by

$$x = -\frac{1}{\epsilon} + \cdots$$

in agreement with (2.41). To determine more terms in the expansion of the second root, we try

$$x = -\frac{1}{\epsilon} + x_0 + \cdots \tag{2.45}$$

Substituting (2.45) into (2.34) yields

$$\epsilon \left(-\frac{1}{\epsilon} + x_0 + \cdots\right)^2 - \frac{1}{\epsilon} + x_0 + \cdots + 1 = 0$$

or

$$\epsilon \left(\frac{1}{\epsilon^2} - \frac{2x_0}{\epsilon} + x_0^2 + \cdots\right) - \frac{1}{\epsilon} + x_0 + 1 + \cdots = 0$$

or

$$-2x_0 + x_0 + 1 + O(\epsilon) = 0$$

Hence, $x_0 = 1$ and (2.45) becomes

$$x = -\frac{1}{\epsilon} + 1 + \cdots$$

in agreement with (2.41).

Alternatively, once ν has been determined, we view (2.42) as a transformation from x to y. Then, putting $x = y/\epsilon$ in (2.34) yields

$$y^2 + y + \epsilon = 0 \tag{2.46}$$

which can be solved to determine both roots because ϵ does not multiply the highest order.

2.2. Cubic Equations

In this section, we also consider three examples. The roots of the first example can be expressed in powers of the small parameter ϵ, the roots of the second example need to be expressed in fractional powers of ϵ, and some of the roots of the third start with inverse powers of ϵ.

EXAMPLE 1
We consider the equation

$$x^3 - (6 + \epsilon)x^2 + (11 + 2\epsilon)x - 6 + \epsilon^2 = 0 \tag{2.47}$$

We try an expansion in powers of ϵ as

$$x = x_0 + \epsilon x_1 + \cdots \tag{2.48}$$

Substituting (2.48) into (2.47) gives

$$(x_0 + \epsilon x_1 + \cdots)^3 - (6 + \epsilon)(x_0 + \epsilon x_1 + \cdots)^2 + (11 + 2\epsilon)(x_0 + \epsilon x_1 + \cdots)$$
$$- 6 + \epsilon^2 = 0$$

or

$$x_0^3 + 3\epsilon x_0^2 x_1 - (6 + \epsilon)(x_0^2 + 2\epsilon x_0 x_1) + (11 + 2\epsilon)(x_0 + \epsilon x_1) - 6 + \epsilon^2 + \cdots = 0$$

or

$$x_0^3 + 3\epsilon x_0^2 x_1 - 6x_0^2 - 12\epsilon x_0 x_1 - \epsilon x_0^2 + 11x_0 + 11\epsilon x_1 + 2\epsilon x_0 - 6 + \cdots = 0$$

Collecting coefficients of equal powers of ϵ gives

$$x_0^3 - 6x_0^2 + 11x_0 - 6 + \epsilon(3x_0^2 x_1 - 12x_0 x_1 + 11x_1 - x_0^2 + 2x_0) + \cdots = 0$$

where terms up to $O(\epsilon)$ have been retained, consistent with the order of the assumed expansion. Equating each of the coefficients of ϵ^0 and ϵ to zero yields

$$x_0^3 - 6x_0^2 + 11x_0 - 6 = 0 \tag{2.49}$$

$$3x_0^2 x_1 - 12x_0 x_1 + 11x_1 - x_0^2 + 2x_0 = 0 \tag{2.50}$$

Equation (2.49) can be rewritten in factored form as

$$(x_0 - 1)(x_0 - 2)(x_0 - 3) = 0$$

Hence,

$$x_0 = 1 \quad \text{or} \quad 2 \quad \text{or} \quad 3$$

It follows from (2.50) that

$$(3x_0^2 - 12x_0 + 11)x_1 = x_0^2 - 2x_0$$

whose solution is

$$x_1 = \frac{x_0^2 - 2x_0}{3x_0^2 - 12x_0 + 11} \qquad (2.51)$$

When $x_0 = 1$, it follows from (2.51) that $x_1 = -\frac{1}{2}$. Hence, one of the roots is given by

$$x = 1 - \tfrac{1}{2}\epsilon + \cdots$$

When $x_0 = 2$, it follows from (2.51) that $x_1 = 0$. Hence, a second root is given by

$$x = 2 + (0)\epsilon + \cdots$$

When $x_0 = 3$, it follows from (2.51) that $x_1 = \frac{3}{2}$. Hence, the third root is given by

$$x = 3 + \tfrac{3}{2}\epsilon + \cdots$$

Thus, in this case all roots go in powers of ϵ.

EXAMPLE 2

As a second example, we consider

$$x^3 - (4 + \epsilon)x^2 + (5 - 2\epsilon)x - 2 + \epsilon^2 = 0 \qquad (2.52)$$

Again, let us try an expansion in the form

$$x = x_0 + \epsilon x_1 + \cdots \qquad (2.53)$$

Substituting (2.53) into (2.52) gives

$$(x_0 + \epsilon x_1 + \cdots)^3 - (4 + \epsilon)(x_0 + \epsilon x_1 + \cdots)^2 + (5 - 2\epsilon)(x_0 + \epsilon x_1 + \cdots)$$
$$- 2 + \epsilon^2 = 0$$

or

$$x_0^3 - 4x_0^2 + 5x_0 - 2 + \epsilon(3x_0^2 x_1 - 8x_0 x_1 - x_0^2 + 5x_1 - 2x_0) + \cdots = 0$$

Equating coefficients of like powers of ϵ, we have

$$x_0^3 - 4x_0^2 + 5x_0 - 2 = 0 \qquad (2.54)$$
$$3x_0^2 x_1 - 8x_0 x_1 - x_0^2 + 5x_1 - 2x_0 = 0 \qquad (2.55)$$

To solve (2.54), we factor its left-hand side and obtain

$$(x_0 - 1)^2 (x_0 - 2) = 0$$

which yields

CUBIC EQUATIONS

$$x_0 = 1, \quad 1, \quad \text{and} \quad 2$$

To solve (2.55) for x_1, we first rewrite it as

$$(3x_0^2 - 8x_0 + 5) x_1 = x_0^2 + 2x_0$$

Hence,

$$x_1 = \frac{x_0^2 + 2x_0}{3x_0^2 - 8x_0 + 5} \tag{2.56}$$

When $x_0 = 2$, it follows from (2.56) that $x_1 = 8$. Hence, one of the roots is given by

$$x = 2 + 8\epsilon + \cdots \tag{2.57}$$

When $x_0 = 1$, it follows from (2.56) that $x_1 = \infty$, indicating that the assumed form of the expansion is wrong.

To determine a valid expansion when $x_0 = 1$, we change the form of the expansion (2.53) to the following:

$$x = 1 + \epsilon^\nu x_1 + \epsilon^{2\nu} x_2 + \cdots \quad \nu > 0 \tag{2.58}$$

and determine ν in the course of analysis. Putting (2.58) in (2.52), we have

$$(1 + \epsilon^\nu x_1 + \epsilon^{2\nu} x_2 + \cdots)^3 - (4 + \epsilon)(1 + \epsilon^\nu x_1 + \epsilon^{2\nu} x_2 + \cdots)^2$$
$$+ (5 - 2\epsilon)(1 + \epsilon^\nu x_1 + \epsilon^{2\nu} x_2 + \cdots) - 2 + \cdots = 0$$

or

$$1 + 3\epsilon^\nu x_1 + 3\epsilon^{2\nu} x_2 + 3\epsilon^{2\nu} x_1^2 - 4 - 8\epsilon^\nu x_1 - 8\epsilon^{2\nu} x_2 - 4\epsilon^{2\nu} x_1^2$$
$$- \epsilon - 2\epsilon^{1+\nu} x_1 + 5 + 5\epsilon^\nu x_1 + 5\epsilon^{2\nu} x_2 - 2\epsilon - 2\epsilon^{1+\nu} x_1 - 2 + \cdots = 0$$

Hence,

$$-x_1^2 \epsilon^{2\nu} - 3\epsilon + \cdots = 0 \tag{2.59}$$

In order that the dominant terms in (2.59) balance each other, 2ν must be equal to 1 or $\nu = \frac{1}{2}$, and $x_1 = \pm\sqrt{3}i$. Hence, it follows from (2.58) that the second and third roots are given by

$$x = 1 \pm \epsilon^{1/2} \sqrt{3}i + \cdots$$

This example illustrates the fact that difficulties arise whenever the assumed form of the expansion is not correct. But once the form is corrected, a consistent solution is obtained. This is typical of perturbation problems.

EXAMPLE 3
As a third example, we consider

$$\epsilon x^3 + x + 2 + \epsilon = 0 \tag{2.60}$$

42 ALGEBRAIC EQUATIONS

in which the small parameter multiplies the highest power of x. As $\epsilon \to 0$, (2.60) reduces to

$$x + 2 = 0$$

and hence, we assume that one of the roots has the form

$$x = x_0 + \epsilon x_1 + \cdots \qquad (2.61)$$

Substituting (2.61) into (2.60) yields

$$\epsilon(x_0 + \epsilon x_1 + \cdots)^3 + x_0 + \epsilon x_1 + \cdots + 2 + \epsilon = 0$$

or

$$x_0 + 2 + \epsilon(x_1 + x_0^3 + 1) + \cdots = 0$$

Equating coefficients of like powers of ϵ, we have

$$x_0 + 2 = 0$$
$$x_1 + x_0^3 + 1 = 0$$

Hence, $x_0 = -2$ and $x_1 = 7$. Therefore, one of the roots is given by

$$x = -2 + 7\epsilon + \cdots$$

To determine the other roots, we note that they tend to ∞ as $\epsilon \to 0$ because ϵ multiplies the highest order. Hence, to determine expansions for these roots, we assume that their leading terms have the form

$$x = \frac{y}{\epsilon^\nu} + \cdots \qquad \nu > 0 \qquad (2.62)$$

Substituting (2.62) into (2.60), we have

$$\epsilon^{1-3\nu} y^3 + \epsilon^{-\nu} y + 2 + \cdots = 0 \qquad (2.63)$$

In order that the dominant parts in (2.63) balance each other,

$$1 - 3\nu = -\nu \quad \text{or} \quad \nu = \tfrac{1}{2}$$
$$y^3 + y = 0$$

Hence,

$$y = 0 \quad i \quad -i$$

The case $y = 0$ corresponds to the first root, and hence, it is discarded here.

To determine an improved approximation to the second and third roots, we use the above information and seek expansions in the form

$$x = \frac{y}{\epsilon^{1/2}} + x_0 + \cdots \qquad (2.64)$$

where $y = i$ or $-i$. Substituting (2.64) into (2.60) gives

$$\epsilon\left(\frac{y^3}{\epsilon^{3/2}} + \frac{3y^2 x_0}{\epsilon} + \cdots\right) + \frac{y}{\epsilon^{1/2}} + x_0 + \cdots + 2 + \epsilon = 0$$

or

$$\epsilon^{-1/2}(y^3 + y) + 3y^2 x_0 + x_0 + 2 + \cdots = 0$$

Equating coefficients of like powers of ϵ, we have

$$y^3 + y = 0$$
$$3y^2 x_0 + x_0 + 2 = 0$$

Hence, as before, $y = \pm i$ and

$$x_0 = -\frac{2}{3y^2 + 1} = 1$$

Therefore, the second and third roots are given by

$$x = \pm \frac{i}{\epsilon^{1/2}} + 1 + \cdots$$

2.3. Higher-Order Equations

In this section, we consider higher-order equations and concentrate on the case in which the small parameter multiplies the highest power of the unknown. Specifically, we consider

$$\epsilon x^n = x^m + a_{m-1} x^{m-1} + a_{m-2} x^{m-2} + \cdots + a_1 x + a_0 \qquad (2.65)$$

where the a_s are independent of ϵ and x, n and m are integers, and $n > m$. As $\epsilon \to 0$, (2.65) reduces to

$$x^m + a_{m-1} x^{m-1} + a_{m-2} x^{m-2} + \cdots + a_1 x + a_0 = 0 \qquad (2.66)$$

which has the roots α_s, where $s = 1, 2, 3, \ldots, m$. To improve upon these roots, we let

$$x = x_0 + \epsilon x_1 + \cdots \qquad (2.67)$$

in (2.65) and obtain

$$\epsilon(x_0 + \epsilon x_1 + \cdots)^n = (x_0 + \epsilon x_1 + \cdots)^m + a_{m-1}(x_0 + \epsilon x_1 + \cdots)^{m-1}$$
$$+ a_{m-2}(x_0 + \epsilon x_1 + \cdots)^{m-2} + \cdots + a_1(x_0 + \epsilon x_1 + \cdots) + a_0$$

or

44 ALGEBRAIC EQUATIONS

$$x_0^m + a_{m-1}x_0^{m-1} + a_{m-2}x_0^{m-2} + \cdots + a_1x_0 + a_0 + \epsilon[mx_0^{m-1}$$
$$+ (m-1)a_{m-1}x_0^{m-2} + (m-2)a_{m-2}x_0^{m-3} + \cdots + a_1]x_1 - \epsilon x_0^n + O(\epsilon^2) = 0$$

Equating coefficients of like powers of ϵ, we have

$$x_0^m + a_{m-1}x_0^{m-1} + a_{m-2}x_0^{m-2} + \cdots + a_1x_0 + a_0 = 0 \qquad (2.68)$$

$$[mx_0^{m-1} + (m-1)a_{m-1}x_0^{m-2} + (m-2)a_{m-2}x_0^{m-3} + \cdots + a_1]x_1 = x_0^n \qquad (2.69)$$

Equation (2.68) is the same as (2.66), and hence, has the roots $x_0 = \alpha_s$, where $s = 1, 2, 3, \ldots, m$. Then, it follows from (2.69) that

$$x_1 = \alpha_s^n [m\alpha_s^{m-1} + (m-1)a_{m-1}\alpha_s^{m-2} + \cdots + a_1]^{-1}$$

Hence,

$$x = \alpha_s + \epsilon \alpha_s^n [m\alpha_s^{m-1} + (m-1)a_{m-1}\alpha_s^{m-2} + \cdots + a_1]^{-1} + \cdots \qquad (2.70)$$

We should note that (2.70) breaks down whenever the term inside the brackets vanishes. This corresponds to a multiple root of (2.68). In this case, the expansion goes in fractional powers of ϵ and one needs to follow the procedure used in Example 2 of the preceding section.

To determine the remaining $n - m$ roots, we note that they tend to ∞ as $\epsilon \to 0$ because ϵ multiplies the highest power of x. Then, we assume expansions for them in the form

$$x = \frac{y}{\epsilon^\nu} + x_0 + \cdots \qquad \nu > 0 \qquad (2.71)$$

Substituting (2.71) into (2.65), we have

$$\epsilon \left(\frac{y^n}{\epsilon^{n\nu}} + \frac{ny^{n-1}x_0}{\epsilon^{(n-1)\nu}} + \cdots \right) = \frac{y^m}{\epsilon^{m\nu}} + \frac{my^{m-1}x_0}{\epsilon^{(m-1)\nu}} + \cdots + \frac{a_{m-1}y^{m-1}}{\epsilon^{(m-1)\nu}}$$
$$+ \frac{(m-1)a_{m-1}y^{m-2}x_0}{\epsilon^{(m-2)\nu}} + \cdots \qquad (2.72)$$

Extracting the dominant terms, we have

$$\epsilon^{(1-n\nu)}y^n = \epsilon^{-m\nu}y^m$$

Hence,

$$(1 - n\nu) = -m\nu \quad \text{so that} \quad \nu = \frac{1}{n-m} \qquad (2.73)$$

$$y^n = y^m \qquad (2.74)$$

Equation (2.74) has 0 as a root with a multiplicity of m and

$$y^{n-m} = 1 = e^{2ir\pi}$$

where $r = 1, 2, 3, \ldots, (n - m)$. Hence,

$$y = \omega, \omega^2, \ldots, \omega^k \qquad \omega = \exp\left(\frac{2i\pi}{n-m}\right) \qquad (2.75)$$

where $k = n - m$. We discard the root $y = 0$ because it corresponds to the first s roots.

Using (2.73) and (2.74), we rewrite (2.72) as

$$ny^{n-1} x_0 \epsilon^v = my^{m-1} x_0 \epsilon^v + a_{m-1} y^{m-1} \epsilon^v + \cdots$$

Hence, equating the coefficients of ϵ^v on both sides yields

$$ny^{n-1} x_0 = my^{m-1} x_0 + a_{m-1} y^{m-1}$$

Hence,

$$x_0 = \frac{a_{m-1} y^{m-1}}{ny^{n-1} - my^{m-1}} = \frac{a_{m-1}}{ny^{n-m} - m} = \frac{a_{m-1}}{n - m} \qquad (2.76)$$

Therefore, the last $n - m$ roots are given by

$$x = \frac{\omega^r}{\epsilon^v} + \frac{a_{m-1}}{n-m} + \cdots \qquad r = 1, 2, \ldots, n - m \qquad (2.77)$$

where v and ω are defined by (2.73) and (2.75).

2.4. Transcendental Equations

We consider the roots of Bessel's function $J_0(x)$ for large x, that is, we consider the roots of the transcendental equation

$$J_0(x) = 0$$

when x is large. In Chapter 13, we determine an expansion for $J_0(x)$ as $x \to \infty$. It follows from (13.141) that

$$J_0(x) \sim \sqrt{\frac{2}{\pi x}} \left[u \cos\left(x - \tfrac{1}{4}\pi\right) + v \sin\left(x - \tfrac{1}{4}\pi\right) \right] \qquad \text{as} \qquad x \to \infty \qquad (2.78)$$

where u and v are defined in (13.129) and (13.130) as

$$u(x) = 1 - \frac{1^2 \cdot 3^2}{4^2 \cdot 2^2 \cdot 2! x^2} + \frac{1^2 \cdot 3^2 \cdot 5^2 \cdot 7^2}{4^4 \cdot 2^4 \cdot 4! x^4} + \cdots \qquad (2.79)$$

$$v(x) = \frac{1}{4 \cdot 2x} - \frac{1^2 \cdot 3^2 \cdot 5^2}{4^3 \cdot 2^3 \cdot 3! x^3} + \cdots \qquad (2.80)$$

46 ALGEBRAIC EQUATIONS

Setting $J_0(x) = 0$ in (2.78), we have

$$u \cos(x - \tfrac{1}{4}\pi) = -v \sin(x - \tfrac{1}{4}\pi)$$

or

$$\cot(x - \tfrac{1}{4}\pi) = -\frac{v}{u} \tag{2.81}$$

It follows from (2.79) and (2.80) that

$$\begin{aligned}\frac{v}{u} &= \left(\frac{1}{8x} - \frac{75}{1024x^3} + \cdots\right)\left(1 - \frac{9}{128x^2} + \cdots\right)^{-1} \\ &= \left(\frac{1}{8x} - \frac{75}{1024x^3} + \cdots\right)\left(1 + \frac{9}{128x^2} + \cdots\right) \\ &= \frac{1}{8x} - \frac{33}{512x^3} + \cdots \end{aligned} \tag{2.82}$$

where use has been made of the binomial theorem. Putting (2.82) into (2.81), we find that the large roots of $J_0(x)$ are governed by

$$\cot(x - \tfrac{1}{4}\pi) = -\frac{1}{8x} + \frac{33}{512x^3} + \cdots \tag{2.83}$$

Since x is large, the right-hand side of (2.83) can be neglected for the first approximation. The result is

$$\cot(x - \tfrac{1}{4}\pi) = 0$$

Hence,

$$x - \tfrac{1}{4}\pi = (n + \tfrac{1}{2})\pi \quad \text{or} \quad x = (n + \tfrac{3}{4})\pi \tag{2.84}$$

where n is an integer, which must be large in order that x be large. As shown below, even $n = 0$ yields an incredibly accurate result.

To determine an improved approximation to x, we put

$$x - \tfrac{1}{4}\pi = (n + \tfrac{1}{2})\pi + \delta \quad \text{or} \quad x = (n + \tfrac{3}{4})\pi + \delta \tag{2.85}$$

in (2.83) and obtain

$$\cot[(n + \tfrac{1}{2})\pi + \delta] = -\frac{1}{2\pi(4n + 3) + 8\delta} + \frac{33}{[2\pi(4n + 3) + 8\delta]^3} + \cdots \tag{2.86}$$

Using trigonometric identities, we rewrite the left-hand side as

$$\cot[(n + \tfrac{1}{2})\pi + \delta] = \frac{\cot(n + \tfrac{1}{2})\pi \cot\delta - 1}{\cot(n + \tfrac{1}{2})\pi + \cot\delta} = \frac{-1}{\cot\delta}$$

$$= -\tan\delta = -(\delta + \tfrac{1}{3}\delta^3 + \cdots)$$

Hence, (2.86) becomes

$$\delta + \tfrac{1}{3}\delta^3 = \frac{1}{2\pi(4n+3)+8\delta} - \frac{33}{[2\pi(4n+3)+8\delta]^3} + \cdots \quad (2.87)$$

which is an algebraic equation for δ.

We take $[2\pi(4n+3)]^{-1} = \epsilon$ as a small parameter and rewrite (2.87) as

$$\delta + \tfrac{1}{3}\delta^3 = \frac{\epsilon}{1+8\epsilon\delta} - \frac{33\epsilon^3}{(1+8\epsilon\delta)^3} + \cdots$$

or

$$\delta + \tfrac{1}{3}\delta^3 = \epsilon(1 - 8\epsilon\delta) - 33\epsilon^3 + \cdots \quad (2.88)$$

Next, we try the following expansion for δ:

$$\delta = \epsilon\delta_1 + \epsilon^2\delta_2 + \epsilon^3\delta_3 + \cdots \quad (2.89)$$

Substituting (2.89) into (2.88), we have

$$\epsilon\delta_1 + \epsilon^2\delta_2 + \epsilon^3\delta_3 + \cdots + \tfrac{1}{3}\epsilon^3\delta_1^3 = \epsilon - 8\epsilon^3\delta_1 - 33\epsilon^3 + \cdots$$

where only terms up to $O(\epsilon^3)$ have been retained. Equating coefficients of like powers of ϵ, we obtain

$$\delta_1 = 1 \qquad \delta_2 = 0 \qquad \delta_3 + \tfrac{1}{3}\delta_1^3 = -8\delta_1 - 33$$

Hence, $\delta_3 = -41\tfrac{1}{3}$. Therefore,

$$x = (n + \tfrac{3}{4})\pi + \epsilon - 41\tfrac{1}{3}\epsilon^3 + \cdots$$

or

$$x = (n + \tfrac{3}{4})\pi + \frac{1}{2\pi(4n+3)} - \frac{31}{6\pi^3(4n+3)^3} + \cdots \quad (2.90)$$

Table 2-1 compares the approximate solution (2.90) with the tabulated roots of $J_0(x)$. The agreement is incredible even for the lowest root, which is approximately 2.40482, a not very large number. The disagreement in the lowest mode is in the fourth significant figure and the error is approximately 0.07%. The accuracy improves as the root number increases. For the fourth root, the perturbation expansion agrees with the tabulated value to seven significant figures.

TABLE 2-1. Comparison of Approximate and Tabulated Roots of Bessel's Function of Order Zero

Root No.	1	2	3	4	5	6	7
Perturbation	2.40308	5.52004	8.65372	11.79153	14.93092	18.07106	21.21164
Tabulated	2.40482	5.52008	8.65373	11.79153	14.93092	18.07106	21.21164

48 ALGEBRAIC EQUATIONS

Exercises

2.1. For small ϵ, determine two terms in the expansion of each root of the following equations:

(a) $x^3 - (2 + \epsilon)x^2 - (1 - \epsilon)x + 2 + 3\epsilon = 0$
(b) $x^3 - (3 + \epsilon)x - 2 + \epsilon = 0$
(c) $x^3 + (3 - 2\epsilon)x^2 + (3 + \epsilon)x + 1 - 2\epsilon = 0$
(d) $x^4 + (2 - 3\epsilon)x^3 - (2 - \epsilon)x - 1 + 4\epsilon = 0$
(e) $x^4 + (4 - \epsilon)x^3 + (6 + 2\epsilon)x^2 + (4 + \epsilon)x + 1 - \epsilon^2 = 0$

2.2. For small ϵ, determine two terms in the expansion of each root of the following equations:

(a) $\epsilon(u^3 + u^2) + 4u^2 - 3u - 1 = 0$
(b) $\epsilon u^3 + u - 2 = 0$
(c) $\epsilon u^3 + (u - 2)^2 = 0$
(d) $u^2 - u - 2 + \frac{1}{2}\epsilon(u^3 + 2u + 3) = 0$
(e) $\epsilon u^4 + u^3 - 2u^2 - u + 2 = 0$
(f) $\epsilon u^4 - u^3 + 3u - 2 = 0$
(g) $\epsilon u^4 - u^2 + 3u - 2 = 0$
(h) $\epsilon u^4 + u^2 - 3u + 2 = 0$
(i) $\epsilon u^4 - u^2 + 2u - 1 = 0$
(j) $\epsilon u^4 + u^2 - 2u + 1 = 0$
(k) $\epsilon(u^4 + u^3) - u^2 + 3u - 2 = 0$
(l) $\epsilon(u^5 + u^4 - 2u^3) + 2u^2 - 3u + 1 = 0$
(m) $\epsilon(u^5 + u^4 - 2u^3) - 4u^2 + 4u - 1 = 0$
(n) $\epsilon^2 u^6 - \epsilon u^4 - u^3 + 2u^2 + u - 2 = 0$

2.3. For small ϵ, determine two-term expansions for the solutions of

(a) $s - \dfrac{\epsilon}{3s^2} - \dfrac{3\epsilon^2}{10s^4} = 0$

(b) $1 - \dfrac{\epsilon}{3\sqrt{s}} + \dfrac{21\epsilon^2}{5\sqrt{s}} = 0$

2.4. Determine two-term expansions for the large roots of

(a) $x \tan x = 1$
(b) $x \cot x = 1$

2.5. Differentiate (2.78) and use (2.79) and (2.80) to show that the large roots of $J_0'(x) = 0$ are approximately given by

$$\tan(x - \tfrac{1}{4}\pi) = -\frac{3}{8x} + \cdots$$

Then, show that

$$x = (n + \tfrac{1}{4})\pi - \frac{3}{2\pi(4n+1)} + \cdots$$

Compare this result with those tabulated for the first seven roots.

2.6. The asymptotic expansion of Bessel's function of second kind of order zero is

$$Y_0(x) \sim \sqrt{\frac{2}{\pi x}} \left[\sin(x - \tfrac{1}{4}\pi) - \frac{1}{8x} \cos(x - \tfrac{1}{4}\pi) \right] \quad \text{as} \quad x \to \infty$$

Show that the roots of $Y_0(x) = 0$ are given by

$$x = (n + \tfrac{1}{4})\pi + \frac{1}{2\pi(4n+1)} + \cdots$$

Compare this result with those tabulated for the first seven roots.

2.7. Using the asymptotic expansion of $Y_0(x)$ in the preceding exercise, show that the roots of $Y_0'(x) = 0$ are given by

$$x = (n + \tfrac{3}{4})\pi - \frac{3}{2\pi(4n+3)} + \cdots$$

Compare this result with those tabulated for the first seven roots.

2.8. The asymptotic expansion of $J_\nu(x)$ is

$$J_\nu(x) \sim \sqrt{\frac{2}{\pi x}} \left[\cos(x - \tfrac{1}{2}\nu\pi - \tfrac{1}{4}\pi) - \frac{4\nu^2 - 1}{8x} \sin(x - \tfrac{1}{2}\nu\pi - \tfrac{1}{4}\pi) \right] \quad \text{as} \quad x \to \infty$$

(a) Show that the roots of $J_\nu(x) = 0$ are given by

$$x = (n + \tfrac{3}{4} + \tfrac{1}{2}\nu)\pi - \frac{4\nu^2 - 1}{2\pi(4n + 3 + 2\nu)} + \cdots$$

(b) Show that the roots of $J_\nu'(x) = 0$ are given by

$$x = (n + \tfrac{1}{4} + \tfrac{1}{2}\nu)\pi - \frac{3 + 4\nu^2}{2\pi(4n + 1 + 2\nu)} + \cdots$$

(c) Compare these results with those obtained in Exercise 2.5 and Section 2.4 for $\nu = 0$.

(d) Compare these results with those tabulated for the first seven roots when $\nu = 1$.

2.9. The asymptotic expansion of $Y_\nu(x)$ is

$$Y_\nu(x) \sim \sqrt{\frac{2}{\pi x}} \left[\sin(x - \tfrac{1}{2}\nu\pi - \tfrac{1}{4}\pi) + \frac{4\nu^2 - 1}{8x} \cos(x - \tfrac{1}{2}\nu\pi - \tfrac{1}{4}\pi) \right] \quad \text{as} \quad x \to \infty$$

50 ALGEBRAIC EQUATIONS

(a) Show that the roots of $Y_\nu(x) = 0$ are given by

$$x = (n + \tfrac{1}{4} + \tfrac{1}{2}\nu) - \frac{4\nu^2 - 1}{2\pi(4n + 1 + 2\nu)} + \cdots$$

Compare this result with that for Y_0 in Exercise 2.6.

(b) Show that the roots of $Y'_\nu(x) = 0$ are given by

$$x = (n + \tfrac{3}{4} + \tfrac{1}{2}\nu)\pi - \frac{3 + 4\nu^2}{2\pi(4n + 3 + 2\nu)} + \cdots$$

Compare this result with that for Y_0 in Exercise 2.7.

(c) Compare these results with those tabulated for the first seven roots when $\nu = 1$.

CHAPTER 3

Integrals

There are many differential and difference equations whose solutions cannot be expressed in terms of elementary functions but can be expressed in the form of integrals. Among the many methods that can be used to represent the solutions of differential equations as integrals, we mention the Laplace and Fourier transforms. Before we discuss methods of determining approximations of integrals, we show how to represent the solution of a simple differential equation as an integral. Other examples are given in Section 13.5 and Exercises 13.17 and 13.18.

We consider the general solution of the following first-order linear ordinary-differential equation:

$$y' + y = \frac{1}{x} \qquad (3.1)$$

Multiplying (3.1) by the integrating factor exp (x) gives

$$y'e^x + ye^x = \frac{e^x}{x} \qquad (3.2)$$

which can be rewritten as

$$\frac{d}{dx}(ye^x) = \frac{e^x}{x} \qquad (3.3)$$

Integrating both sides yields

$$ye^x = \int_{x_0}^{x} \frac{e^\tau}{\tau} d\tau + c \qquad (3.4)$$

where τ is a dummy variable of integration, x_0 is an arbitrary limit of integration, and c is a constant. If $y(1) = a$, then

$$ae = \int_{x_0}^{1} \frac{e^\tau}{\tau} d\tau + c$$

or

$$c = ae - \int_{x_0}^{1} \frac{e^\tau}{\tau} d\tau$$

Substituting for c in (3.4) gives

$$ye^x = \int_{x_0}^{x} \frac{e^\tau}{\tau} d\tau + ae - \int_{x_0}^{1} \frac{e^\tau}{\tau} d\tau$$

$$= ae + \int_{x_0}^{x} \frac{e^\tau}{\tau} d\tau + \int_{1}^{x_0} \frac{e^\tau}{\tau} d\tau$$

$$= ae + \int_{1}^{x} \frac{e^\tau}{\tau} d\tau$$

Hence,

$$y = ae^{1-x} + e^{-x} \int_{1}^{x} \frac{e^\tau}{\tau} d\tau \qquad (3.5)$$

In this chapter, we discuss a number of methods for determining approximations to integrals such as the one in (3.5). These methods include expansions of the integrands, integration by parts, Laplace's method, the method of stationary phase, and the method of steepest descent. These methods are described by applying them to specific examples.

3.1. Expansion of Integrands

In this section, we consider four examples.

EXAMPLE 1
As a first example, we consider the value of the integral

$$I(\epsilon) = \int_{0}^{1} \sin \epsilon x^2 \, dx \qquad (3.6)$$

for small ϵ. Expanding the integrand in a Taylor series gives

$$\sin \epsilon x^2 = \sum_{n=1}^{\infty} \frac{(-1)^{n+1}(\epsilon x^2)^{2n-1}}{(2n-1)!} = \epsilon x^2 - \frac{1}{6} \epsilon^3 x^6 + \frac{1}{120} \epsilon^5 x^{10} + O(\epsilon^7)$$

$$(3.7)$$

Using the ratio test in (3.7) yields

$$\lim_{n\to\infty} \frac{n\text{th term}}{(n-1)\text{th term}} = \lim_{n\to\infty} \frac{(-1)^{n+1}(\epsilon x^2)^{2n-1}(2n-3)!}{(2n-1)!(-1)^n(\epsilon x^2)^{2n-3}}$$

$$= \lim_{n\to\infty} \frac{-(\epsilon x^2)^2}{(2n-1)(2n-2)} = 0$$

Hence, the series (3.7) converges for all values of ϵx^2. Since $|x| \leq 1$ and ϵ is small, the remainder term in (3.7) is $O(\epsilon^7)$ for all values of x. Substituting (3.7) into (3.6) and integrating term by term, we obtain

$$I(\epsilon) = \sum_{n=1}^{\infty} \frac{(-1)^{n+1}\epsilon^{2n-1}}{(2n-1)!} \int_0^1 x^{4n-2} dx = \sum_{n=1}^{\infty} \frac{(-1)^{n+1}\epsilon^{2n-1}}{(2n-1)!(4n-1)}$$

$$= \frac{1}{3}\epsilon - \frac{1}{42}\epsilon^3 + \frac{1}{1320}\epsilon^5 + O(\epsilon^7) \tag{3.8}$$

EXAMPLE 2

As a second example, we consider the complete elliptic integral of the first kind

$$I(m) = \int_0^{(1/2)\pi} \frac{d\theta}{\sqrt{1 - m \sin^2 \theta}} \tag{3.9}$$

for small m. Using the binomial theorem, we write

$$(1 - m \sin^2 \theta)^{-1/2} = 1 + \tfrac{1}{2} m \sin^2 \theta + \frac{(-\tfrac{1}{2})(-\tfrac{3}{2})}{2!}(-m \sin^2 \theta)^2$$

$$+ \frac{(-\tfrac{1}{2})(-\tfrac{3}{2})(-\tfrac{5}{2})}{3!}(-m \sin^2 \theta)^3 + \frac{(-\tfrac{1}{2})(-\tfrac{3}{2})(-\tfrac{5}{2})(-\tfrac{7}{2})}{4!}$$

$$\times (-m \sin^2 \theta)^4 + O(m^5) \tag{3.10}$$

Using the ratio test yields

$$\lim_{n\to\infty} \frac{n\text{th term}}{(n-1)\text{th term}}$$

$$= \lim_{n\to\infty} \frac{(-\tfrac{1}{2})(-\tfrac{3}{2})(-\tfrac{5}{2})\cdots\left(-\frac{2n-3}{2}\right)(n-2)!(-m \sin^2 \theta)^{n-1}}{(-\tfrac{1}{2})(-\tfrac{3}{2})(-\tfrac{5}{2})\cdots\left(-\frac{2n-5}{2}\right)(n-1)!(-m \sin^2 \theta)^{n-2}}$$

$$= \lim_{n\to\infty} \frac{(2n-3)m \sin^2 \theta}{2(n-1)} = m \sin^2 \theta$$

TABLE 3-1. Variation of I_a/I_e with m

m	0	0.2	0.4	0.5	0.6	0.7	0.8
I_a/I_e	1	1.0000	0.99917	0.99720	0.99216	0.98043	0.95382

Hence, expansion (3.10) converges for all values of θ such that $m \sin^2 \theta < 1$. Since $\sin^2 \theta \leq 1$ and m is small, the remainder in (3.10) is $O(m^5)$ for all values of θ.

Substituting (3.10) into (3.9) and integrating term by term, we obtain

$$I(m) = \int_0^{(1/2)\pi} d\theta + \tfrac{1}{2} m \int_0^{(1/2)\pi} \sin^2 \theta \, d\theta + \tfrac{3}{8} m^2 \int_0^{(1/2)\pi} \sin^4 \theta \, d\theta$$

$$+ \frac{5}{16} m^3 \int_0^{(1/2)\pi} \sin^6 \theta \, d\theta + \frac{35}{128} m^4 \int_0^{(1/2)\pi} \sin^8 \theta \, d\theta + O(m^5)$$

(3.11)

We note that

$$\int_0^{(1/2)\pi} \sin^{2n} \theta \, d\theta = \frac{(2n)! \pi}{(n!)^2 2^{2n+1}} \tag{3.12}$$

Hence, (3.9) becomes

$$I(m) = \tfrac{1}{2} \pi \left[1 + \tfrac{1}{4} m + \tfrac{9}{64} m^2 + \tfrac{25}{256} m^3 + \tfrac{1225}{16384} m^4 + O(m^5) \right] \tag{3.13}$$

Table 3-1 shows the ratio I_a/I_e as a function of m, where I_a is the approximate expression (3.13) and I_e is the exact value of I as tabulated on page 608 of Abramowitz and Stegun (1964). As $m \to 0$, $I_a/I_e \to 1$. When $m \leq 0.5$, the error incurred in representing I_e by I_a is less than 0.28%. For $m = 0.7$, the error incurred in representing I_e by I_a is less than 2%. Thus, I_a is a good approximation of I_e for small values of m.

EXAMPLE 3

As a third example, we consider the integral

$$I(x) = \int_0^x t^{-3/4} e^{-t} \, dt \tag{3.14}$$

for small x. We expand the exponential in a Taylor series as

EXPANSION OF INTEGRANDS 55

$$e^{-t} = \sum_{n=0}^{\infty} \frac{(-1)^n t^n}{n!} = 1 - t + \tfrac{1}{2} t^2 - \tfrac{1}{6} t^3 + \tfrac{1}{24} t^4 + O(t^5) \qquad (3.15)$$

Using the ratio test for the series in (3.15), we have

$$\lim_{n \to \infty} \frac{n\text{th term}}{(n-1)\text{th term}} = \lim_{n \to \infty} \frac{(-1)^n t^n (n-1)!}{n! (-1)^{n-1} t^{n-1}} = \lim_{n \to \infty} -\frac{t}{n} = 0 \qquad (3.16)$$

for all t. Hence, the series (3.15) converges for all values of t. Moreover, since t is small, the order of error in (3.15) is uniform.

Substituting (3.15) into (3.14) and integrating term by term, we have

$$I(x) = \sum_{n=0}^{\infty} \frac{(-1)^n}{n!} \int_0^x t^{n-(3/4)} \, dt = \sum_{n=0}^{\infty} \frac{(-1)^n x^{n+(1/4)}}{n! \, (n + \tfrac{1}{4})}$$

$$= 4x^{1/4} - \tfrac{4}{5} x^{5/4} + \tfrac{2}{9} x^{9/4} - \tfrac{2}{39} x^{13/4} + O(x^{17/4}) \qquad (3.17)$$

EXAMPLE 4

As a last example, we consider the error integral

$$I(x) = \int_x^{\infty} e^{-t^2} \, dt \qquad (3.18)$$

for small x. Expanding the integrand in a Taylor series, we have

$$e^{-t^2} = \sum_{n=0}^{\infty} \frac{(-1)^n t^{2n}}{n!} \qquad (3.19)$$

which converges for all values of t as shown in the preceding example. However, no finite number of terms can represent $\exp(-t^2)$ for all t as shown in Figure 5-2. On the other hand,

$$e^{-t^2} = \sum_{n=0}^{N-1} \frac{(-1)^n t^{2n}}{n!} + O(t^{2N}) \qquad (3.20)$$

for small t. Hence, we represent (3.18) in the following alternate form:

$$I(x) = \int_0^{\infty} e^{-t^2} \, dt - \int_0^x e^{-t^2} \, dt \qquad (3.21)$$

To evaluate the first integral, we note that

$$I_1 = \int_0^{\infty} e^{-u^2} \, du = \int_0^{\infty} e^{-v^2} \, dv \qquad (3.22)$$

Hence,

$$I_1^2 = \left(\int_0^\infty e^{-u^2}\, du\right)\left(\int_0^\infty e^{-v^2}\, dv\right) = \int_0^\infty \int_0^\infty e^{-(u^2+v^2)}\, du\, dv \quad (3.23)$$

We change the integration variables from the Cartesian coordinates u and v to the polar coordinates r and θ. Thus,

$$du\, dv = r\, dr\, d\theta \quad (3.24)$$

and the limit of integration on θ goes from 0 to $\frac{1}{2}\pi$ and that on r goes from 0 to ∞. Hence, (3.23) becomes

$$I_1^2 = \int_0^\infty \int_0^{(1/2)\pi} re^{-r^2}\, dr\, d\theta = \left(\int_0^{(1/2)\pi} d\theta\right)\left(\int_0^\infty re^{-r^2}\, dr\right)$$

$$= \tfrac{1}{2}\pi\left(-\tfrac{1}{2}e^{-r^2}\right)\Big|_0^\infty = \tfrac{1}{4}\pi$$

so that

$$I_1 = \int_0^\infty e^{-x^2}\, dx = \tfrac{1}{2}\sqrt{\pi} \quad (3.25)$$

Substituting (3.19) into the second integral in (3.21), integrating term by term, and using (3.25), we obtain

$$I(x) = \tfrac{1}{2}\sqrt{\pi} - \sum_{n=0}^\infty \frac{(-1)^n}{n!}\int_0^x t^{2n}\, dt = \tfrac{1}{2}\sqrt{\pi} - \sum_{n=0}^\infty \frac{(-1)^n x^{2n+1}}{n!\,(2n+1)}$$

$$= \tfrac{1}{2}\sqrt{\pi} - x + \tfrac{1}{3}x^3 - \tfrac{1}{10}x^5 + \tfrac{1}{42}x^7 + O(x^9) \quad (3.26)$$

3.2. Integration by Parts

We explain this approach by applying it to five examples. The last example brings out a shortcoming of this approach and leads into Laplace's method and the method of stationary phase.

EXAMPLE 1

As a first example, we consider the incomplete factorial function defined by

$$I(x) = \int_x^\infty \frac{e^{-t}}{t^2}\, dt \quad (3.27)$$

for large x.

The method of integration by parts is based on the identity

$$d(uv) = u\,dv + v\,du \tag{3.28}$$

or

$$u\,dv = d(uv) - v\,du \tag{3.29}$$

If u and v are functions of t, then integrating both sides of (3.29) from $t = t_1$ to $t = t_2$ gives

$$\int_{t_1}^{t_2} u\,dv = uv \Big|_{t_1}^{t_2} - \int_{t_1}^{t_2} v\,du \tag{3.30}$$

To apply (3.30), we need to express the quantity under the integral sign in (3.27) as $u\,dv$, that is,

$$\frac{e^{-t}}{t^2}\,dt = u\,dv \tag{3.31}$$

Usually, u and dv are chosen such that the resulting expression for dv is integrable. Moreover, they are chosen such that the successive terms in the expansion of $I(x)$ decrease in order. To illustrate these points, we try two choices. First, we let

$$u = e^{-t} \qquad dv = \frac{dt}{t^2} \tag{3.32}$$

Hence,

$$du = -e^{-t}dt \qquad v = -\frac{1}{t} \tag{3.33}$$

Substituting (3.32) and (3.33) into (3.30) yields

$$\int_x^\infty \frac{e^{-t}}{t^2}\,dt = \int_x^\infty u\,dv = -\frac{e^{-t}}{t}\bigg|_x^\infty - \int_x^\infty \frac{e^{-t}}{t}\,dt$$

or

$$\int_x^\infty \frac{e^{-t}}{t^2}\,dt = \frac{e^{-x}}{x} - \int_x^\infty \frac{e^{-t}}{t}\,dt \tag{3.34}$$

To continue the process, we put

$$u = e^{-t} \qquad dv = \frac{dt}{t} \tag{3.35}$$

58 INTEGRALS

Hence,

$$du = -e^{-t} dt \quad v = \ln t \tag{3.36}$$

Then,

$$\int_x^\infty \frac{e^{-t}}{t} dt = \int_x^\infty u\, dv = e^{-t} \ln t \Big|_x^\infty + \int_x^\infty e^{-t} \ln t\, dt \tag{3.37}$$

Since

$$\lim_{t \to \infty} e^{-t} \ln t = \lim_{t \to \infty} \frac{\ln t}{e^t} = \lim_{t \to \infty} \frac{1}{te^t} = 0$$

according to l'Hospital's rule, (3.37) becomes

$$\int_x^\infty \frac{e^{-t}}{t} dt = -e^{-x} \ln x + \int_x^\infty e^{-t} \ln t\, dt \tag{3.38}$$

Substituting (3.38) into (3.34) yields

$$\int_x^\infty \frac{e^{-t}}{t^2} dt = \frac{e^{-x}}{x} + e^{-x} \ln x - \int_x^\infty e^{-t} \ln t\, dt \tag{3.39}$$

We note that the second term on the right-hand side of (3.39) is much bigger than the first term as $x \to \infty$. Therefore, the above choices (3.32) and (3.35) do not yield an asymptotic expansion.

Second, we let

$$u = \frac{1}{t^2} \quad dv = e^{-t} dt \tag{3.40}$$

Hence,

$$du = -\frac{2}{t^3} dt \quad v = -e^{-t} \tag{3.41}$$

Substituting (3.40) and (3.41) into (3.30) yields

$$\int_x^\infty \frac{e^{-t}}{t^2} dt = \int_x^\infty u\, dv = -\frac{e^{-t}}{t^2} \Big|_x^\infty - 2\int_x^\infty \frac{e^{-t}}{t^3} dt$$

or

$$\int_x^\infty \frac{e^{-t}}{t^2} dt = \frac{e^{-x}}{x^2} - 2\int_x^\infty \frac{e^{-t}}{t^3} dt \tag{3.42}$$

INTEGRATION BY PARTS 59

To continue the process further, we let

$$u = \frac{1}{t^3} \qquad dv = e^{-t}\, dt \qquad (3.43)$$

from which

$$du = -\frac{3}{t^4}\, dt \qquad v = -e^{-t} \qquad (3.44)$$

Hence,

$$\int_x^\infty \frac{e^{-t}}{t^3}\, dt = -\left.\frac{e^{-t}}{t^3}\right|_x^\infty - 3 \int_x^\infty \frac{e^{-t}}{t^4}\, dt$$

or

$$\int_x^\infty \frac{e^{-t}}{t^3}\, dt = \frac{e^{-x}}{x^3} - 3 \int_x^\infty \frac{e^{-t}}{t^4}\, dt \qquad (3.45)$$

Substituting (3.45) into (3.42) yields

$$\int_x^\infty \frac{e^{-t}}{t^2}\, dt = \frac{e^{-x}}{x^2} - \frac{2e^{-x}}{x^3} + 3! \int_x^\infty \frac{e^{-t}}{t^4}\, dt \qquad (3.46)$$

Continuing the process, we obtain

$$\int_x^\infty \frac{e^{-t}}{t^2}\, dt = \frac{e^{-x}}{x^2} - \frac{2!\,e^{-x}}{x^3} + \frac{3!\,e^{-x}}{x^4} - \frac{4!\,e^{-x}}{x^5} + \cdots + \frac{(-1)^{n-1} n!\, e^{-x}}{x^{n+1}}$$

$$+ (-1)^n (n+1)! \int_x^\infty \frac{e^{-t}}{t^{n+2}}\, dt \qquad (3.47)$$

Since $t^{n+2} \geq x^{n+2}$ when $x \leq t < \infty$,

$$\frac{1}{t^{n+2}} \leq \frac{1}{x^{n+2}}.$$

and

$$\int_x^\infty \frac{e^{-t}}{t^{n+2}}\, dt < \frac{1}{x^{n+2}} \int_x^\infty e^{-t}\, dt = \frac{e^{-x}}{x^{n+2}}$$

Then, (3.47) can be rewritten as

60 INTEGRALS

$$I(x) = e^{-x} \sum_{n=1}^{N} \frac{(-1)^{n-1} n!}{x^{n+1}} + e^{-x} O\left(\frac{1}{x^{N+2}}\right) \quad (3.48)$$

and hence, it is an asymptotic expansion. We note that the series in (3.48) diverges because

$$\lim_{m \to \infty} \frac{m\text{th term}}{(m-1)\text{th term}} = \lim_{m \to \infty} \frac{(-1)^{m-1} m! x^m}{x^{m+1}(-1)^{m-2}(m-1)!} = \lim_{m \to \infty} \frac{-m}{x} = -\infty$$

However, for a fixed N, the error can be made arbitrarily small by increasing x.

EXAMPLE 2

As a second example, we consider

$$I(x) = \int_0^\infty \frac{e^{-t}}{x+t} dt \quad (3.49)$$

for large positive x. As in the preceding example, the choice

$$u = e^{-t} \quad dv = (x+t)^{-1} dt$$

does not lead to an asymptotic expansion. Thus, we let

$$u = (x+t)^{-1} \quad dv = e^{-t} dt \quad (3.50)$$

Hence,

$$du = -(x+t)^{-2} dt \quad v = -e^{-t} \quad (3.51)$$

Substituting (3.50) and (3.51) into (3.30) yields

$$I(x) = \int_0^\infty \frac{e^{-t}}{x+t} dt = -\frac{e^{-t}}{x+t} \Big|_0^\infty - \int_0^\infty \frac{e^{-t}}{(x+t)^2} dt$$

or

$$I(x) = \frac{1}{x} - \int_0^\infty \frac{e^{-t}}{(x+t)^2} dt \quad (3.52)$$

To continue the process, we let

$$u = (x+t)^{-2} \quad dv = e^{-t} dt \quad (3.53)$$

Hence,

$$du = -2(x+t)^{-3} dt \quad v = -e^{-t} \quad (3.54)$$

Substituting (3.53) and (3.54) into (3.30) yields

$$\int_0^\infty \frac{e^{-t}}{(x+t)^2} dt = -\frac{e^{-t}}{(x+t)^2} \bigg|_0^\infty - 2\int_0^\infty \frac{e^{-t}}{(x+t)^3} dt = \frac{1}{x^2} - 2\int_0^\infty \frac{e^{-t}}{(x+t)^3} dt \quad (3.55)$$

Then, (3.52) becomes

$$I(x) = \frac{1}{x} - \frac{1}{x^2} + 2\int_0^\infty \frac{e^{-t}}{(x+t)^3} dt \quad (3.56)$$

Continuing the process, we have

$$I(x) = \frac{1}{x} - \frac{1}{x^2} + \frac{2!}{x^3} - \frac{3!}{x^4} + \cdots + \frac{(-1)^{n-1}(n-1)!}{x^n}$$

$$+ (-1)^n n! \int_0^\infty \frac{e^{-t}}{(x+t)^{n+1}} dt \quad (3.57)$$

Since x and t are positive,

$$\frac{1}{(x+t)^{n+1}} \leq \frac{1}{x^{n+1}}$$

and hence,

$$\int_0^\infty \frac{e^{-t}}{(x+t)^{n+1}} dt < \frac{1}{x^{n+1}} \int_0^\infty e^{-t} dt = \frac{1}{x^{n+1}}$$

Therefore,

$$I(x) = \sum_{n=1}^N \frac{(-1)^{n-1}(n-1)!}{x^n} + O\left(\frac{1}{x^{N+1}}\right) \quad (3.58)$$

and hence, it is an asymptotic expansion. We note that the series in (3.58) diverges because

$$\lim_{m \to \infty} \frac{m\text{th term}}{(m-1)\text{th term}} = \lim_{m \to \infty} \frac{(-1)^{m-1}(m-1)! x^{m-1}}{x^m (-1)^{m-2}(m-2)!} = \lim_{m \to \infty} \frac{-(m-1)}{x} = -\infty$$

However, for a fixed N, the remainder can be made arbitrarily small by increasing x.

EXAMPLE 3
As a third example, we consider the Laplace integral

$$I(x) = \int_0^\infty e^{-xt} f(t) dt \quad (3.59)$$

62 INTEGRALS

for large positive x when $f(t)$ is analytic (i.e., all its derivatives exist in the interval of interest) and the integral exists. Such integrals occur in the solution of differential equations by using the Laplace transform. We let

$$u = f(t) \qquad dv = e^{-xt}\, dt \qquad (3.60)$$

so that

$$du = f'(t)\, dt \qquad v = -\frac{e^{-xt}}{x} \qquad (3.61)$$

Had we chosen $u = \exp(-xt)$ and $dv = f(t)\, dt$, we would have found the resulting expression not to be an asymptotic expansion. Substituting (3.60) and (3.61) into (3.30) yields

$$\int_0^\infty e^{-xt} f(t)\, dt = -\frac{e^{-xt}}{x} f(t)\Big|_0^\infty + \frac{1}{x}\int_0^\infty e^{-xt} f'(t)\, dt$$

or

$$I(x) = \frac{f(0)}{x} + \frac{1}{x}\int_0^\infty e^{-xt} f'(t)\, dt \qquad (3.62)$$

Continuing the process, we put

$$u = f'(t) \qquad dv = e^{-xt}\, dt \qquad (3.63)$$

so that

$$du = f''(t)\, dt \qquad v = -\frac{e^{-xt}}{x} \qquad (3.64)$$

Substituting (3.63) and (3.64) into (3.30) and then into (3.62) gives

$$I(x) = \frac{f(0)}{x} - \frac{e^{-xt}}{x^2} f'(t)\Big|_0^\infty + \frac{1}{x^2}\int_0^\infty e^{-xt} f''(t)\, dt$$

or

$$I(x) = \frac{f(0)}{x} + \frac{f'(0)}{x^2} + \frac{1}{x^2}\int_0^\infty e^{-xt} f''(t)\, dt \qquad (3.65)$$

Continuing the process, we obtain

$$I(x) = \frac{f(0)}{x} + \frac{f'(0)}{x^2} + \frac{f''(0)}{x^3} + \frac{f'''(0)}{x^4} + \cdots + \frac{f^{(n)}(0)}{x^{n+1}}$$

$$+ \frac{1}{x^{n+1}} \int_0^\infty e^{-xt} f^{(n+1)}(t)\, dt \qquad (3.66)$$

Thus,

$$I(x) = \sum_{n=0}^{N} \frac{f^{(n)}(0)}{x^{n+1}} + O\left(\frac{1}{x^{N+2}}\right) \qquad (3.67)$$

and hence, it is an asymptotic expansion.

EXAMPLE 4

As a fourth example, we consider the Fourier integral

$$I(\alpha) = \int_0^\infty e^{i\alpha t} f(t)\, dt \qquad (3.68)$$

for large positive α when $f(t)$ is an analytic function of t. It is assumed that the function $f(t)$ decays properly at ∞ so that the integral (3.68) exists. Such integrals occur in the solutions of differential equations by using the Fourier transform. To integrate (3.68) by parts, we put

$$u = f(t) \qquad dv = e^{i\alpha t}\, dt \qquad (3.69)$$

so that

$$du = f'(t)\, dt \qquad v = \frac{e^{i\alpha t}}{i\alpha} \qquad (3.70)$$

Substituting (3.69) and (3.70) into (3.30) gives

$$\int_0^\infty e^{i\alpha t} f(t)\, dt = \frac{e^{i\alpha t}}{i\alpha} f(t) \Big|_0^\infty - \frac{1}{i\alpha} \int_0^\infty e^{i\alpha t} f'(t)\, dt$$

or

$$I(\alpha) = -\frac{f(0)}{i\alpha} - \frac{1}{i\alpha} \int_0^\infty e^{i\alpha t} f'(t)\, dt \qquad (3.71)$$

Continuing the process as in the preceding example, we obtain

$$I(\alpha) = \sum_{n=0}^{N} \frac{f^{(n)}(0)}{(-i\alpha)^{n+1}} + O\left(\frac{1}{\alpha^{N+2}}\right) \qquad (3.72)$$

which is an asymptotic expansion.

64 INTEGRALS

EXAMPLE 5

As a last example, we consider the generalized Laplace integral

$$I(x) = \int_a^b e^{xh(t)} f(t) \, dt \quad b > a \tag{3.73}$$

for large positive x when $h(t)$ and $f(t)$ are differentiable functions of t. To integrate (3.73) by parts, we need to express the quantity under the integral sign in (3.73) as $u \, dv$. In this case, we cannot put

$$u = f(t) \quad dv = e^{xh(t)} \, dt \tag{3.74}$$

because the last expression is not integrable in simple terms. To obtain an integrable expression, we modify (3.74) and put

$$u = \frac{f(t)}{h'(t)} \quad dv = e^{xh(t)} h'(t) \, dt \tag{3.75}$$

so that

$$du = \left[\frac{f(t)}{h'(t)}\right]' dt \quad v = \frac{e^{xh(t)}}{x} \tag{3.76}$$

Substituting (3.75) and (3.76) into (3.30) yields

$$\int_a^b e^{xh(t)} f(t) \, dt = \frac{e^{xh(t)} f(t)}{xh'(t)} \bigg|_a^b - \frac{1}{x} \int_a^b e^{xh(t)} \left[\frac{f(t)}{h'(t)}\right]' dt$$

or

$$I(x) = \frac{e^{xh(b)} f(b)}{xh'(b)} - \frac{e^{xh(a)} f(a)}{xh'(a)} - \frac{1}{x} \int_a^b e^{xh(t)} \left[\frac{f(t)}{h'(t)}\right]' dt \tag{3.77}$$

Continuing the process, we find that

$$I(x) \sim \frac{e^{xh(b)} f(b)}{xh'(b)} - \frac{e^{xh(a)} f(a)}{xh'(a)} \quad \text{as} \quad x \to \infty \tag{3.78}$$

In writing (3.77) and (3.78), we assumed that $f(t)$ and $h(t)$ are differentiable and that $h'(t) \neq 0$ in $[a, b]$. If $h(t)$ is *stationary* (i.e., $h' = 0$) at any point in $[a, b]$, the above process breaks down because the integral in (3.77) does not exist. If $h'(t) \neq 0$, (3.78) shows that only the immediate neighborhoods of the end points contribute to the integral. Moreover, if $h(a) \neq h(b)$, only the immediate neighborhood of the end point with the larger value of h contributes to the asymptotic development of the integral. This suggests that only the immediate neighborhood of the point $t = c$ in $[a, b]$ corresponding to the maximum value of $h(t)$ contributes to the asymptotic expansion of the integral, irrespective of whether $t = c$ is an interior or an end point. This is the central idea of Laplace's method. If the maximum value of $h(t)$ occurs at the boundary, say $t = a$, and if it is nonstationary, then one can replace the upper limit in (3.73) with c provided that $h(t)$ has no stationary values in $[a, c]$, integrate the result by parts, and obtain the asymptotic development of $I(x)$. If the maximum value of $h(t)$ corresponds to a stationary point, the method of integration by parts fails. In the next section, we discuss Laplace's method for determining the asymptotic expansions of integrals such as (3.73) without restricting $f(t)$ to be differentiable and the maximum value of $h(t)$ to occur at a nonstationary point.

3.3. Laplace's Method

In this section, we consider integrals of the general form

$$I(x) = \int_a^b e^{xh(t)} f(t) \, dt \tag{3.79}$$

for real $h(t)$ and large positive x. We assume that the integral in (3.79) exists (i.e., it has a finite value). As discussed in the preceding section, according to Laplace, only the immediate neighborhood of the point corresponding to the maximum value of $h(t)$ in $[a, b]$ contributes to the asymptotic expansion of $I(x)$. Laplace devised a general method for determining the asymptotic development of integrals having the form (3.79). In this section, we explain Laplace's method using six examples.

EXAMPLE 1
As a first example, we consider the special case

$$I(x) = \int_0^{10} \frac{e^{-xt}}{1+t} \, dt \tag{3.80}$$

To determine an asymptotic development for $I(x)$ for large x, we expand $(1 + t)^{-1}$ in powers of t, integrate the resulting expansion term by term, and replace the upper limit by ∞. This is justified below.

It follows from the binomial theorem that

$$(1 + t)^{-1} = 1 - t + t^2 - t^3 + \cdots = \sum_{n=0}^{\infty} (-1)^n t^n \tag{3.81}$$

Using the ratio test, we have

$$\lim_{n \to \infty} \frac{n\text{th term}}{(n-1)\text{th term}} = \lim_{n \to \infty} \frac{(-1)^n t^n}{(-1)^{n-1} t^{n-1}} = \lim_{n \to \infty} (-t) = -t$$

Hence, the series (3.81) converges only for $|t| < 1$. Consequently, substituting (3.81) into (3.80) and integrating the result term by term from $t = 0$ to $t = 10$ appears to be unjustified because (3.81) is valid only for $|t| < 1$. To circumvent this difficulty, we break the interval of integration into the two intervals $[0, \delta]$ and $[\delta, 10]$, where δ is a small positive number. Thus, we write (3.80) as

$$I(x) = \int_0^\delta \frac{e^{-xt}}{1+t} dt + \int_\delta^{10} \frac{e^{-xt}}{1+t} dt \tag{3.82}$$

Next, we show that the second integral in (3.82) is exponentially small for large positive x. To this end, we note that $1 > (1 + t)^{-1}$ for positive values of t. Hence,

$$\int_\delta^{10} \frac{e^{-xt}}{1+t} dt < \int_\delta^{10} e^{-xt} dt = \left. \frac{e^{-xt}}{-x} \right|_\delta^{10} = -\frac{1}{x}(e^{-10x} - e^{-\delta x})$$

As $x \to \infty$, $\exp(-10x)$ tends to zero much faster than any power of x^{-1}. Moreover, for finite values of δ and as $x \to \infty$, $\exp(-\delta x)$ tends to zero much faster than any power of x^{-1}. Therefore, the second integral in (3.82) tends exponentially to zero and

$$I(x) = \int_0^\delta \frac{e^{-xt}}{1+t} dt + \text{exponentially small terms} \tag{3.83}$$

as $x \to \infty$. Thus, only the immediate neighborhood of $t = 0$ contributes to the asymptotic development of (3.80).

Substituting (3.81) into (3.83) and integrating term by term, we obtain

$$I(x) = \int_0^\delta e^{-xt} \left(\sum_{n=0}^\infty (-1)^n t^n \right) dt = \sum_{n=0}^\infty (-1)^n \int_0^\delta t^n e^{-xt} dt \tag{3.84}$$

To evaluate the last integral in (3.84), we introduce the transformation $\tau = xt$ so that $dt = d\tau/x$. Hence,

$$\int_0^\delta t^n e^{-xt} dt = \frac{1}{x^{n+1}} \int_0^{\delta x} \tau^n e^{-\tau} d\tau \tag{3.85}$$

A repeated integration by parts of the last integral in (3.85) yields

$$\int_0^\delta t^n e^{-xt} dt = \frac{1}{x^{n+1}} \int_0^{\delta x} \tau^n e^{-\tau} d\tau = -\frac{1}{x^{n+1}} [\tau^n + n\tau^{n-1} + n(n-1)\tau^{n-2}$$
$$+ \cdots + \tfrac{1}{2} n! \tau^2 + n!\tau + n!] e^{-\tau} \Big|_0^{\delta x}$$

or

$$\int_0^\delta t^n e^{-xt}\, dt = \frac{n!}{x^{n+1}} - e^{-\delta x}\left[\frac{\delta^n}{x} + \frac{n\delta^{n-1}}{x^2} + \frac{n(n-1)\delta^{n-2}}{x^3}\right.$$
$$\left. + \cdots + \frac{n!\delta^2}{2x^{n-1}} + \frac{n!\delta}{x^n} + \frac{n!}{x^{n+1}}\right] \quad (3.86)$$

Since $\exp(-\delta x)$ tends to zero as $x \to \infty$ much faster than any power of x^{-1}, the term multiplying $\exp(-\delta x)$ in (3.86) is much smaller than any power of x^{-1}. Hence,

$$\int_0^\delta t^n e^{-xt}\, dt = \frac{n!}{x^{n+1}} + \text{EST} \quad (3.87)$$

where EST stands for exponentially small terms. This result is independent of the value of δ including ∞. In fact,

$$\int_0^\infty t^n e^{-xt}\, dt = \frac{n!}{x^{n+1}}$$

This completes justification of the above stated procedure.

Substituting (3.87) into (3.84) gives

$$I(x) = \sum_{n=0}^\infty \frac{(-1)^n n!}{x^{n+1}} \quad (3.88)$$

where the exponentially small terms have been omitted. Using the ratio test, we have

$$\lim_{n\to\infty} \frac{n\text{th term}}{(n-1)\text{th term}} = \lim_{n\to\infty} \frac{(-1)^n n!\, x^n}{x^{n+1}(-1)^{n-1}(n-1)!} = \lim_{n\to\infty} \frac{-n}{x} = -\infty$$

Hence, the series (3.88) diverges. Therefore, the equality sign cannot be used and it must be replaced with an asymptotic sign. Thus, we write

$$I(x) \sim \sum_{n=0}^\infty \frac{(-1)^n n!}{x^{n+1}} \quad \text{as} \quad x \to \infty \quad (3.89)$$

This result could have been obtained by integration by parts.

The procedure stated above is usually referred to as Watson's lemma, which gives the full asymptotic development of integrals of the form

$$I(x) = \int_0^b f(t) e^{-xt}\, dt \quad (3.90)$$

68 INTEGRALS

where $f(t)$ is continuous on the interval $[0, b]$, has the asymptotic expansion

$$f(t) \sim \sum_{m=1}^{\infty} a_m t^{m\beta - 1} \quad \text{as} \quad t \to 0 \quad (3.91a)$$

with β being positive for the integral to converge, and if $b = +\infty$

$$f(t) < K e^{\alpha t} \quad (3.91b)$$

with K and α being positive numbers independent of t. Then, Watson's lemma states that the full asymptotic expansion of $I(x)$ is obtained by substituting (3.91a) into (3.90), integrating the result term by term, and replacing the upper limit with ∞, that is,

$$I(x) \sim \sum_{m=1}^{\infty} a_m \int_0^{\infty} t^{m\beta - 1} e^{-xt} \, dt$$

or

$$I(x) \sim \sum_{m=1}^{\infty} a_m \Gamma(m\beta)/x^{m\beta} \quad (3.92)$$

EXAMPLE 2

In the preceding example, $h(t) = -t$ has its maximum at the lower limit (i.e., $t = 0$). Since $h'(0) = -1$, this maximum is not a relative maximum. As a second example, we consider a case in which $h(t)$ has a relative maximum at $t = 0$. Specifically, we consider

$$I(x) = \int_0^5 \frac{e^{-xt^2}}{1+t} \, dt \quad (3.93)$$

In this case, the method of integration by parts would fail as discussed in Example 5 of the preceding section. For large positive x, only the immediate neighborhood of $t = 0$ contributes to the asymptotic development of $I(x)$ because $h(t) = -t^2$ has its maximum there. Using Watson's lemma, we determine an asymptotic expansion for $I(x)$ by substituting the expansion (3.81) of $(1 + t)^{-1}$ into (3.93), integrating the result term by term, and replacing the upper limit by ∞. The result is

$$I(x) \sim \sum_{n=0}^{\infty} (-1)^n \int_0^{\infty} t^n e^{-xt^2} \, dt \quad (3.94)$$

We denote the integrals in (3.94) by

$$I_n = \int_0^\infty t^n e^{-xt^2} \, dt \qquad (3.95)$$

We start with

$$I_0 = \int_0^\infty e^{-xt^2} \, dt \qquad (3.96)$$

If we let $\sqrt{x}\, t = \tau$, then $dt = d\tau/\sqrt{x}$ and (3.96) becomes

$$I_0 = \frac{1}{\sqrt{x}} \int_0^\infty e^{-\tau^2} \, d\tau \qquad (3.97)$$

Using (3.25), we rewrite (3.97) as

$$I_0 = \int_0^\infty e^{-xt^2} \, dt = \frac{\sqrt{\pi}}{2\sqrt{x}} \qquad (3.98)$$

Differentiating (3.98) with respect to x, we have

$$-\int_0^\infty t^2 e^{-xt^2} \, dt = -\frac{\sqrt{\pi}}{4x^{3/2}}$$

Hence,

$$I_2 = \int_0^\infty t^2 e^{-xt^2} \, dt = \frac{\sqrt{\pi}}{4x^{3/2}} \qquad (3.99)$$

Differentiating (3.99) with respect to x, we have

$$-\int_0^\infty t^4 e^{-xt^2} \, dt = -\frac{3\sqrt{\pi}}{8x^{5/2}}$$

Hence,

$$I_4 = \int_0^\infty t^4 e^{-xt^2} \, dt = \frac{3\sqrt{\pi}}{8x^{5/2}} \qquad (3.100)$$

Continuing the process, we find that

$$I_{2m} = \tfrac{1}{2}\sqrt{\pi}\,(-1)^m \frac{d^m}{dx^m}(x^{-1/2}) \qquad (3.101)$$

To determine the I_n for odd n, we start with

70 INTEGRALS

$$I_1 = \int_0^\infty t e^{-xt^2} dt \qquad (3.102)$$

We let $xt^2 = \tau$ so that $2xt\, dt = d\tau$. Hence,

$$I_1 = \frac{1}{2x} \int_0^\infty e^{-\tau} d\tau = -\frac{1}{2x} e^{-\tau} \Big|_0^\infty = \frac{1}{2x} \qquad (3.103)$$

Differentiating (3.103) with respect to x yields

$$-\int_0^\infty t^3 e^{-xt^2} dt = -\frac{1}{2x^2}$$

Hence,

$$I_3 = \int_0^\infty t^3 e^{-xt^2} dt = \frac{1}{2x^2} \qquad (3.104)$$

Differentiating (3.104) with respect to x yields

$$-\int_0^\infty t^5 e^{-xt^2} dt = -\frac{1}{x^3}$$

Hence,

$$I_5 = \int_0^\infty t^5 e^{-xt^2} dt = \frac{1}{x^3} \qquad (3.105)$$

Continuing the process, we find that

$$I_{2m+1} = \tfrac{1}{2}(-1)^m \frac{d^m}{dx^m}(x^{-1}) \qquad (3.106)$$

Using (3.101) and (3.106), we rewrite (3.94) as

$$I(x) \sim \tfrac{1}{2}\sqrt{\pi} \sum_{m=0}^\infty (-1)^m \frac{d^m}{dx^m}(x^{-1/2}) - \tfrac{1}{2} \sum_{m=0}^\infty (-1)^m \frac{d^m}{dx^m}(x^{-1})$$

$$= \frac{\sqrt{\pi}}{2x^{1/2}} - \frac{1}{2x} + \frac{\sqrt{\pi}}{4x^{3/2}} - \frac{1}{2x^2} + \frac{3\sqrt{\pi}}{8x^{5/2}} - \frac{1}{x^3} + \cdots \qquad (3.107)$$

THE GAMMA FUNCTION

Alternatively, we can express (3.95) in terms of the gamma function defined by

$$\Gamma(z) = \int_0^\infty \tau^{z-1} e^{-\tau} d\tau \qquad (3.108)$$

where z must be greater than zero; otherwise, the integral does not exist. To relate (3.95) to the gamma function, we introduce the transformation

$$\tau = xt^2 \qquad d\tau = 2xt\, dt \qquad (3.109)$$

so that (3.95) becomes

$$I_n = \tfrac{1}{2} x^{-(n+1)/2} \int_0^\infty \tau^{(n-1)/2} e^{-\tau} d\tau = \tfrac{1}{2} x^{-(n+1)/2} \Gamma\left(\frac{n+1}{2}\right) \qquad (3.110)$$

Substituting (3.110) into (3.94) yields

$$I(x) \sim \sum_{n=0}^\infty \frac{(-1)^n \Gamma\left(\dfrac{n+1}{2}\right)}{2 x^{(n+1)/2}} \qquad (3.111)$$

To show that (3.111) is the same as (3.107), we use the following recursion formula:

$$\Gamma(z) = (z-1)\Gamma(z-1) \qquad (3.112)$$

for $z > 1$. To prove this formula, we use the method of integration by parts and let

$$u = \tau^{z-1} \qquad dv = e^{-\tau} d\tau$$

so that

$$du = (z-1)\tau^{z-2} d\tau \qquad v = -e^{-\tau}$$

Then, it follows from (3.30) that

$$\Gamma(z) = -\tau^{z-1} e^{-\tau} \Big|_0^\infty + (z-1) \int_0^\infty \tau^{z-2} e^{-\tau} d\tau \qquad (3.113)$$

The integral in (3.113) exists and equals $\Gamma(z-1)$ only for $z > 1$. Then, (3.112) follows from (3.113). If $\Gamma(z)$ is tabulated for $0 \leq z < 1$ only, the recursion formula (3.112) can be used to determine $\Gamma(z)$ for $z > 1$. For example,

$$\Gamma(5.3) = 4.3\Gamma(4.3) = (4.3)(3.3)\Gamma(3.3) = (4.3)(3.3)(2.3)\Gamma(2.3)$$
$$= (4.3)(3.3)(2.3)(1.3)\Gamma(1.3) = (4.3)(3.3)(2.3)(1.3)(.3)\Gamma(.3)$$

The integral defining the gamma function can be integrated analytically whenever z is a positive integer or a multiple of $\tfrac{1}{2}$. When z is a positive integer, the recursion formula (3.112) gives

72 INTEGRALS

$$\Gamma(z) = (z-1)\Gamma(z-1) = (z-1)(z-2)\Gamma(z-2)$$
$$= (z-1)(z-2)(z-3) \cdots (3)(2)(1)\Gamma(1) \qquad (3.114)$$

Putting $z = 1$ in (3.108), we have

$$\Gamma(1) = \int_0^\infty e^{-\tau}\, d\tau = -e^{-\tau}\Big|_0^\infty = 1 \qquad (3.115)$$

Hence, it follows from (3.114) that

$$\Gamma(z) = (z-1)! \qquad (3.116)$$

if z is a positive integer. Therefore, if $n = 2m + 1$ where m is a positive integer

$$\Gamma\left(\frac{n+1}{2}\right) = \Gamma\left(\frac{2m+1+1}{2}\right) = \Gamma(m+1) = m!$$

Then, it follows from (3.110) that

$$I_{2m+1} = \tfrac{1}{2}\, m!\, x^{-(m+1)} \qquad (3.117)$$

in agreement with (3.106).

If $z = m + \tfrac{1}{2}$ where m is a positive integer, it follows from (3.112) that

$$\Gamma(m+\tfrac{1}{2}) = (m-\tfrac{1}{2})\,\Gamma(m-\tfrac{1}{2}) = (m-\tfrac{1}{2})(m-\tfrac{3}{2})\,\Gamma(m-\tfrac{3}{2})$$
$$= (m-\tfrac{1}{2})(m-\tfrac{3}{2})(m-\tfrac{5}{2}) \cdots (\tfrac{5}{2})(\tfrac{3}{2})(\tfrac{1}{2})\,\Gamma(\tfrac{1}{2}) \qquad (3.118)$$

Putting $z = \tfrac{1}{2}$ in (3.108), we have

$$\Gamma(\tfrac{1}{2}) = \int_0^\infty \tau^{-1/2}\, e^{-\tau}\, d\tau \qquad (3.119)$$

With the substitution $\tau = y^2$, (3.119) becomes

$$\Gamma(\tfrac{1}{2}) = 2\int_0^\infty e^{-y^2}\, dy \qquad (3.120)$$

But the integral in (3.120) is $\tfrac{1}{2}\sqrt{\pi}$ according to (3.25); hence

$$\Gamma(\tfrac{1}{2}) = \sqrt{\pi} \qquad (3.121)$$

Therefore, (3.118) becomes

$$\Gamma(m+\tfrac{1}{2}) = (m-\tfrac{1}{2})(m-\tfrac{3}{2})(m-\tfrac{5}{2}) \cdots (\tfrac{5}{2})(\tfrac{3}{2})(\tfrac{1}{2})\sqrt{\pi} \qquad (3.122)$$

if m is a positive integer. Consequently, if $n = 2m$ where m is an integer,

$$\Gamma\left(\frac{n+1}{2}\right) = \Gamma\left(\frac{2m+1}{2}\right) = \Gamma(m + \tfrac{1}{2})$$

$$= (m - \tfrac{1}{2})(m - \tfrac{3}{2})(m - \tfrac{5}{2}) \cdots (\tfrac{5}{2})(\tfrac{3}{2})(\tfrac{1}{2})\sqrt{\pi} \qquad (3.123)$$

Then, it follows from (3.110) that

$$I_{2m} = \tfrac{1}{2}(m - \tfrac{1}{2})(m - \tfrac{3}{2})(m - \tfrac{5}{2}) \cdots (\tfrac{5}{2})(\tfrac{3}{2})(\tfrac{1}{2})\sqrt{\pi}\, x^{-[m+(1/2)]} \qquad (3.124)$$

in agreement with (3.101).

EXAMPLE 3

As a third example, we consider the following integral, which is slightly more general than the preceding example:

$$I(x) = \int_a^b f(t)\, e^{xh(t)}\, dt \qquad b > a \qquad (3.125)$$

where

$$h'(a) = 0 \qquad h''(a) < 0 \qquad (3.126)$$

$$f(t) = f_0 (t - a)^\lambda \qquad \text{as} \qquad t \to a \qquad (3.127)$$

and f_0 is a constant. The conditions (3.126) imply that $h(t)$ has a relative maximum at $t = a$. We assume that this maximum is also an absolute maximum as shown in Figure 3-1. The constant λ must be greater than -1 for the integral (3.125) to exist. Only the immediate neighborhood of $t = a$ contributes to the asymptotic development of $I(x)$ because $h(t)$ has its maximum there. Thus, the upper limit can be replaced with $a + \delta$, where δ is a small positive number. The result is

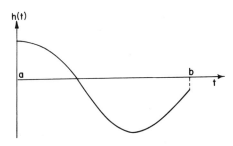

Figure 3-1. A function with a relative and absolute maximum at $t = a$.

74 INTEGRALS

$$I(x) \sim \int_a^{a+\delta} f(t) e^{xh(t)} dt \qquad (3.128)$$

To determine the principal part of $I(x)$, we expand $h(t)$ in a Taylor series about $t = a$, that is,

$$h(t) = h(a) + h'(a)(t - a) + \tfrac{1}{2} h''(a) (t - a)^2 + \cdots$$

or

$$h(t) = h(a) + \tfrac{1}{2} h''(a) (t - a)^2 + \cdots \qquad (3.129)$$

because $h'(a) = 0$. Substituting (3.127) and (3.129) into (3.128), we have

$$I(x) \sim f_0 e^{xh(a)} \int_a^{a+\delta} (t - a)^\lambda \, e^{(1/2)xh''(a)(t-a)^2} dt \qquad (3.130)$$

Since $h''(a) < 0$, δ can be replaced with ∞ because only the immediate neighborhood of $t = a$ contributes to the asymptotic development of $I(x)$. Thus, we rewrite (3.130) as

$$I(x) \sim f_0 e^{xh(a)} \int_a^\infty (t - a)^\lambda \, e^{(1/2)xh''(a)(t-a)^2} dt \qquad (3.131)$$

The integral in (3.131) can be expressed in terms of the gamma function. To this end, we let

$$-\tfrac{1}{2} xh''(a) (t - a)^2 = \tau \qquad (3.132)$$

and obtain

$$I(x) \sim \frac{1}{2} \left[\frac{2}{-xh''(a)} \right]^{(\lambda+1)/2} f_0 e^{xh(a)} \int_0^\infty \tau^{(\lambda-1)/2} e^{-\tau} d\tau \qquad (3.133)$$

or

$$I(x) \sim \frac{1}{2} \left[\frac{2}{-xh''(a)} \right]^{(\lambda+1)/2} f_0 e^{xh(a)} \Gamma\left(\frac{\lambda + 1}{2}\right) \quad \text{as} \quad x \to \infty \qquad (3.134)$$

according to (3.108).

To determine the higher-order terms in the development of $I(x)$, we need to keep the higher-order terms in $f(t)$ and the higher-order terms in $h(t)$.

EXAMPLE 4

As a fourth example, we consider the preceding example with $h(t)$ having an inflection point at $t = a$ as shown in Figure 3-2. Thus,

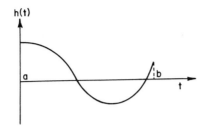

Figure 3-2. A function with an inflection and absolute maximum at $t = a$.

$$h'(a) = h''(a) = 0 \quad \text{and} \quad h'''(a) < 0 \qquad (3.135)$$

Since only the immediate neighborhood of $t = a$ contributes to the asymptotic development of (3.125) for large x, the upper limit can be replaced with $a + \delta$, where δ is a small positive number, as in (3.128).

To determine the principal part of (3.128), we expand $h(t)$ in a Taylor series about $t = a$, that is,

$$h(t) = h(a) + h'(a)(t-a) + \frac{1}{2!}h''(a)(t-a)^2 + \frac{1}{3!}h'''(a)(t-a)^3 + \cdots$$

or

$$h(t) = h(a) + \tfrac{1}{6} h'''(a)(t-a)^3 + \cdots \qquad (3.136)$$

on account of (3.135). Substituting (3.127) and (3.136) into (3.128) and replacing δ with ∞, we obtain

$$I(x) \sim f_0 e^{xh(a)} \int_a^\infty (t-a)^\lambda \, e^{(1/6)xh'''(a)(t-a)^3} \, dt \qquad (3.137)$$

The integral in (3.137) can be expressed in terms of the gamma function. To accomplish this, we let

$$-\tfrac{1}{6} xh'''(a)(t-a)^3 = \tau \qquad (3.138)$$

and obtain

$$I(x) \sim \frac{1}{3} \left[\frac{6}{-xh'''(a)} \right]^{(\lambda+1)/3} f_0 e^{xh(a)} \int_0^\infty \tau^{(\lambda-2)/3} e^{-\tau} \, d\tau \qquad (3.139)$$

Hence,

$$I(x) \sim \frac{1}{3} \left[\frac{6}{-xh'''(a)} \right]^{(\lambda+1)/3} f_0 e^{xh(a)} \, \Gamma\!\left(\frac{\lambda+1}{3}\right) \quad \text{as} \quad x \to \infty \qquad (3.140)$$

according to (3.108).

76 INTEGRALS

EXAMPLE 5

In all the preceding examples, $h(t)$ has its maximum at an end point. Here, we consider a case in which $h(t)$ has its maximum at a point inside the interval of integration as shown in Figure 3-3. Specifically, we consider

$$I(x) = \int_a^b f(t) e^{xh(t)} dt \qquad b > a \qquad (3.141)$$

where x is large

$$h'(c) = 0 \qquad h''(c) < 0 \qquad a < c < b \qquad (3.142)$$

$$f(t) \sim f_0 (t-c)^\lambda \qquad \text{as} \qquad t \to c \qquad (3.143)$$

and $\lambda > -1$ for the integral to exist. To determine the principal part of $I(x)$ for large x, we expand $h(t)$ in a Taylor series as

$$h(t) = h(c) + \tfrac{1}{2} h''(c) (t-c)^2 + \cdots \qquad (3.144)$$

because $h'(c) = 0$. Then, we have

$$I(x) \sim f_0 e^{xh(c)} \int_{c-\delta}^{c+\delta} (t-c)^\lambda e^{(1/2)xh''(c)(t-c)^2} dt \qquad (3.145)$$

where δ is a positive number. Since only the immediate neighborhood of $t = c$ contributes to $I(x)$, δ can be replaced with ∞. The result is

$$I(x) \sim f_0 e^{xh(c)} \int_{-\infty}^{\infty} (t-c)^\lambda e^{(1/2)xh''(c)(t-c)^2} dt \qquad (3.146)$$

Using the transformation

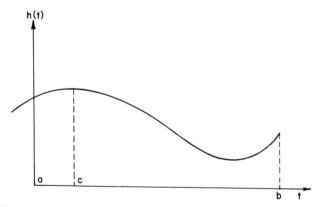

Figure 3-3. A function with an absolute maximum at an interior point.

LAPLACE'S METHOD 77

$$-\tfrac{1}{2} x h''(c) (t - c)^2 = \tau^2 \qquad (3.147)$$

we express (3.146) as

$$I(x) \sim f_0 \left[\frac{-2}{xh''(c)}\right]^{(\lambda+1)/2} e^{xh(c)} \int_{-\infty}^{\infty} \tau^\lambda e^{-\tau^2} d\tau \quad \text{as} \quad x \to \infty \qquad (3.148)$$

If λ is an odd integer, the integral in (3.148) vanishes, and one needs to determine the next term in the asymptotic development. If λ is an even integer,

$$\int_{-\infty}^{\infty} \tau^\lambda e^{-\tau^2} d\tau = 2 \int_{0}^{\infty} \tau^\lambda e^{-\tau^2} d\tau \qquad (3.149)$$

Using the transformation $\tau^2 = \theta$, (3.149) becomes

$$\int_{-\infty}^{\infty} \tau^\lambda e^{-\tau^2} d\tau = \int_{0}^{\infty} \theta^{(\lambda-1)/2} e^{-\theta} d\theta = \Gamma\left(\frac{\lambda+1}{2}\right) \qquad (3.150)$$

according to (3.108). Hence, (3.148) becomes

$$I(x) \sim f_0 \left[\frac{-2}{xh''(c)}\right]^{(\lambda+1)/2} e^{xh(c)} \Gamma\left(\frac{\lambda+1}{2}\right) \quad \text{as} \quad x \to \infty \qquad (3.151)$$

EXAMPLE 6
In all preceding examples,

$$f(t) \sim f_0 (t - c)^\lambda \qquad \lambda > -1 \qquad (3.152)$$

where c is the location of the maximum of $h(t)$. As a last example, we consider a case in which $f(t) \to 0$ as $t \to c$ faster than any power of $t - c$ so that it cannot be represented as in (3.152). Specifically, we consider

$$I(x) = \int_{a}^{b} e^{-1/(t-a)} e^{-x(t-a)^2} dt \qquad a < b \qquad (3.153)$$

for large positive x. Since $f(t) = \exp\left[-(t - a)^{-1}\right]$ tends to zero much faster than any power of $(t - a)$, the contribution to the integral from the immediate neighborhood of $t = a$ is exponentially small. Consequently, application of Watson's lemma directly to (3.153) does not yield its asymptotic development.

To determine the principal part of the asymptotic development of $I(x)$, we cannot separate $\exp\left[-(t - a)^{-1}\right]$ from $\exp\left[-x(t - a)^2\right]$. Rather, we should combine them and rewrite (3.153) as

$$I(x) = \int_{a}^{b} e^{h(x, t)} dt \qquad (3.154)$$

where

$$h(x, t) = -(t - a)^{-1} - x(t - a)^2 \qquad (3.155)$$

The stationary values of $h(x, t)$ are located where

$$\frac{dh}{dt} = 0 = \frac{1}{(t - a)^2} - 2x(t - a) \qquad (3.156)$$

Solving (3.156) yields

$$t = a + (2x)^{-1/3} \qquad (3.157)$$

for the location of the maximum of $h(x, t)$. This location is a function of x, in contrast with the preceding examples. Thus, to determine the asymptotic expansion of the integral, we first need to transform the variable of integration so that the maximum of the exponent is independent of x. Letting

$$t - a = x^{-1/3} s \qquad (3.158)$$

we rewrite (3.153) as

$$I(x) = \frac{1}{x^{1/3}} \int_0^{(b-a)x^{1/3}} e^{-x^{1/3}[s^2 + (1/s)]} ds \qquad (3.159)$$

The maximum value of $h(s) = -(s^2 + s^{-1})$ in (3.159) occurs at $s = 2^{-1/3}$. Hence, only the immediate neighborhood of this point contributes to the asymptotic expansion of $I(x)$ if b is greater than the location of the maximum of h. Thus, we expand $h(s)$ in a Taylor series about this point and obtain

$$h(s) = h(2^{-1/3}) - 3(s - 2^{-1/3})^2 + \cdots \qquad (3.160)$$

Then, we let

$$3x^{1/3} (s - 2^{-1/3})^2 = \tau^2 \qquad (3.161)$$

in (3.159) and replace the upper and lower limits by $+\infty$ and $-\infty$, respectively. The result is

$$I(x) \sim \frac{e^{-(3/2)(2x)^{1/3}}}{\sqrt{3x}} \int_{-\infty}^{\infty} e^{-\tau^2} d\tau$$

or

$$I(x) \sim \left(\frac{\pi}{3x}\right)^{1/2} e^{-(3/2)(2x)^{1/3}} \qquad (3.162)$$

on account of (3.25).

3.4. The Method of Stationary Phase

In this section, we consider the generalized Fourier integral

$$I(\alpha) = \int_a^b f(t) e^{i\alpha h(t)} dt \quad b > a \qquad (3.163)$$

for large positive α when $f(t)$ and $h(t)$ are real and the integral exists. The integrand is a complex number expressed in polar form with $f(t)$ and $\alpha h(t)$ being its amplitude and argument (phase). If $f(t)$ and $h(t)$ are continuously differentiable, one may be tempted to use the method of integration by parts by letting

$$u = \frac{f(t)}{h'(t)} \quad dv = e^{i\alpha h(t)} h'(t) dt$$

so that

$$du = \left[\frac{f(t)}{h'(t)}\right]' dt \quad v = \frac{1}{i\alpha} e^{i\alpha h(t)}$$

Then, it follows from (3.163) that

$$I(\alpha) = \frac{f(t) e^{i\alpha h(t)}}{i\alpha h'(t)} \bigg|_a^b - \frac{1}{i\alpha} \int_a^b \left[\frac{f(t)}{h'(t)}\right]' e^{i\alpha h(t)} dt \qquad (3.164)$$

Continuing the process further, one can write

$$I(\alpha) = \frac{i}{\alpha}\left[\frac{f(a) e^{i\alpha h(a)}}{h'(a)} - \frac{f(b) e^{i\alpha h(b)}}{h'(b)}\right] + O\left(\frac{1}{\alpha^2}\right) \qquad (3.165)$$

As in the case of the generalized Laplace integral (Example 5 of Section 3.2), the method of integration by parts fails if $h'(t)$ vanishes at any point in the interval $[a, b]$. If $h'(t) \neq 0$ in $[a, b]$, (3.165) shows that only the immediate neighborhoods of the end points contribute to the asymptotic development of $I(\alpha)$. The rapid oscillations of $\exp[i\alpha h(t)]$ tend to cancel the contributions to the integral except in the neighborhoods of the end points, as shown in Figure 3-4. Moreover, both ends contribute to the asymptotic expansion, in contrast with the generalized Laplace integral in which only the end with the larger value of h contributes to the asymptotic expansion. If $h'(t)$ vanishes in the interval (i.e., the phase has stationary points), the contribution to the asymptotic expansion of the integral arises from the immediate neighborhoods of the end and stationary points, with the major contribution arising from the neighborhoods of the stationary points, as evident from Figure 3-5. The rapid oscillations of $\exp[i\alpha h(t)]$ shown in Figure 3-5 tend to cancel contributions to the integral except in the neighborhoods of the end and stationary points. Figure 3-5 shows

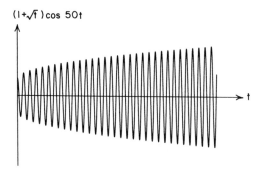

Figure 3-4. A function with rapid oscillations.

clearly that there is less cancellation from the neighborhood of a stationary point than from the neighborhood of an end point. Hence, the leading terms to the asymptotic expansion of Fourier integrals arise from the neighborhoods of stationary points. In the absence of stationary points, the method of integration by parts yields a good approximation to the integral. As in (3.165), the principal contribution is $O(\alpha^{-1})$. In the presence of stationary points, Stokes developed the so-called method of *stationary phase* to determine the contribution of the neighborhood of a stationary point $t = c$ to the asymptotic development of the integral by expanding $f(t)$ and $h(t)$ in powers of $t - c$. As shown below, the principal contribution from the neighborhood of a stationary point is $O(\alpha^{-1/2})$, and hence, only the stationary points contribute to the leading term in the asymptotic expansion of $I(\alpha)$. Next, we describe the method of stationary phase by applying it to four examples.

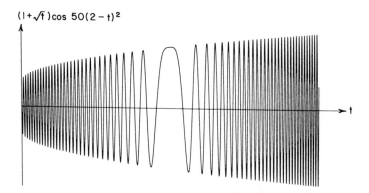

Figure 3-5. A function with a stationary point.

EXAMPLE 1

We begin with a case in which $h(t)$ has a stationary point at $t = a$ corresponding to a maximum or a minimum and it has no other stationary points. Moreover, we assume that $f(a)$ is finite. According to the Stokes method of stationary phase, only the immediate neighborhood of $t = a$ contributes to the leading term in the asymptotic expansion of $I(\alpha)$. Hence,

$$I(\alpha) \sim \int_a^{a+\delta} f(t) e^{i\alpha h(t)} dt \qquad (3.166)$$

where δ is a small positive number. Thus, to the first approximation, $f(t)$ can be replaced by $f(a)$ and $h(t)$ can be expanded in a Taylor series as

$$h(t) = h(a) + \tfrac{1}{2} h''(a)(t-a)^2 + \cdots$$

because $h'(a) = 0$. Then, (3.166) can be rewritten as

$$I(\alpha) \sim f(a) e^{i\alpha h(a)} \int_a^{a+\delta} e^{(1/2)i\alpha h''(a)(t-a)^2} dt \qquad (3.167)$$

Since only the immediate neighborhood of $t = a$ contributes to the integral, δ can be replaced with ∞. Letting $t - a = z$, we rewrite (3.167) as

$$I(\alpha) \sim f(a) e^{i\alpha h(a)} \int_0^\infty e^{(1/2)i\alpha h''(a)z^2} dz \qquad (3.168)$$

To evaluate the integral in (3.168), we appeal to Cauchy's theorem, which states that if the derivative of a function $F(z)$ of a complex variable z exists and is continuous inside and on a closed curve C (i.e., F is analytic), then

$$\int_C F(z) dz = 0 \qquad (3.169)$$

where the integration is carried around the closed curve C. The basic idea is to choose C in such a way that the original Fourier integral is transformed into a Laplace integral, that is, the dominant part of the integrand is a real decaying exponential at ∞. In the case under consideration

$$F(z) = \exp\left[\tfrac{1}{2} i\alpha h''(a) z^2\right] \qquad (3.170)$$

and its derivative exists for all values of z. Hence, Cauchy's theorem applies and we have (3.169). To apply Cauchy's theorem, we take the closed curve C to consist of the real axis x, a line making $45°$ to the real axis, and an eighth of a circle with the radius R, as shown in Figure 3-6. Hence,

82 INTEGRALS

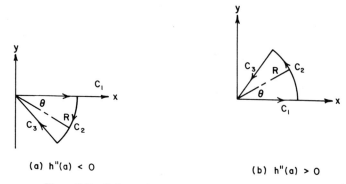

(a) $h''(a) < 0$ (b) $h''(a) > 0$

Figure 3-6. Deformation of the contour of integration.

$$\int_{C_1} F(z)\,dz + \int_{C_2} F(z)\,dz + \int_{C_3} F(z)\,dz = 0 \tag{3.171}$$

But on C_2

$$z = x + iy = R\cos\theta + iR\sin\theta = Re^{i\theta}$$

so that

$$z^2 = R^2 e^{2i\theta} = R^2\cos 2\theta + iR^2\sin 2\theta$$

Hence, in order that the contribution of C_2 as $R \to \infty$ vanishes, we must choose θ to be positive or negative depending on whether $h''(a)$ is positive or negative, respectively. Moreover, in order that the integral be converted into a Laplace integral, the angle of rotation must be $\frac{1}{4}\pi$ or $-\frac{1}{4}\pi$ depending on whether $h''(a)$ is positive or negative, respectively. Hence, it follows from Figure 3.6b that when $h''(a) > 0$

$$\int_{C_2} F(z)\,dz = iR \int_0^{(1/4)\pi} e^{-(1/2)\alpha h''(a)R^2 \sin 2\theta + (1/2)i\alpha h''(a)R^2 \cos 2\theta + i\theta}\,d\theta$$

$$\tag{3.172}$$

Then,

$$\left| \int_{C_2} F(z)\,dz \right| \leqslant R \int_0^{(1/4)\pi} e^{-(1/2)\alpha h''(a)R^2 \sin 2\theta}\,d\theta$$

$$= R \int_0^{\epsilon} e^{-(1/2)\alpha h''(a)R^2 \sin 2\theta}\,d\theta + R \int_{\epsilon}^{(1/4)\pi} e^{-(1/2)\alpha h''(a)R^2 \sin 2\theta}\,d\theta \tag{3.173}$$

THE METHOD OF STATIONARY PHASE

where ϵ is a small positive number. Since $h''(a) > 0$ and $\sin 2\theta > 0$ in the interval $[\epsilon, \frac{1}{4}\pi]$, the integrand tends to zero uniformly, and hence, the last integral in (3.173) tends to zero there as $R \to \infty$. To estimate the integral over the interval $[0, \epsilon]$, we approximate $\sin 2\theta$ by 2θ and obtain

$$R \int_0^\epsilon e^{-(1/2)\alpha h''(a)R^2 \sin 2\theta} d\theta \approx R \int_0^\epsilon e^{-\alpha h''(a)R^2 \theta} d\theta$$

$$= \frac{1 - e^{-\alpha h''(a)R^2 \epsilon}}{\alpha R h''(a)} \to 0 \quad \text{as} \quad R \to \infty$$

Therefore, the integral over the contour C_2 vanishes and it follows from (3.171) that

$$\int_{C_1} F(z) \, dz = -\int_{C_3} F(z) \, dz \quad \text{as} \quad R \to \infty \qquad (3.174)$$

Along the curve C_1, $z = x$ so that $z^2 = x^2$, whereas along the curve C_3, $z = r \exp(\frac{1}{4} i\pi)$ so that $z^2 = r^2 \exp(\frac{1}{2} i\pi) = ir^2$, where r is the distance from the origin to any point on C_3. Then, substituting (3.170) into (3.174) yields

$$\int_0^\infty e^{(1/2)i\alpha h''(a)x^2} dx = -e^{(1/4)i\pi} \int_\infty^0 e^{-(1/2)\alpha h''(a)r^2} dr$$

$$= e^{(1/4)i\pi} \int_0^\infty e^{-(1/2)\alpha h''(a)r^2} dr \qquad (3.175)$$

Equation (3.175) shows that the Fourier integral in the neighborhood of a stationary point has been converted into a Laplace integral by rotating the contour of integration from the real x axis by the angle $\frac{1}{4}\pi$. Making the substitution

$$\sqrt{\tfrac{1}{2}\alpha h''(a)} \, r = \tau$$

in (3.175), we obtain

$$\int_0^\infty e^{(1/2)i\alpha h''(a)x^2} dx = \frac{\sqrt{2} \, e^{(1/4)i\pi}}{\sqrt{\alpha h''(a)}} \int_0^\infty e^{-\tau^2} d\tau = \frac{\sqrt{\pi} \, e^{(1/4)i\pi}}{\sqrt{2\alpha h''(a)}}$$

$$(3.176)$$

on account of (3.25). Using (3.176) in (3.168) yields

$$I(\alpha) \sim \frac{\sqrt{\pi} f(a) \, e^{i\alpha h(a) + (1/4)i\pi}}{\sqrt{2\alpha h''(a)}} \qquad (3.177)$$

84 INTEGRALS

As mentioned above, the contribution from the neighborhood of a stationary point is $O(\alpha^{-1/2})$ compared with the $O(\alpha^{-1})$ contribution from an end point.

When $h''(a) < 0$, the contour of integration needs to be rotated by the angle $-\frac{1}{4}\pi$ so that

$$z = re^{-(1/4)i\pi} \qquad z^2 = r^2 e^{-(1/2)i\pi} = -ir^2$$

Then, (3.168) becomes

$$I(\alpha) \sim f(a) e^{i\alpha h(a) - (1/4)i\pi} \int_0^\infty e^{(1/2)\alpha h''(a) r^2} \, dr \qquad (3.178)$$

which is a Laplace integral because $h''(a) < 0$. Making the substitution

$$\sqrt{-\tfrac{1}{2}\alpha h''(a)}\, r = \tau$$

and using (3.25), we obtain from (3.178) that

$$I(\alpha) \sim \frac{\sqrt{\pi}\, f(a) e^{i\alpha h(a) - (1/4)i\pi}}{\sqrt{-2\alpha h''(a)}} \qquad \text{as} \qquad \alpha \to \infty \qquad (3.179)$$

In general, one needs to rotate the real axis by an angle θ so that $i\alpha h''(a) z^2$ in (3.168) becomes a negative real number. Since

$$z = re^{i\theta} \qquad z^2 = r^2 e^{2i\theta} = r^2 \cos 2\theta + ir^2 \sin 2\theta$$

it follows that $\cos 2\theta = 0$ or $\theta = \pm \tfrac{1}{4}\pi$. Then, $z^2 = ir^2 \sin 2\theta$ and

$$i\alpha h''(a) z^2 = -\alpha r^2 h''(a) \sin 2\theta$$

Consequently, when $h''(a) > 0$, one should take $\theta = \tfrac{1}{4}\pi$, whereas when $h''(a) < 0$, one should take $\theta = -\tfrac{1}{4}\pi$ so that the integral in (3.168) becomes a Laplace integral.

EXAMPLE 2

As a second example, we consider a case in which $h(t)$ has a stationary point at $t = c$ where $a < c < b$. We assume that $h(t)$ has no other stationary points and that $f(c)$ is finite. Hence, according to Stokes' method of stationary phase, the leading term in the asymptotic expansion of (3.163) arises from the immediate neighborhood of $t = c$ and we write

$$I(\alpha) \sim \int_{c-\delta}^{c+\delta} f(t) e^{i\alpha h(t)} \, dt \qquad (3.180)$$

where δ is a positive small number. Hence, $f(t)$ can be replaced with $f(c)$. Moreover, expanding $h(t)$ in a Taylor series around $t = c$, we have

$$h(t) = h(c) + \tfrac{1}{2} h''(c) (t - c)^2 + \cdots \qquad (3.181)$$

because $h'(c) = 0$. Substituting (3.181) into (3.180), replacing $f(t)$ with $f(c)$ and $t - c$ with z, and replacing δ with ∞, we obtain

$$I(\alpha) \sim f(c) \, e^{i\alpha h(c)} \int_{-\infty}^{\infty} e^{(1/2)i\alpha h''(c)z^2} \, dz \tag{3.182}$$

To evaluate the integral in (3.182) when $h''(c) > 0$, we rotate the contour of integration by the angle $\frac{1}{4}\pi$, so that $z = r \exp\left(\frac{1}{4}i\pi\right)$. The result is

$$I(\alpha) \sim f(c) e^{i\alpha h(c) + (1/4)i\pi} \int_{-\infty}^{\infty} e^{-(1/2)\alpha h''(c)r^2} \, dr$$

We make the substitution

$$\sqrt{\tfrac{1}{2}\alpha h''(c)} \, r = \tau$$

and obtain

$$I(\alpha) \sim \frac{\sqrt{2} \, f(c) e^{i\alpha h(c) + (1/4)i\pi}}{\sqrt{\alpha h''(c)}} \int_{-\infty}^{\infty} e^{-\tau^2} \, d\tau$$

Hence

$$I(\alpha) \sim \frac{\sqrt{2\pi} \, f(c) \, e^{i\alpha h(c) + (1/4)i\pi}}{\sqrt{\alpha h''(c)}} \quad \text{as} \quad \alpha \to \infty \tag{3.183}$$

When $h''(c) < 0$, we rotate the contour of integration by the angle $-\frac{1}{4}\pi$, make the substitution

$$\sqrt{-\tfrac{1}{2}\alpha h''(c)} \, r = \tau$$

and obtain

$$I(\alpha) \sim \frac{\sqrt{2\pi} \, f(c) \, e^{i\alpha h(c) - (1/4)i\pi}}{\sqrt{-\alpha h''(c)}} \quad \text{as} \quad \alpha \to \infty \tag{3.184}$$

EXAMPLE 3

As a third example, we consider a case in which $h(t)$ has a stationary point at $t = a$, no other stationary points, and

$$h'(a) = h''(a) = \cdots = h^{(n-1)}(a) = 0$$

but $h^{(n)}(a) \neq 0$. We assume that $f(a)$ is finite. We expand $h(t)$ in a Taylor series and obtain

$$h(t) = h(a) + \frac{1}{n!} h^{(n)}(a) (t - a)^n + \cdots \tag{3.185}$$

Then, it follows from (3.163) that to the first approximation

$$I(\alpha) \sim f(a) e^{i\alpha h(a)} \int_a^\infty e^{i\alpha h^{(n)}(a)(t-a)^n/n!} \, dt \qquad (3.186)$$

Letting $t - a = z$, we rewrite (3.186) as

$$I(\alpha) \sim f(a) e^{i\alpha h(a)} \int_0^\infty e^{i\alpha h^{(n)}(a) z^n/n!} \, dz \qquad (3.187)$$

To evaluate the integral in (3.187), we convert it into a Laplace integral by rotating the contour of integration by the angle $\pi/2n$ when $h^{(n)}(a) > 0$ so that $z = r \exp(i\pi/2n)$. The result is

$$I(\alpha) \sim f(a) e^{i\alpha h(a) + i\pi/2n} \int_0^\infty e^{-\alpha h^{(n)}(a) r^n/n!} \, dr \qquad (3.188)$$

The integral in (3.188) can be expressed in terms of the gamma function by making the substitution

$$\frac{\alpha h^{(n)}(a) r^n}{n!} = s$$

and obtaining

$$\int_0^\infty e^{-\alpha h^{(n)}(a) r^n/n!} \, dr = \frac{1}{n} \left[\frac{n!}{\alpha h^{(n)}(a)} \right]^{1/n} \int_0^\infty s^{(1/n)-1} e^{-s} \, ds$$

$$= \left[\frac{n!}{\alpha h^{(n)}(a)} \right]^{1/n} \frac{\Gamma\left(\frac{1}{n}\right)}{n} \qquad (3.189)$$

according to (3.108). Hence, (3.188) can be rewritten as

$$I(\alpha) \sim \left[\frac{n!}{\alpha h^{(n)}(a)} \right]^{1/n} \frac{f(a) \Gamma\left(\frac{1}{n}\right)}{n} e^{i\alpha h(a) + i\pi/2n} \quad \text{as} \quad \alpha \to \infty \qquad (3.190)$$

When $h^{(n)}(a) < 0$, the contour of integration needs to be rotated from the real axis by the angle $-\pi/2n$. Then following steps similar to those above, one obtains

$$I(\alpha) \sim \left[\frac{n!}{-\alpha h^{(n)}(a)} \right]^{1/n} \frac{f(a) \Gamma\left(\frac{1}{n}\right)}{n} e^{i\alpha h(a) - i\pi/2n} \quad \text{as} \quad \alpha \to \infty \qquad (3.191)$$

THE METHOD OF STATIONARY PHASE 87

We note from (3.190) that the leading term in the contribution of a neighborhood of a stationary point to the asymptotic expansion of (3.163) is $O(\alpha^{-1/n})$, where n corresponds to the order of the lowest derivative of h that does not vanish at the stationary point. Therefore, if $h(t)$ has many stationary points in $[a, b]$, the leading term to the asymptotic expansion of (3.163) arises from the neighborhood of the stationary point corresponding to the largest value of n. If more than one stationary point corresponds to the largest value of n, the leading term in the asymptotic expansion of (3.163) can be obtained as the summation of the leading contributions from the neighborhoods of these points.

EXAMPLE 4

In all preceding examples, $f(t)$ is finite at the stationary point. In this example, we consider the preceding example when

$$f(t) \sim f_0 (t-a)^\lambda \quad \text{as} \quad t \to a \tag{3.192}$$

where λ must be greater than -1 for the integral (3.163) to exist. Substituting for $h(t)$ and $f(t)$ from (3.185) and (3.192) into (3.163) and replacing b with ∞, we obtain

$$I(\alpha) \sim f_0 e^{i\alpha h(a)} \int_a^\infty (t-a)^\lambda e^{i\alpha h^{(n)}(a)(t-a)^n/n!} \, dt \tag{3.193}$$

where λ is assumed to be less than $n - 1$ for (3.193) to exist. Letting $t - a = z$, we rewrite (3.193) as

$$I(\alpha) \sim f_0 e^{i\alpha h(a)} \int_0^\infty z^\lambda e^{i\alpha h^{(n)}(a) z^n/n!} \, dz \tag{3.194}$$

As before, when $h^{(n)}(a) > 0$, we transform (3.194) into a Laplace integral by rotating the contour of integration from the real axis by the angle $\pi/2n$ so that $z = r \exp(i\pi/2n)$. The result is

$$I(\alpha) \sim f_0 e^{i\alpha h(a) + i(\lambda+1)\pi/2n} \int_0^\infty r^\lambda e^{-\alpha h^{(n)}(a) r^n/n!} \, dr \tag{3.195}$$

The integral in (3.195) can be expressed in terms of the gamma function by making the substitution

$$\frac{\alpha h^{(n)}(a) r^n}{n!} = s$$

and obtaining

$$\int_0^\infty r^\lambda e^{-\alpha h^{(n)}(a) r^n / n!} \, dr = \frac{1}{n} \left[\frac{n!}{\alpha h^{(n)}(a)} \right]^{(\lambda+1)/n} \int_0^\infty s^{-1+(\lambda+1)/n} e^{-s} \, ds$$

$$= \left[\frac{n!}{\alpha h^{(n)}(a)} \right]^{(\lambda+1)/n} \frac{\Gamma\left(\frac{\lambda+1}{n}\right)}{n} \qquad (3.196)$$

Substituting (3.196) into (3.195) yields

$$I(\alpha) \sim \left[\frac{n!}{\alpha h^{(n)}(a)} \right]^{(\lambda+1)/n} \frac{f_0 \Gamma\left(\frac{\lambda+1}{n}\right)}{n} e^{i\alpha h(a) + i(\lambda+1)\pi/2n} \quad \text{as} \quad \alpha \to \infty$$

(3.197)

When $h^{(n)}(a) < 0$, rotating the contour of integration by the angle $-\pi/2n$ and following steps similar to those above, we obtain

$$I(\alpha) \sim \left[\frac{n!}{-\alpha h^{(n)}(a)} \right]^{(\lambda+1)/n} \frac{f_0 \Gamma\left(\frac{\lambda+1}{n}\right)}{n} e^{i\alpha h(a) - i(\lambda+1)\pi/2n} \quad \text{as} \quad \alpha \to \infty$$

(3.198)

We note that (3.197) and (3.198) tend to (3.190) and (3.191) as $\lambda \to 0$. We note from (3.198) that if $\lambda + 1 > n$, then the leading contribution to the asymptotic expansion of the integral arises from the end point $t = b$.

3.5. The Method of Steepest Descent

So far, we considered only integrals in which the arguments of the exponents in the integrands are either purely real (Laplace integrals) or purely imaginary (Fourier integrals). In this section, we consider integrals in which the arguments of the exponents in the integrands are complex. Thus, we consider integrals of the form

$$I(\alpha) = \int_C f(z) e^{\alpha h(z)} \, dz \qquad (3.199)$$

where α is a large real positive number, C is a contour of integration in the complex z plane, and $f(z)$ and $h(z)$ are analytic functions of z, regular in a region of the z plane that contains the contour of integration. A function $h(z)$ is called *analytic* in the domain D of the z plane if it is defined and has a derivative at every point in the domain D. A function $h(z)$ that is analytic in a region D, except for a finite number of points, is said to be *meromorphic* in D. The exceptional points are called the *singularities* of the function $h(z)$.

A function $h(z)$ of a complex variable $z = x + iy$ is *differentiable* at the point z_0 if

$$\lim_{\Delta z \to 0} \frac{h(z_0 + \Delta z) - h(z_0)}{\Delta z} \qquad (3.200)$$

exists and is independent of the choice of Δz. The limit is called the derivative of $h(z)$ at z_0 and it is usually denoted by $h'(z_0)$ or $dh(z_0)/dz$. Putting $z = x + iy$ in $h(z)$, one can usually separate the result into real and imaginary parts and obtain

$$h(z) = h(x + iy) = \phi(x, y) + i\psi(x, y) \qquad (3.201)$$

Substituting (3.201) into (3.200) and taking $\Delta z = \Delta x$, we have

$$\frac{dh}{dz}(z_0) = \lim_{\Delta x \to 0} \frac{\phi(x_0 + \Delta x, y_0) - \phi(x_0, y_0)}{\Delta x}$$

$$+ i \lim_{\Delta x \to 0} \frac{\psi(x_0 + \Delta x, y_0) - \psi(x_0, y_0)}{\Delta x}$$

Hence,

$$\frac{dh}{dz} = \frac{\partial \phi}{\partial x} + i \frac{\partial \psi}{\partial x} \qquad \text{at} \quad z = z_0 \qquad (3.202)$$

Substituting (3.201) into (3.200) and taking $\Delta z = i\Delta y$, we have

$$\frac{dh}{dz}(z_0) = \lim_{\Delta y \to 0} \frac{\phi(x_0, y_0 + \Delta y) - \phi(x_0, y_0)}{i\Delta y} + \lim_{\Delta y \to 0} \frac{\psi(x_0, y_0 + \Delta y) - \psi(x_0, y_0)}{\Delta y}$$

Hence,

$$\frac{dh}{dz} = \frac{\partial \psi}{\partial y} - i \frac{\partial \phi}{\partial y} \qquad \text{at} \quad z = z_0 \qquad (3.203)$$

Since the derivative must be independent of the choice of Δz for a function $h(z)$ to be differentiable, the expressions in (3.202) and (3.203) must be the same. Thus,

$$\frac{\partial \phi}{\partial x} + i \frac{\partial \psi}{\partial x} = \frac{\partial \psi}{\partial y} - i \frac{\partial \phi}{\partial y} \qquad (3.204)$$

Separating real and imaginary parts in (3.204) yields the so-called *Cauchy-Riemann equations*

$$\frac{\partial \phi}{\partial x} = \frac{\partial \psi}{\partial y} \qquad (3.205a)$$

90 INTEGRALS

$$\frac{\partial \phi}{\partial y} = -\frac{\partial \psi}{\partial x} \quad (3.205b)$$

Eliminating ψ from (3.205) by cross differentiation yields

$$\frac{\partial^2 \phi}{\partial x^2} + \frac{\partial^2 \phi}{\partial y^2} = 0 \quad (3.206)$$

whereas eliminating ϕ from (3.205) by cross differentiation yields

$$\frac{\partial^2 \psi}{\partial x^2} + \frac{\partial^2 \psi}{\partial y^2} = 0 \quad (3.207)$$

To determine the asymptotic development of $I(\alpha)$, we use the analyticity of the integrand and appeal to Cauchy's theorem (3.169) to deform the contour of integration C into a new contour C' on which either ψ or ϕ is constant, thereby transforming the integral into either a Fourier or a Laplace integral. Then, the asymptotic development can be determined by using either the method of stationary phase or Laplace's method. It is preferable to transform the integral into a Laplace integral (i.e., constant ψ) because the full asymptotic development of a Laplace integral arises only from the immediate neighborhood of the point where ϕ is the largest on C'. In contrast, the full asymptotic development of a Fourier integral depends, in general, on the end points as well as all stationary points of ψ on C'.

We note that *constant-phase contours are steepest descent and ascent contours*. To show this, we appeal to the concept of gradient in elementary calculus. The gradient of $\phi(x,y)$ is defined by

$$\nabla \phi = \left(\frac{\partial \phi}{\partial x}, \frac{\partial \phi}{\partial y} \right) \quad (3.208)$$

and the derivative of ϕ in the direction defined by the unit vector $\mathbf{n}$ is

$$\frac{d\phi}{dn} = \nabla \phi \cdot \mathbf{n} \quad (3.209)$$

Hence, the maximum value of $d\phi/dn$ occurs when $\mathbf{n} = \nabla\phi/|\nabla\phi|$. Thus, the steepest ascent contours are parallel to $\nabla\phi$, whereas the steepest descent contours are parallel to $-\nabla\phi$. It follows from the Cauchy-Riemann equations (3.205) that

$$\nabla \phi \cdot \nabla \psi = 0 \quad (3.210)$$

so that $\nabla\phi$ is perpendicular to $\nabla\psi$ and the derivative of ψ in the direction $\nabla\phi$ is zero. Thus, ψ is constant on contours whose tangents are parallel to $\nabla\phi$, showing that constant-phase contours correspond to steepest ascent and descent contours.

We should note that $\phi(x, y)$ can never have a true maximum or minimum except at a singularity because if, for example,

$$\frac{\partial^2 \phi}{\partial x^2} < 0 \quad \text{then} \quad \frac{\partial^2 \phi}{\partial y^2} > 0 \qquad (3.211)$$

on account of (3.206). However, the surface $\phi(x, y)$ may possess a flat spot at which

$$\frac{\partial \phi}{\partial x} = \frac{\partial \phi}{\partial y} = 0 \qquad (3.212)$$

Such a spot is called a *saddle point* because the surface $\phi(x,y)$ resembles a saddle or a mountain pass or a col, as shown in Figure 3-7. By the Cauchy-Riemann equations (3.205), we see that (3.212) implies that

$$\frac{\partial \psi}{\partial x} = \frac{\partial \psi}{\partial y} = 0 \qquad (3.213)$$

Thus, a saddle point of $\phi(x,y)$ is also a saddle point of $\psi(x,y)$ as well as a point where $h'(z) = 0$. If $z = z_0$ is a saddle point and if $h'(z_0) = h''(z_0) = \cdots = h^{(m)}(z_0) = 0$, we call $z = z_0$ a *saddle point of order* $m + 1$.

Through a saddle point z_0, there are two or more *level curves* (curves of constant ϕ), separating the neighborhood of the saddle point into sectors. Moreover, through a saddle point, there are two or more constant-phase contours (curves of

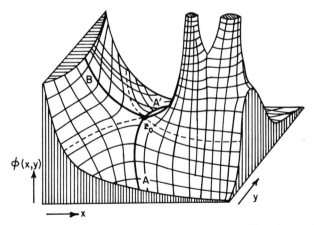

Figure 3-7. Topography of the surface $\phi = \mathrm{Re}\,h(z)$ near the saddle point z_0, for a typical function $h(z)$. The heavy solid curves follow the centers of the ridges and valleys from the saddle point, and the dashed curves follow level contours, $\phi = \phi(x_0, y_0) = $ constant. The curve AA' is the path of steepest descent.

constant ψ), which are the steepest paths through the saddle point. Some of these steepest paths descend, whereas the others ascend.

Although the forms of the constant-level and -phase paths passing through a saddle may be complicated in the whole z plane, it is easy to find their forms near a saddle point. If the saddle point $z = z_0$ is of order m

$$h(z) \approx h(z_0) + \frac{1}{m!} h^{(m)}(z_0)(z - z_0)^m \qquad (3.214)$$

Hence, if

$$\frac{1}{m!} h^{(m)}(z_0) = Ke^{i\chi} \qquad z - z_0 = re^{i\theta} \qquad (3.215)$$

then,

$$h(z) \approx h(z_0) + Kr^m e^{i(\chi + m\theta)} = \phi + i\psi$$

or

$$\phi = \phi_0 + Kr^m \cos(\chi + m\theta) \qquad \psi = \psi_0 + Kr^m \sin(\chi + m\theta) \qquad (3.216)$$

where ϕ_0 and ψ_0 are the values of ϕ and ψ at the saddle point. Thus, the level curves $\phi = \phi_0$ passing through the saddle point are approximately given by

$$\cos(\chi + m\theta) = 0 \quad \text{or} \quad \chi + m\theta = (n + \tfrac{1}{2})\pi \qquad (3.217)$$

where $n = 1, 2, 3, \ldots, 2m$. Equation (3.217) provides the $2m$ constant level curves $\theta = [(n + \tfrac{1}{2})\pi - \chi]/m$ that divide the neighborhood of z_0 into m hills and valleys. It follows from (3.216) that the constant-phase (steepest) contours $\psi = \psi_0$ are approximately given by

$$\sin(\chi + m\theta) = 0 \quad \text{or} \quad \chi + m\theta = n\pi \qquad (3.218)$$

where $n = 1, 2, 3, \ldots, 2m$. Equation (3.218) provides the $2m$ steepest contours $\theta = (n\pi - \chi)/m$, m contours of steepest descent, and m contours of steepest ascent.

In the simplest case, the saddle point is of order 2 so that there are two steepest descent contours and two steepest ascent contours, as shown in Figure 3-8. Moreover, there are four constant level curves, separating the neighborhood of the saddle point into two hills and two valleys, as shown in Figure 3-8. For a saddle point of order 3, there are six constant-level contours, separating the neighborhood of the saddle point into three hills and three valleys. Moreover, there are three steepest descent contours in the valleys and three steepest ascent contours in the hills, as shown in Figure 3-9.

The above discussion shows that an effective method of determining the asymptotic development of integrals whose end points lie in two different

THE METHOD OF STEEPEST DESCENT 93

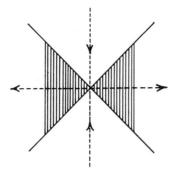

Figure 3-8. Topography of the surface $\phi = \text{Re}h(z)$ near a saddle point of order 2 when $\chi = 0$. The valleys are shaded, the steepest curves dotted, the solid lines are the level curves, and the arrows indicate the direction in which ϕ decreases.

valleys is the *method of steepest descent* developed by Riemann and Debye. It consists of deforming the contour of integration C to a new contour C' such that

1. *the contour of integration passes through a zero of $h'(z)$*;
2. *the imaginary part ψ of $h(z)$ is constant on the contour*;
3. *the contour is that of steepest descent.*

If the restrictions on the deformed contour are such that it passes through more than one saddle point, each will make its contribution to the integral with the main contribution arising from the one corresponding to the largest ϕ. If $h'(z)$ does not vanish, the contour of integration is chosen to satisfy the second and

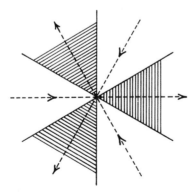

Figure 3-9. Topography of the surface $\phi = \text{Re}h(z)$ near a saddle point of order 3 when $\chi = 0$. The valleys are shaded, the steepest curves dotted, the solid lines are the level curves, and the arrows indicate the direction in which ϕ decreases.

94 INTEGRALS

third conditions only. Next, the method of steepest descent is described by its application to three examples.

EXAMPLE 1

As a first example, we consider the following integral representation of Bessel's function of the first kind and zeroth order:

$$J_0(\alpha) = \frac{1}{\pi} \int_{-1}^{1} \frac{e^{i\alpha z}}{\sqrt{1-z^2}} dz \quad \text{for large } \alpha \qquad (3.219)$$

Integration by parts fails in this case because the choice

$$u = e^{i\alpha z} \qquad dv = (1-z^2)^{-1/2} dz$$

leads to a nonasymptotic expansion, whereas the choice

$$u = (1-z^2)^{-1/2} \qquad dv = e^{i\alpha z} dz$$

leads to singular expressions at $z = \pm 1$.

To determine an approximate expression for $J_0(\alpha)$ for large α by using the method of steepest descent, we deform the contour of integration into a constant-phase contour. We note that $h(z) = iz$ so that the phase ψ at $z = 1$ is 1, whereas the phase at $z = -1$ is -1. Thus, the contour cannot be continuously deformed into a single contour along which the phase is constant. However, we can deform the contour into one that consists of three line segments: C_1, which runs up from -1 to $-1 + iY$ along a straight line parallel to the y axis; C_2, which runs parallel to the x axis from $-1 + iY$ to $1 + iY$; and C_3, which runs down from $1 + iY$ to 1 along a straight line parallel to the y axis, as shown in Figure 3-10. By Cauchy's theorem (3.169),

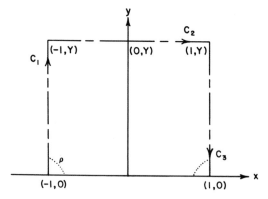

Figure 3-10. Contour deformation for (3.219).

THE METHOD OF STEEPEST DESCENT

$$\int_{C_1+C_2+C_3} \frac{e^{i\alpha z}}{\sqrt{1-z^2}} dz = \int_{-1}^{1} \frac{e^{i\alpha z}}{\sqrt{1-z^2}} dz \qquad (3.220)$$

Strictly speaking, the contour of integration must be deformed as shown by the dotted lines to avoid the so-called branch singularities at $x = \pm 1$. However, in the limit as $\rho \to 0$ the integral tends to that given in (3.220). Hence, we will not worry about such border branch points in what follows. As $Y \to \infty$, the integral along C_2 vanishes because the integrand vanishes uniformly there. Then,

$$J_0(\alpha) = \frac{1}{\pi} \int_{-1}^{-1+i\infty} \frac{e^{i\alpha z}}{\sqrt{1-z^2}} dz + \frac{1}{\pi} \int_{1+i\infty}^{1} \frac{e^{i\alpha z}}{\sqrt{1-z^2}} dz \qquad (3.221)$$

Making the substitution $z = -1 + iy$ in the first integral and $z = 1 + iy$ in the second integral, we rewrite (3.221) as

$$J_0(\alpha) = \frac{ie^{-i\alpha}}{\pi} \int_0^\infty \frac{e^{-\alpha y}}{\sqrt{2iy+y^2}} dy + \frac{ie^{i\alpha}}{\pi} \int_\infty^0 \frac{e^{-\alpha y} dy}{\sqrt{-2iy+y^2}}$$

or

$$J_0(\alpha) = \frac{ie^{-i\alpha-(1/4)i\pi}}{\sqrt{2}\,\pi} \int_0^\infty e^{-\alpha y} y^{-1/2} (1 - \tfrac{1}{2} iy)^{-1/2} dy$$

$$- \frac{ie^{i\alpha+(1/4)i\pi}}{\sqrt{2}\,\pi} \int_0^\infty e^{-\alpha y} y^{-1/2} (1 + \tfrac{1}{2} iy)^{-1/2} dy \qquad (3.222)$$

The integrals in (3.222) are Laplace integrals and only the immediate neighborhood of $y = 0$ contributes to their asymptotic developments for large α. Thus, using Watson's lemma, we expand the nonexponential parts of the integrands for small y and integrate the results term by term. The leading term in the asymptotic development of $J_0(\alpha)$ is

$$J_0(\alpha) \sim \frac{e^{-i\alpha+(1/4)i\pi}}{\sqrt{2}\,\pi} \int_0^\infty y^{-1/2} e^{-\alpha y} dy + \frac{e^{i\alpha-(1/4)i\pi}}{\sqrt{2}\,\pi} \int_0^\infty y^{-1/2} e^{-\alpha y} dy$$

which, with the substitution $\alpha y = t$, becomes

$$J_0(\alpha) \sim \frac{1}{\sqrt{2\alpha}\,\pi} [e^{-i\alpha+(1/4)i\pi} + e^{i\alpha-(1/4)i\pi}] \int_0^\infty t^{-1/2} e^{-t} dt \qquad (3.223)$$

Using (3.121) and the identity

$$(e^{i\theta} + e^{-i\theta}) = 2\cos\theta$$

96 INTEGRALS

we rewrite (3.223) as

$$J_0(\alpha) \sim \sqrt{\frac{2}{\alpha\pi}} \cos(\alpha - \tfrac{1}{4}\pi) \quad \text{as} \quad \alpha \to \infty \tag{3.224}$$

EXAMPLE 2

As a second example, we consider the Airy integral

$$Ai(\alpha) = \frac{1}{\pi}\int_0^\infty \cos(\tfrac{1}{3}s^3 + \alpha s)\,ds \quad \text{for} \quad \alpha \gg 1 \tag{3.225}$$

To transform (3.225) into an integral having the form (3.199), we introduce the transformation $s = \alpha^{1/2} z$ and obtain

$$Ai(\alpha) = \frac{\alpha^{1/2}}{\pi}\int_0^\infty \cos[\alpha^{3/2}(\tfrac{1}{3}z^3 + z)]\,dz = \frac{\alpha^{1/2}}{2\pi}\int_0^\infty \{e^{i\alpha^{3/2}[(1/3)z^3 + z]}$$

$$+ e^{-i\alpha^{3/2}[(1/3)z^3 + z]}\}\,dz = \frac{\alpha^{1/2}}{2\pi}\int_0^\infty e^{i\alpha^{3/2}[(1/3)z^3 + z]}\,dz$$

$$+ \frac{\alpha^{1/2}}{2\pi}\int_0^\infty e^{-i\alpha^{3/2}[(1/3)z^3 + z]}\,dz = \frac{\alpha^{1/2}}{2\pi}\int_0^\infty e^{i\alpha^{3/2}[(1/3)z^3 + z]}\,dz$$

$$- \frac{\alpha^{1/2}}{2\pi}\int_0^{-\infty} e^{i\alpha^{3/2}[(1/3)z^3 + z]}\,dz$$

Hence,

$$Ai(\alpha) = \frac{\alpha^{1/2}}{2\pi}\int_{-\infty}^\infty e^{i\alpha^{3/2}[(1/3)z^3 + z]}\,dz \tag{3.226}$$

Integration by parts yields a trivial solution because it forces the expansion to have the form of a power series in α^{-1} whereas, as shown below, the expansion has an exponential factor that tends to zero faster than any power of α^{-1}.

To determine the asymptotic development of $Ai(\alpha)$, we use the method of steepest descent. Here,

$$h(z) = i(\tfrac{1}{3}z^3 + z) \quad h'(z) = i(z^2 + 1)$$

and the saddle points are the zeros of $h'(z) = 0$, that is, the points $z = \pm i$. At the saddle points,

$$h(\pm i) = i(\mp \tfrac{1}{3}i \pm i) = \mp \tfrac{2}{3}$$

Hence, Im $h(\pm i) = 0$. Putting $z = x + iy$ in the expression for h yields

$$h(z) = i[\tfrac{1}{3}(x + iy)^3 + x + iy] = y(\tfrac{1}{3}y^2 - x^2 - 1) + ix(\tfrac{1}{3}x^2 - y^2 + 1) \tag{3.227}$$

Since Im $h = 0$ at the saddle points, it follows from (3.227) that the steepest paths passing through the saddle points are given by Im $h = 0$, that is,

$$x(\tfrac{1}{3}x^2 - y^2 + 1) = 0 \tag{3.228}$$

Equation (3.228) is a cubic equation consisting of the imaginary axis and of the two branches of the hyperbola

$$\tfrac{1}{3}x^2 - y^2 + 1 = 0 \tag{3.229}$$

These paths are shown in Figure 3-11 with arrows indicating the directions of decreasing Re $h(z)$. Therefore, to apply the method of steepest descent, we deform the contour of integration from the real axis into the contour C_1

1. which passes through the saddle point $z = i$,
2. which is a constant-phase contour, and
3. which is a steepest-descent contour from the saddle point.

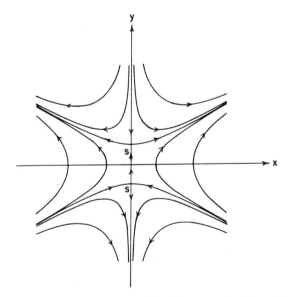

Figure 3-11. Constant phase contours including the steepest paths passing through the saddle points for the Airy function; the arrows indicate the direction in which Re$h(z)$ decreases.

The deformation is possible since the integral decays exponentially in the region between the contours as $z \to \infty$. Hence, we replace (3.226) with

$$Ai(\alpha) = \frac{\alpha^{1/2}}{2\pi} \int_{C_1} e^{i\alpha^{3/2}[(1/3)z^3 + z]} \, dz \qquad (3.230)$$

Along C_1, (3.229) holds so that $x^2 = 3(y^2 - 1)$ and (3.227) becomes

$$h(z) = y(2 - \tfrac{8}{3} y^2) \quad \text{for} \quad y \geq 1 \qquad (3.231)$$

Moreover, $h(z)$ has its maximum at $z = i$ (i.e., $y = 1$). Therefore, the integral in (3.230) is a generalized Laplace integral and its asymptotic development can be obtained by using Watson's lemma. To this end, we make the substitution

$$\alpha^{3/2} h(z) = i\alpha^{3/2} (\tfrac{1}{3} z^3 + z) = \alpha^{3/2} h(i) - \tau^2 = -\tfrac{2}{3} \alpha^{3/2} - \tau^2 \qquad (3.232)$$

Hence,

$$i\alpha^{3/2}(z^2 + 1) \, dz = -2\tau \, d\tau \qquad (3.233)$$

and (3.230) becomes

$$Ai(\alpha) \sim \frac{ie^{-(2/3)\alpha^{3/2}}}{\pi \alpha} \int_{-\infty}^{\infty} \frac{\tau e^{-\tau^2}}{z^2 + 1} \, d\tau \qquad (3.234)$$

Next, we need to expand $(z^2 + 1)^{-1}$ for small τ, and hence, we need to determine z as a function of τ. To this end, we expand the left-hand side of (3.232) in a Taylor series around $z = i$ and obtain

$$(z - i)^2 - \tfrac{1}{3} i(z - i)^3 = \alpha^{-3/2} \tau^2$$

whose solution for small τ can be expressed in the form

$$z = i + \alpha^{-3/4} \tau + \cdots \qquad (3.235)$$

Thus,

$$(z^2 + 1)^{-1} = (2i\alpha^{-3/4} \tau)^{-1} \quad \text{as} \quad \tau \to 0 \qquad (3.236)$$

Substituting (3.236) into (3.234) leads to the following expression for the leading term in the asymptotic expansion of $Ai(\alpha)$:

$$Ai(\alpha) \sim \frac{e^{-(2/3)\alpha^{3/2}}}{2\sqrt{\pi} \, \alpha^{1/4}} \quad \text{as} \quad \alpha \to \infty \qquad (3.237)$$

It should be noted that detailed tracing of the steepest descent contours is seldom necessary because the asymptotic development of an integral wholy arises from the immediate neighborhoods of the points where ϕ is greatest. These points could be interior or end points. In order that ϕ be greatest at an

THE METHOD OF STEEPEST DESCENT

interior point, it is necessary that such a point be a saddle point. Thus, in the present example, it is sufficient to trace a very short segment of steepest descent contour of integration through the saddle point. Since $z = i$ is a saddle point, the exponent in (3.230) near this point can be approximated as

$$i\alpha^{3/2}(\tfrac{1}{3} z^3 + z) = -\tfrac{2}{3} \alpha^{3/2} - \alpha^{3/2} (z - i)^2 + \cdots \qquad (3.238)$$

and the direction of the contour will be such that $\alpha^{3/2}(z - i)^2 = \tau^2$ is real and positive. Hence, for the leading term, we substitute (3.238) into (3.230), take the limits of integration from $-\infty$ to ∞, and obtain

$$Ai(\alpha) \sim \frac{e^{-(2/3)\alpha^{3/2}}}{2\pi\alpha^{1/4}} \int_{-\infty}^{\infty} e^{-\tau^2} \, d\tau$$

Hence,

$$Ai(\alpha) \sim \frac{e^{-(2/3)\alpha^{3/2}}}{2\sqrt{\pi}\,\alpha^{1/4}} \qquad \text{as} \qquad \alpha \to \infty \qquad (3.239)$$

on account of (3.25).

EXAMPLE 3

As a final example, we consider

$$I(\alpha) = \int_0^\infty e^{\alpha(3\lambda z - z^3)} \, dz \qquad (3.240)$$

for large positive α where λ is a complex number that is independent of α. The saddle points are given by

$$h'(z) = \frac{d}{dz}(3\lambda z - z^3) = 0 \quad \text{or} \quad \lambda - z^2 = 0$$

Hence, they are located at $z = \pm \lambda^{1/2}$. If we let $\lambda = \sigma^2 \exp(2i\nu)$, then the saddle points are located at $z = \pm \sigma \exp(i\nu)$. Since $h''(z) = -6z$ is different from zero at the saddle points, the saddle points are of order two.

To determine the asymptotic development of (3.240), we use the analyticity of the integrand and deform the contour of integration into the contour C that passes through the origin and the saddle point $z = \sigma \exp(i\nu)$ and that is a steepest-descent contour (see Figure 3-12). Thus, we rewrite (3.240) as

$$I(\alpha) = \int_0^\infty e^{\alpha(3\lambda z - z^3)} \, dz = \int_C e^{\alpha(3\lambda z - z^3)} \, dz \qquad (3.241)$$

As mentioned earlier, only a very short segment $\hat{C}$ of the contour C passing

100 INTEGRALS

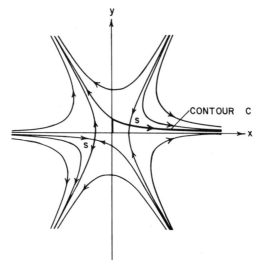

Figure 3-12. Deformed contour C of integration for $\sigma = 1$ and $\nu = 22.5°$.

through the saddle point contributes to the asymptotic development of $I(\alpha)$. Hence,

$$I(\alpha) \sim \int_{\hat{C}} e^{\alpha(3\lambda z - z^3)}\, dz \qquad (3.242)$$

Since $z = \sigma \exp(i\nu)$ is a saddle point, the exponent in (3.242) near this point can be expanded as

$$\alpha(3\lambda z - z^3) = 2\alpha\sigma^3\, e^{3i\nu} - 3\alpha\sigma\, e^{i\nu}(z - \sigma e^{i\nu})^2 + \cdots \qquad (3.243)$$

and the direction of the path $\hat{C}$ will be such that

$$3\alpha\sigma\, e^{i\nu}(z - \sigma e^{i\nu})^2 = \tau^2 \qquad (3.244)$$

where τ^2 is real and positive. Hence,

$$z - \sigma e^{i\nu} = \pm \frac{e^{-(1/2)i\nu}}{\sqrt{3\alpha\sigma}}\, \tau \qquad (3.245)$$

Thus, there are two possible choices for the direction of the path of integration, namely $-\tfrac{1}{2}\nu$ and $\pi - \tfrac{1}{2}\nu$. At this point, one needs to inspect the behaviors of ϕ and ψ over the complex plane (Figure 3-12) in order to decide the sense in which the path passes through the saddle point. Figure 3-12 shows that the direction of the path of integration is $-\tfrac{1}{2}\nu$, so that we must take the positive sign in (3.245) because τ goes from a negative to a positive value on the path. These values are taken to be $-\infty$ and ∞ according to Watson's lemma. Then, substituting (3.245) into (3.242), we obtain for the leading term

$$I(\alpha) \sim \frac{\exp\left[-\frac{1}{2}i\nu + 2\alpha\sigma^3 e^{3i\nu}\right]}{\sqrt{3\sigma\alpha}} \int_{-\infty}^{\infty} e^{-\tau^2} d\tau$$

or

$$I(\alpha) \sim \sqrt{\frac{\pi}{3\sigma\alpha}} \exp\left[-\frac{1}{2}i\nu + 2\alpha\sigma^3 e^{3i\nu}\right] \quad \text{as} \quad \alpha \to \infty \quad (3.246)$$

Exercises

3.1. Show that as $\epsilon \to 0$

$$\int_0^1 \frac{\sin \epsilon t}{t} dt \sim \epsilon - \frac{1}{18}\epsilon^3 + \frac{1}{600}\epsilon^5$$

3.2. Show that as $x \to 0$

$$\int_0^x t^{-3/4} e^{-t} dt \sim 4x^{1/4} - \frac{4}{5}x^{5/4} + \frac{2}{9}x^{9/4}$$

3.3. Show that as $x \to \infty$

(a) $\displaystyle\int_x^\infty \frac{e^{-t}}{t} dt \sim e^{-x}\left[\frac{1}{x} - \frac{1!}{x^2} + \frac{2!}{x^3} - \frac{3!}{x^4}\right]$

(b) $\displaystyle\int_x^\infty \frac{e^{-t}}{t^n} dt \sim \frac{e^{-x}}{x^n}\left[1 - \frac{n}{x} + \frac{n(n+1)}{x^2} - \frac{n(n+1)(n+2)}{x^3}\right]$

3.4. Show that as $x \to \infty$

$$\int_x^\infty e^{-t} t^{\lambda-1} dt \sim x^\lambda e^{-x}\left[\frac{1}{x} + \frac{\lambda-1}{x^2} + \frac{(\lambda-1)(\lambda-2)}{x^3}\right]$$

3.5. Show that as $x \to \infty$

$$\int_x^\infty e^{-t^2} dt \sim e^{-x^2}\left[\frac{1}{2x} - \frac{1}{2^2 x^3} + \frac{1\cdot 3}{2^3 x^5} - \frac{1\cdot 3\cdot 5}{2^4 x^7}\right]$$

3.6. Consider the complete elliptic integral of the second kind

$$I(m) = \int_0^{(1/2)\pi} \sqrt{1 - m\sin^2\theta}\, d\theta$$

Show that

102 INTEGRALS

$$I(m) = \frac{1}{2}\pi\left[1 - \frac{1}{4}m - \frac{3}{64}m^2 - \frac{5}{256}m^3 - \frac{175}{16384}m^4 + \cdots\right]$$

Compare this result with those tabulated in Abramowitz and Stegun.

3.7. Show that as $x \to \infty$

$$\int_x^\infty \frac{\cos t}{t}\,dt \sim \left(-\frac{1}{x} + \frac{2!}{x^3} - \frac{4!}{x^5}\right)\sin x + \left(\frac{1}{x^2} - \frac{3!}{x^4} + \frac{5!}{x^6}\right)\cos x$$

3.8. Show that as $x \to \infty$

(a) $\displaystyle\int_x^\infty \frac{\cos t}{\sqrt{t}}\,dt \sim \frac{1}{\sqrt{x}}(f\cos x - g\sin x)$

(b) $\displaystyle\int_x^\infty \frac{\sin t}{\sqrt{t}}\,dt \sim \frac{1}{\sqrt{x}}(f\sin x + g\cos x)$

where

$$f \sim \frac{1}{2x} - \frac{1\cdot 3\cdot 5}{(2x)^3} + \frac{1\cdot 3\cdot 5\cdot 7\cdot 9}{(2x)^5}$$

$$g \sim 1 - \frac{1\cdot 3}{(2x)^2} + \frac{1\cdot 3\cdot 5\cdot 7}{(2x)^4}$$

3.9. Show that as $x \to \infty$

(a) $\displaystyle\int_x^\infty \frac{\cos(t-x)}{t}\,dt \sim \frac{1}{x^2} - \frac{3!}{x^4} + \frac{5!}{x^6}$

(b) $\displaystyle\int_x^\infty \frac{\sin(t-x)}{t}\,dt \sim \frac{1}{x} - \frac{2!}{x^3} + \frac{4!}{x^5}$

3.10. Show that as $x \to \infty$

(a) $\displaystyle\int_x^\infty \frac{dt}{t^2 \ln t} \sim \frac{1}{x \ln x}$

(b) $\displaystyle\int_2^x \frac{dt}{t \ln \ln t} \sim \frac{\ln x}{\ln \ln x}$

3.11. Show that as $x \to \infty$

(a) $\displaystyle\int_x^\infty \frac{t^{1/2}}{1+t^2}\,dt \sim 2x^{-1/2} - \frac{2}{5}x^{-5/2} + \frac{2}{9}x^{-9/2}$

(b) $\displaystyle\int_0^x (t^3 + t^2)^{1/2}\,dt \sim \frac{2}{5}x^{5/2} + \frac{1}{3}x^{3/2}$

(c) $\displaystyle\int_a^x \frac{dt}{\ln t} \sim \frac{x}{\ln x}$

3.12. Show that as $x \to \infty$

(a) $\displaystyle\int_0^1 e^{-xt} \ln(2+t)\, dt \sim \frac{\ln 2}{x}$

(b) $\displaystyle\int_0^1 e^{-xt} \ln(1+t)\, dt \sim \frac{1}{x^2}$

(c) $\displaystyle\int_0^1 e^{-xt} \sin t\, dt \sim \frac{1}{x^2}$

(d) $\displaystyle\int_0^1 e^{-(x/t)+t+xt}\, dt \sim \frac{e}{2x}$

3.13. Show that as $\omega \to \infty$

$$\int_0^\infty \frac{e^{-x}}{\omega + x + x\sqrt{\omega}}\, dx \sim \frac{1}{\omega} - \frac{1}{\omega^{3/2}}$$

3.14. Show that as $\omega \to \infty$

(a) $\displaystyle\int_0^\infty e^{-\omega(x^2+2x)}(1+x)^{5/2}\, dx \sim \frac{1}{2\omega}$

(b) $\displaystyle\int_0^\infty e^{-\omega(x^2+2x)} \ln(2+x)\, dx \sim \frac{\ln 2}{2\omega}$

(c) $\displaystyle\int_0^\infty e^{-\omega(x^2+2x)} \ln(1+x)\, dx \sim \frac{1}{4\omega^2}$

(d) $\displaystyle\int_0^\infty \frac{e^{-\omega(x^2+2x)}}{\sqrt{x+3x^2}}\, dx \sim \sqrt{\frac{\pi}{2\omega}}$

3.15. Show that as $\omega \to \infty$

$$\int_a^b (x-a)^\lambda e^{-\omega h(x)}\, dx \sim \frac{e^{-\omega h(a)} \Gamma(\lambda + 1)}{[\omega h'(a)]^{\lambda + 1}}$$

where $\lambda > -1$, $b > a$, and $h(x) > h(a)$.

3.16. Show that as $\omega \to \infty$

$$\int_1^\infty e^{-\omega x^2} x^{5/2} \ln(1+x)\, dx \sim \frac{e^{-\omega} \ln 2}{2\omega}$$

104 INTEGRALS

$$\int_1^\infty e^{-\omega x^2} x^{5/2} \ln x \, dx \sim \frac{e^{-\omega}}{4\omega^2}$$

3.17. Show that as $\omega \to \infty$

(a) $\displaystyle\int_0^\infty \frac{e^{-\omega x^2}}{\sqrt{x+x^2}} dx \sim \frac{\Gamma(\frac{1}{4})}{2\omega^{1/4}}$

(b) $\displaystyle\int_0^\infty e^{-\omega x^2} x^{5/2} \ln(1+x) \, dx \sim \frac{\Gamma(\frac{9}{4})}{2\omega^{9/4}}$

3.18. Show that as $\omega \to \infty$

(a) $\displaystyle\int_{-\infty}^\infty e^{-\omega x^2} \ln(2+x^2) \, dx \sim \frac{\sqrt{\pi} \ln 2}{\sqrt{\omega}}$

(b) $\displaystyle\int_{-\infty}^\infty e^{-\omega x^2} \ln(1+x^2) \, dx \sim \frac{\sqrt{\pi}}{2\omega^{3/2}}$

3.19. Show that as $x \to \infty$

(a) $\displaystyle\int_1^2 e^{-x[t+(1/t)]} dt \sim \frac{\sqrt{\pi}}{2\sqrt{x}} e^{-2x}$

(b) $\displaystyle\int_1^2 e^{-x[t+(1/t)]} \ln(1+t) \, dt \sim \frac{\sqrt{\pi} \ln 2}{2\sqrt{x}} e^{-2x}$

(c) $\displaystyle\int_1^2 e^{-x[t+(1/t)]} \ln t \, dt \sim \frac{e^{-2x}}{2x}$

(d) $\displaystyle\int_1^2 \frac{e^{-x[t+(1/t)]}}{\sqrt{t^2-1}} dt \sim \frac{\Gamma(\frac{1}{4})}{2\sqrt{2}x^{1/4}} e^{-2x}$

3.20. Show that as $x \to \infty$

(a) $\displaystyle\int_0^1 \frac{e^{-xt^n}}{1+t} dt \sim \frac{\Gamma\left(\frac{1}{n}\right)}{nx^{1/n}}$

(b) $\displaystyle\int_0^1 e^{-xt^n} \ln(1+t) \, dt \sim \frac{\Gamma\left(\frac{2}{n}\right)}{nx^{2/n}}$

(c) $\displaystyle\int_0^1 \frac{e^{-xt^n}}{\sqrt{t}} dt \sim \frac{\Gamma\left(\frac{1}{2n}\right)}{nx^{1/2n}}$

3.21. Show that as $\alpha \to \infty$

$$Ai(-\alpha) = \frac{1}{\pi} \int_0^\infty \cos\left(\tfrac{1}{3} t^3 - \alpha t\right) dt \sim \frac{1}{\sqrt{\pi}\alpha^{1/4}} \sin\left(\tfrac{2}{3} \alpha^{3/2} + \tfrac{1}{4}\pi\right)$$

3.22. Show that as $\alpha \to \infty$

$$\int_0^\infty e^{i\alpha[(1/3)t^3 + t]} dt \sim \frac{i}{\alpha}$$

3.23. Show that as $x \to \infty$

$$J_0(x) = \frac{1}{\pi} \int_0^\pi \cos(x \cos\theta) \, d\theta \sim \sqrt{\frac{2}{\pi x}} \cos\left(x - \tfrac{1}{4}\pi\right)$$

3.24. Show that as $\alpha \to \infty$

(a) $\displaystyle \int_0^1 e^{i\alpha t^3} dt \sim \frac{\Gamma(\tfrac{1}{3})e^{(1/6)i\pi}}{3\alpha^{1/3}}$

(b) $\displaystyle \int_0^1 \frac{e^{i\alpha t^3}}{\sqrt{t}} dt \sim \frac{\Gamma(\tfrac{1}{6})e^{(1/12)i\pi}}{3\alpha^{1/6}}$

(c) $\displaystyle \int_0^1 e^{i\alpha t^3} \ln(1 + t) \, dt \sim \frac{\Gamma(\tfrac{2}{3})e^{(1/3)i\pi}}{3\alpha^{2/3}}$

(d) $\displaystyle \int_0^1 e^{i\alpha t^3} \ln(2 + t) \, dt \sim \frac{\Gamma(\tfrac{1}{3})e^{(1/6)i\pi} \ln 2}{3\alpha^{1/3}}$

3.25. Show that as $x \to \infty$

(a) $\displaystyle K_0(x) = \int_1^\infty \frac{e^{-xt}}{\sqrt{t^2 - 1}} dt \sim \sqrt{\frac{\pi}{2x}} e^{-x}$

(b) $\displaystyle H_0^{(1)}(x) = -\frac{2}{\pi} \int_1^\infty \frac{e^{ixt}}{\sqrt{1 - t^2}} dt \sim \sqrt{\frac{2}{\pi x}} e^{i[x - (1/4)\pi]}$

(c) $\displaystyle J_0(x) = \frac{2}{\pi} \int_1^\infty \frac{\sin xt}{\sqrt{t^2 - 1}} dt \sim \sqrt{\frac{2}{\pi x}} \cos\left(x - \tfrac{1}{4}\pi\right)$

(d) $\displaystyle Y_0(x) = -\frac{2}{\pi} \int_1^\infty \frac{\cos xt}{\sqrt{t^2 - 1}} dt \sim \sqrt{\frac{2}{\pi x}} \sin\left(x - \tfrac{1}{4}\pi\right)$

(e) $\displaystyle I_n(x) = \frac{1}{\pi} \int_0^\pi e^{x \cos\theta} \cos n\theta \, d\theta \sim \frac{e^x}{\sqrt{2\pi x}}$

3.26. Show that as $z \to \infty$

$$\Gamma(z) = \int_0^\infty t^{z-1} e^{-t} dt \sim \sqrt{\frac{2\pi}{z}} z^z e^{-z}$$

106 INTEGRALS

3.27. Show that as $x \to \infty$

$$\int_{-1}^{1} e^{ixt} (1 - t^2)^{n-1/2} \, dt \sim \left(\frac{2}{x}\right)^{n+1/2} \Gamma(n + \tfrac{1}{2}) \cos [x - \tfrac{1}{2} \pi (n + \tfrac{1}{2})]$$

3.28. Show that as $\omega \to \infty$

$$\int_0^1 \ln t \, e^{i\omega t} \, dt \sim -\frac{i \ln \omega}{\omega} - \frac{i\gamma + \tfrac{1}{2}\pi}{\omega}$$

where $\gamma = -\displaystyle\int_0^\infty e^{-t} \ln t \, dt$.

3.29. Show that as $\alpha \to \infty$

$$\int_0^\infty e^{\alpha[z - (1/3)iz^3]} \, dz \sim \frac{\sqrt{\pi} \exp[-\tfrac{1}{8} i\pi + \tfrac{2}{3} \alpha e^{-(1/4)i\pi}]}{\alpha^{1/2}}$$

3.30. Consider the integral representation of the Legendre polynomial of order n

$$P_n(\mu) = \frac{1}{\pi} \int_0^\pi [\mu + \sqrt{\mu^2 - 1} \cos \theta]^n \, d\theta$$

Show that

$$P_n(\mu) \sim \frac{1}{\sqrt{2n\pi}} \frac{(\mu + \sqrt{\mu^2 - 1})^{n+1/2}}{(\mu^2 - 1)^{1/4}} \quad \text{as} \quad n \to \infty$$

3.31. Consider the Airy function $Bi(z)$ of the second kind defined by

$$Bi(z) = \frac{1}{\pi} \int_0^\infty [e^{-(1/3)t^3 + zt} + \sin(\tfrac{1}{3} t^3 + zt)] \, dt$$

(a) As $z \to \infty$, use Laplace's method for the first integral and integration by parts for the second integral and show that

$$Bi(z) \sim \frac{e^{(2/3)z^{3/2}}}{\sqrt{\pi} \, z^{1/4}}$$

(b) Replace z by $-\alpha$ in the above integrals; then, use the Laplace's method for the first integral and the method of stationary phase for the second integral, and obtain

$$Bi(-z) \sim \frac{1}{\sqrt{\pi} \, \alpha^{1/4}} \cos(\tfrac{2}{3} \alpha^{3/2} + \tfrac{1}{4} \pi)$$

CHAPTER 4

The Duffing Equation

The free oscillations of many conservative systems having a single degree of freedom are governed by an equation of the form

$$\frac{d^2x^*}{dt^{*2}} + f(x^*) = 0 \qquad (4.1)$$

where f is a nonlinear function of x^*. Here, d^2x^*/dt^{*2} is the acceleration of the system, whereas $f(x^*)$ is the restoring force. Let $x^* = x_0^*$ be an equilibrium position of the system; then, $f(x_0^*) = 0$. Moreover, assume that f is an analytic function at $x^* = x_0^*$; then, it can be expanded in a Taylor series as

$$f(x^*) = k_1(x^* - x_0^*) + k_2(x^* - x_0^*)^2 + k_3(x^* - x_0^*)^3 + \cdots \qquad (4.2)$$

where

$$k_n = \frac{1}{n!} \frac{d^n f^*}{dx^{*n}}(x_0^*)$$

Hence, (4.1) can be rewritten as

$$\frac{d^2x^*}{dt^{*2}} + k_1(x^* - x_0^*) + k_2(x^* - x_0^*)^2 + k_3(x^* - x_0^*)^3 + \cdots = 0 \qquad (4.3)$$

Equation (4.3) describes the motion of the system in the neighborhood of the equilibrium position. It is convenient to introduce the transformation $u^* = x^* - x_0^*$ so that (4.3) becomes

$$\frac{d^2u^*}{dt^{*2}} + k_1 u^* + k_2 u^{*2} + k_3 u^{*3} + \cdots = 0 \qquad (4.4)$$

Most of this chapter is devoted to the following special case of (4.4):

$$\frac{d^2u^*}{dt^{*2}} + k_1 u^* + k_3 u^{*3} = 0 \qquad (4.5)$$

where $k_1 > 0$ and k_3 may be positive or negative. Equation (4.5) is usually called the *Duffing equation*. As mentioned in Chapter 1, one should make it a practice to write the governing equations in dimensionless form before solving them. To this end, we choose a characteristic length U^* of the motion and a characteristic time T^* and let

$$t = \frac{t^*}{T^*} \qquad u = \frac{u^*}{U^*}$$

Using the chain rule, we have

$$\frac{d}{dt^*} = \frac{d}{dt}\frac{dt}{dt^*} = \frac{1}{T^*}\frac{d}{dt}$$

$$\frac{d^2}{dt^{*2}} = \frac{1}{T^{*2}}\frac{d^2}{dt^2}$$

Then, (4.5) becomes

$$\ddot{u} + k_1 T^{*2} u + k_3 T^{*2} U^{*2} u^3 = 0 \qquad (4.6)$$

It is convenient to choose T^* so that $k_1 T^{*2} = 1$ and let $\epsilon = k_3 T^{*2} U^{*2} = k_3 U^{*2}/k_1$. Hence, (4.6) can be rewritten as

$$\ddot{u} + u + \epsilon u^3 = 0 \qquad (4.7)$$

We note that ϵ is a dimensionless quantity, and it is a measure of the strength of the nonlinearity. As initial conditions, we take

$$u(0) = x_0 \qquad \dot{u}(0) = \dot{x}_0 \qquad (4.8)$$

The solution u of our problem is a function of the independent variable t and the parameter ϵ. Hence, we write $u = u(t; \epsilon)$, where the parameter ϵ is separated from the independent variable t by a semicolon. In the next section, we determine a straightforward approximation to (4.7) and (4.8) for small but finite ϵ. This expansion is nonuniform for large times. Then, an exact solution is obtained in Section 4.2, and it is used to show that the frequency ω of the system is a function of ϵ (i.e., the nonlinearity). This fact is used in Section 4.3 to determine a uniform expansion by expanding both u and ω in powers of ϵ, this is, the Lindstedt-Poincaré technique. In Section 4.4, we introduce the expansions for u and ω into the straightforward expansion and render it uniform or normal, this is, the method of renormalization. In Section 4.5, we describe the method of multiple scales, in Section 4.6, we describe the method of variation of parameters (method of special perturbations), and in Section 4.7, we describe the method of averaging.

4.1. The Straightforward Expansion

When $\epsilon = 0$, (4.7) reduces to

$$\ddot{u} + u = 0 \tag{4.9}$$

whose general solution is

$$u_0 = a_0 \cos(t + \beta_0) \tag{4.10}$$

where a_0 and β_0 are arbitrary constants. When ϵ is small but different from zero, the general solution of (4.7) is no longer given by (4.10), and a correction must be added to it. We try a correction in the form of a power series in ϵ, that is, we let

$$u(t; \epsilon) = u_0(t) + \epsilon u_1(t) + \epsilon^2 u_2(t) + \epsilon^3 u_3(t) + \cdots \tag{4.11}$$

Here, we restrict our discussion to the first term in the correction series. Thus, we seek an approximate solution in the form

$$u(t; \epsilon) = u_0(t) + \epsilon u_1(t) + O(\epsilon^2) \tag{4.12}$$

Since only one term is kept in the correction series, we call (4.12) a first-order expansion.

We substitute (4.12) into (4.7) and obtain

$$\ddot{u}_0 + \epsilon \ddot{u}_1 + O(\epsilon^2) + u_0 + \epsilon u_1 + O(\epsilon^2) + \epsilon [u_0 + \epsilon u_1 + O(\epsilon^2)]^3 = 0 \tag{4.13}$$

Using the binomial theorem to expand the last term, we have

$$[u_0 + \epsilon u_1 + O(\epsilon^2)]^3 = u_0^3 + 3u_0^2 [\epsilon u_1 + O(\epsilon^2)] + 3u_0 [\epsilon u_1 + O(\epsilon^2)]^2$$
$$+ [\epsilon u_1 + O(\epsilon^2)]^3 = u_0^3 + 3\epsilon u_0^2 u_1 + O(\epsilon^2) \tag{4.14}$$

Substituting (4.14) into (4.13) and collecting coefficients of equal powers of ϵ, we obtain

$$\ddot{u}_0 + u_0 + \epsilon(\ddot{u}_1 + u_1 + u_0^3) + O(\epsilon^2) = 0 \tag{4.15}$$

We note that since we are only interested in terms up to $O(\epsilon)$, we need only the terms that are independent of ϵ from the quantity inside the brackets in (4.13). This fact can be used to minimize the algebra. Setting $\epsilon = 0$ in (4.15) gives

$$\ddot{u}_0 + u_0 = 0 \tag{4.16}$$

Then, (4.15) becomes

$$\epsilon(\ddot{u}_1 + u_1 + u_0^3) + O(\epsilon^2) = 0 \tag{4.17}$$

Dividing (4.17) by ϵ yields

110 THE DUFFING EQUATION

$$\ddot{u}_1 + u_1 + u_0^3 + O(\epsilon) = 0 \tag{4.18}$$

Setting $\epsilon = 0$ in (4.18) gives

$$\ddot{u}_1 + u_1 + u_0^3 = 0 \tag{4.19}$$

Comparing (4.16) and (4.19) with (4.15), we note that the former equations can be obtained by simply setting each of the coefficients of ϵ equal to zero in (4.15). This is how one usually derives the equations governing u_0 and u_1. Moreover, we note that (4.16) and (4.19) have to be solved in succession. One first solves (4.16) for u_0, substitutes the result into (4.19), and then solves for u_1.

The general solution of (4.16) can be written as

$$u_0 = a_0 \cos(t + \beta_0) \tag{4.20}$$

where a_0 and β_0 are arbitrary constants. Substituting for u_0 into (4.19) yields

$$\ddot{u}_1 + u_1 = -a_0^3 \cos^3(t + \beta_0) \tag{4.21}$$

Equation (4.21) is an inhomogeneous equation whose general solution consists of the sum of a homogeneous solution and a particular solution (Appendix B). To determine a particular solution, we find it convenient to express the inhomogeneous term in a Fourier series using the trigonometric identity (A18)

$$\cos^3 \theta = \tfrac{1}{4} \cos 3\theta + \tfrac{3}{4} \cos \theta$$

and rewrite (4.21) as

$$\ddot{u}_1 + u_1 = -\tfrac{3}{4} a_0^3 \cos(t + \beta_0) - \tfrac{1}{4} a_0^3 \cos(3t + 3\beta_0) \tag{4.22}$$

The homogeneous solution of (4.22) can be expressed as

$$u_{1_h} = a_1 \cos(t + \beta_1) \tag{4.23}$$

where a_1 and β_1 are arbitrary constants. Since (4.22) is linear, one can use the principle of superposition and determine particular solutions as the sum of two particular solutions corresponding to the two inhomogeneous terms. That is, one determines a particular solution by adding two particular solutions of the following two equations:

$$\ddot{u}_1 + u_1 = -\tfrac{3}{4} a_0^3 \cos(t + \beta_0) \tag{4.24}$$

$$\ddot{u}_1 + u_1 = -\tfrac{1}{4} a_0^3 \cos(3t + 3\beta_0) \tag{4.25}$$

A particular solution of (4.24) is (B75 and B76)

$$u_{1_p}^{(1)} = -\tfrac{3}{8} a_0^3 t \sin(t + \beta_0) \tag{4.26}$$

A particular solution of (4.25) is (B68 and B69)

$$u_{1_p}^{(2)} = \tfrac{1}{32} a_0^3 \cos(3t + 3\beta_0) \tag{4.27}$$

Hence, according to the principle of superposition, a particular solution of (4.22) is

$$u_{1_p} = -\tfrac{3}{8} a_0^3 t \sin(t + \beta_0) + \tfrac{1}{32} a_0^3 \cos(3t + 3\beta_0) \tag{4.28}$$

Therefore, the general solution of (4.22) is

$$u_1 = a_1 \cos(t + \beta_1) - \tfrac{3}{8} a_0^3 t \sin(t + \beta_0) + \tfrac{1}{32} a_0^3 \cos(3t + 3\beta_0) \tag{4.29}$$

Substituting for u_0 and u_1 from (4.20) and (4.29), respectively, into (4.12) yields the following first-order expansion for the general solution of (4.7):

$$u = a_0 \cos(t + \beta_0) + \epsilon [a_1 \cos(t + \beta_1) - \tfrac{3}{8} a_0^3 t \sin(t + \beta_0)$$
$$+ \tfrac{1}{32} a_0^3 \cos(3t + 3\beta_0)] + \cdots \tag{4.30}$$

where a_0, a_1, β_0, and β_1 are arbitrary constants. We started with a second-order equation that can satisfy two initial conditions but it appears that we ended up with four arbitrary constants. It turns out that the constants a_0, a_1, β_0, and β_1 are not arbitrary and that the two initial conditions (4.8) are enough to determine them. To see this, we substitute (4.30) into (4.8) and obtain

$$x_0 = a_0 \cos \beta_0 + \epsilon(a_1 \cos \beta_1 + \tfrac{1}{32} a_0^3 \cos 3\beta_0) + \cdots \tag{4.31}$$

$$\dot{x}_0 = -a_0 \sin \beta_0 - \epsilon(a_1 \sin \beta_1 + \tfrac{3}{8} a_0^3 \sin \beta_0 + \tfrac{3}{32} a_0^3 \sin 3\beta_0) + \cdots \tag{4.32}$$

Transposing all terms to one side and equating the coefficients of the powers of ϵ to zero in (4.31) and (4.32) is equivalent to equating the coefficients of like powers of ϵ on both sides of these equations. The results are

Order ϵ^0

$$x_0 = a_0 \cos \beta_0 \tag{4.33}$$

$$\dot{x}_0 = -a_0 \sin \beta_0 \tag{4.34}$$

Order ϵ

$$a_1 \cos \beta_1 = -\tfrac{1}{32} a_0^3 \cos 3\beta_0 \tag{4.35}$$

$$a_1 \sin \beta_1 = -\tfrac{3}{8} a_0^3 \sin \beta_0 - \tfrac{3}{32} a_0^3 \sin 3\beta_0 \tag{4.36}$$

Squaring and adding (4.33) and (4.34) yields

$$x_0^2 + \dot{x}_0^2 = a_0^2 \cos^2 \beta_0 + a_0^2 \sin^2 \beta_0 = a_0^2$$

or

$$a_0 = (x_0^2 + \dot{x}_0^2)^{1/2} \tag{4.37}$$

Then, solving (4.33) and (4.34) for β_0 gives

112 THE DUFFING EQUATION

$$\beta_0 = -\sin^{-1}\left[\frac{\dot{x}_0}{(x_0^2 + \dot{x}_0^2)^{1/2}}\right] = \cos^{-1}\left[\frac{x_0}{(x_0^2 + \dot{x}_0^2)^{1/2}}\right] \quad (4.38)$$

Similarly, solving (4.35) and (4.36) yields

$$a_1 = \tfrac{1}{32} a_0^3 [\cos^2 3\beta_0 + 9(\sin 3\beta_0 + 4 \sin \beta_0)^2]^{1/2} \quad (4.39)$$

$$\beta_1 = -\sin^{-1}\frac{3a_0^3(4\sin\beta_0 + \sin 3\beta_0)}{32 a_1} = \cos^{-1}\left[-\frac{a_0^3 \cos 3\beta_0}{32 a_1}\right] \quad (4.40)$$

Thus, once a_0 and β_0 are calculated from (4.37) and (4.38), a_1 and β_1 can be calculated from (4.39) and (4.40).

Let us return to (4.30) and note that

$a_0 \cos(t + \beta_0) + \epsilon a_1 \cos(t + \beta_1)$

$= a_0 \cos t \cos \beta_0 - a_0 \sin t \sin \beta_0 + \epsilon a_1 \cos t \cos \beta_1 - \epsilon a_1 \sin t \sin \beta_1$

$= (a_0 \cos \beta_0 + \epsilon a_1 \cos \beta_1) \cos t - (a_0 \sin \beta_0 + \epsilon a_1 \sin \beta_1) \sin t$

$= a \cos t \cos \beta - a \sin t \sin \beta$

$= a \cos(t + \beta) \quad (4.41)$

where

$$a \cos \beta = a_0 \cos \beta_0 + \epsilon a_1 \cos \beta_1 \quad (4.42)$$

$$a \sin \beta = a_0 \sin \beta_0 + \epsilon a_1 \sin \beta_1 \quad (4.43)$$

It follows from (4.42) and (4.43) that

$$a_0 = a + O(\epsilon) \qquad \beta_0 = \beta + O(\epsilon) \quad (4.44)$$

Using (4.41) and (4.44), we rewrite (4.30) as

$u = a \cos(t + \beta) + \epsilon \{ -\tfrac{3}{8} [a + O(\epsilon)]^3 t \sin[t + \beta + O(\epsilon)]$

$\quad + \tfrac{1}{32} [a + O(\epsilon)]^3 \cos[3t + 3\beta + O(\epsilon)] \} + \cdots$

Hence,

$$u = a \cos(t + \beta) + \epsilon a^3 [-\tfrac{3}{8} t \sin(t + \beta) + \tfrac{1}{32} \cos(3t + 3\beta)] + \cdots \quad (4.45)$$

We could have obtained this solution directly from (4.16) and (4.19) as follows: The solution of (4.16) is taken in the form

$$u_0 = a \cos(t + \beta) \quad (4.46)$$

Then, (4.19) becomes

$$\ddot{u}_1 + u_1 = -a^3 \cos^3(t + \beta) = -\tfrac{3}{4} a^3 \cos(t + \beta) - \tfrac{1}{4} a^3 \cos(3t + 3\beta) \quad (4.47)$$

Now, in writing down the solution of (4.47), we take only the particular solution. That is, we write the solution of (4.47) as

$$u_1 = -\tfrac{3}{8} a^3 t \sin(t + \beta) + \tfrac{1}{32} a^3 \cos(3t + 3\beta) \qquad (4.48)$$

Substituting for u_0 and u_1 from (4.46) and (4.48) into (4.12), we obtain (4.45). Thus, in solving problems of this kind, one has two choices. First, one may include the homogeneous solution at each order and consider the arbitrary constants to be independent of ϵ. Second, one may disregard the homogeneous solution at all orders except the first and consider the arbitrary constants to be dependent upon ϵ. The latter choice is used in most of this book.

Returning to (4.45), we find that to the first approximation

$$u = a \cos(t + \beta)$$

and that its first correction is

$$-\tfrac{3}{8} a^3 \epsilon t \sin(t + \beta) + \tfrac{1}{32} \epsilon a^3 \cos(3t + 3\beta)$$

We note that this correction is small, as it is supposed to be, only when ϵt is small compared with unity. When ϵt is $O(1)$, the term that is supposed to be a small correction becomes the order of the main term. Moreover, when $\epsilon t > O(1)$, the "small-correction" term becomes larger than the main term. Hence, the straightforward expansion (4.45) is valid only for times such that $\epsilon t < O(1)$, that is, $t < O(\epsilon^{-1})$. Consequently, we say such expansions are *nonuniform* or *breakdown* for long times and we call such expansions *pedestrian expansions*. The reason for the breakdown of this expansion is the presence of the term $t \sin(t + \beta)$, a product of algebraic and circular terms. Such terms are called *mixed-secular terms* in the astronomy literature. The word secular is derived from the French word siècle, which means a century. This designation is the result of ϵ being very small in astronomical applications and ϵt becomes appreciable after very long times, the order of a century. Thus, for the expansion to be uniform, the corrections must be free of secular terms. In the next section, we obtain and examine the exact solution in order to determine the source of the secular terms. In subsequent sections, we develop methods that avoid secular terms, and hence, yield uniform expansions.

4.2. Exact Solution

Equation (4.7) belongs to the class of second-order equations from which the first derivative is absent. This class of equations can be integrated by making the change of variable $\dot{u} = v$ and changing the independent variable from t to u. To this end, we have

$$\ddot{u} = \dot{v} = \frac{dv}{dt} = \frac{dv}{du}\frac{du}{dt} = v\frac{dv}{du} \qquad (4.49)$$

Then, (4.7) becomes

114 THE DUFFING EQUATION

$$v\frac{dv}{du} + u + \epsilon u^3 = 0 \tag{4.50}$$

which, upon separation of variables, becomes

$$v\,dv = -(u + \epsilon u^3)\,du \tag{4.51}$$

Integrating (4.51) gives

$$\tfrac{1}{2} v^2 = h - (\tfrac{1}{2} u^2 + \tfrac{1}{4} \epsilon u^4) = h - F(u) \tag{4.52}$$

where h is a constant of integration. Since $v = \dot u$, $\tfrac{1}{2} v^2$ is proportional to the kinetic energy of the system. Since $u + \epsilon u^3$ is proportional to the restoring force,

$$F(u) = \int (u + \epsilon u^3)\,du = \tfrac{1}{2} u^2 + \tfrac{1}{4} \epsilon u^4$$

is proportional to the potential energy of the system. Hence, h is proportional to the total energy of the system. For a given h, (4.52) yields an integral of the motion in the uv plane, which is called the *phase plane*. It can be used to delineate the qualitative characteristics of the motion.

To construct some of the integral curves in the phase plane, we draw first the curve $F(u)$ in Figure 4-1. When $\epsilon > 0$, $F(u)$ has one stationary point, namely $u = 0$. It corresponds to a minimum of $F(u)$. On the other hand, when $\epsilon < 0$, $F(u)$ has stationary points at $u = 0$ and $u = \pm|\epsilon|^{-1/2}$. The first point corresponds to a minimum of $F(u)$, whereas the other two points correspond to maxima of $F(u)$. Also shown in Figure 4-1 are three horizontal lines corresponding to different values of h. It follows from (4.52) that

$$v = \pm\sqrt{2}\,[h - F(u)]^{1/2} \tag{4.53}$$

Thus, there are real values for v, and hence, there are real motions, if and only if $h \geq F(u)$. Moreover, the motion is symmetric about the u axis. Since $\dot u = v$, a representative point on an integral curve moves clockwise as t increases.

When $\epsilon > 0$, there are real motions only when $h \geq h_0 = 0$. When $h = h_0$, the integral curve consists of a single point, which is referred to as *center*. When $h > h_0$, the integral curve consists of a closed trajectory, which corresponds to a *periodic motion*. Thus, all possible motions are periodic.

When $\epsilon < 0$, there are real motions for all values of h. Figure 4-1b shows that, when $h = h_0$, the integral curve consists of the origin, a center, and two trajectories (branches), one opening to the right and one opening to the left. When $h = h_2$, the integral curve consists of two trajectories passing through the points: $u = |\epsilon|^{-1/2}$ and $v = 0$, and $u = -|\epsilon|^{-1/2}$ and $v = 0$. These points are referred to as *saddle points*, and they correspond to the maxima of $F(u)$ and are labeled by the letter S. The trajectories that pass through the saddle points are usually referred to as *separatrices*. When $h < h_0$, the integral curve consists of two trajectories, one opening to the right and one opening to the left. When $h_0 < h < h_2$, the integral curve consists of three trajectories, one closed trajectory surrounding the

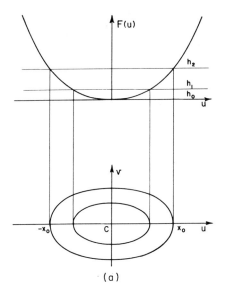

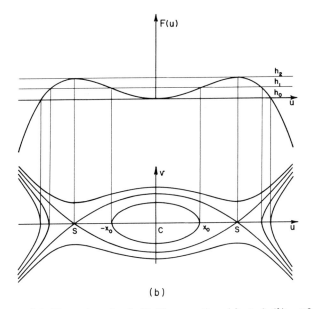

Figure 4-1. Phase plane for the Duffing equation: (a) $\epsilon > 0$; (b) $\epsilon < 0$.

115

116 THE DUFFING EQUATION

origin, one trajectory opening to the right, and one trajectory opening to the left. Thus, periodic motions exist only when $h_0 < h < h_2$ and when the initial conditions restrict the motion to the closed branch of the integral curve.

Next, we integrate (4.52). Using the initial conditions (4.8) in (4.52) and recalling that $v = \dot{u}$, we have

$$\tfrac{1}{2} \dot{x}_0^2 = h - (\tfrac{1}{2} x_0^2 + \tfrac{1}{4} \epsilon x_0^4)$$

Hence,

$$h = \tfrac{1}{2} \dot{x}_0^2 + \tfrac{1}{2} x_0^2 + \tfrac{1}{4} \epsilon x_0^4 \tag{4.54}$$

Since $v = \dot{u}$, it follows from (4.52) that

$$\dot{u} = \pm(2h - u^2 - \tfrac{1}{2} \epsilon u^4)^{1/2} \tag{4.55}$$

where the positive and negative signs correspond to motions above and below the u axis, respectively. Upon separation of variables, (4.55) becomes

$$dt = \pm \frac{du}{(2h - u^2 - \tfrac{1}{2} \epsilon u^4)^{1/2}} \tag{4.56}$$

Integrating (4.56) gives

$$\int_0^t dt = \pm \int_{x_0}^u \frac{du}{(2h - u^2 - \tfrac{1}{2} \epsilon u^4)^{1/2}}$$

Hence,

$$t = \pm \int_{x_0}^u \frac{du}{(2h - u^2 - \tfrac{1}{2} \epsilon u^4)^{1/2}} \tag{4.57}$$

Next, we specialize this solution to the case of periodic motions.

Figure 4-1 shows that a closed trajectory (i.e., periodic motion) intersects the u axis at two points. Denote these intersections by $u = -x_0$ and $u = x_0$. Substituting $u = \pm x_0$ and $v = \dot{u}_0 = 0$ into (4.54) and solving for h, we obtain

$$h = \tfrac{1}{2} x_0^2 + \tfrac{1}{4} \epsilon x_0^4$$

Hence,

$$2h - u^2 - \tfrac{1}{2} \epsilon u^4 = x_0^2 + \tfrac{1}{2} \epsilon x_0^4 - u^2 - \tfrac{1}{2} \epsilon u^4$$
$$= (x_0^2 - u^2)(1 + \tfrac{1}{2} \epsilon x_0^2 + \tfrac{1}{2} \epsilon u^2) \tag{4.58}$$

Then, (4.57) becomes

$$t = \pm \int_{x_0}^u \frac{du}{(x_0^2 - u^2)^{1/2} (1 + \tfrac{1}{2} \epsilon x_0^2 + \tfrac{1}{2} \epsilon u^2)^{1/2}} \tag{4.59}$$

which can be put in a standard elliptic form by introducing the transformation

$$u = -x_0 \cos \theta \tag{4.60}$$

Then,

$$du = x_0 \sin \theta \, d\theta$$

Moreover, $u = \pm x_0$ corresponds to $\theta_0 = 0$ and π. Hence, (4.59) becomes

$$t = \pm \int_{\theta_0}^{\theta} \frac{d\theta}{(1 + \tfrac{1}{2} \epsilon x_0^2 + \tfrac{1}{2} \epsilon x_0^2 \cos^2 \theta)^{1/2}}$$

or

$$t = \pm \frac{1}{\sqrt{1 + \epsilon x_0^2}} \int_{\theta_0}^{\theta} \frac{d\theta}{\sqrt{1 - m \sin^2 \theta}} \tag{4.61}$$

where

$$m = \frac{\epsilon x_0^2}{2(1 + \epsilon x_0^2)} \tag{4.62}$$

Since the closed trajectory is symmetric with respect to the u axis, the time needed for the representative point to move from $u = -x_0$ to $u = x_0$ is one-half the period T. But $u = -x_0$ corresponds to $\theta = 0$ and $u = x_0$ corresponds to $\theta = \pi$. Hence, it follows from (4.61) that

$$T = \frac{2}{\sqrt{1 + \epsilon x_0^2}} \int_0^{\pi} \frac{d\theta}{\sqrt{1 - m \sin^2 \theta}} \tag{4.63}$$

Since the integral in (4.63) over the interval $[0, \tfrac{1}{2}\pi]$ is the same as that over the interval $[\tfrac{1}{2}\pi, \pi]$, (4.63) can be rewritten as

$$T = \frac{4}{\sqrt{1 + \epsilon x_0^2}} \int_0^{1/2 \pi} \frac{d\theta}{\sqrt{1 - m \sin^2 \theta}} \tag{4.64}$$

The integral (4.64) is called a complete elliptic integral of the first kind, and it is tabulated. The variation of T with ϵx_0^2 is shown in Table 4-1. It is clear from

TABLE 4-1. Exact Periods of the Periodic Motions of the Duffing Equation

				m			
	0	0.02	0.04	0.06	0.08	0.10
ϵx_0^2	0	0.042	0.087	0.136	0.190	0.25
T	6.283	6.187	6.088	5.986	5.879	5.767

118 THE DUFFING EQUATION

TABLE 4-2. Variation of T_a/T with ϵx_0^2

ϵx_0^2	0	0.042	0.087	0.136	0.190	0.25
T_a/T	1.000	1.0004	1.002	1.004	1.007	1.013

Table 4-1 that the period is a function of ϵx_0^2, and hence, the nonlinearity. Since the angular frequency $\omega = 2\pi/T$, it is clear also that ω is a function of the nonlinearity. This explains the breakdown of the straightforward expansion (4.45), which forces the angular frequency of the system to be unity, irrespective of the nonlinearity. This also suggests that any uniform expansion must account for the fact that the angular frequency is a function of the nonlinearity. The earliest technique that uses this fact seems to be the Lindstedt-Poincaré technique, which is discussed in the next section.

When ϵ is small, m is also small. Then, an approximation T_a to T can be obtained by expanding the integrand in (4.64). The result is

$$T_a = \frac{4}{\sqrt{1 + \epsilon x_0^2}} \int_0^{1/2\pi} (1 + \tfrac{1}{2} m \sin^2 \theta + \cdots) d\theta$$

$$= \frac{4}{\sqrt{1 + \epsilon x_0^2}} (\tfrac{1}{2}\pi + \tfrac{1}{8} m\pi + \cdots) = \frac{2\pi}{\sqrt{1 + \epsilon x_0^2}} \left[1 + \frac{\epsilon x_0^2}{8(1 + \epsilon x_0^2)} + \cdots \right]$$

$$= 2\pi(1 - \tfrac{1}{2} \epsilon x_0^2 + \cdots)(1 + \tfrac{1}{8} \epsilon x_0^2 + \cdots)$$

Hence,

$$T_a = 2\pi(1 - \tfrac{3}{8} \epsilon x_0^2 + \cdots) \tag{4.65}$$

Table 4-2 shows that the two term expansion for the period is close to the exact value. The error is only 1.3% for values of ϵx_0^2 as high as 0.25.

4.3. The Lindstedt-Poincaré Technique

As discussed in the preceding section, the breakdown in the straightforward expansion is due to its failure to account for the nonlinear dependence of the frequency on the nonlinearity. Thus, any expansion that does not account for a nonlinear frequency is doomed to failure. A number of techniques that yield uniformly valid expansions have been developed. Four of these techniques are discussed in this chapter. We start with the Lindstedt-Poincaré technique in this section.

To account for the nonlinear dependence of the frequency, we explicitly exhibit the frequency ω of the system in the differential equation. To this end, we introduce the transformation

$$\tau = \omega t \tag{4.66}$$

where ω is a constant that depends on ϵ. Then, we need to change the independent variable from t to τ. Using the chain rule, we transform the derivatives according to

$$\frac{d}{dt} = \frac{d\tau}{dt}\frac{d}{d\tau} = \omega \frac{d}{d\tau}$$

$$\frac{d^2}{dt^2} = \omega \frac{d^2}{dt\,d\tau} = \omega \frac{d\tau}{dt}\frac{d^2}{d\tau^2} = \omega^2 \frac{d^2}{d\tau^2}$$

Hence, (4.7) becomes

$$\omega^2 u'' + u + \epsilon u^3 = 0 \tag{4.67}$$

where the prime indicates the derivative with respect to τ. We note that the actual frequency of the system now appears explicitly in the equation. Until now, both u and ω are unknowns. We seek approximate solutions for them in the form of power series in terms of ϵ. That is, we let

$$u = u_0(\tau) + \epsilon u_1(\tau) + \cdots \tag{4.68}$$

$$\omega = 1 + \epsilon \omega_1 + \cdots \tag{4.69}$$

We note that the first term in the expansion of ω is the linear frequency; in this case, the linear frequency is unity. The corrections to the linear frequency are determined in the course of the analysis by requiring the expansion of u to be uniform for all τ.

Substituting (4.68) and (4.69) into (4.67) gives

$$(1 + \epsilon\omega_1 + \cdots)^2 (u_0'' + \epsilon u_1'' + \cdots) + u_0 + \epsilon u_1 + \cdots + \epsilon(u_0 + \epsilon u_1 + \cdots)^3 = 0$$

which, upon expansion, becomes

$$u_0'' + 2\epsilon\omega_1 u_0'' + \epsilon u_1'' + \cdots + u_0 + \epsilon u_1 + \cdots + \epsilon u_0^3 + \cdots = 0$$

or

$$u_0'' + u_0 + \epsilon(u_1'' + u_1 + u_0^3 + 2\omega_1 u_0'') + \cdots = 0 \tag{4.70}$$

Equating each of the coefficients of ϵ^0 and ϵ to zero yields

$$u_0'' + u_0 = 0 \tag{4.71}$$

$$u_1'' + u_1 = -u_0^3 - 2\omega_1 u_0'' \tag{4.72}$$

The general solution of (4.71) is

$$u_0 = a \cos(\tau + \beta) \tag{4.73}$$

where a and β are constants. Then, (4.72) becomes

$$u_1'' + u_1 = -a^3 \cos^3(\tau + \beta) + 2\omega_1 a \cos(\tau + \beta)$$

or

$$u_1'' + u_1 = (2\omega_1 a - \tfrac{3}{4} a^3) \cos(\tau + \beta) - \tfrac{1}{4} a^3 \cos(3\tau + 3\beta) \tag{4.74}$$

The particular solution of (4.74) is (B69 and B76)

$$u_1 = \tfrac{1}{2}(2\omega_1 a - \tfrac{3}{4} a^3)\tau \sin(\tau + \beta) + \tfrac{1}{32} a^3 \cos(3\tau + 3\beta) \tag{4.75}$$

We note that the particular solution of u_1 contains a mixed-secular term, which makes the expansion nonuniform. For a uniform expansion, we cannot permit such secular terms to appear in $u_1, u_2, u_3, \cdots$. In contrast with the straightforward expansion (4.45), where the secular term cannot be annihilated unless $a = 0$ (i.e., trivial solution for u), in this case, we still have to choose the parameter ω_1. Thus, we choose it to annihilate the secular term. To this end, we set the coefficient of the secular term zero. That is,

$$2\omega_1 a - \tfrac{3}{4} a^3 = 0 \tag{4.76}$$

Then, (4.75) becomes

$$u_1 = \tfrac{1}{32} a^3 \cos(3\tau + 3\beta) \tag{4.77}$$

Disregarding the trivial case $a = 0$, we find that (4.76) is satisfied if

$$\omega_1 = \tfrac{3}{8} a^2 \tag{4.78}$$

We note that, to determine the condition (4.76) for the elimination of the secular term from u_1, we do not need to determine the particular solution first as done above. Instead, we only need to inspect the inhomogeneous terms in (4.74) governing u_1 and choose the coefficient of $\cos(\tau + \beta)$ to be zero because this is the term that produces the secular term in u_1.

Substituting (4.73) and (4.77) into (4.68) yields

$$u = a \cos(\tau + \beta) + \tfrac{1}{32} \epsilon a^3 \cos(3\tau + 3\beta) + \cdots \tag{4.79}$$

Substituting for ω_1 from (4.78) into (4.69) yields

$$\omega = 1 + \tfrac{3}{8} \epsilon a^2 + \cdots \tag{4.80}$$

Since $\tau = \omega t$, we rewrite (4.79) as

$$u = a \cos[(1 + \tfrac{3}{8} \epsilon a^2)t + \beta] + \tfrac{1}{32} \epsilon a^3 \cos[3(1 + \tfrac{3}{8} \epsilon a^2)t + 3\beta] + \cdots \tag{4.81}$$

Thus, the expansion (4.81) is uniform to first-order because secular terms do not appear in it and the correction term (the term proportional to ϵ) is small compared with the first term.

Next, we compare the period obtained in this section by using the Lindstedt-Poincaré technique with the approximate period (4.65) obtained in the preceding section by expanding the integrand in the exact solution. Since the value

THE METHOD OF RENORMALIZATION 121

$u = -x_0$ is the intersection of the closed trajectory with the u axis, $\dot{u} = 0$ when $u = -x_0$. Using these values as initial conditions in (4.81), we have

$$-x_0 = a \cos \beta + \tfrac{1}{32} \epsilon a^3 \cos 3\beta + \cdots \tag{4.82}$$

It follows from (4.82) that $a^2 = x_0^2$ because $\beta = 0$ for $\dot{u}(0) = 0$. Hence, (4.80) becomes

$$\omega = 1 + \tfrac{3}{8} \epsilon x_0^2 + \cdots$$

But $T_a = 2\pi/\omega$, therefore

$$T_a = 2\pi (1 + \tfrac{3}{8} \epsilon x_0^2 + \cdots)^{-1} = 2\pi (1 - \tfrac{3}{8} \epsilon x_0^2) + \cdots \tag{4.83}$$

which is in full agreement with (4.65).

4.4. The Method of Renormalization

Instead of introducing the transformation (4.66) into the differential equation and carrying out another expansion as in the preceding section (the Lindstedt-Poincaré technique), we introduce this transformation into the nonuniform straightforward expansion (4.45). It follows from (4.66) and (4.69) that

$$t = \omega^{-1} \tau = (1 + \epsilon \omega_1 + \cdots)^{-1} \tau = \tau (1 - \epsilon \omega_1 + \cdots) = \tau - \epsilon \omega_1 \tau + \cdots \tag{4.84}$$

Then, (4.45) becomes

$$u = a \cos(\tau + \beta - \epsilon \omega_1 \tau + \cdots) + \epsilon a^3 \left[-\tfrac{3}{8}(\tau - \epsilon \omega_1 \tau + \cdots) \right.$$
$$\left. \cdot \sin(\tau + \beta - \epsilon \omega_1 \tau + \cdots) + \tfrac{1}{32} \cos(3\tau + 3\beta - 3\epsilon \omega_1 \tau + \cdots) \right] + \cdots \tag{4.85}$$

Using Taylor series expansions, we have

$$\cos(\tau + \beta - \epsilon \omega_1 \tau + \cdots) = \cos(\tau + \beta) + \epsilon \omega_1 \tau \sin(\tau + \beta) + \cdots \tag{4.86a}$$

$$\sin(\tau + \beta - \epsilon \omega_1 \tau + \cdots) = \sin(\tau + \beta) - \epsilon \omega_1 \tau \cos(\tau + \beta) + \cdots \tag{4.86b}$$

$$\cos(3\tau + 3\beta - 3\epsilon \omega_1 \tau + \cdots) = \cos(3\tau + 3\beta) + 3\epsilon \omega_1 \tau \sin(3\tau + 3\beta) + \cdots \tag{4.86c}$$

Using these expansions, we rewrite (4.85) as

$$u = a \cos(\tau + \beta) + \epsilon \left[(\omega_1 a - \tfrac{3}{8} a^3) \tau \sin(\tau + \beta) + \tfrac{1}{32} a^3 \cos(3\tau + 3\beta) \right] + \cdots \tag{4.87}$$

In contrast with the straightforward expansion, in which the mixed-secular term cannot be annihilated unless $a = 0$, corresponding to the trivial solution,

122 THE DUFFING EQUATION

we have built in (4.87) the parameter ω_1, which can be chosen to annihilate the mixed-secular terms. Thus, setting the coefficient of the mixed-secular term to zero, we have

$$\omega_1 a - \tfrac{3}{8} a^3 = 0 \qquad (4.88)$$

Then, (4.87) reduces to

$$u = a \cos(\tau + \beta) + \tfrac{1}{32} \epsilon a^3 \cos(3\tau + 3\beta) + \cdots \qquad (4.89)$$

Disregarding the trivial case $a = 0$, we find that (4.88) is satisfied when $\omega_1 = \tfrac{3}{8} a^2$, in agreement with (4.78) obtained by using the Lindstedt-Poincaré technique. Comparing (4.89) with (4.79) we see that they are identical. Thus, the present technique yields the same expansion as the Lindstedt-Poincaré technique. Since the transformation (4.66) and (4.69) is introduced into the nonuniform straightforward expansion (4.45), which is then made uniform, the present technique is usually referred to as a *uniformization* or a *renormalization procedure*.

4.5. The Method of Multiple Scales

We return to the uniform expansion (4.81) obtained by using the Lindstedt-Poincaré technique and rewrite it as

$$u = a \cos(t + \beta + \tfrac{3}{8}\epsilon t a^2) + \tfrac{1}{32} \epsilon a^3 \cos(3t + 3\beta + \tfrac{9}{8}\epsilon t a^2) + \cdots \qquad (4.90)$$

We note from (4.90) that the functional dependence of u on t and ϵ is not disjoint because u depends on the combination ϵt as well as on the individual t and ϵ. Thus, in place of $u = u(t; \epsilon)$, we write $u = \hat{u}(t, \epsilon t; \epsilon)$. Carrying out the expansion (4.90) to higher order, we find that u, besides the individual t and ϵ, depends on the combinations $\epsilon t, \epsilon^2 t, \epsilon^3 t, \cdots$. Hence, we write

$$u(t; \epsilon) = \hat{u}(t, \epsilon t, \epsilon^2 t, \epsilon^3 t, \cdots; \epsilon)$$

or

$$u(t; \epsilon) = \hat{u}(T_0, T_1, T_2, T_3, \cdots; \epsilon) \qquad (4.91)$$

where the T_n are defined by

$$T_0 = t \quad T_1 = \epsilon t \quad T_2 = \epsilon^2 t \quad T_3 = \epsilon^3 t, \cdots \qquad (4.92)$$

We note that the T_n represent *different time scales* because ϵ is a small parameter. For example, if $\epsilon = \tfrac{1}{60}$, variations on the scale T_0 can be observed on the second arm of a watch, variations on the scale T_1 can be observed on the minute arm of a watch, and variations on the scale T_2 can be observed on the hour arm of a watch. Thus, T_0 represents a fast scale, T_1 represents a slower scale, T_2 represents an even slower scale, and so on. Since the dependence of u on t and ϵ

THE METHOD OF MULTIPLE SCALES   123

occurs on different scales, we imagine that we have a watch and attempt to observe the behavior of u on the different scales of the watch.

Thus, instead of determining u as a function of t, we determine u as a function of $T_0, T_1, T_2, \cdots$. To this end, we change the independent variable in the original equation (4.7) from t to $T_0, T_1, T_2 \cdots$. Using the chain rule, we have

$$\frac{d}{dt} = \frac{\partial}{\partial T_0} + \epsilon \frac{\partial}{\partial T_1} + \epsilon^2 \frac{\partial}{\partial T_2} + \cdots \tag{4.93}$$

$$\frac{d^2}{dt^2} = \frac{\partial^2}{\partial T_0^2} + 2\epsilon \frac{\partial^2}{\partial T_0 \partial T_1} + \epsilon^2 \left(2 \frac{\partial^2}{\partial T_0 \partial T_2} + \frac{\partial^2}{\partial T_1^2} \right) + \cdots \tag{4.94}$$

Hence, (4.7) becomes

$$\frac{\partial^2 u}{\partial T_0^2} + 2\epsilon \frac{\partial^2 u}{\partial T_0 \partial T_1} + \epsilon^2 \left(2 \frac{\partial^2 u}{\partial T_0 \partial T_2} + \frac{\partial^2 u}{\partial T_1^2} \right) + u + \epsilon u^3 + \cdots = 0 \tag{4.95}$$

We note that we have replaced the original ordinary-differential equation by a partial-differential equation. Consequently, it appears that the problem has been complicated. This is true, but experience with this method has shown that the disadvantages of introducing this complication are far outweighed by the advantages. Not only does this method provide a uniform expansion, it also provides all the various nonlinear resonance phenomena, as we shall see in subsequent chapters.

We seek a uniform approximate solution to (4.95) in the form

$$u = u_0(T_0, T_1, T_2, \cdots) + \epsilon u_1(T_0, T_1, T_2, \cdots) + \cdots \tag{4.96}$$

Substituting for u from (4.96) into (4.95) gives

$$\frac{\partial^2 u_0}{\partial T_0^2} + \epsilon \frac{\partial^2 u_1}{\partial T_0^2} + 2\epsilon \frac{\partial^2 u_0}{\partial T_0 \partial T_1} + u_0 + \epsilon u_1 + \epsilon u_0^3 + \cdots = 0 \tag{4.97}$$

Equating each of the coefficients of ϵ^0 and ϵ to zero, we have

$$\frac{\partial^2 u_0}{\partial T_0^2} + u_0 = 0 \tag{4.98}$$

$$\frac{\partial^2 u_1}{\partial T_0^2} + u_1 = -2 \frac{\partial^2 u_0}{\partial T_0 \partial T_1} - u_0^3 \tag{4.99}$$

The general solution of (4.98) can be written as

$$u_0 = a(T_1, T_2, \cdots) \cos [T_0 + \beta(T_1, T_2, \cdots)] \tag{4.100}$$

We note that a and β are not constants but functions of the slow scales T_1, $T_2, \cdots$ because u_0 is a function of $T_0, T_1, T_2, \cdots$ and the derivatives in (4.98) are with respect to T_0. The functional dependence of a and β on $T_1, T_2, \cdots$

124 THE DUFFING EQUATION

is not known at this level of approximation; it is determined at subsequent levels of approximation by eliminating the secular terms.

Next, we substitute (4.100) into (4.99) and obtain

$$\frac{\partial^2 u_1}{\partial T_0^2} + u_1 = -2 \frac{\partial^2}{\partial T_0 \partial T_1} [a \cos(T_0 + \beta)] - a^3 \cos^3(T_0 + \beta)$$

$$= 2 \frac{\partial a}{\partial T_1} \sin(T_0 + \beta) + 2a \frac{\partial \beta}{\partial T_1} \cos(T_0 + \beta)$$

$$- \tfrac{3}{4} a^3 \cos(T_0 + \beta) - \tfrac{1}{4} a^3 \cos(3T_0 + 3\beta)$$

or

$$\frac{\partial^2 u_1}{\partial T_0^2} + u_1 = 2 \frac{\partial a}{\partial T_1} \sin(T_0 + \beta) + \left(2a \frac{\partial \beta}{\partial T_1} - \tfrac{3}{4} a^3\right) \cos(T_0 + \beta)$$

$$- \tfrac{1}{4} a^3 \cos(3T_0 + 3\beta) \qquad (4.101)$$

The right-hand side contains terms that produce secular terms in u_1. For a uniform expansion, these terms must be eliminated. This is accomplished by setting each of the coefficients of $\sin(T_0 + \beta)$ and $\cos(T_0 + \beta)$ equal to zero. The result is

$$\frac{\partial a}{\partial T_1} = 0 \qquad (4.102)$$

$$2a \frac{\partial \beta}{\partial T_1} - \tfrac{3}{4} a^3 = 0 \qquad (4.103)$$

Then, the particular solution of (4.101) becomes

$$u_1 = \tfrac{1}{32} a^3 \cos(3T_0 + 3\beta) \qquad (4.104)$$

The solution of (4.102) is $a = a(T_2, T_3, \cdots)$. Then, if $a \neq 0$, (4.103) can be rewritten as

$$\frac{\partial \beta}{\partial T_1} = \tfrac{3}{8} a^2$$

whose solution is

$$\beta = \tfrac{3}{8} a^2 T_1 + \beta_0(T_2, T_3, \cdots) \qquad (4.105)$$

Substituting for u_0 and u_1 from (4.100) and (4.104) into (4.96), we have

$$u = a \cos(T_0 + \beta) + \tfrac{1}{32} \epsilon a^3 \cos(3T_0 + 3\beta) + \cdots \qquad (4.106)$$

Substituting for β from (4.105) into (4.106) and recalling that $a = a(T_2, T_3, \cdots)$, we obtain

$$u = a(T_2, T_3, \cdots) \cos [T_0 + \tfrac{3}{8} T_1 a^2(T_2, T_3, \cdots) + \beta_0(T_2, T_3, \cdots)]$$
$$+ \tfrac{1}{32} \epsilon a^3(T_2, T_3, \cdots) \cos [3T_0 + \tfrac{9}{8} T_1 a^2(T_2, T_3, \cdots)$$
$$+ 3\beta_0(T_2, T_3, \cdots)] + \cdots \qquad (4.107)$$

If we stop the expansion as in (4.107), a and β can be considered constants to within the order of the error indicated. This is so because

$$a(T_2, T_3, \cdots) = a(\epsilon^2 t, \epsilon^3 t, \cdots)$$
$$= a(0, 0, \cdots) + \frac{\partial a}{\partial T_2} \epsilon^2 t + \cdots$$
$$= \hat{a} + O(\epsilon^2 t)$$
$$\beta_0(T_2, T_3, \cdots) = \beta_0(\epsilon^2 t, \epsilon^3 t, \cdots)$$
$$= \beta_0(0, 0, \cdots) + \frac{\partial \beta_0}{\partial T_0} \epsilon^2 t + \cdots$$
$$= \hat{\beta}_0 + O(\epsilon^2 t)$$

Thus, replacing a and β_0 by the constants $\hat{a}$ and $\hat{\beta}_0$ in (4.107), we have

$$u = \hat{a} \cos (T_0 + \tfrac{3}{8} T_1 \hat{a}^2 + \hat{\beta}_0)$$
$$+ \tfrac{1}{32} \epsilon \hat{a}^3 \cos (3T_0 + \tfrac{9}{8} T_1 \hat{a}^2 + 3\hat{\beta}_0) + O(\epsilon^2 t) \qquad (4.108)$$

In terms of the original variable t, (4.108) can be expressed as

$$u = \hat{a} \cos (t + \tfrac{3}{8} \epsilon t \hat{a}^2 + \hat{\beta}_0) + \tfrac{1}{32} \epsilon \hat{a}^3 \cos (3t + \tfrac{9}{8} \epsilon t \hat{a}^2$$
$$+ 3\hat{\beta}_0) + O(\epsilon^2 t) \qquad (4.109)$$

in agreement with the expansion (4.90) obtained by either the Lindstedt-Poincaré technique or the method of renormalization.

Inspecting (4.109), we find that the error is $O(1)$, and hence, the order of the first term when $t = O(\epsilon^{-2})$. Thus, (4.109) is not valid for $t \geq O(\epsilon^{-2})$. Moreover, if $t = O(\epsilon^{-1})$, the error is $O(\epsilon)$, and hence, the order of the second term. Therefore an expansion that is valid when $t = O(\epsilon^{-1})$ consists of the first term only. That is,

$$u = \hat{a} \cos (t + \tfrac{3}{8} \epsilon t \hat{a}^2 + \hat{\beta}_0) + O(\epsilon) \qquad (4.110)$$

for all times up to $O(\epsilon^{-1})$. This means that to determine a uniform first-order expansion, we need only to eliminate the terms that produce secular terms from the equation for u_1 without actually solving for u_1, thereby determining the dependence of u_0 on T_1. Similarly, in determining a first-order uniform expansion by using either the Lindstedt-Poincaré technique or the method of renormalization, we need only to eliminate the secular terms from the equation describing u_1, and hence, determine ω_1. Furthermore, in the higher approximations, we let

126 THE DUFFING EQUATION

$$u = \sum_{n=0}^{N-1} \epsilon^n u_n(T_0, T_1, \cdots, T_N) + O(\epsilon^N) \qquad (4.111)$$

That is, if we are after an Nth-order expansion, we include the scales $T_0, T_1, \cdots, T_N$ but we do not include the term $O(\epsilon^N)$.

Before closing this section, we present an alternative representation of the solutions of the perturbation equations; namely, we represent the solution of (4.98) in a complex form rather than the real form (4.100). To this end, we use the fact that (A22)

$$\cos \theta = \tfrac{1}{2}(e^{i\theta} + e^{-i\theta})$$

Thus, (4.100) can be rewritten as

$$u_0 = \tfrac{1}{2} a [e^{i(T_0 + \beta)} + e^{-i(T_0 + \beta)}]$$
$$= \tfrac{1}{2} a e^{i(T_0 + \beta)} + \tfrac{1}{2} a e^{-i(T_0 + \beta)}$$
$$= \tfrac{1}{2} a e^{i\beta} e^{iT_0} + \tfrac{1}{2} a e^{-i\beta} e^{-iT_0}$$

or

$$u_0 = A e^{iT_0} + \overline{A} e^{-iT_0} \qquad (4.112)$$

where $\overline{A}$ is the complex conjugate of A and

$$A = \tfrac{1}{2} a e^{i\beta} \qquad (4.113)$$

For a first-order expansion, we consider A to be a function of T_1 only.

Substituting (4.112) into (4.99), we have

$$\frac{\partial^2 u_1}{\partial T_0^2} + u_1 = -2i \frac{\partial A}{\partial T_1} e^{iT_0} + 2i \frac{\partial \overline{A}}{\partial T_1} e^{-iT_0} - (A e^{iT_0} + \overline{A} e^{-iT_0})^3 \qquad (4.114)$$

Expanding the cubic term in (4.114) and collecting coefficients of harmonics, we obtain

$$\frac{\partial^2 u_1}{\partial T_0^2} + u_1 = -\left(2i \frac{\partial A}{\partial T_1} + 3A^2 \overline{A}\right) e^{iT_0} + \left(2i \frac{\partial \overline{A}}{\partial T_1} - 3\overline{A}^2 A\right) e^{-iT_0}$$
$$- A^3 e^{3iT_0} - \overline{A}^3 e^{-3iT_0} \qquad (4.115)$$

We note that the terms proportional to $\exp(iT_0)$ and $\exp(-iT_0)$ produce secular terms in the particular solution of u_1. Thus, for a uniform expansion, each of the coefficients of $\exp(iT_0)$ and $\exp(-iT_0)$ must vanish. That is,

$$2i \frac{\partial A}{\partial T_1} + 3A^2 \overline{A} = 0 \qquad (4.116)$$

$$2i \frac{\partial \overline{A}}{\partial T_1} - 3\overline{A}^2 A = 0 \qquad (4.117)$$

We remind the reader that, for a uniform first-order solution, we do not need to solve for u_1 and it is sufficient to inspect equation (4.115) and eliminate the terms that produce secular terms in u_1.

Comparing (4.116) and (4.117), we find that they are not independent because taking the complex conjugate of (4.116) leads to (4.117). Hence, if one of them is satisfied, the other is automatically satisfied. To analyze (4.116), we replace A by its polar form (4.113). The result is

$$2i\left(\frac{1}{2}\frac{\partial a}{\partial T_1}e^{i\beta} + \frac{1}{2}ai\frac{\partial \beta}{\partial T_1}e^{i\beta}\right) + 3 \cdot \frac{a^2}{4}e^{2i\beta} \cdot \frac{a}{2}e^{-i\beta} = 0$$

or

$$i\frac{\partial a}{\partial T_1}e^{i\beta} - a\frac{\partial \beta}{\partial T_1}e^{i\beta} + \tfrac{3}{8}a^3 e^{i\beta} = 0$$

or

$$i\frac{\partial a}{\partial T_1} - a\frac{\partial \beta}{\partial T_1} + \tfrac{3}{8}a^3 = 0 \qquad (4.118)$$

We recall the fact that a complex number vanishes if and only if its real and imaginary parts vanish independently. Since a and β are real, the independent vanishing of the real and imaginary parts in (4.118) yields

$$\frac{\partial a}{\partial T_1} = 0 \qquad (4.119)$$

$$a\frac{\partial \beta}{\partial T_1} - \tfrac{3}{8}a^3 = 0 \qquad (4.120)$$

in agreement with (4.102) and (4.103) obtained above by expressing the solution in real form. Comparing the complex with the real representation, we find it more convenient to use the complex form. Thus, the complex form is used in the remainder of the book.

4.6. Variation of Parameters

When $\epsilon = 0$, the solution of (4.7) can be written as

$$u = a \cos(t + \beta) \qquad (4.121)$$

where a and β are constants, which are sometimes referred to as *parameters*. It follows from (4.121) that

$$\dot{u} = -a \sin(t + \beta) \qquad (4.122)$$

When $\epsilon \neq 0$, we assume that the solution of (4.7) is still given by (4.121) but with time-varying a and β. In other words, we consider (4.121) as a transformation from $u(t)$ to $a(t)$ and $\beta(t)$. This is why this approach is called the *method of variation of parameters*. Using this view, we note that we have two equations, namely (4.7) and (4.121), for the three unknowns $u(t)$, $a(t)$, and $\beta(t)$. Hence, we have the freedom of imposing a third condition (third equation). This condition is arbitrary except that it must be independent of (4.7) and (4.121). This arbitrariness can be used to advantage, namely to produce a simple and convenient transformation. Out of all possible conditions, we choose to impose the condition (4.122), thereby assuming that u as well as $\dot{u}$ have the same form as the linear case. This condition leads to a convenient transformation because it leads to a set of first-order rather than second-order equations for $a(t)$ and $\beta(t)$.

Differentiating (4.121) with respect to t and recalling that a and β are functions of t, we have

$$\dot{u} = -a \sin(t+\beta) + \dot{a} \cos(t+\beta) - a\dot{\beta} \sin(t+\beta) \tag{4.123}$$

Comparing (4.123) with (4.122), we conclude that

$$\dot{a} \cos(t+\beta) - a\dot{\beta} \sin(t+\beta) = 0 \tag{4.124}$$

Differentiating (4.122) with respect to t, we obtain

$$\ddot{u} = -a \cos(t+\beta) - \dot{a} \sin(t+\beta) - a\dot{\beta} \cos(t+\beta) \tag{4.125}$$

Substituting for u and $\ddot{u}$ from (4.121) and (4.125) into (4.7), we have

$$\dot{a} \sin(t+\beta) + a\dot{\beta} \cos(t+\beta) = \epsilon a^3 \cos^3(t+\beta) \tag{4.126}$$

We note that (4.124) and (4.126) constitute a system of two first-order equations for $\dot{a}$ and $\dot{\beta}$. They can be simplified further. To this end, we multiply (4.124) by $\cos(t+\beta)$ and (4.126) by $\sin(t+\beta)$, add the results, recall (A1), and obtain

$$\dot{a} = \epsilon a^3 \sin(t+\beta) \cos^3(t+\beta) \tag{4.127}$$

Substituting for $\dot{a}$ into (4.124) and solving for $\dot{\beta}$, we obtain

$$\dot{\beta} = \epsilon a^2 \cos^4(t+\beta) \tag{4.128}$$

if $a \neq 0$. Thus, the original second-order equation (4.7) for $u(t)$ has been replaced by the two first-order equations (4.127) and (4.128) for $a(t)$ and $\beta(t)$. We emphasize that no approximations have been made in deriving (4.127) and (4.128).

Comparing the transformed equations (4.127) and (4.128) with the original equation (4.7), we find that the transformed equations are more nonlinear than the original equation. Then, the question arises what is the value of this transformation? The answer to this question depends on the value of ϵ. If ϵ is small, the major parts of a and β vary more slowly than u with t as shown in Figure

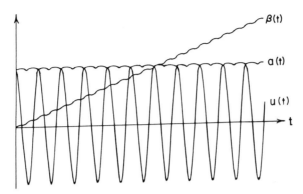

Figure 4-2. The variation of a, β, and u with t for $a(0) = 1.5$, $\beta(0) = 0.0$, and $\epsilon = 0.05$.

4-2. This fact can be used to advantage analytically and numerically. The analytical advantage is utilized in the method of averaging (Section 4.7). Numerically, it is advantageous to solve the transformed equations instead of the original equation because a large step size can be used in the integration. This is the reason why astronomers use the method of variation of parameters to determine the equations describing the parameters of the orbits. Then, they numerically solve the variational equations rather than the original equations. Usually, astronomers and celestial mechanicians refer to this approach as the "special method of perturbations."

4.7. The Method of Averaging

Using the trigonometric identities

$$\sin\phi \cos^3\phi = \tfrac{1}{4}\sin 2\phi + \tfrac{1}{8}\sin 4\phi$$
$$\cos^4\phi = \tfrac{3}{8} + \tfrac{1}{2}\cos 2\phi + \tfrac{1}{8}\cos 4\phi$$

we rewrite (4.127) and (4.128) as

$$\dot{a} = \tfrac{1}{8}\epsilon a^3(2\sin 2\phi + \sin 4\phi) \qquad (4.129)$$

$$\dot{\beta} = \tfrac{1}{8}\epsilon a^2(3 + 4\cos 2\phi + \cos 4\phi) \qquad (4.130)$$

where $\phi = t + \beta$. Since $-1 \leq \sin n\phi \leq 1$ and $-1 \leq \cos n\phi \leq 1$, $\dot{a} = O(\epsilon)$ and $\dot{\beta} = O(\epsilon)$ if a is bounded. Thus, the major parts of a and β are slowly varying functions of time if ϵ is small, as shown in Figure 4.2. Hence, they change very little during the time interval π (the period of the circular functions) and, to the first approximation, they can be considered constant in the interval $[0, \pi]$.

We average both sides of (4.129) and (4.130) over the interval $[0, \pi]$ and obtain

130 THE DUFFING EQUATION

$$\frac{1}{\pi}\int_0^\pi \dot{a}\, dt = \frac{1}{8\pi}\epsilon \int_0^\pi a^3(2\sin 2\phi + \sin 4\phi)\, dt \tag{4.131}$$

$$\frac{1}{\pi}\int_0^\pi \dot{\beta}\, dt = \frac{1}{8\pi}\epsilon \int_0^\pi a^2(3 + 4\cos 2\phi + \cos 4\phi)\, dt \tag{4.132}$$

Since a and β can be considered constant in the interval $[0, \pi]$, $a, \dot{a}$, and $\dot{\beta}$ can be taken outside the integral signs in (4.131) and (4.132). The result is

$$\dot{a} = \frac{1}{8\pi}\epsilon a^3 \int_0^\pi (2\sin 2\phi + \sin 4\phi)\, dt \tag{4.133}$$

$$\dot{\beta} = \tfrac{3}{8}\epsilon a^2 + \frac{1}{8\pi}\epsilon a^2 \int_0^\pi (4\cos 2\phi + \cos 4\phi)\, dt \tag{4.134}$$

Next, we change the integration variable from t to $\phi = t + \beta$. Hence, $d\phi = dt$ because β can be considered constant in $[0, \pi]$. Substituting this change of variable into (4.133) and (4.134) gives

$$\dot{a} = \frac{1}{8\pi}\epsilon a^3 \int_\beta^{\pi+\beta} (2\sin 2\phi + \sin 4\phi)\, d\phi$$

$$= -\frac{1}{8\pi}\epsilon a^3 (\cos 2\phi + \tfrac{1}{4}\cos 4\phi)\Big|_\beta^{\pi+\beta} = 0 \tag{4.135}$$

$$\dot{\beta} = \tfrac{3}{8}\epsilon a^2 + \frac{1}{8\pi}\epsilon a^2 \int_\beta^{\pi+\beta} (4\cos 2\phi + \cos 4\phi)\, d\phi$$

$$= \tfrac{3}{8}\epsilon a^2 + \frac{1}{8\pi}\epsilon a^2 (2\sin 2\phi + \tfrac{1}{4}\sin 4\phi)\Big|_\beta^{\pi+\beta} = \tfrac{3}{8}\epsilon a^2 \tag{4.136}$$

This averaging technique is usually referred to as the *van der Pol method* or the *Krylov-Bogoliubov method*.

It follows from (4.135) that

$$a = a_0 = \text{constant} \tag{4.137}$$

Then, it follows from (4.136) that

$$\beta = \tfrac{3}{8}\epsilon a_0^2 t + \beta_0 \tag{4.138}$$

where β_0 is a constant. Substituting for a and β into (4.121), we obtain to the first approximation

$$u = a_0 \cos\left[\left(1 + \tfrac{3}{8}\epsilon a_0^2\right)t + \beta_0\right] \tag{4.139}$$

in agreement with the solutions obtained by using the Lindstedt-Poincaré technique, the method of renormalization, and the method of multiple scales.

Before closing this section, we note that one can arrive at the final results in (4.135) and (4.136) without going through the averaging process. We note that the right-hand sides of (4.129) and (4.130) are the sum of two groups of terms— a group that is a linear combination of fast varying terms and a group that is a linear combination of slowly varying terms. Then, to the first approximation, $\dot{a}$ in (4.129) is equal to the slowly varying group on its right-hand side, which is zero. Moreover, to the first approximation, $\dot{\beta}$ in (4.130) is equal to the slowly varying group on its right-hand side, which is $\tfrac{3}{8}\epsilon a^2$.

Exercises

4.1. Use the method of renormalization to render the following expansions uniformly valid:

(a) $u(t; \epsilon) = a \cos(\omega_0 t + \theta) + \epsilon a^3 t \sin(\omega_0 t + \theta) + O(\epsilon^2)$
(b) $u(t; \epsilon) = a \cos(\omega_0 t + \theta) + \epsilon[a^2 t \sin(\omega_0 t + \theta) + (1 - a^2)at \cos(\omega_0 t + \theta)] + O(\epsilon^2)$

4.2. Consider the equation

$$\ddot{u} + \omega_0^2 u = \epsilon \dot{u}^2 u \quad \epsilon \ll 1$$

(a) Determine a two-term straightforward expansion and discuss its uniformity.
(b) Render this expansion uniformly valid by using the method of renormalization.
(c) Determine a first-order uniform expansion by using the Lindstedt-Poincaré technique.
(d) Use the method of multiple scales to determine a first-order uniform expansion.
(e) Use the method of averaging to determine a first-order uniform expansion.

4.3. Consider the equation

$$\ddot{u} + 4u + \epsilon u^2 \ddot{u} = 0$$

(a) Determine a two-term straightforward expansion and discuss its uniformity.
(b) Render this expansion uniformly valid by using the method of renormalization.
(c) Determine a first-order uniform expansion by using the Lindstedt-Poincaré technique.
(d) Use the method of multiple scales to determine a first-order uniform expansion.
(e) Use the method of averaging to determine a first-order uniform expansion.

4.4. Consider the equation

132 THE DUFFING EQUATION

$$\ddot{u} + \frac{\omega^2 u}{1 + u^2} = 0$$

(a) Determine a two-term straightforward expansion for small but finite u and discuss its uniformity.
(b) Render this expansion uniformly valid by using the method of renormalization.
(c) Determine a first-order uniform expansion by using the Lindstedt-Poincaré technique.
(d) Use the method of multiple scales to determine a first-order uniform expansion.
(e) Use the method of averaging to determine a first-order uniform expansion.

4.5. Consider the equation

$$\ddot{u} + \omega_0^2 u = \epsilon u^5 \qquad \epsilon \ll 1$$

(a) Determine a two-term straightforward expansion and discuss its uniformity.
(b) Render this expansion uniformly valid by using the method of renormalization.
(c) Determine a first-order uniform expansion by using the Lindstedt-Poincaré technique.
(d) Use the method of multiple scales to determine a first-order uniform expansion.
(e) Use the method of averaging to determine a first-order uniform expansion.

4.6. The motion of a simple pendulum is governed by

$$\ddot{\theta} + \frac{g}{\ell} \sin \theta = 0$$

(a) Expand for small θ and keep up to cubic terms.
(b) Determine a first-order uniform expansion for small but finite θ.

4.7. Consider the equation

$$\ddot{\theta} = \Omega^2 \sin \theta \cos \theta - \frac{g}{R} \sin \theta$$

(a) Expand for small θ and keep up to cubic terms.
(b) Determine a first-order uniform expansion for small but finite θ.

4.8. The motion of a particle on a rotating parabola is governed by

$$(1 + 4p^2 x^2)\ddot{x} + \Lambda x + 4p^2 \dot{x}^2 x = 0$$

where p and Λ are constants. Determine a first-order uniform expansion for small but finite x.

4.9. Consider the equation

$$\left(1 + \frac{u^2}{1 - u^2}\right)\ddot{u} + \frac{u \dot{u}^2}{(1 - u^2)^2} + \omega_0^2 u + \frac{g}{\ell} \frac{u}{\sqrt{1 - u^2}} = 0$$

(a) Expand for small u and keep up to cubic terms.
(b) Determine a first-order uniform expansion for small but finite u.

4.10. Consider the equation

$$(l^2 + r^2 - 2rl \cos \theta)\ddot{\theta} + rl \sin \theta \dot{\theta}^2 + gl \sin \theta = 0$$

where g, r, and l are constants. Determine a first-order expansion for small but finite θ. Expand first for small θ and keep up to cubic terms.

4.11. Consider the equation

$$(\tfrac{1}{12} l^2 + r^2 \theta^2)\ddot{\theta} + r^2 \theta \dot{\theta}^2 + gr\theta \cos \theta = 0$$

where r, l, and g are constants. Determine a first-order uniform expansion for small but finite θ.

4.12. Consider the equation

$$m\ddot{x} + kx(x^2 + l^2)^{-1/2}[(x^2 + l^2)^{1/2} - \tfrac{1}{2} l] = 0$$

Determine a first-order uniform expansion for small but finite x.

4.13. Expand the integrand in (4.64) up to $O(m^2)$ and obtain

$$T_a = \frac{2\pi}{\sqrt{1 + \epsilon x_0^2}}(1 + \tfrac{1}{4} m + \tfrac{9}{64} m^2 + \cdots)$$

Then, express T_a in terms of m as

$$T_a^{(1)} = 2\pi\sqrt{1 - 2m}\,(1 + \tfrac{1}{4}m + \tfrac{9}{64}m^2 + \cdots)$$

and in terms of x_0 as

$$T^{(2)} = 2\pi(1 - \tfrac{3}{8}\epsilon x_0^2 + \tfrac{57}{256}\epsilon^2 x_0^4 + \cdots)$$

Show that $T_a^{(1)}$ is more accurate than $T_a^{(2)}$ by comparing them with the tabulated values of the exact solution (4.64). Note that $T_a^{(1)}$ can be obtained from $T_a^{(2)}$ by using the Euler transformation $m = \epsilon x_0^2/2(1 + \epsilon x_0^2)$. Often, this transformation extends the validity of an asymptotic series.

CHAPTER 5

The Linear Oscillator

In contrast with the preceding chapter, in which we discussed conservative systems, in this chapter, we discuss systems with damping. To describe the techniques with the minimum amount of algebra, we use the simplest possible equation with damping, namely the equation governing the free oscillations of a particle with mass m connected to a rigid support by a spring with constant k and a dashpot with coefficient μ, as shown in Figure 5-1. The governing equation can be written as follows:

$$m\frac{d^2u^*}{dt^{*2}} + \mu\frac{du^*}{dt^*} + ku^* = 0 \tag{5.1}$$

In the absence of damping, the system has the angular frequency $\omega_0 = \sqrt{k/m}$.
As before, we introduce dimensionless variables

$$t = \omega_0 t^* \quad \text{and} \quad u = u^*/u_0^*$$

where u_0^* is any characteristic displacement such as the initial displacement. Then, (5.1) becomes

$$\ddot{u} + 2\epsilon\dot{u} + u = 0 \tag{5.2}$$

where $\epsilon = \frac{1}{2}\mu/\sqrt{km}$ is a measure of the ratio of the damping force to the restoring force of the spring and dots denote the derivatives with respect to t. In this chapter, we consider the general solution for (5.2) for small ϵ.

We start in the following section by the straightforward expansion and discuss its nonuniformity. In Section 5.2, we investigate the exact solution to exhibit the source of nonuniformity. In Section 5.3, we show that the Lindstedt-Poincaré technique may lead to the trivial solution. In Section 5.4, we show how the method of multiple scales leads to a uniform expansion. Finally, in Section 5.5, we use the method of averaging to determine a uniform first approximation.

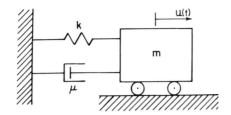

Figure 5-1. A mass restrained by a spring and a damper.

5.1. The Straightforward Expansion

We seek a second-order straightforward expansion in powers of ϵ in the form

$$u(t; \epsilon) = u_0(t) + \epsilon u_1(t) + \epsilon^2 u_2(t) + \cdots \quad (5.3)$$

Substituting (5.3) into (5.2) yields

$$\ddot{u}_0 + \epsilon \ddot{u}_1 + \epsilon^2 \ddot{u}_2 + \cdots + 2\epsilon(\dot{u}_0 + \epsilon \dot{u}_1 + \epsilon^2 \dot{u}_2 + \cdots) + u_0 + \epsilon u_1 + \epsilon^2 u_2 + \cdots = 0$$

Collecting coefficients of equal powers of ϵ gives

$$\ddot{u}_0 + u_0 + \epsilon(\ddot{u}_1 + u_1 + 2\dot{u}_0) + \epsilon^2(\ddot{u}_2 + u_2 + 2\dot{u}_1) + \cdots = 0$$

Equating coefficients of like powers of ϵ to zero yields

$$\ddot{u}_0 + u_0 = 0 \quad (5.4)$$

$$\ddot{u}_1 + u_1 = -2\dot{u}_0 \quad (5.5)$$

$$\ddot{u}_2 + u_2 = -2\dot{u}_1 \quad (5.6)$$

These equations can be solved sequentially for u_0, u_1, and u_2.

The general solution of (5.4) can be written as

$$u_0 = a \cos(t + \beta) \quad (5.7)$$

where a and β are constants. Then, (5.5) becomes

$$\ddot{u}_1 + u_1 = 2a \sin(t + \beta) \quad (5.8)$$

As discussed in Section 4.1, we do not need the homogeneous solution of (5.8) if we allow a and β to be functions of ϵ when enforcing the initial conditions. Then, we write the solution for u_1 as the particular solution only. It follows from (B81) and (B82) that the solution of (5.8) is

$$u_1 = -at \cos(t + \beta) \quad (5.9)$$

Substituting for u_1 into (5.6) yields

$$\ddot{u}_2 + u_2 = 2a \cos(t + \beta) - 2at \sin(t + \beta) \quad (5.10)$$

136 THE LINEAR OSCILLATOR

The particular solution of (5.10) can be obtained using operational calculus, or variation of parameters, or the method of undetermined coefficients. The result is (B69 and B76)

$$u_2 = \tfrac{1}{2} at^2 \cos(t + \beta) + \tfrac{1}{2} at \sin(t + \beta) \tag{5.11}$$

Substituting for u_0, u_1, and u_2 from (5.7), (5.9), and (5.11), respectively, into (5.3) gives

$$u = a \cos(t + \beta) - \epsilon at \cos(t + \beta) + \tfrac{1}{2} \epsilon^2 a [t^2 \cos(t + \beta) + t \sin(t + \beta)] + \cdots \tag{5.12}$$

This straightforward or pedestrian expansion is not valid when $t \geqslant O(\epsilon^{-1})$ due to the presence of the mixed-secular terms. We note that the secular terms become more compounded at higher orders. The first approximation has a linear secular term, whereas the second approximation has a quadratic secular term. Carrying out the expansion to nth-order, one finds that the nth approximation contains an nth secular term t^n. Next, we discuss the exact solution and investigate the source of the secular terms.

5.2. Exact Solution

To exhibit the source of nonuniformity and motivate the methods that yield uniform expansions, we investigate the exact solution to (5.2). Since (5.2) is a linear differential equation with constant coefficients, it has solutions in the form

$$u = c \exp(\lambda t) \tag{5.13}$$

where c and λ are constants, which may be complex. Substituting (5.13) into (5.2) yields

$$(\lambda^2 + 2\epsilon\lambda + 1) c \exp(\lambda t) = 0$$

Hence, for a nontrivial solution

$$\lambda^2 + 2\epsilon\lambda + 1 = 0 \tag{5.14}$$

or

$$\lambda = -\epsilon \pm \sqrt{\epsilon^2 - 1} \tag{5.15}$$

When $\epsilon < 1$, it is convenient to write (5.15) as

$$\lambda = -\epsilon \pm i\sqrt{1 - \epsilon^2} \tag{5.16}$$

Then, the general solution of (5.2) can be written as

$$u = c_1 \exp[-\epsilon t + i\sqrt{1 - \epsilon^2}\, t] + c_2 \exp[-\epsilon t - i\sqrt{1 - \epsilon^2}\, t] \tag{5.17}$$

Since we are interested in real solutions, the constant c_2 must be the complex conjugate of c_1. To write (5.17) in terms of circular functions, we express c_1 and c_2 in the polar forms

$$c_1 = \tfrac{1}{2} a \exp(i\beta) \qquad c_2 = \tfrac{1}{2} a \exp(-i\beta)$$

where a and β are real constants. Then, (5.17) becomes

$$u = ae^{-\epsilon t} \cos(\sqrt{1 - \epsilon^2}\, t + \beta) \qquad (5.18)$$

To exhibit the source of the nonuniformity, we determine the first three terms in the expansion of (5.18) for small ϵ and compare the result with the straightforward expansion (5.12) obtained directly from the differential equation. To this end, we expand the exponential and circular functions in the exact solution separately. Using Taylor series, we have

$$e^{-\epsilon t} = 1 - \epsilon t + \frac{1}{2!}\epsilon^2 t^2 - \frac{1}{3!}\epsilon^3 t^3 + \cdots = \sum_{n=0}^{\infty} \frac{1}{n!}(-\epsilon t)^n \qquad (5.19)$$

To expand the circular function, we first expand the radical, using the binomial theorem, that is,

$$\sqrt{1-\epsilon^2} = (1-\epsilon^2)^{1/2} = 1 - \tfrac{1}{2}\epsilon^2 + \frac{(\tfrac{1}{2})(-\tfrac{1}{2})}{2!}\epsilon^4 - \frac{(\tfrac{1}{2})(-\tfrac{1}{2})(-\tfrac{3}{2})}{3!}\epsilon^6 + \cdots$$

$$= 1 - \tfrac{1}{2}\epsilon^2 - \tfrac{1}{8}\epsilon^4 + \cdots \qquad (5.20)$$

Then,

$$\cos(\sqrt{1-\epsilon^2}\, t + \beta) = \cos[t + \beta - \tfrac{1}{2}\epsilon^2 t - \tfrac{1}{8}\epsilon^4 t + \cdots]$$

$$= \cos(t+\beta)\cos(\tfrac{1}{2}\epsilon^2 t + \tfrac{1}{8}\epsilon^4 t + \cdots) + \sin(t+\beta)\sin(\tfrac{1}{2}\epsilon^2 t + \tfrac{1}{8}\epsilon^4 t + \cdots)$$

But

$$\cos\delta = 1 - \frac{1}{2!}\delta^2 + \frac{1}{4!}\delta^4 + \cdots = \sum_{n=0}^{\infty} \frac{(-1)^n \delta^{2n}}{(2n)!} \qquad (5.21)$$

$$\sin\delta = \delta - \frac{1}{3!}\delta^3 + \frac{1}{5!}\delta^5 + \cdots = \sum_{n=1}^{\infty} \frac{(-1)^{n+1}\delta^{2n-1}}{(2n-1)!} \qquad (5.22)$$

Hence, to $O(\epsilon^4)$

$$\cos(\sqrt{1-\epsilon^2}\, t + \beta) = \cos(t+\beta)[1 - \tfrac{1}{2}(\tfrac{1}{2}\epsilon^2 t + \cdots)^2]$$

$$+ \sin(t+\beta)[\tfrac{1}{2}\epsilon^2 t + \tfrac{1}{8}\epsilon^4 t + \cdots] + \cdots$$

$$= (1 - \tfrac{1}{8}\epsilon^4 t^2)\cos(t+\beta) + (\tfrac{1}{2}\epsilon^2 t + \tfrac{1}{8}\epsilon^4 t)\sin(t+\beta) + \cdots$$

Therefore,

138 THE LINEAR OSCILLATOR

$$\cos(\sqrt{1-\epsilon^2}\, t + \beta) = \cos(t+\beta) + \tfrac{1}{2}\epsilon^2 t \sin(t+\beta)$$
$$+ \tfrac{1}{8}\epsilon^4 [t \sin(t+\beta) - t^2 \cos(t+\beta)] + \cdots \quad (5.23)$$

To compare with (5.12), we need the expansion of (5.18) to $O(\epsilon^2)$. Substituting for the exponential and circular functions from (5.19) and (5.23) into (5.18) and keeping terms up to $O(\epsilon^2)$, we obtain

$$u = a(1 - \epsilon t + \tfrac{1}{2}\epsilon^2 t^2 + \cdots)[\cos(t+\beta) + \tfrac{1}{2}\epsilon^2 t \sin(t+\beta) + \cdots] = a\cos(t+\beta)$$
$$- \epsilon a t \cos(t+\beta) + \tfrac{1}{2}\epsilon^2 a [t^2 \cos(t+\beta) + t \sin(t+\beta)] + \cdots \quad (5.24)$$

in agreement with (5.12) obtained directly from the differential equation.

The above development shows that, in arriving at the straightforward expansion, we had to expand the exponential function as in (5.19) and the circular function as in (5.23). Using the ratio test in (5.19), we have

$$\lim_{n\to\infty} \frac{n\text{th term}}{(n-1)\text{th term}} = \lim_{n\to\infty} \frac{(-\epsilon t)^n (n-1)!}{n! (-\epsilon t)^{n-1}} = \lim_{n\to\infty} \frac{-\epsilon t}{n} = 0$$

Hence, the series (5.19) converges for all values of ϵt. However, Figure 5-2 shows that $\exp(-\epsilon t)$ cannot be approximated uniformly by a finite number of terms in the series for all values of ϵt. Therefore, any expansion procedure that is based on approximating $\exp(-\epsilon t)$ by a finite number of terms in a series expansion in ϵt is doomed to fail for large t.

In expanding the circular function, we had to use the three expansions (5.20) through (5.22). Using the ratio test for the series (5.20), we have

$$\lim_{n\to\infty} \frac{n\text{th term}}{(n-1)\text{th term}} = \lim_{n\to\infty} \frac{(\tfrac{1}{2})(-\tfrac{1}{2})(-\tfrac{3}{2})\cdots\left(-\dfrac{2n-1}{2}\right)(-\epsilon^2)^{n+1} n!}{(n+1)!(\tfrac{1}{2})(-\tfrac{1}{2})(-\tfrac{3}{2})\cdots\left(-\dfrac{2n-3}{2}\right)(-\epsilon^2)^n}$$

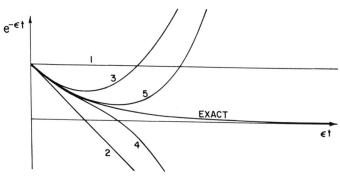

Figure 5-2. Approximation of an exponential function by a finite number of terms in a Taylor series.

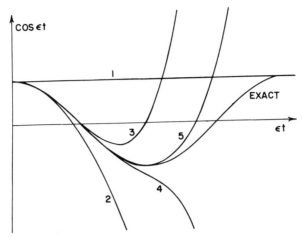

Figure 5-3. Approximation of a cosine function by a finite number of terms in a Taylor series.

$$= \lim_{n \to \infty} \frac{\epsilon^2 (2n-1)}{2n+2} = \epsilon^2$$

Hence, (5.20) converges for all $\epsilon^2 < 1$. Thus, the expansion (5.20) is uniform for small ϵ. Using the ratio test for the series (5.21), we have

$$\lim_{n \to \infty} \frac{n\text{th term}}{(n-1)\text{th term}} = \lim_{n \to \infty} \frac{(-1)^n \, \delta^{2n} \, (2n-2)!}{(2n)! \, (-1)^{n-1} \, \delta^{2n-2}} = \lim_{n \to \infty} \frac{-\delta^2}{2n(2n-1)} = 0$$

Hence, the series (5.21) converges for all values of δ. However, Figure 5-3 shows that $\cos \delta$ cannot be approximated uniformly by a finite number of terms for all values of δ. Similarly, one can show that $\sin \delta$ cannot be approximated uniformly by a finite number of terms for all values of δ. Therefore, any expansion that is based on approximating $\cos \delta$ and/or $\sin \delta$ by a finite number of terms in a series expansion in δ is doomed to fail for large δ.

In summary, the failure of the straightforward expansion for large t is due to the expansion of $\exp(-\epsilon t)$ and $\cos(\sqrt{1-\epsilon^2}\, t + \beta)$ in powers of ϵ. We note that the straightforward expansion (5.23) forces the frequency to be unity, which is independent of the damping. In fact, the presence of the damping changes the frequency from 1 to $\sqrt{1-\epsilon^2}$. Thus, any expansion procedure that does not account for the dependence of the frequency on ϵ will fail for large t. Next, we show that the Lindstedt-Poincaré technique may lead to the trivial solution.

5.3. The Lindstedt-Poincaré Technique

To account for the fact that the frequency is a function of ϵ, we let $\tau = \omega t$ in (5.2) and obtain

140 THE LINEAR OSCILLATOR

$$\omega^2 u'' + 2\epsilon\omega u' + u = 0 \tag{5.25}$$

where the prime indicates the derivative with respect to τ. Next, we try expanding u and ω in powers of ϵ, that is,

$$u = u_0(\tau) + \epsilon u_1(\tau) + \epsilon^2 u_2(\tau) + \cdots \tag{5.26}$$

$$\omega = 1 + \epsilon\omega_1 + \epsilon^2\omega_2 + \cdots \tag{5.27}$$

Note that the first term in (5.27) is unity, which is the unperturbed (undamped) frequency. Substituting (5.26) and (5.27) into (5.25) gives

$$(1 + \epsilon\omega_1 + \epsilon^2\omega_2 + \cdots)^2 (u_0'' + \epsilon u_1'' + \epsilon^2 u_2'' + \cdots) + 2\epsilon(1 + \epsilon\omega_1 + \epsilon^2\omega_2 + \cdots)$$
$$\cdot (u_0' + \epsilon u_1' + \epsilon^2 u_2' + \cdots) + u_0 + \epsilon u_1 + \epsilon^2 u_2 + \cdots = 0$$

Expanding for small ϵ and collecting coefficients of equal powers of ϵ, we obtain

$$u_0'' + u_0 + \epsilon[u_1'' + u_1 + 2\omega_1 u_0'' + 2u_0']$$
$$+ \epsilon^2[u_2'' + u_2 + 2\omega_2 u_0'' + \omega_1^2 u_0'' + 2\omega_1 u_1'' + 2\omega_1 u_0' + 2u_1'] + \cdots = 0$$

Equating coefficients of like powers of ϵ yields the following equations that can be solved sequentially for u_0, u_1, and u_2:

$$u_0'' + u_0 = 0 \tag{5.28}$$

$$u_1'' + u_1 = -2\omega_1 u_0'' - 2u_0' \tag{5.29}$$

$$u_2'' + u_2 = -2\omega_2 u_0'' - \omega_1^2 u_0'' - 2\omega_1 u_1'' - 2\omega_1 u_0' - 2u_1' \tag{5.30}$$

The general solution of (5.28) can be written as

$$u_0 = a \cos(\tau + \beta) \tag{5.31}$$

where a and β are constants. Then, (5.29) becomes

$$u_1'' + u_1 = 2a\omega_1 \cos(\tau + \beta) + 2a \sin(\tau + \beta) \tag{5.32}$$

To eliminate the secular terms from the particular solution for u_1, we need to annihilate the right-hand side of (5.32). Since τ is a variable, this condition demands that each of the coefficients of $\sin(\tau + \beta)$ and $\cos(\tau + \beta)$ must vanish independently. That is,

$$2a\omega_1 = 0 \quad \text{and} \quad a = 0 \tag{5.33}$$

Hence, (5.32) becomes

$$u_1'' + u_1 = 0 \tag{5.34}$$

Equations (5.33) cannot be satisfied simultaneously unless $a = 0$. But if $a = 0$, $u_0 = 0$ according to (5.31), then $u_1 = 0$ if the homogeneous solution is not included. Substituting for $u_0 = 0$ into (5.30) yields

$$u_2'' + u_2 = 0 \tag{5.35}$$

whose particular solution is $u_2 = 0$. Thus, we have ended up with the trivial solution.

Had we included the homogeneous solution of (5.34), we would have obtained

$$u_1 = a_1 \cos(\tau + \beta_1) \tag{5.36}$$

where a_1 and β_1 are arbitrary constants. Then, (5.35) would have been replaced by

$$u_2'' + u_2 = 2\omega_1 a_1 \cos(\tau + \beta_1) + 2a_1 \sin(\tau + \beta_1) \tag{5.37}$$

Again, the condition for the elimination of secular terms demands that $a_1 = 0$. Consequently, only the trivial solution is obtained.

The above development shows that the Lindstedt-Poincaré technique leads to the trivial solution only. Similarly, the method of renormalization yields the trivial solution of this problem only. The reason for the failure of these techniques to yield a uniform nontrivial solution is our insistence on a uniform solution having a constant amplitude as in (5.31). Since the amplitude is $a \exp(-\epsilon t)$ according to the exact solution (5.18), the only constant-amplitude uniform solution is the one attained after a long time (i.e., steady state). Therefore, although the Lindstedt-Poincaré technique and the method of renormalization are effective in determining periodic solutions, they are incapable of determining transient responses.

In this example, this shortcoming of the Lindstedt-Poincaré technique and the method of renormalization can be remedied if one permits the ω_n to be complex. To show this, we express the solution of (5.28) in the complex form

$$u_0 = Ae^{i\tau} + \overline{A}e^{-i\tau} \tag{5.38}$$

where A is a complex constant. Then, (5.29) becomes

$$u_1'' + u_1 = 2(\omega_1 - i)Ae^{i\tau} + cc \tag{5.39}$$

Eliminating the terms that produce secular terms from (5.39), we obtain $\omega_1 = i$ and (5.39) reduces to (5.34). Hence, $u_1 = 0$. Substituting for u_0, u_1, and ω_1 into (5.30) gives

$$u_2'' + u_2 = (2\omega_2 + 1)Ae^{i\tau} + cc \tag{5.40}$$

The condition for the elimination of secular terms leads to $\omega_2 = -\frac{1}{2}$, and hence $u_2 = 0$. Therefore,

$$\tau = \omega t = (1 + i\epsilon - \tfrac{1}{2}\epsilon^2 + \cdots)t \tag{5.41}$$

and

$$\begin{aligned} u &= A \exp\left[i(1 + i\epsilon - \tfrac{1}{2}\epsilon^2 + \cdots)t\right] + cc \\ &= A \exp(-\epsilon t) \exp\left[i(1 - \tfrac{1}{2}\epsilon^2)t\right] + cc + \cdots \end{aligned} \tag{5.42}$$

Letting $A = \frac{1}{2}a \exp(i\beta)$, where a and β are real, we rewrite (5.42) as

$$u = ae^{-\epsilon t} \cos[(1 - \tfrac{1}{2}\epsilon^2)t + \beta + \cdots] \tag{5.43}$$

in agreement with the exact solution (5.18) to $O(\epsilon^2)$.

We should note that the success of the preceding modification of the Lindstedt-Poincaré technique is accidental. In fact, this modified technique may lead to an erroneous solution, as shown in Section 6.2.

5.4. The Method of Multiple Scales

Using (5.20), we rewrite (5.18) as

$$u = ae^{-\epsilon t} \cos(t + \beta - \tfrac{1}{2}\epsilon^2 t - \tfrac{1}{8}\epsilon^4 t + \cdots) \tag{5.44}$$

Thus, $u(t; \epsilon) = u(t, \epsilon t, \epsilon^2 t, \epsilon^4 t, \cdots)$ and this problem is well-suited for the application of the method of multiple scales. To determine a second-order expansion, we need only the three scales $T_0 = t$, $T_1 = \epsilon t$, and $T_2 = \epsilon^2 t$. Then, using the chain rule we transform the time derivatives according to (4.93) and (4.94). However, here we streamline the notation and write

$$\frac{d}{dt} = D_0 + \epsilon D_1 + \epsilon^2 D_2 + \cdots \tag{5.45a}$$

$$\frac{d^2}{dt^2} = (D_0 + \epsilon D_1 + \epsilon^2 D_2 + \cdots)^2$$

$$= D_0^2 + 2\epsilon D_0 D_1 + \epsilon^2(2D_0 D_2 + D_1^2) + \cdots \tag{5.45b}$$

where

$$D_n = \frac{\partial}{\partial T_n} \tag{5.46}$$

Then, (5.2) becomes

$$[D_0^2 + 2\epsilon D_0 D_1 + \epsilon^2(2D_0 D_2 + D_1^2) + \cdots]u + 2\epsilon(D_0 + \epsilon D_1 + \epsilon^2 D_2 + \cdots)u + u = 0 \tag{5.47}$$

Thus, the original ordinary-differential equation (5.2) is transformed into the partial-differential equation (5.47)

We seek a solution for (5.47) in powers of ϵ in the form

$$u = u_0(T_0, T_1, T_2) + \epsilon u_1(T_0, T_1, T_2) + \epsilon^2 u_2(T_0, T_1, T_2) + \cdots \tag{5.48}$$

Substituting (5.48) into (5.47) and collecting coefficients of equal powers of ϵ, we have

THE METHOD OF MULTIPLE SCALES 143

$$D_0^2 u_0 + u_0 + \epsilon[D_0^2 u_1 + u_1 + 2D_0 D_1 u_0 + 2D_0 u_0] + \epsilon^2[D_0^2 u_2 + u_2 + 2D_0 D_2 u_0$$
$$+ D_1^2 u_0 + 2D_0 D_1 u_1 + 2D_0 u_1 + 2D_1 u_0] + \cdots = 0 \quad (5.49)$$

Equating each of the coefficients of ϵ to zero yields

$$D_0^2 u_0 + u_0 = 0 \quad (5.50)$$

$$D_0^2 u_1 + u_1 = -2D_0 D_1 u_0 - 2D_0 u_0 \quad (5.51)$$

$$D_0^2 u_2 + u_2 = -2D_0 D_2 u_0 - D_1^2 u_0 - 2D_0 D_1 u_1 - 2D_0 u_1 - 2D_1 u_0 \quad (5.52)$$

As discussed in Section 4.5, it is convenient to express the solution of (5.50) in the complex form

$$u_0 = A(T_1, T_2)e^{iT_0} + \bar{A}(T_1, T_2)e^{-iT_0} \quad (5.53)$$

Then, (5.51) becomes

$$D_0^2 u_1 + u_1 = -2i(D_1 A + A)e^{iT_0} + cc \quad (5.54)$$

where cc stands for the complex conjugate of the preceding terms; in this case, it stands for $2i(D_1 \bar{A} + \bar{A}) \exp(-iT_0)$. Thus, this notation serves to display long expressions efficiently, and hence, it is used frequently.

The condition for the elimination of secular terms from u_1 demands the vanishing of each of the coefficients of $\exp(iT_0)$ and $\exp(-iT_0)$ independently. The vanishing of the coefficient of $\exp(iT_0)$ yields

$$D_1 A + A = 0 \quad (5.55)$$

The vanishing of the coefficient of $\exp(-iT_0)$ yields the complex conjugate of (5.55), and hence, it does not yield a new condition. Once (5.55) is satisfied, its complex conjugate is automatically satisfied. With (5.55), (5.54) becomes

$$D_0^2 u_1 + u_1 = 0 \quad (5.56)$$

In keeping up with our approach of not including the homogeneous solutions except in the lowest-order problem, we write the solution of (5.56) as

$$u_1 = 0 \quad (5.57)$$

because (5.56) is homogeneous. The solution of (5.55) is

$$A = B(T_2)e^{-T_1} \quad (5.58)$$

where $B(T_2)$, the constant of integration as far as the derivative with respect to T_1 is concerned, is a function of T_2, and it is determined by eliminating the terms that produce secular terms in the second-order problem, the problem for u_2.

Substituting for u_0 and u_1 into (5.52), we have

$$D_0^2 u_2 + u_2 = -(2iD_2 A + D_1^2 A + 2D_1 A)e^{iT_0} + cc \quad (5.59)$$

144 THE LINEAR OSCILLATOR

The condition for the elimination of secular terms from u_2 demands the vanishing of the coefficient of $\exp(iT_0)$. That is,

$$2iD_2 A + D_1^2 A + 2D_1 A = 0 \qquad (5.60)$$

Substituting for A from (5.58) into (5.60) yields

$$2iD_2 Be^{-T_1} - Be^{-T_1} = 0$$

or

$$2iD_2 B - B = 0 \qquad (5.61)$$

whose solution is

$$B = ce^{-(1/2)iT_2} \qquad (5.62)$$

where c is a complex constant because B is a function of T_2 only.

Putting B in (5.58), we have

$$A = ce^{-T_1 - (1/2)iT_2} \qquad (5.63)$$

Then, (5.53) becomes

$$u_0 = ce^{-T_1 + i[T_0 - (1/2)T_2]} + \bar{c}e^{-T_1 - i[T_0 - (1/2)T_2]}$$

Expressing c in the polar form $\tfrac{1}{2} a \exp(i\beta)$, we rewrite u_0 as

$$u_0 = \tfrac{1}{2} a e^{-T_1} e^{i[T_0 - (1/2)T_2 + \beta]} + \tfrac{1}{2} a e^{-T_1} e^{-i[T_0 - (1/2)T_2 + \beta]}$$

$$= \tfrac{1}{2} a e^{-T_1} \{ e^{i[T_0 - (1/2)T_2 + \beta]} + e^{-i[T_0 - (1/2)T_2 + \beta]} \}$$

$$= a e^{-T_1} \cos(T_0 - \tfrac{1}{2} T_2 + \beta)$$

In terms of the original variable, u_0 becomes

$$u_0 = a e^{-\epsilon t} \cos(t - \tfrac{1}{2}\epsilon^2 t + \beta) \qquad (5.64)$$

Substituting for u_0 and u_1 from (5.64) and (5.57) into (5.48), we obtain

$$u = a e^{-\epsilon t} \cos(t - \tfrac{1}{2}\epsilon^2 t + \beta) + O(\epsilon^2) \qquad (5.65)$$

which is uniform and in agreement with (5.44) to $O(\epsilon^2)$. Therefore, the method of multiple scales is effective in determining the transient response as well as determining the approximation to the frequency of the system.

5.5. The Method of Averaging

When $\epsilon = 0$, the solution of (5.2) can be written as

$$u = a \cos(t + \beta) \qquad (5.66)$$

where a and β are constants. Differentiating (5.66) with respect to t yields

$$\dot{u} = -a \sin(t + \beta) \qquad (5.67)$$

THE METHOD OF AVERAGING 145

When $\epsilon \neq 0$, we still represent the solution by (5.66) subject to the constraint (5.67) but with time varying a and β.

Since $a = a(t)$ and $\beta = \beta(t)$, differentiating (5.66) with respect to t gives

$$\dot{u} = \dot{a} \cos(t + \beta) - a(1 + \dot{\beta}) \sin(t + \beta)$$
$$= -a \sin(t + \beta) + \dot{a} \cos(t + \beta) - a\dot{\beta} \sin(t + \beta) \qquad (5.68)$$

Comparing (5.67) and (5.68), we conclude that

$$\dot{a} \cos(t + \beta) - a\dot{\beta} \sin(t + \beta) = 0 \qquad (5.69)$$

Differentiating (5.67) with respect to t gives

$$\ddot{u} = -\dot{a} \sin(t + \beta) - a(1 + \dot{\beta}) \cos(t + \beta) \qquad (5.70)$$

Substituting (5.66), (5.67), and (5.70) into (5.2), we obtain

$$-\dot{a} \sin(t + \beta) - a \cos(t + \beta) - a\dot{\beta} \cos(t + \beta) - 2\epsilon a \sin(t + \beta) + a \cos(t + \beta) = 0$$

or

$$\dot{a} \sin(t + \beta) + a\dot{\beta} \cos(t + \beta) = -2\epsilon a \sin(t + \beta) \qquad (5.71)$$

Multiplying (5.69) by $\cos(t + \beta)$ and (5.71) by $\sin(t + \beta)$, adding the results, and noting that $\sin^2 \theta + \cos^2 \theta = 1$, we obtain

$$\dot{a} = -2\epsilon a \sin^2(t + \beta) = -\epsilon a + \epsilon a \cos(2t + 2\beta) \qquad (5.72)$$

Substituting for $\dot{a}$ into (5.69) gives

$$-2\epsilon a \sin^2(t + \beta) \cos(t + \beta) - a\dot{\beta} \sin(t + \beta) = 0 \qquad (5.73)$$

When $a \neq 0$, it follows from (5.73) that

$$\dot{\beta} = -2\epsilon \sin(t + \beta) \cos(t + \beta) = -\epsilon \sin(2t + 2\beta) \qquad (5.74)$$

For small ϵ, we have two approaches as discussed in Section 4.6. First, we can average (5.72) and (5.74) over the period of the circular functions on the right-hand sides of these equations to determine a first approximation. Second, we can keep only the slowly varying terms in (5.72) and (5.74). Applying the latter approach, we obtain to the first approximation

$$\dot{a} = -\epsilon a \qquad (5.75)$$

$$\dot{\beta} = 0 \qquad (5.76)$$

The solutions of (5.75) and (5.76) are

$$a = a_0 e^{-\epsilon t} \quad \text{and} \quad \beta = \beta_0 \qquad (5.77)$$

where a_0 and β_0 are constants. Then, it follows from (5.66) that

$$u = a_0 e^{-\epsilon t} \cos(t + \beta_0) \qquad (5.78)$$

146 THE LINEAR OSCILLATOR

which is in agreement with (5.44) up to $O(\epsilon)$. Therefore, the first approximation of the method of averaging is effective in determining the transient response, but it is incapable of determining the corrections to the frequency, which are of higher order than the first. Thus, to determine these corrections, one needs to carry out the solutions of (5.72) and (5.74) to higher order; this is accomplished by using the generalized method of averaging (Section 7.6).

Exercises

5.1. Consider the equation

$$\ddot{u} + 2\epsilon\mu\dot{u} + u + \epsilon u^3 = 0 \quad \epsilon \ll 1$$

Use the methods of multiple scales and averaging to determine a first-order uniform expansion for u.

5.2. Consider the equation

$$\ddot{u} + \omega_0^2 u + \epsilon \dot{u}^3 = 0 \quad \epsilon \ll 1$$

Use the methods of multiple scales and averaging to determine a first-order uniform expansion for u.

5.3. Consider the following equation:

$$\ddot{u} + \omega_0^2 u + 2\epsilon\mu u^2 \dot{u} + \epsilon u^3 = 0$$

Show that to the first approximation

$$u = a \cos(\omega_0 t + \beta) + O(\epsilon)$$

and determine the equations governing a and β by using the methods of multiple scales and averaging.

5.4. Use the methods of multiple scales and averaging to determine a first-order uniform expansion for the general solution of

$$\ddot{\theta} + \omega^2 \sin\theta + \frac{4\sin^2\theta}{1 + 4(1 - \cos\theta)}\dot{\theta} = 0$$

for small but finite θ.

5.5. Consider the equation

$$\ddot{u} + \omega_0^2 u + \frac{\mu\dot{u}}{1 - u^2} = 0$$

Determine a first-order uniform expansion for small u.

5.6. Use the methods of multiple scales and averaging to determine a first-order uniform expansion for

$$\ddot{u} + u + \epsilon\dot{u}^5 = 0 \quad \epsilon \ll 1$$

CHAPTER 6

Self-Excited Oscillators

In contrast with the preceding chapter, which deals with systems possessing positive damping, we consider in this chapter systems with negative damping. Specifically, we consider self-excited systems having a single degree of freedom. Such systems are governed by equations of the form

$$m\frac{d^2u^*}{dt^{*2}} + ku^* = \mu f^*\left(u^*, \frac{du^*}{dt^*}\right)\frac{du^*}{dt^*} \qquad (6.1)$$

where μ is a positive parameter and f^* is positive for small u^*.

To simplify the algebra, we consider the following special equation:

$$m\frac{d^2u^*}{dt^{*2}} + ku^* = \mu\left[1 - \alpha\left(\frac{du^*}{dt^*}\right)^2\right]\frac{du^*}{dt^*} \qquad (6.2)$$

where α is positive. This equation is usually called the *Rayleigh equation*. As discussed earlier, we need to express (6.2) in dimensionless form before carrying out the analysis. To this end, we use a characteristic displacement u_0^* and the linear natural frequency $\omega_0 = \sqrt{k/m}$, as reference quantities, and define the following dimensionless variables without the asterisks:

$$u = \frac{u^*}{u_0^*} \qquad t = t^*\sqrt{k/m}$$

Then, (6.2) becomes

$$\ddot{u} + u = \epsilon\left(1 - \frac{\alpha u_0^{*2}k}{m}\dot{u}^2\right)\dot{u} \qquad (6.3)$$

where $\epsilon = \mu/\sqrt{km}$. We choose u_0^* so that $\alpha u_0^{*2}k = \frac{1}{3}m$ and (6.2) can be put in the standard form

$$\ddot{u} + u = \epsilon(\dot{u} - \tfrac{1}{3}\dot{u}^3) \qquad (6.4)$$

Differentiating (6.4) with respect to t gives

147

148 SELF-EXCITED OSCILLATORS

$$\ddot{u} + u = \epsilon(\dot{u} - \dot{u}^2 \dot{u}) \qquad (6.5)$$

If we let $\dot{u} = v$, we can rewrite (6.5) in the form

$$\dot{v} + v = \epsilon(1 - v^2)\dot{v} \qquad (6.6)$$

This equation is usually called the *van der Pol equation*.

In this chapter, we describe techniques of determining approximate solutions of (6.4), and hence, of (6.6) for small ϵ. We start by determining a straightforward expansion and discussing its uniformity. We show that neither the Lindstedt-Poincaré technique nor the method of renormalization is capable of yielding the transient response. Then, we show that the methods of multiple scales and averaging can yield the transient response.

6.1. The Straightforward Expansion

We seek a first-order straightforward expansion for the solution of (6.4) in the form

$$u(t; \epsilon) = u_0(t) + \epsilon u_1(t) + \cdots \qquad (6.7)$$

Substituting (6.7) into (6.4) yields

$$\ddot{u}_0 + \epsilon \ddot{u}_1 + \cdots + u_0 + \epsilon u_1 + \cdots = \epsilon(\dot{u}_0 + \epsilon \dot{u}_1 + \cdots) - \tfrac{1}{3} \epsilon(\dot{u}_0 + \epsilon \dot{u}_1 + \cdots)^3 \qquad (6.8)$$

Using the binomial expansion theorem and keeping terms up to $O(\epsilon)$ in (6.8), we have

$$\ddot{u}_0 + u_0 + \epsilon(\ddot{u}_1 + u_1) + \cdots = \epsilon(\dot{u}_0 - \tfrac{1}{3} \dot{u}_0^3) + \cdots \qquad (6.9)$$

Equating coefficients of like powers of ϵ on both sides of (6.9) gives

$$\ddot{u}_0 + u_0 = 0 \qquad (6.10)$$

$$\ddot{u}_1 + u_1 = \dot{u}_0 - \tfrac{1}{3} \dot{u}_0^3 \qquad (6.11)$$

which can be solved sequentially for u_0 and u_1.

The general solution of (6.10) can be written as

$$u_0 = a \cos(t + \beta) \qquad (6.12)$$

where a and β are constants. Then, (6.11) becomes

$$\ddot{u}_1 + u_1 = -a \sin(t + \beta) + \tfrac{1}{3} a^3 \sin^3(t + \beta) \qquad (6.13)$$

Using the trigonometric identity (A16)

$$\sin^3 \theta = \tfrac{3}{4} \sin \theta - \tfrac{1}{4} \sin 3\theta$$

we rewrite (6.13) as

$$\ddot{u}_1 + u_1 = (\tfrac{1}{4} a^2 - 1)a \sin(t + \beta) - \tfrac{1}{12} a^3 \sin(3t + 3\beta) \qquad (6.14)$$

Since (6.14) is linear, a particular solution can be obtained as the sum of two particular solutions corresponding to the two inhomogeneous terms. That is, a particular solution of (6.14) can be obtained as the sum of any particular solution of

$$\ddot{u}_1^{(1)} + u_1^{(1)} = (\tfrac{1}{4} a^2 - 1)a \sin(t + \beta) \tag{6.15}$$

and any particular solution of

$$\ddot{u}_1^{(2)} + u_1^{(2)} = -\tfrac{1}{12} a^3 \sin(3t + 3\beta) \tag{6.16}$$

A particular solution of (6.15) is (B81 and B82)

$$u_{1_p}^{(1)} = \tfrac{1}{2}(1 - \tfrac{1}{4} a^2) at \cos(t + \beta) \tag{6.17}$$

whereas a particular solution of (6.16) is (B68 and B69)

$$u_{1_p}^{(2)} = \tfrac{1}{96} a^3 \sin(3t + 3\beta) \tag{6.18}$$

Hence,

$$u_{1_p} = u_{1_p}^{(1)} + u_{2_p}^{(2)} = \tfrac{1}{2}(1 - \tfrac{1}{4} a^2) at \cos(t + \beta) + \tfrac{1}{96} a^3 \sin(3t + 3\beta) \tag{6.19}$$

Since the solution of the homogeneous equation (6.14) is not needed if a and β are considered functions of ϵ according to Section 4.1, the solution of (6.14) is given by (6.19). Substituting (6.12) and (6.19) into (6.7), we obtain

$$u = a \cos(t + \beta) + \epsilon[\tfrac{1}{2}(1 - \tfrac{1}{4} a^2) at \cos(t + \beta) + \tfrac{1}{96} a^3 \sin(3t + 3\beta)] + \cdots \tag{6.20}$$

This straightforward expansion is nonuniform for $t \geqslant O(\epsilon^{-1})$ because the correction term is the order or larger than the first term owing to the presence of the mixed-secular term. This nonuniformity is illustrated in Figure 6-1, which compares (6.20) with solutions obtained by numerically integrating (6.4). Initially, the straightforward and numerical solutions are in agreement. But as t increases, the analytical solution deviates more and more from the numerical solution, which approaches a periodic solution having an amplitude of approximately two, irrespective of the initial conditions. The periodic solution approached by the numerical solution is called a *limit cycle*. Figure 6-2 shows the numerical solutions in the phase plane (uv plane, where $v = \dot{u}$) for several values of ϵ and initial conditions. When ϵ is small, the limit cycle has an amplitude of approximately two, irrespective of the initial conditions.

We should note that the mixed-secular term in (6.20) disappears if

$$(1 - \tfrac{1}{4} a^2)a = 0 \tag{6.21}$$

Excluding the trivial case $a = 0$, (6.21) is satisfied and the mixed-secular term disappears from (6.20) if $a = \pm 2$. If the amplitude is defined to be positive, the mixed-secular term disappears from (6.20) if $a = 2$ and (6.20) becomes

$$u = 2 \cos(t + \beta) + \tfrac{1}{12} \epsilon \sin(3t + 3\beta) + \cdots \tag{6.22}$$

which is periodic and to first-order has an amplitude of 2. This is the limit cycle.

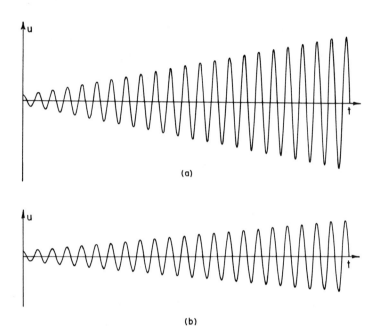

Figure 6-1. Comparison of straightforward expansion (a) with exact solution (b) for $u(0) = 0.5$, $\dot{u}(0) = 0$, and $\epsilon = 0.1$.

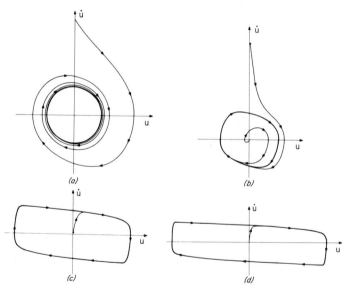

Figure 6-2. Phase planes for Rayleigh's equation: (a) $\epsilon = 0.01$; (b) $\epsilon = 0.1$; (c) $\epsilon = 1$; (d) $\epsilon = 10.0$.

6.2. The Method of Renormalization

In this section, we apply the method of renormalization to (6.20). Thus, we let

$$\tau = \omega t = (1 + \epsilon\omega_1 + \cdots)t \tag{6.23}$$

Hence,

$$t = \tau(1 + \epsilon\omega_1 + \cdots)^{-1} = \tau - \epsilon\omega_1\tau + \cdots \tag{6.24}$$

Substituting (6.24) into (6.20), we have

$$u = a\cos(\tau + \beta - \epsilon\omega_1\tau + \cdots) + \epsilon[\tfrac{1}{2}(1 - \tfrac{1}{4}a^2)a(\tau - \epsilon\omega_1\tau + \cdots)$$
$$\cdot \cos(\tau + \beta - \epsilon\omega_1\tau + \cdots) + \tfrac{1}{96}a^3\sin(3\tau + 3\beta - 3\epsilon\omega_1\tau + \cdots)] + \cdots \tag{6.25}$$

Using Taylor series expansions such as those in (4.86), we rewrite (6.25) as

$$u = a\cos(\tau + \beta) + \epsilon[\omega_1 a\tau \sin(\tau + \beta) + \tfrac{1}{2}(1 - \tfrac{1}{4}a^2)a\tau \cos(\tau + \beta)$$
$$+ \tfrac{1}{96}a^3\sin(3\tau + 3\beta)] + \cdots \tag{6.26}$$

If ω_1 is real, the elmination of the secular terms from (6.26) demands that

$$\omega_1 a = 0 \tag{6.27}$$

$$(1 - \tfrac{1}{4}a^2)a = 0 \tag{6.28}$$

Since $a \neq 0$ for a nontrivial solution, it follows from (6.27) and (6.28) that $\omega_1 = 0$ and $a = 2$. (The case $a = -2$ can be disregarded if the amplitude is defined to be positive.) Then, it follows from (6.23) that $\tau = t + O(\epsilon^2)$ and from (6.26) that

$$u = 2\cos(t + \beta) + \tfrac{1}{12}\epsilon \sin(3t + 3\beta) + \cdots \tag{6.29}$$

which is the limit cycle.

The question arises whether permitting the ω_n to be complex will lead to a uniform expansion as in Section 5.3. To answer this question, we express (6.26) in complex form. To this end, we note that

$$\cos\theta = \tfrac{1}{2}(e^{i\theta} + e^{-i\theta}) \quad \sin\theta = -\frac{i}{2}(e^{i\theta} - e^{-i\theta})$$

Then, we rewrite (6.26) as

$$u = \tfrac{1}{2}ae^{i(\tau+\beta)} + \epsilon\left\{\tfrac{1}{2}[\tfrac{1}{2}(1 - \tfrac{1}{4}a^2) - i\omega_1]a\tau e^{i(\tau+\beta)} - \frac{i}{192}a^3 e^{3i(\tau+\beta)}\right\}$$
$$+ cc + \cdots \tag{6.30}$$

Eliminating the secular term in (6.30) demands that

$$\omega_1 = -\tfrac{1}{2}i(1 - \tfrac{1}{4}a^2) \tag{6.31}$$

152 SELF-EXCITED OSCILLATORS

Then,
$$u = \tfrac{1}{2} a e^{i(\tau + \beta)} + cc + \cdots \tag{6.32}$$

It follows from (6.23) and (6.31) that
$$\tau = t - \tfrac{1}{2} i\epsilon(1 - \tfrac{1}{4} a^2)t + \cdots \tag{6.33}$$

Hence, (6.32) becomes
$$u = \tfrac{1}{2} a \exp \{i[t - \tfrac{1}{2} i\epsilon(1 - \tfrac{1}{4} a^2)t + \beta + \cdots]\} + cc + \cdots$$
$$= \tfrac{1}{2} a \exp [\tfrac{1}{2} \epsilon(1 - \tfrac{1}{4} a^2)t] \exp [i(t + \beta)] + cc + \cdots$$

Therefore,
$$u = a \exp [\tfrac{1}{2} \epsilon(1 - \tfrac{1}{4} a^2)t] \cos (t + \beta) + \cdots \tag{6.34}$$

Equation (6.34) shows that $u \to \infty$ as $t \to \infty$ if $|a| < 2$ and that $u \to 0$ as $t \to \infty$ if $|a| > 2$. This result is erroneous because Figures 6-1 and 6-2 show that the numerical solutions of (6.4) approach approximately two, irrespective of the initial conditions, and hence, the value of a. Hence, the modification of the Lindstedt-Poincaré technique or the method of renormalization by allowing the ω_n to be complex may lead to erroneous results as in this case. Therefore, one should avoid applying either of these techniques to determine other than periodic solutions.

6.3. The Method of Multiple Scales

To determine a uniform first-order expansion for the solution of (6.4) by using the method of multiple scales, we introduce the scales $T_0 = t$ and $T_1 = \epsilon t$. Then, the derivatives with respect to t transform into
$$\frac{d}{dt} = D_0 + \epsilon D_1 + \cdots$$
$$\frac{d^2}{dt^2} = D_0^2 + 2\epsilon D_0 D_1 + \cdots$$

where $D_n = \partial/\partial T_n$. Hence, (6.4) becomes
$$D_0^2 u + 2\epsilon D_0 D_1 u + u = \epsilon[D_0 u - \tfrac{1}{3}(D_0 u)^3] + \cdots \tag{6.35}$$

We seek a solution of (6.35) in the form
$$u = u_0(T_0, T_1) + \epsilon u_1(T_0, T_1) + \cdots \tag{6.36}$$

Substituting (6.36) into (6.35) and equating the coefficients of ϵ^0 and ϵ on both sides, we obtain
$$D_0^2 u_0 + u_0 = 0 \tag{6.37}$$

THE METHOD OF MULTIPLE SCALES 153

$$D_0^2 u_1 + u_1 = -2D_0 D_1 u_0 + D_0 u_0 - \tfrac{1}{3}(D_0 u_0)^3 \qquad (6.38)$$

As before, the general solution of (6.37) is expressed in the following complex form:

$$u_0 = A(T_1)e^{iT_0} + \bar{A}(T_1)e^{-iT_0} \qquad (6.39)$$

Then, (6.38) becomes

$$D_0^2 u_1 + u_1 = -2iA'e^{iT_0} + 2i\bar{A}'e^{-iT_0} + iAe^{iT_0} - i\bar{A}e^{-iT_0} - \tfrac{1}{3}(iAe^{iT_0} - i\bar{A}e^{-iT_0})^3$$

or

$$D_0^2 u_1 + u_1 = -i(2A' - A + A^2\bar{A})e^{iT_0} + \tfrac{1}{3}iA^3 e^{3iT_0} + cc \qquad (6.40)$$

Eliminating the secular terms from u_1 demands that

$$2A' - A + A^2\bar{A} = 0 \qquad (6.41)$$

As before, we express A in the polar form

$$A = \tfrac{1}{2} a e^{i\beta} \qquad (6.42)$$

where a and β are real functions of T_1. Then, (6.39) becomes

$$u_0 = \tfrac{1}{2} a e^{i(T_0 + \beta)} + \tfrac{1}{2} a e^{-i(T_0 + \beta)}$$

or

$$u_0 = a \cos(T_0 + \beta) \qquad (6.43)$$

Substituting (6.42) into (6.41) gives

$$a'e^{i\beta} + ia\beta' e^{i\beta} - \tfrac{1}{2} a e^{i\beta} + \tfrac{1}{8} a^3 e^{i\beta} = 0$$

Dividing by the factor $\exp(i\beta)$ and separating real and imaginary parts, we obtain

$$a' = \tfrac{1}{2} a - \tfrac{1}{8} a^3 \qquad (6.44)$$

$$\beta' = 0 \qquad (6.45)$$

The solution of (6.45) is

$$\beta = \beta_0 = \text{constant} \qquad (6.46)$$

The solution of (6.44) can be obtained by separation of variables, that is,

$$dT_1 = \frac{8 da}{4a - a^3} = \frac{8 da}{a(2-a)(2+a)} \qquad (6.47)$$

Expressing the right-hand side of (6.47) in partial fractions, we obtain

$$dT_1 = \frac{2 da}{a} + \frac{da}{2-a} - \frac{da}{2+a} \qquad (6.48)$$

Integrating (6.48) yields

154 SELF-EXCITED OSCILLATORS

$$T_1 + c = 2 \log a - \log |2 - a| - \log (2 + a)$$

or

$$T_1 + c = \log \frac{a^2}{|4 - a^2|}$$

where c is a constant. Hence,

$$\frac{a^2}{4 - a^2} = e^{T_1 + c} = e^{\epsilon t + c} \qquad (6.49)$$

Solving (6.49) for a^2 gives

$$a^2 = \frac{4 \exp(\epsilon t + c)}{1 + \exp(\epsilon t + c)} = \frac{4}{1 + \exp(-\epsilon t - c)} \qquad (6.50)$$

Substituting (6.46) and (6.50) into (6.43) and setting $T_0 = t$, we have

$$u_0 = 2[1 + e^{-\epsilon t - c}]^{-1/2} \cos(t + \beta_0) \qquad (6.51)$$

Substituting (6.51) into (6.36), we obtain the following first-order expansion for the general solution of (6.4):

$$u = 2[1 + \exp(-\epsilon t - c)]^{-1/2} \cos(t + \beta_0) + \cdots \qquad (6.52)$$

Using the initial conditions

$$u(0) = a_0 \qquad \dot{u}(0) = 0 \qquad (6.53)$$

we find from (6.52) that

$$a_0 = 2[1 + e^{-c}]^{-1/2} \cos \beta_0 \qquad (6.54)$$

$$0 = -2[1 + e^{-c}]^{-1/2} \sin \beta_0 + O(\epsilon) \qquad (6.55)$$

It follows from (6.55) that $\sin \beta_0 = O(\epsilon)$ or $\beta_0 = 0 + O(\epsilon)$. Then, it follows from (6.54) that

$$a_0^2 = 4[1 + e^{-c}]^{-1} \qquad (6.56)$$

Solving (6.56) for $\exp(-c)$ gives

$$e^{-c} = \frac{4}{a_0^2} - 1 \qquad (6.57)$$

Therefore, (6.52) becomes

$$u = 2\left[1 + \left(\frac{4}{a_0^2} - 1\right) e^{-\epsilon t}\right]^{-1/2} \cos t + \cdots \qquad (6.58)$$

Equation (6.58) shows that

$$u \to 2 \cos t + O(\epsilon) \qquad (6.59)$$

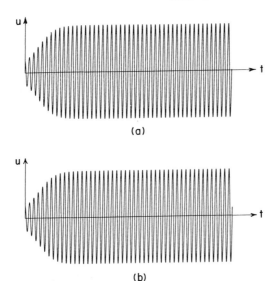

Figure 6-3. Comparison of the approximate solution (b) with the exact solution (a) for $u(0) = 0.5$, $\dot{u}(0) = 0$, and $\epsilon = 0.1$.

as $t \to \infty$, irrespective of the value of a_0 as long as it is different from zero. This result is in agreement with the numerical solutions in Figures 6-1 and 6-3. Figure 6-3 shows that (6.58) is in good agreement with the numerical solutions of (6.4).

Equation (6.58) clearly shows that u does not have the form $a \cos(t + \beta)$ where a is an exponential function of time. The Lindstedt-Poincaré technique and the method of renormalization, even when ω is complex, force a to be an exponential function. Thus, these approaches cannot be expected to yield a good approximation of the solution. On the other hand, with the method of multiple scales, changing the original ordinary-differential equation into a system of partial-differential equations permits enough generality in the form of the solution to obtain an excellent approximation.

6.4. The Method of Averaging

As in Sections 4.6 and 5.5, to apply the method of averaging to (6.4), we need to use the method of variation of parameters and introduce the transformation

$$u(t; \epsilon) = a(t) \cos[t + \beta(t)] \qquad (6.60)$$

$$\dot{u}(t; \epsilon) = -a(t) \sin[t + \beta(t)] \qquad (6.61)$$

As before, we differentiate (6.60) with respect to t and obtain

$$\dot{u} = -a\sin(t+\beta) + \dot{a}\cos(t+\beta) - a\dot{\beta}\sin(t+\beta) \qquad (6.62)$$

It follows from (6.61) and (6.62) that

$$\dot{a}\cos(t+\beta) - a\dot{\beta}\sin(t+\beta) = 0 \qquad (6.63)$$

Differentiating (6.61) with respect to t gives

$$\ddot{u} = -a\cos(t+\beta) - \dot{a}\sin(t+\beta) - a\dot{\beta}\cos(t+\beta) \qquad (6.64)$$

Substituting (6.60), (6.61), and (6.64) into (6.4) yields

$$\dot{a}\sin(t+\beta) + a\dot{\beta}\cos(t+\beta) = \epsilon a\sin(t+\beta) - \tfrac{1}{3}\epsilon a^3 \sin^3(t+\beta) \qquad (6.65)$$

Solving (6.63) and (6.65) for $\dot{a}$ and $\dot{\beta}$ gives the following variational equations:

$$\dot{a} = \epsilon[a\sin(t+\beta) - \tfrac{1}{3}a^3 \sin^3(t+\beta)]\sin(t+\beta) \qquad (6.66)$$

$$a\dot{\beta} = \epsilon[a\sin(t+\beta) - \tfrac{1}{3}a^3 \sin^3(t+\beta)]\cos(t+\beta) \qquad (6.67)$$

Using the trigonometric identities

$$\sin^2\theta = \tfrac{1}{2} - \tfrac{1}{2}\cos 2\theta$$

$$\sin^4\theta = \tfrac{1}{8}(\cos 4\theta - 4\cos 2\theta + 3)$$

$$\sin\theta\cos\theta = \tfrac{1}{2}\sin 2\theta$$

$$\sin^3\theta\cos\theta = \tfrac{1}{8}(2\sin 2\theta - \sin 4\theta)$$

We rewrite (6.66) and (6.67) as

$$\dot{a} = \epsilon a[\tfrac{1}{2}(1-\tfrac{1}{4}a^2) - \tfrac{1}{2}(1-\tfrac{1}{3}a^2)\cos(2t+2\beta) - \tfrac{1}{24}a^2\cos(4t+4\beta)] \qquad (6.68)$$

$$\dot{\beta} = \epsilon[\tfrac{1}{2}(1-\tfrac{1}{6}a^2)\sin(2t+2\beta) + \tfrac{1}{24}a^2\sin(4t+4\beta)] \qquad (6.69)$$

The assumption $a \neq 0$ is used in arriving at (6.69). To the first approximation, we keep only the slowly varying terms, that is, the terms independent of the circular functions. Hence, to the first approximation, (6.68) and (6.69) are replaced by the following averaged equations:

$$\dot{a} = \tfrac{1}{2}\epsilon a(1 - \tfrac{1}{4}a^2) \qquad (6.70)$$

$$\dot{\beta} = 0 \qquad (6.71)$$

which are in full agreement with (6.44) and (6.45) obtained by using the method of multiple scales.

Figure 6-4 compares the solutions of the variational equations (6.68) and (6.69) with the solutions of the averaged equation (6.70) for the initial conditions

$$a(0) = a_0 \qquad \beta(0) = 0 \qquad (6.72)$$

It is clear that the solutions of the averaged equations are averages of the solutions of the variational equations.

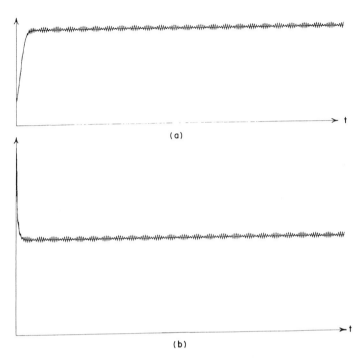

Figure 6-4. Comparison of solutions of the variational equations (6.68) and (6.69) with those of the averaged equation (6.70) for $\epsilon = 0.3$: (a) $a(0) = 0.5$, $\beta(0) = 0$; (b) $a(0) = 4.0$, $\beta(0) = 0$.

Exercises

6.1. Consider van der Pol's equation

$$\ddot{u} + u = \epsilon(1 - u^2)\dot{u}$$

(a) Determine two terms in the straightforward expansion and discuss its uniformity.
(b) Use the method of renormalization to render the straightforward expansion uniformly valid.
(c) Use the methods of multiple scales and averaging to determine a first-order uniform expansion.
(d) Compare the results in (b) and (c) and indicate the limitations of the method of renormalization.

6.2. Use the methods of multiple scales and averaging to determine a first-order uniform expansion including the transient response of the solution of

$$\ddot{u} + \omega_0^2 u = \epsilon[\dot{u} - \dot{u}^3 + \dot{u}^2 u] \qquad \epsilon \ll 1$$

158 SELF-EXCITED OSCILLATORS

6.3. Consider the equation

$$\ddot{x} + x + \dot{x} - \tfrac{1}{2}(\dot{x} - |\dot{x}|)\delta(x - x_0) = 0$$

where x_0 is a constant and δ is the Dirac delta function. Determine a first-order uniform expansion for small but finite x.

6.4. Consider the equations

$$\ddot{u} + \omega_0^2 u = 2\epsilon[(1 - v)\dot{u} - \dot{v}u]$$

$$\dot{v} + v = u^2$$

Determine a first-order uniform expansion for u and v.

6.5. Use the methods of averaging and multiple scales to determine a first-order uniform expansion for

$$\ddot{u} + u = \epsilon(1 - u^4)\dot{u}$$

when $\epsilon \ll 1$.

6.6. Use the methods of multiple scales and averaging to determine a first-order uniform expansion for

$$\ddot{u} + u - \epsilon(1 - u^2)\dot{u} + \epsilon u^3 = 0$$

when $\epsilon \ll 1$.

CHAPTER 7

Systems with Quadratic and Cubic Nonlinearities

We consider the free oscillations of a particle of mass m under the action of gravity and restrained by a nonlinear spring, as shown in Figure 7-1. The equation of motion is

$$m \frac{d^2 x^*}{dt^{*2}} + f(x^*) = mg \qquad (7.1)$$

where g is the gravitational acceleration and $f(x^*)$ is the restraining force of the spring. We assume that $f(x^*)$ is a cubic function of x^*, that is

$$f(x^*) = k_1 x^* + k_3 x^{*3} \qquad (7.2)$$

where $k_1 > 0$. Substituting (7.2) into (7.1) gives

$$m \frac{d^2 x^*}{dt^{*2}} + k_1 x^* + k_3 x^{*3} = mg \qquad (7.3)$$

The equilibrium positions x_s^* can be obtained from (7.3) by dropping the acceleration term. The result is

$$k_1 x_s^* + k_3 x_s^{*3} = mg \qquad (7.4)$$

In this chapter, we investigate small oscillations about one of the equilibrium positions. To this end, we let

$$x^* = x_s^* + u^* \qquad (7.5)$$

in (7.3) and obtain

$$m \frac{d^2 u^*}{dt^{*2}} + k_1 (x_s^* + u^*) + k_3 (x_s^* + u^*)^3 = mg \qquad (7.6)$$

Expanding the cubic term and using (7.4), we rewrite (7.6) as

$$m \frac{d^2 u^*}{dt^{*2}} + (k_1 + 3 k_3 x_s^{*2}) u^* + 3 k_3 x_s^* u^{*2} + k_3 u^{*3} = 0 \qquad (7.7)$$

160 SYSTEMS WITH QUADRATIC AND CUBIC NONLINEARITIES

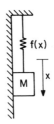

Figure 7-1. Mass restrained by a nonlinear spring in the presence of gravity force.

As before, we introduce the following dimensionless quantities:

$$u = u^*/x_s^* \qquad t = \omega t^*$$

where

$$\omega = \sqrt{(k_1 + 3k_3 x_s^{*2})/m}$$

is the linear natural frequency, which is assumed to be real. (We note that when the mass-spring system is oriented horizontally, the natural frequency is $\sqrt{k_1/m}$.) Then, (7.7) becomes

$$\ddot{u} + u + 3\alpha u^2 + \alpha u^3 = 0 \qquad (7.8)$$

where $\alpha = k_3 x_s^{*2}/m\omega^2$. In what follows, we assume that $\alpha = O(1)$. In contrast with the Duffing equation, (7.8) contains a quadratic as well as cubic term. Instead of (7.8), we consider a slightly more general equation, namely

$$\ddot{u} + u + \alpha_2 u^2 + \alpha_3 u^3 = 0 \qquad (7.9)$$

where α_2 and α_3 are constants.

In the next section, we determine a second-order straightforward expansion to the solutions of (7.9) for small but finite amplitudes. We render this expansion uniform in Sections 7.2 and 7.3 by using the method of renormalization and the Lindstedt-Poincaré technique. In Section 7.4, we determine a uniform second-order expansion by using the method of multiple scales. In Section 7.5, we show that the first approximation of the method of averaging yields an incomplete solution. In Section 7.6, we introduce the generalized method of averaging and obtain a uniform second-order expansion to (7.9). Finally, in Section 7.7, we introduce the Krylov-Bogoliubov-Mitropolsky technique.

7.1. The Straightforward Expansion

To carry out a straightforward expansion for small but finite amplitudes for (7.9), we need to introduce a small parameter because none appear explicitly in this equation. To this end, we seek an expansion in the form

$$u = \epsilon u_1(t) + \epsilon^2 u_2(t) + \epsilon^3 u_3(t) + \cdots \qquad (7.10)$$

where ϵ is a small dimensionless parameter that is a measure of the amplitude of oscillation. It can be used as a *bookkeeping* or *crutching device* and set equal to unity if the amplitude is taken to be small as described below.

Substituting (7.10) into (7.9) gives

$$\epsilon \ddot{u}_1 + \epsilon^2 \ddot{u}_2 + \epsilon^3 \ddot{u}_3 + \cdots + \epsilon u_1 + \epsilon^2 u_2 + \epsilon^3 u_3 + \cdots + \alpha_2 (\epsilon u_1 + \epsilon^2 u_2 + \epsilon^3 u_3 + \cdots)^2$$
$$+ \alpha_3 (\epsilon u_1 + \epsilon^2 u_2 + \epsilon^3 u_3 + \cdots)^3 = 0 \quad (7.11)$$

Using the binomial theorem to expand the terms in parentheses in (7.11) and keeping terms to $O(\epsilon^3)$ only, we obtain

$$\epsilon(\ddot{u}_1 + u_1) + \epsilon^2(\ddot{u}_2 + u_2 + \alpha_2 u_1^2) + \epsilon^3(\ddot{u}_3 + u_3 + 2\alpha_2 u_1 u_2 + \alpha_3 u_1^3) + \cdots = 0$$
$$(7.12)$$

Equating each of the coefficients of ϵ to zero in (7.12) yields

$$\ddot{u}_1 + u_1 = 0 \quad (7.13)$$

$$\ddot{u}_2 + u_2 = -\alpha_2 u_1^2 \quad (7.14)$$

$$\ddot{u}_3 + u_3 = -2\alpha_2 u_1 u_2 - \alpha_3 u_1^3 \quad (7.15)$$

which can be solved sequentially for u_1, u_2, and u_3.

The general solution of (7.13) can be expressed as

$$u_1 = a \cos(t + \beta) \quad (7.16)$$

where a and β are constants. Then, (7.14) becomes

$$\ddot{u}_2 + u_2 = -\alpha_2 a^2 \cos^2(t + \beta) = -\tfrac{1}{2} \alpha_2 a^2 - \tfrac{1}{2} \alpha_2 a^2 \cos(2t + 2\beta) \quad (7.17)$$

As before, we do not include the solution of the homogeneous problem for u_2. Moreover, a particular solution for (7.17) can be obtained as the sum of two particular solutions, one for each of the following equations:

$$\ddot{u}_2^{(1)} + u_2^{(1)} = -\tfrac{1}{2} \alpha_2 a^2 \quad (7.18)$$

$$\ddot{u}_2^{(2)} + u_2^{(2)} = -\tfrac{1}{2} \alpha_2 a^2 \cos(2t + 2\beta) \quad (7.19)$$

A particular solution of (7.18) is

$$u_{2p}^{(1)} = -\tfrac{1}{2} \alpha_2 a^2 \quad (7.20)$$

whereas a particular solution of (7.19) is (B68 and B69)

$$u_{2p}^{(2)} = \tfrac{1}{6} \alpha_2 a^2 \cos(2t + 2\beta) \quad (7.21)$$

Hence,

$$u_2 = u_{2p}^{(1)} + u_{2p}^{(2)} = -\tfrac{1}{2} \alpha_2 a^2 + \tfrac{1}{6} \alpha_2 a^2 \cos(2t + 2\beta) \quad (7.22)$$

Substituting (7.16) and (7.22) into (7.15) gives

$$\ddot{u}_3 + u_3 = -2\alpha_2 a \cos(t + \beta)[-\tfrac{1}{2} \alpha_2 a^2 + \tfrac{1}{6} \alpha_2 a^2 \cos(2t + 2\beta)] - \alpha_3 a^3 \cos^3(t + \beta)$$

or

$$\ddot{u}_3 + u_3 = (\tfrac{5}{6}\alpha_2^2 - \tfrac{3}{4}\alpha_3)a^3 \cos(t+\beta) - (\tfrac{1}{4}\alpha_3 + \tfrac{1}{6}\alpha_2^2)a^3 \cos(3t+3\beta) \quad (7.23)$$

Since (7.23) is linear,

$$u_3 = u_{3p}^{(1)} + u_{3p}^{(2)} \quad (7.24)$$

where $u_{3p}^{(1)}$ and $u_{3p}^{(2)}$ are particular solutions of the following equations:

$$\ddot{u}_3^{(1)} + u_3^{(1)} = (\tfrac{5}{6}\alpha_2^2 - \tfrac{3}{4}\alpha_3)a^3 \cos(t+\beta) \quad (7.25)$$

$$\ddot{u}_3^{(2)} + u_3^{(2)} = -(\tfrac{1}{4}\alpha_3 + \tfrac{1}{6}\alpha_2^2)a^3 \cos(3t+3\beta) \quad (7.26)$$

It follows (B69 and B76) that

$$u_{3p}^{(1)} = (\tfrac{5}{12}\alpha_2^2 - \tfrac{3}{8}\alpha_3)a^3 t \sin(t+\beta) \quad (7.27)$$

$$u_{3p}^{(2)} = \tfrac{1}{8}(\tfrac{1}{4}\alpha_3 + \tfrac{1}{6}\alpha_2^2)a^3 \cos(3t+3\beta) \quad (7.28)$$

Substituting (7.27) and (7.28) into (7.24) yields

$$u_3 = (\tfrac{5}{12}\alpha_2^2 - \tfrac{3}{8}\alpha_3)a^3 t \sin(t+\beta) + \tfrac{1}{8}(\tfrac{1}{4}\alpha_3 + \tfrac{1}{6}\alpha_2^2)a^3 \cos(3t+3\beta) \quad (7.29)$$

Substituting (7.16), (7.22), and (7.29) into (7.10) yields the following third-order [up to $O(\epsilon^3)$] straightforward expansion:

$$u = \epsilon a \cos(t+\beta) + \tfrac{1}{6}\alpha_2 \epsilon^2 a^2 [\cos(2t+2\beta) - 3]$$
$$+ \epsilon^3 a^3 [(\tfrac{5}{12}\alpha_2^2 - \tfrac{3}{8}\alpha_3) t \sin(t+\beta)$$
$$+ \tfrac{1}{8}(\tfrac{1}{4}\alpha_3 + \tfrac{1}{6}\alpha_2^2) \cos(3t+3\beta)] + \cdots \quad (7.30)$$

We note that the dependence of u on ϵ and a is in the combination ϵa. Thus, one can set $\epsilon = 1$ and consider a as the perturbation parameter. The straightforward expansion (7.30) breaks down for $t \geq O(\epsilon^{-1}a^{-1})$ because the second correction term is the same order or larger than the first correction term, owing to the presence of the mixed-secular term. Next, we use the method of renormalization to render (7.30) uniform.

7.2. The Method of Renormalization

As discussed in Sections 4.3 and 4.4, we can construct a uniform expansion for (7.9) either by applying the Lindstedt-Poincaré technique to the differential equation or by applying the method of renormalization to (7.30). In either case, we introduce a transformation $\tau = \omega t$, where ω is the nonlinear frequency of (7.9). Moreover, we expand ω in a power series of ϵ with the first term being the linear frequency (in this case it is unity). Thus, we write

$$\omega = 1 + \epsilon \omega_1 + \epsilon^2 \omega_2 + \cdots \quad (7.31)$$

We note that we included terms up to $O(\epsilon^2)$ in (7.31) because the secular term

THE METHOD OF RENORMALIZATION 163

appears at $O(\epsilon^3)$ in (7.30). Then, we have

$$t = (1 + \epsilon\omega_1 + \epsilon^2\omega_2 + \cdots)^{-1}\tau$$
$$= [1 - (\epsilon\omega_1 + \epsilon^2\omega_2 + \cdots) + (\epsilon\omega_1 + \epsilon^2\omega_2 + \cdots)^2]\tau$$

Keeping terms up to $O(\epsilon^2)$ gives

$$t = \tau - \epsilon\omega_1\tau + \epsilon^2(\omega_1^2 - \omega_2)\tau + \cdots \tag{7.32}$$

Substituting (7.32) into (7.30) yields

$$u = \epsilon a \cos\left[\tau + \beta - \epsilon\omega_1\tau + \epsilon^2(\omega_1^2 - \omega_2)\tau + \cdots\right]$$
$$+ \tfrac{1}{6}\alpha_2\epsilon^2 a^2 \left\{\cos\left[2\tau + 2\beta - 2\epsilon\omega_1\tau + 2\epsilon^2(\omega_1^2 - \omega_2)\tau + \cdots\right] - 3\right\}$$
$$+ \epsilon^3 a^3 \left\{(\tfrac{5}{12}\alpha_2^2 - \tfrac{3}{8}\alpha_3)[\tau - \epsilon\omega_1\tau + \epsilon^2(\omega_1^2 - \omega_2)\tau + \cdots] \sin\left[\tau + \beta - \epsilon\omega_1\tau\right.\right.$$
$$\left.\left.+ \epsilon^2(\omega_1^2 - \omega_2)\tau + \cdots\right] + \tfrac{1}{8}(\tfrac{1}{4}\alpha_3 + \tfrac{1}{6}\alpha_2^2)\cos\left[3\tau + 3\beta - 3\epsilon\omega_1\tau\right.\right.$$
$$\left.\left.+ 3\epsilon^2(\omega_1^2 - \omega_2)\tau + \cdots\right]\right\} + \cdots \tag{7.33}$$

Next, we expand (7.33) for small ϵ keeping τ fixed. To this end, we need to expand the circular functions in (7.33). Using Taylor series, we obtain

$$\cos\left[\tau + \beta - \epsilon\omega_1\tau + \epsilon^2(\omega_1^2 - \omega_2)\tau + \cdots\right]$$
$$= \cos(\tau + \beta) + [\epsilon\omega_1\tau - \epsilon^2(\omega_1^2 - \omega_2)\tau]\sin(\tau + \beta)$$
$$- \frac{1}{2!}[\epsilon\omega_1\tau - \epsilon^2(\omega_1^2 - \omega_2)\tau + \cdots]^2 \cos(\tau + \beta)$$
$$= \cos(\tau + \beta) + \epsilon\omega_1\tau \sin(\tau + \beta) - \epsilon^2[(\omega_1^2 - \omega_2)\tau \sin(\tau + \beta)$$
$$+ \tfrac{1}{2}\omega_1^2\tau^2 \cos(\tau + \beta)] + \cdots \tag{7.34}$$

$$\cos\left[2\tau + 2\beta - 2\epsilon\omega_1\tau + 2\epsilon^2(\omega_1^2 - \omega_2)\tau + \cdots\right]$$
$$= \cos(2\tau + 2\beta) + [2\epsilon\omega_1\tau - 2\epsilon^2(\omega_1^2 - \omega_2)\tau]\sin(2\tau + 2\beta) + \cdots$$
$$= \cos(2\tau + 2\beta) + 2\epsilon\omega_1\tau \sin(2\tau + 2\beta) + \cdots \tag{7.35}$$

$$\sin\left[\tau + \beta - \epsilon\omega_1\tau + \epsilon^2(\omega_1^2 - \omega_2)\tau + \cdots\right] = \sin(\tau + \beta) + \cdots \tag{7.36}$$

$$\cos\left[3\tau + 3\beta - 3\epsilon\omega_1\tau + 3\epsilon^2(\omega_1^2 - \omega_2)\tau + \cdots\right] = \cos(3\tau + 3\beta) + \cdots \tag{7.37}$$

Substituting (7.34) through (7.37) into (7.33) and keeping terms up to $O(\epsilon^3)$ only, we obtain

$$u = \epsilon a \cos(\tau + \beta) + \epsilon^2[\omega_1 a\tau \sin(\tau + \beta) + \tfrac{1}{6}\alpha_2 a^2 \cos(2\tau + 2\beta) - \tfrac{1}{2}\alpha_2 a^2]$$
$$+ \epsilon^3\left\{-\tfrac{1}{2}\omega_1^2 a\tau^2 \cos(\tau + \beta) + [\tfrac{5}{12}\alpha_2^2 - \tfrac{3}{8}\alpha_3)a^3 - a(\omega_1^2 - \omega_2)]\tau \sin(\tau + \beta)\right.$$
$$\left.+ \tfrac{1}{3}\alpha_2\omega_1 a^2\tau \sin(2\tau + 2\beta) + \tfrac{1}{8}(\tfrac{1}{4}\alpha_3 + \tfrac{1}{6}\alpha_2^2)a^3 \cos(3\tau + 3\beta)\right\} + \cdots \tag{7.38}$$

Eliminating the secular term at $O(\epsilon^2)$ demands that $\omega_1 = 0$ if $a \neq 0$. Then, all secular terms are eliminated at $O(\epsilon^3)$ except one, which is eliminated if

$$(\tfrac{5}{12}\alpha_2^2 - \tfrac{3}{8}\alpha_3)a^3 + a\omega_2 = 0 \tag{7.39}$$

Hence,

$$\omega_2 = -(\tfrac{5}{12}\alpha_2^2 - \tfrac{3}{8}\alpha_3)a^2 \tag{7.40}$$

With the secular terms eliminated, (7.38) becomes

$$u = \epsilon a \cos(\tau + \beta) + \tfrac{1}{6}\epsilon^2 a^2 \alpha_2 [\cos(2\tau + 2\beta) - 3] + \cdots \tag{7.41}$$

where

$$\tau = \omega t = [1 + (\tfrac{3}{8}\alpha_3 - \tfrac{5}{12}\alpha_2^2)\epsilon^2 a^2]t + \cdots \tag{7.42}$$

We note that the secular term at $O(\epsilon^3)$ in (7.38) is used to determine ω_2. The remaining part is bounded and hence it is not included in (7.41) because the expansion is ended to this order, and the error is $O(\epsilon^3)$ for all $t \leqslant O(\epsilon^{-1})$. This point is discussed in more detail in Section 4.5.

Returning to (7.30), we note that the first secular term appears at $O(\epsilon^3)$. Consequently, we could have concluded that $\omega_1 = 0$ before carrying out the expansion because the term $\epsilon\omega_1$ in (7.31) creates secular terms at $O(\epsilon^2)$ and not at $O(\epsilon^3)$, as needed to eliminate the secular term from (7.30). Had we used this fact, we would have shortened the algebra considerably. Next, we show that application of the Lindstedt-Poincaré technique to this problem involves less algebra than that needed in applying the method of renormalization.

7.3. The Lindstedt-Poincaré Technique

Introducing the transformation $\tau = \omega t$ in (7.9) gives

$$\omega^2 u'' + u + \alpha_2 u^2 + \alpha_3 u^3 = 0 \tag{7.43}$$

where the prime denotes the derivative with respect to τ. Next, we expand ω and u in powers of ϵ as

$$u = \epsilon u_1(\tau) + \epsilon^2 u_2(\tau) + \epsilon^3 u_3(\tau) + \cdots \tag{7.44}$$

$$\omega = 1 + \epsilon\omega_1 + \epsilon^2\omega_2 + \cdots \tag{7.45}$$

As discussed in the preceding chapters, the first term in the expansion of ω is the linear frequency, which is unity in this case. Substituting (7.44) and (7.45) into (7.43), we have

$$(1 + \epsilon\omega_1 + \epsilon^2\omega_2 + \cdots)^2(\epsilon u_1'' + \epsilon^2 u_2'' + \epsilon^3 u_3'' + \cdots) + \epsilon u_1 + \epsilon^2 u_2 + \epsilon^3 u_3 + \cdots$$
$$+ \alpha_2(\epsilon u_1 + \epsilon^2 u_2 + \epsilon^3 u_3 + \cdots)^2 + \alpha_3(\epsilon u_1 + \epsilon^2 u_2 + \epsilon^3 u_3 + \cdots)^3 = 0$$

Using the binomial theorem to expand the exponentiated quantities and keeping terms to $O(\epsilon^3)$, we obtain

$$(1 + 2\epsilon\omega_1 + \epsilon^2\omega_1^2 + 2\epsilon^2\omega_2)(\epsilon u_1'' + \epsilon^2 u_2'' + \epsilon^3 u_3'') + \epsilon u_1 + \epsilon^2 u_2 + \epsilon^3 u_3$$
$$+ \alpha_2(\epsilon^2 u_1^2 + 2\epsilon^3 u_1 u_2) + \alpha_3 \epsilon^3 u_1^3 + \cdots = 0$$

Multiplying the first two terms and equating the coefficients of like powers of ϵ to zero yields

$$u_1'' + u_1 = 0 \tag{7.46}$$

$$u_2'' + u_2 = -2\omega_1 u_1'' - \alpha_2 u_1^2 \tag{7.47}$$

$$u_3'' + u_3 = -2\omega_1 u_2'' - (\omega_1^2 + 2\omega_2)u_1'' - 2\alpha_2 u_1 u_2 - \alpha_3 u_1^3 \tag{7.48}$$

The general solution of (7.46) can be expressed as

$$u_1 = a \cos(\tau + \beta) \tag{7.49}$$

where a and β are constants. Then, (7.47) becomes

$$u_2'' + u_2 = 2\omega_1 a \cos(\tau + \beta) - \alpha_2 a^2 \cos^2(\tau + \beta)$$

or

$$u_2'' + u_2 = 2\omega_1 a \cos(\tau + \beta) - \tfrac{1}{2}\alpha_2 a^2 - \tfrac{1}{2}\alpha_2 a^2 \cos(2\tau + 2\beta) \tag{7.50}$$

Eliminating the secular terms from u_2 demands that $\omega_1 = 0$. Then, the solution of (7.50) can be obtained as in Section 7.1. The result is

$$u_2 = -\tfrac{1}{2}\alpha_2 a^2 + \tfrac{1}{6}\alpha_2 a^2 \cos(2\tau + 2\beta) \tag{7.51}$$

Substituting (7.49) and (7.51) into (7.48) and using the fact that $\omega_1 = 0$, we obtain

$$u_3'' + u_3 = 2\omega_2 a \cos(\tau + \beta) - 2\alpha_2 a \cos(\tau + \beta)[-\tfrac{1}{2}\alpha_2 a^2$$
$$+ \tfrac{1}{6}\alpha_2 a^2 \cos(2\tau + 2\beta)] - \alpha_3 a^3 \cos^3(\tau + \beta) \tag{7.52}$$

Using trigonometric identities as in Section 7.1, we rewrite (7.52) as

$$u_3'' + u_3 = (2\omega_2 a - \tfrac{3}{4}\alpha_3 a^3 + \tfrac{5}{6}\alpha_2^2 a^3) \cos(\tau + \beta)$$
$$- (\tfrac{1}{4}\alpha_3 + \tfrac{1}{6}\alpha_2^2) a^3 \cos(3\tau + 3\beta) \tag{7.53}$$

Eliminating the secular terms from (7.53) demands that

$$2\omega_2 a - \tfrac{3}{4}\alpha_3 a^3 + \tfrac{5}{6}\alpha_2^2 a^3 = 0$$

or

$$\omega_2 = \tfrac{3}{8}\alpha_3 a^2 - \tfrac{5}{12}\alpha_2^2 a^2 \tag{7.54}$$

Substituting (7.49) and (7.51) into (7.44) yields (7.41), whereas substituting (7.54) into (7.45), using $\tau = \omega t$, and recalling that $\omega_1 = 0$ yields (7.42). Thus, the Lindstedt-Poincaré technique produced an expansion that is in full agreement with that obtained by using the method of renormalization with less algebra.

7.4. The Method of Multiple Scales

In this section, we determine a third-order uniform expansion by using the method of multiple scales. We note that to third-order we need the three scales $T_0 = t$, $T_1 = \epsilon t$, and $T_2 = \epsilon^2 t$. Then, the time derivatives become

$$\frac{d}{dt} = D_0 + \epsilon D_1 + \epsilon^2 D_2 + \cdots \quad (7.55)$$

$$\frac{d^2}{dt^2} = D_0^2 + 2\epsilon D_0 D_1 + \epsilon^2(D_1^2 + 2D_0 D_2) + \cdots \quad (7.56)$$

where $D_n = \partial/\partial T_n$. Using (7.56), we transform (7.9) into

$$D_0^2 u + 2\epsilon D_0 D_1 u + \epsilon^2(D_1^2 u + 2D_0 D_2 u) + u + \alpha_2 u^2 + \alpha_3 u^3 + \cdots = 0 \quad (7.57)$$

We seek an approximate solution to (7.57) in the form

$$u = \epsilon u_1(T_0, T_1, T_2) + \epsilon^2 u_2(T_0, T_1, T_2) + \epsilon^3 u_3(T_0, T_1, T_2) + \cdots \quad (7.58)$$

Substituting (7.58) into (7.57) and equating each of the coefficients of ϵ to zero, we obtain

$$D_0^2 u_1 + u_1 = 0 \quad (7.59)$$

$$D_0^2 u_2 + u_2 = -2D_0 D_1 u_1 - \alpha_2 u_1^2 \quad (7.60)$$

$$D_0^2 u_3 + u_3 = -D_1^2 u_1 - 2D_0 D_2 u_1 - 2D_0 D_1 u_2 - 2\alpha_2 u_1 u_2 - \alpha_3 u_1^3 \quad (7.61)$$

The solution of (7.59) can be expressed in the following form:

$$u_1 = A(T_1, T_2) e^{iT_0} + \overline{A}(T_1, T_2) e^{-iT_0} \quad (7.62)$$

Then, (7.60) becomes

$$D_0^2 u_2 + u_2 = -2iD_1 A e^{iT_0} + 2iD_1 \overline{A} e^{-iT_0} - \alpha_2(A^2 e^{2iT_0} + 2A\overline{A} + \overline{A}^2 e^{-2iT_0}) \quad (7.63)$$

Eliminating the secular terms from u_2 demands that

$$D_1 A = 0 \quad \text{or} \quad A = A(T_2) \quad (7.64)$$

Then, the solution of (7.63) is taken to consist of its particular solution only, which can be obtained by using the principle of superposition, as in the preceding section. The result is

$$u_2 = \tfrac{1}{3} \alpha_2 A^2 e^{2iT_0} + \tfrac{1}{3} \alpha_2 \overline{A}^2 e^{-2iT_0} - 2\alpha_2 A\overline{A} \quad (7.65)$$

Substituting (7.62), (7.64), and (7.65) into (7.61) yields

$$D_0^2 u_3 + u_3 = -2iA' e^{iT_0} + 2i\overline{A}' e^{-iT_0} - 2\alpha_2(A e^{iT_0} + \overline{A} e^{-iT_0})(\tfrac{1}{3}\alpha_2 A^2 e^{2iT_0}$$
$$+ \tfrac{1}{3}\alpha_2 \overline{A}^2 e^{-2iT_0} - 2\alpha_2 A\overline{A}) - \alpha_3(A e^{iT_0} + \overline{A} e^{-iT_0})^3 \quad (7.66)$$

where the prime indicates the derivative with respect to T_2. In arriving at (7.66),

THE METHOD OF MULTIPLE SCALES 167

we used (7.64) so that $D_1 u_1 = D_1 u_2 = 0$. Using the binomial theorem, we rewrite (7.66) as

$$D_0^2 u_3 + u_3 = [-2iA' + \tfrac{10}{3}\alpha_2^2 A^2 \bar{A} - 3\alpha_3 A^2 \bar{A}]e^{iT_0}$$
$$- (\tfrac{2}{3}\alpha_2^2 + \alpha_3)A^3 e^{3iT_0} + cc \qquad (7.67)$$

Eliminating the secular terms from u_3 demands that

$$-2iA' + (\tfrac{10}{3}\alpha_2^2 - 3\alpha_3)A^2 \bar{A} = 0 \qquad (7.68)$$

Expressing A in the polar form

$$A = \tfrac{1}{2} a e^{i\beta} \qquad (7.69)$$

where a and β are real, we rewrite (7.68) as

$$-ia' e^{i\beta} + a\beta' e^{i\beta} + (\tfrac{5}{12}\alpha_2^2 - \tfrac{3}{8}\alpha_3)a^3 e^{i\beta} = 0$$

or

$$-ia' + a\beta' + (\tfrac{5}{12}\alpha_2^2 - \tfrac{3}{8}\alpha_3)a^3 = 0 \qquad (7.70)$$

Separating real and imaginary parts in (7.70) yields

$$a' = 0 \qquad (7.71)$$

$$a\beta' = (\tfrac{3}{8}\alpha_3 - \tfrac{5}{12}\alpha_2^2)a^3 \qquad (7.72)$$

It follows from (7.71) that $a = a_0 =$ constant. Then, it follows from (7.72) that

$$\beta = (\tfrac{3}{8}\alpha_3 - \tfrac{5}{12}\alpha_2^2)a_0^2 T_2 + \beta_0 \qquad (7.73)$$

where β_0 is a constant and a_0 is assumed to be different from zero.

Substituting (7.69) into (7.62) and (7.65) and recalling that $T_0 = t$, we obtain

$$u_1 = a \cos(t + \beta) \qquad (7.74)$$

$$u_2 = \tfrac{1}{6}\alpha_2 a^2 \cos(2t + 2\beta) - \tfrac{1}{2}\alpha_2 a^2 \qquad (7.75)$$

Then, (7.58) becomes

$$u = \epsilon a \cos(t + \beta) + \tfrac{1}{6}\epsilon^2 a^2 \alpha_2 [\cos(2t + 2\beta) - 3] + \cdots \qquad (7.76)$$

Using (7.73) in (7.76) and the facts that $a = a_0$ and $T_2 = \epsilon^2 t$, we rewrite (7.76) as

$$u = \epsilon a_0 \cos(\omega t + \beta_0) + \tfrac{1}{6}\epsilon^2 a_0^2 \alpha_2 [\cos(2\omega t + 2\beta_0) - 3] + \cdots \qquad (7.77)$$

where

$$\omega = 1 + (\tfrac{3}{8}\alpha_3 - \tfrac{5}{12}\alpha_2^2)\epsilon^2 a_0^2 + \cdots \qquad (7.78)$$

which is in agreement with (7.41) and (7.42) obtained by using the method of renormalization and the Lindstedt-Poincaré method.

168 SYSTEMS WITH QUADRATIC AND CUBIC NONLINEARITIES

Next, we apply the first approximation of the method of averaging to (7.9).

7.5. The Method of Averaging

As before, we use the method of variation of parameters to transform the dependent variable from u to a and β, where

$$u = \epsilon a \cos(t + \beta) \tag{7.79}$$

$$\dot{u} = -\epsilon a \sin(t + \beta) \tag{7.80}$$

and ϵ is a small dimensionless parameter that is a measure of the amplitude of oscillation. For the nonlinear problem, a and β are variables. Differentiating (7.79) with respect to t yields

$$\dot{u} = -\epsilon a \sin(t + \beta) + \epsilon \dot{a} \cos(t + \beta) - \epsilon a \dot{\beta} \sin(t + \beta) \tag{7.81}$$

Comparing (7.80) and (7.81), we conclude that

$$\dot{a} \cos(t + \beta) - a\dot{\beta} \sin(t + \beta) = 0 \tag{7.82}$$

Differentiating (7.80) with respect to t yields

$$\ddot{u} = -\epsilon a \cos(t + \beta) - \epsilon \dot{a} \sin(t + \beta) - \epsilon a \dot{\beta} \cos(t + \beta) \tag{7.83}$$

Substituting (7.79) and (7.83) into (7.9), we have

$$\dot{a} \sin(t + \beta) + a\dot{\beta} \cos(t + \beta) = \alpha_2 \epsilon a^2 \cos^2(t + \beta) + \alpha_3 \epsilon^2 a^3 \cos^3(t + \beta) \tag{7.84}$$

Solving (7.82) and (7.84) for $\dot{a}$ and $\dot{\beta}$, we obtain

$$\dot{a} = \alpha_2 \epsilon a^2 \sin(t + \beta) \cos^2(t + \beta) + \alpha_3 \epsilon^2 a^3 \sin(t + \beta) \cos^3(t + \beta) \tag{7.85}$$

$$\dot{\beta} = \alpha_2 \epsilon a \cos^3(t + \beta) + \alpha_3 \epsilon^2 a^2 \cos^4(t + \beta) \tag{7.86}$$

where a is assumed to be different from zero in arriving at (7.86).

Since a is small, $\dot{a}$ and $\dot{\beta}$ are slowly varying functions of t. Then, one might attempt to average (7.85) and (7.86) for a first approximation. To this end, one uses trigonometric identities to rewrite (7.85) and (7.86) as

$$\dot{a} = \tfrac{1}{4} \alpha_2 \epsilon a^2 [\sin(t + \beta) + \sin(3t + 3\beta)] + \tfrac{1}{8} \alpha_3 \epsilon^2 a^3 [2 \sin(2t + 2\beta) + \sin(4t + 4\beta)] \tag{7.87}$$

$$\dot{\beta} = \tfrac{1}{4} \alpha_2 \epsilon a [3 \cos(t + \beta) + \cos(3t + 3\beta)] + \tfrac{1}{8} \alpha_3 \epsilon^2 a^2 [\cos(4t + 4\beta) + 4 \cos(2t + 2\beta) + 3] \tag{7.88}$$

Keeping the slowly varying parts on the right-hand side of (7.87) and (7.88), we obtain

$$\dot{a} = 0 \tag{7.89}$$

THE GENERALIZED METHOD OF AVERAGING 169

$$\dot{\beta} = \tfrac{3}{8}\alpha_3\epsilon^2 a^2 \qquad (7.90)$$

Whereas (7.89) agrees with (7.71), (7.90) does not agree with (7.72) obtained by using the method of multiple scales. There is a term $\tfrac{5}{12}\alpha_2^2\epsilon^2 a^2$ missing from (7.90). Following the details of the solution in the preceding section, one finds that this term is the result of the interaction of the first- and second-order approximations. This interaction was not taken into account in arriving at (7.89) and (7.90). To include the effect of this interaction, we need to carry out the solutions of (7.87) and (7.88) to higher order. This is accomplished by using the generalized method of averaging, which is discussed next, or its variant the Krylov-Bogoliubov-Mitropolsky technique, which is discussed in Section 7.7.

7.6. The Generalized Method of Averaging

To apply this method, we introduce the variable

$$\phi = t + \beta \qquad (7.91)$$

and rewrite (7.87) and (7.88) as

$$\dot{a} = \tfrac{1}{4}\alpha_2\epsilon a^2(\sin\phi + \sin 3\phi) + \tfrac{1}{8}\alpha_3\epsilon^2 a^3(2\sin 2\phi + \sin 4\phi) \qquad (7.92)$$

$$\dot{\phi} = 1 + \tfrac{1}{4}\alpha_2\epsilon a(3\cos\phi + \cos 3\phi) + \tfrac{1}{8}\alpha_3\epsilon^2 a^2(\cos 4\phi + 4\cos 2\phi + 3) \qquad (7.93)$$

We seek approximate solutions to (7.92) and (7.93) in the form

$$a = a_0(t) + \epsilon a_1(a_0,\phi_0) + \epsilon^2 a_2(a_0,\phi_0) + \cdots \qquad (7.94)$$

$$\phi = \phi_0(t) + \epsilon\phi_1(a_0,\phi_0) + \epsilon^2 \phi_2(a_0,\phi_0) + \cdots \qquad (7.95)$$

$$\dot{a}_0 = \epsilon A_1(a_0) + \epsilon^2 A_2(a_0) + \cdots \qquad (7.96)$$

$$\dot{\phi}_0 = 1 + \epsilon \Phi_1(a_0) + \epsilon^2 \Phi_2(a_0) + \cdots \qquad (7.97)$$

The functions $a_1, a_2, \ldots$ and $\phi_1, \phi_2, \ldots$ are fast varying functions of ϕ_0, while it follows from (7.96) and (7.97) that a_0, and hence, the A_n and Φ_n are slowly varying functions of t.

Using the chain rule, we write the first derivatives of (7.94) and (7.95) as

$$\dot{a} = \dot{a}_0 + \epsilon\frac{\partial a_1}{\partial a_0}\dot{a}_0 + \epsilon\frac{\partial a_1}{\partial \phi_0}\dot{\phi}_0 + \epsilon^2\frac{\partial a_2}{\partial a_0}\dot{a}_0 + \epsilon^2\frac{\partial a_2}{\partial \phi_0}\dot{\phi}_0 + \cdots \qquad (7.98)$$

$$\dot{\phi} = \dot{\phi}_0 + \epsilon\frac{\partial \phi_1}{\partial a_0}\dot{a}_0 + \epsilon\frac{\partial \phi_1}{\partial \phi_0}\dot{\phi}_0 + \epsilon^2\frac{\partial \phi_2}{\partial a_0}\dot{a}_0 + \epsilon^2\frac{\partial \phi_2}{\partial \phi_0}\dot{\phi}_0 + \cdots \qquad (7.99)$$

Substituting (7.96) and (7.97) into (7.98) and (7.99) and keeping terms up to $O(\epsilon^2)$, we obtain

$$\dot{a} = \epsilon\left[A_1 + \frac{\partial a_1}{\partial \phi_0}\right] + \epsilon^2\left[A_2 + \frac{\partial a_2}{\partial \phi_0} + A_1\frac{\partial a_1}{\partial a_0} + \Phi_1\frac{\partial a_1}{\partial \phi_0}\right] + \cdots \qquad (7.100)$$

170 SYSTEMS WITH QUADRATIC AND CUBIC NONLINEARITIES

$$\dot{\phi} = 1 + \epsilon \left[\Phi_1 + \frac{\partial \phi_1}{\partial \phi_0} \right] + \epsilon^2 \left[\Phi_2 + \frac{\partial \phi_2}{\partial \phi_0} + A_1 \frac{\partial \phi_1}{\partial a_0} + \Phi_1 \frac{\partial \phi_1}{\partial \phi_0} \right] + \cdots \quad (7.101)$$

Next, we need to substitute (7.94) and (7.95) into (7.92) and (7.93) and expand the right-hand sides for small ϵ keeping terms up to $O(\epsilon^2)$. From the right-hand side of (7.92), we have

$$\epsilon a^2 (\sin \phi + \sin 3\phi) = \epsilon (a_0 + \epsilon a_1)^2 [\sin (\phi_0 + \epsilon \phi_1) + \sin (3\phi_0 + 3\epsilon \phi_1)] + \cdots$$

$$= \epsilon (a_0^2 + 2\epsilon a_0 a_1)[\sin \phi_0 + \epsilon \phi_1 \cos \phi_0 + \sin 3\phi_0$$

$$+ 3\epsilon \phi_1 \cos 3\phi_0] + \cdots$$

$$= \epsilon a_0^2 (\sin \phi_0 + \sin 3\phi_0) + 2\epsilon^2 a_0 a_1 (\sin \phi_0 + \sin 3\phi_0)$$

$$+ \epsilon^2 a_0^2 \phi_1 (\cos \phi_0 + 3 \cos 3\phi_0) + \cdots \quad (7.102)$$

$$\epsilon^2 a^3 (2 \sin 2\phi + \sin 4\phi) = \epsilon^2 a_0^3 (2 \sin 2\phi_0 + \sin 4\phi_0) + \cdots \quad (7.103)$$

From the right-hand side of (7.93), we have

$$\epsilon a(3 \cos \phi + \cos 3\phi) = \epsilon (a_0 + \epsilon a_1)[3 \cos (\phi_0 + \epsilon \phi_1) + \cos (3\phi_0 + 3\epsilon \phi_1)] + \cdots$$

$$= \epsilon (a_0 + \epsilon a_1)[3 \cos \phi_0 - 3\epsilon \phi_1 \sin \phi_0 + \cos 3\phi_0$$

$$- 3\epsilon \phi_1 \sin 3\phi_0] + \cdots$$

$$= \epsilon a_0 (3 \cos \phi_0 + \cos 3\phi_0) + \epsilon^2 a_1 (3 \cos \phi_0 + \cos 3\phi_0)$$

$$- 3\epsilon^2 a_0 \phi_1 (\sin \phi_0 + \sin 3\phi_0) + \cdots \quad (7.104)$$

and

$$\epsilon^2 a^2 (\cos 4\phi + 4 \cos 2\phi + 3) = \epsilon^2 a_0^2 (\cos 4\phi_0 + 4 \cos 2\phi_0 + 3) + \cdots \quad (7.105)$$

Substituting (7.100) through (7.105) into (7.92) and (7.93) and equating coefficients of like powers of ϵ, we obtain

$$A_1 + \frac{\partial a_1}{\partial \phi_0} = \tfrac{1}{4} \alpha_2 a_0^2 (\sin \phi_0 + \sin 3\phi_0) \quad (7.106)$$

$$A_2 + \frac{\partial a_2}{\partial \phi_2} + A_1 \frac{\partial a_1}{\partial a_0} + \Phi_1 \frac{\partial a_1}{\partial \phi_0} = \tfrac{1}{2} \alpha_2 a_0 a_1 (\sin \phi_0 + \sin 3\phi_0)$$

$$+ \tfrac{1}{4} \alpha_2 a_0^2 \phi_1 (\cos \phi_0 + 3 \cos 3\phi_0) + \tfrac{1}{8} \alpha_3 a_0^3 (2 \sin 2\phi_0 + \sin 4\phi_0) \quad (7.107)$$

$$\Phi_1 + \frac{\partial \phi_1}{\partial \phi_0} = \tfrac{1}{4} \alpha_2 a_0 (3 \cos \phi_0 + \cos 3\phi_0) \quad (7.108)$$

$$\Phi_2 + \frac{\partial \phi_2}{\partial \phi_0} + A_1 \frac{\partial \phi_1}{\partial \phi_0} + \Phi_1 \frac{\partial \phi_1}{\partial \phi_0} = \tfrac{1}{4} \alpha_2 a_1 (3 \cos \phi_0 + \cos 3\phi_0) - \tfrac{3}{4} \alpha_2 a_0 \phi_1 (\sin \phi_0$$

$$+ \sin 3\phi_0) + \tfrac{1}{8}\alpha_3 a_0^2(\cos 4\phi_0 + 4\cos 2\phi_0 + 3) \quad (7.109)$$

Next, we use the method of separation of variables to separate the fast and slowly varying terms in (7.106) through (7.109). The slowly varying parts of (7.106) and (7.108) yield

$$A_1 = \Phi_1 = 0 \quad (7.110)$$

whereas their fast varying parts yield

$$\frac{\partial a_1}{\partial \phi_0} = \tfrac{1}{4}\alpha_2 a_0^2(\sin \phi_0 + \sin 3\phi_0) \quad (7.111)$$

$$\frac{\partial \phi_1}{\partial \phi_0} = \tfrac{1}{4}\alpha_2 a_0(3\cos \phi_0 + \cos 3\phi_0) \quad (7.112)$$

A particular solution of (7.111) is

$$a_1 = -\tfrac{1}{4}\alpha_2 a_0^2(\cos \phi_0 + \tfrac{1}{3}\cos 3\phi_0) \quad (7.113)$$

whereas a particular solution of (7.112) is

$$\phi_1 = \tfrac{1}{4}\alpha_2 a_0(3\sin \phi_0 + \tfrac{1}{3}\sin 3\phi_0) \quad (7.114)$$

Substituting (7.110), (7.113), and (7.114) into (7.107) and (7.109), we obtain

$$A_2 + \frac{\partial a_2}{\partial \phi_0} = -\tfrac{1}{8}\alpha_2^2 a_0^3(\cos \phi_0 + \tfrac{1}{3}\cos 3\phi_0)(\sin \phi_0 + \sin 3\phi_0)$$

$$+ \tfrac{1}{16}\alpha_2^2 a_0^3(3\sin \phi_0 + \tfrac{1}{3}\sin 3\phi_0)(\cos \phi_0 + 3\cos 3\phi_0)$$

$$+ \tfrac{1}{8}\alpha_3 a_0^3(2\sin 2\phi_0 + \sin 4\phi_0) \quad (7.115)$$

$$\Phi_2 + \frac{\partial \phi_2}{\partial \phi_0} = -\tfrac{1}{16}\alpha_2^2 a_0^2(\cos \phi_0 + \tfrac{1}{3}\cos 3\phi_0)(3\cos \phi_0 + \cos 3\phi_0)$$

$$- \tfrac{3}{16}\alpha_2^2 a_0^2(3\sin \phi_0 + \tfrac{1}{3}\sin 3\phi_0)(\sin \phi_0 + \sin 3\phi_0)$$

$$+ \tfrac{1}{8}\alpha_3 a_0^2(\cos 4\phi_0 + 4\cos 2\phi_0 + 3) \quad (7.116)$$

Using trigonometric identities (Appendix A), we rewrite (7.115) and (7.116) as

$$A_2 + \frac{\partial a_2}{\partial \phi_0} = \tfrac{1}{16} a_0^3 [(4\alpha_3 - \tfrac{9}{2}\alpha_2^2)\sin 2\phi_0 + (2\alpha_3 + \tfrac{10}{3}\alpha_2^2)\sin 4\phi_0 + \tfrac{1}{6}\alpha_2^2 \sin 6\phi_0]$$

$$(7.117)$$

$$\Phi_2 + \frac{\partial \phi_2}{\partial \phi_0} = (\tfrac{3}{8}\alpha_3 - \tfrac{5}{12}\alpha_2^2)a_0^2 + (\tfrac{1}{2}\alpha_3 - \tfrac{3}{16}\alpha_2^2)a_0^2 \cos 2\phi_0$$

$$+ (\tfrac{1}{8}\alpha_3 + \tfrac{1}{4}\alpha_2^2)a_0^2 \cos 4\phi_0 + \tfrac{1}{48}\alpha_2^2 a_0^2 \cos 6\phi_0 \quad (7.118)$$

172 SYSTEMS WITH QUADRATIC AND CUBIC NONLINEARITIES

Since we are seeking an expansion valid to $O(\epsilon^2)$, we do not need to solve for a_2 and ϕ_2. All we need to do is to investigate (7.117) and (7.118) to determine the slowly varying parts, and hence, determine A_2 and Φ_2. Hence,

$$A_2 = 0 \tag{7.119}$$

$$\Phi_2 = (\tfrac{3}{8} \alpha_3 - \tfrac{5}{12} \alpha_2^2) a_0^2 \tag{7.120}$$

Substituting (7.113) and (7.114) into (7.94) and (7.95) yields

$$a = a_0 - \tfrac{1}{4} \alpha_2 \epsilon a_0^2 (\cos \phi_0 + \tfrac{1}{3} \cos 3\phi_0) + \cdots \tag{7.121}$$

$$\phi = \phi_0 + \tfrac{1}{4} \alpha_2 \epsilon a_0 (3 \sin \phi_0 + \tfrac{1}{3} \sin 3\phi_0) + \cdots \tag{7.122}$$

Substituting (7.110), (7.119), and (7.120) into (7.96) and (7.97) gives

$$\dot{a}_0 = 0 \tag{7.123}$$

$$\dot{\phi}_0 = 1 + (\tfrac{3}{8} \alpha_3 - \tfrac{5}{12} \alpha_2^2) \epsilon^2 a_0^2 \tag{7.124}$$

It follows from (7.123) that a_0 = constant, and then it follows from (7.124) that

$$\phi_0 = t + (\tfrac{3}{8} \alpha_3 - \tfrac{5}{12} \alpha_2^2) \epsilon^2 a_0^2 t + \beta_0 \tag{7.125}$$

where β_0 is a constant. Substituting (7.121) and (7.122) into (7.79) gives

$$u = \epsilon [a_0 - \tfrac{1}{4} \alpha_2 \epsilon a_0^2 (\cos \phi_0 + \tfrac{1}{3} \cos 3\phi_0) + \cdots] \cos [\phi_0 + \tfrac{1}{4} \alpha_2 \epsilon a_0 (3 \sin \phi_0 + \tfrac{1}{3} \sin 3\phi_0) + \cdots] \tag{7.126}$$

To compare the present solution with those obtained by using the method of multiple scales and the Lindstedt-Poincaré technique, we need to expand the circular function in (7.126) for small ϵ about ϕ_0. Thus, we write

$$u = \epsilon [a_0 - \tfrac{1}{4} \alpha_2 \epsilon a_0^2 (\cos \phi_0 + \tfrac{1}{3} \cos 3\phi_0)] [\cos \phi_0$$
$$- \tfrac{1}{4} \alpha_2 \epsilon a_0 \sin \phi_0 (3 \sin \phi_0 + \tfrac{1}{3} \sin 3\phi_0)] + \cdots$$
$$= \epsilon a_0 \cos \phi_0 - \tfrac{1}{4} \alpha_2 \epsilon^2 a_0^2 [\cos^2 \phi_0 + \tfrac{1}{3} \cos \phi_0 \cos 3\phi_0$$
$$+ 3 \sin^2 \phi_0 + \tfrac{1}{3} \sin \phi_0 \sin 3\phi_0] + \cdots$$
$$= \epsilon a_0 \cos \phi_0 - \tfrac{1}{4} \alpha_2 \epsilon^2 a_0^2 [\tfrac{1}{2} + \tfrac{1}{2} \cos 2\phi_0 + \tfrac{1}{3} \cos 2\phi_0$$
$$+ \tfrac{3}{2} - \tfrac{3}{2} \cos 2\phi_0] + \cdots$$

or

$$u = \epsilon a_0 \cos \phi_0 + \tfrac{1}{6} \epsilon^2 a_0^2 \alpha_2 (\cos 2\phi_0 - 3) + \cdots \tag{7.127}$$

The expansion represented by (7.125) and (7.127) agrees with the expansion (7.77) and (7.78) obtained by using the method of multiple scales. Comparing the algebra in this case with those in Sections 7.2 and 7.4, we conclude that the methods of renormalization and multiple scales have advantages over the generalized method of averaging.

7.7. The Krylov-Bogoliubov-Mitropolsky Technique

In this section, we describe a variant of the generalized method of averaging, namely the Krylov-Bogoliubov-Mitropolsky technique. It is often referred to as the *asymptotic method*.

When the nonlinear terms are neglected, the solution of (7.9) is

$$u = \epsilon a \cos(t + \beta) \tag{7.128}$$

where a and β are constants and ϵ is a small dimensionless parameter that is a measure of the amplitude. When the nonlinear terms are included, we consider (7.128) to be the first term in an approximate solution of (7.9) but with slowly varying rather than constant a and β. Moreover, we introduce the fast scale $\phi = t + \beta$ and use a to represent the slow variations. Thus, we seek an approximate solution to (7.9) in the form

$$u = \epsilon a \cos\phi + \epsilon^2 u_2(a, \phi) + \epsilon^3 u_3(a, \phi) + \cdots \tag{7.129}$$

Since a and β are slowly varying functions of t, we express them in power series of ϵ in terms of a. Then, we write

$$\dot{a} = \epsilon A_1(a) + \epsilon^2 A_2(a) + \cdots \tag{7.130}$$

$$\dot{\phi} = 1 + \epsilon \Phi_1(a) + \epsilon^2 \Phi_2(a) + \cdots \tag{7.131}$$

Thus, this method can be viewed as a multiple scales procedure with a and ϕ being the scales.

Using the chain rule, we can express the derivatives with respect to t in terms of the new independent variables a and ϕ as

$$\frac{d}{dt} = \dot{a}\frac{\partial}{\partial a} + \dot{\phi}\frac{\partial}{\partial \phi} \tag{7.132}$$

$$\frac{d^2}{dt^2} = \dot{a}^2\frac{\partial^2}{\partial a^2} + \ddot{a}\frac{\partial}{\partial a} + 2\dot{a}\dot{\phi}\frac{\partial^2}{\partial a \partial \phi} + \dot{\phi}^2\frac{\partial^2}{\partial \phi^2} + \ddot{\phi}\frac{\partial}{\partial \phi} \tag{7.133}$$

Differentiating (7.130) with respect to t gives

$$\ddot{a} = \epsilon A_1' \dot{a} + \epsilon^2 A_2' \dot{a} + \cdots \tag{7.134}$$

where the prime denotes the derivative with respect to the argument. Substituting (7.130) into (7.134) gives

$$\ddot{a} = \epsilon A_1'(\epsilon A_1 + \epsilon^2 A_2 + \cdots) + \epsilon^2 A_2'(\epsilon A_1 + \cdots) + \cdots$$

or

$$\ddot{a} = \epsilon^2 A_1 A_1' + O(\epsilon^3) \tag{7.135}$$

Differentiating (7.131) with respect to t yields

$$\ddot{\phi} = \epsilon \Phi_1' \dot{a} + \epsilon^2 \Phi_2' \dot{a} + \cdots \tag{7.136}$$

174 SYSTEMS WITH QUADRATIC AND CUBIC NONLINEARITIES

Substituting (7.130) into (7.136) gives

$$\ddot{\phi} = \epsilon \Phi_1'(\epsilon A_1 + \epsilon^2 A_2 + \cdots) + \epsilon^2 \Phi_2'(\epsilon A_1 + \cdots) + \cdots$$

or

$$\ddot{\phi} = \epsilon^2 A_1 \Phi_1' + O(\epsilon^3) \tag{7.137}$$

Substituting (7.130), (7.131), (7.135), and (7.137) into (7.132) and (7.133), we have

$$\frac{d}{dt} = (\epsilon A_1 + \epsilon^2 A_2 + \cdots)\frac{\partial}{\partial a} + (1 + \epsilon \Phi_1 + \epsilon^2 \Phi_2 + \cdots)\frac{\partial}{\partial \phi}$$

$$\frac{d^2}{dt^2} = (\epsilon A_1 + \epsilon^2 A_2 + \cdots)^2 \frac{\partial^2}{\partial a^2} + (\epsilon^2 A_1 A_1' + \cdots)\frac{\partial}{\partial a}$$

$$+ 2(\epsilon A_1 + \epsilon^2 A_2 + \cdots)(1 + \epsilon \Phi_1 + \epsilon^2 \Phi_2 + \cdots)\frac{\partial^2}{\partial a \partial \phi}$$

$$+ (1 + \epsilon \Phi_1 + \epsilon^2 \Phi_2 + \cdots)^2 \frac{\partial^2}{\partial \phi^2} + (\epsilon^2 A_1 \Phi_1' + \cdots)\frac{\partial}{\partial \phi}$$

Hence,

$$\frac{d}{dt} = \frac{\partial}{\partial \phi} + \epsilon \left[A_1 \frac{\partial}{\partial a} + \Phi_1 \frac{\partial}{\partial \phi} \right] + \epsilon^2 \left[A_2 \frac{\partial}{\partial a} + \Phi_2 \frac{\partial}{\partial \phi} \right] \tag{7.138}$$

$$\frac{d^2}{dt^2} = \frac{\partial^2}{\partial \phi^2} + 2\epsilon \left[\Phi_1 \frac{\partial^2}{\partial \phi^2} + A_1 \frac{\partial^2}{\partial a \partial \phi} \right]$$

$$+ \epsilon^2 \left[(\Phi_1^2 + 2\Phi_2)\frac{\partial^2}{\partial \phi^2} + 2(A_2 + A_1 \Phi_1)\frac{\partial^2}{\partial a \partial \phi} \right.$$

$$\left. + A_1^2 \frac{\partial^2}{\partial a^2} + A_1 A_1' \frac{\partial}{\partial a} + A_1 \Phi_1' \frac{\partial}{\partial \phi} \right] + \cdots \tag{7.139}$$

Then, (7.9) becomes

$$\frac{\partial^2 u}{\partial \phi^2} + 2\epsilon \left[\Phi_1 \frac{\partial^2 u}{\partial \phi^2} + A_1 \frac{\partial^2 u}{\partial a \partial \phi} \right] + \epsilon^2 \left[(\Phi_1^2 + 2\Phi_2)\frac{\partial^2 u}{\partial \phi^2} + 2(A_2 + A_1 \Phi_1)\frac{\partial^2 u}{\partial a \partial \phi} \right.$$

$$\left. + A_1^2 \frac{\partial^2 u}{\partial a^2} + A_1 A_1' \frac{\partial u}{\partial a} + A_1 \Phi_1' \frac{\partial u}{\partial \phi} \right] + u + \alpha_2 u^2 + \alpha_3 u^3 + \cdots = 0 \tag{7.140}$$

Substituting (7.129) into (7.140) and equating coefficients of like powers of ϵ, we obtain

$$\frac{\partial^2 u_2}{\partial \phi^2} + u_2 - 2\Phi_1 a \cos \phi - 2A_1 \sin \phi + \alpha_2 a^2 \cos^2 \phi = 0 \tag{7.141}$$

$$\frac{\partial^2 u_3}{\partial \phi^2} + u_3 + 2\Phi_1 \frac{\partial^2 u_2}{\partial \phi^2} + 2A_1 \frac{\partial^2 u_2}{\partial a \partial \phi} - (\Phi_1^2 + 2\Phi_2) a \cos \phi - 2(A_2 + A_1\Phi_1) \sin \phi$$
$$+ A_1 A_1' \cos \phi - A_1 \Phi_1' a \sin \phi + 2\alpha_2 u_2 a \cos \phi + \alpha_3 a^3 \cos^3 \phi = 0 \quad (7.142)$$

Using the trigonometric identity (A13), we rewrite (7.141) as

$$\frac{\partial^2 u_2}{\partial \phi^2} + u_2 = 2\Phi_1 a \cos \phi + 2A_1 \sin \phi - \tfrac{1}{2} \alpha_2 a^2 - \tfrac{1}{2} \alpha_2 a^2 \cos 2\phi \quad (7.143)$$

Eliminating the secular terms from u_2 demands that

$$\Phi_1 = 0 \quad \text{and} \quad A_1 = 0 \quad (7.144)$$

Then, as before, the solution of (7.143) can be written as

$$u_2 = -\tfrac{1}{2} \alpha_2 a^2 + \tfrac{1}{6} \alpha_2 a^2 \cos 2\phi \quad (7.145)$$

Substituting (7.144) and (7.145) into (7.142) gives

$$\frac{\partial^2 u_3}{\partial \phi^2} + u_3 = 2\Phi_2 a \cos \phi + 2A_2 \sin \phi - 2\alpha_2 a \cos \phi \, [-\tfrac{1}{2} \alpha_2 a^2$$
$$+ \tfrac{1}{6} \alpha_2 a^2 \cos 2\phi] - \alpha_3 a^3 \cos^3 \phi \quad (7.146)$$

Using trigonometric identities (Appendix A), we rewrite (7.146) as

$$\frac{\partial^2 u_3}{\partial \phi^2} + u_3 = (2\Phi_2 - \tfrac{3}{4} \alpha_3 a^2 + \tfrac{5}{6} \alpha_2^2 a^2) a \cos \phi + 2A_2 \sin \phi$$
$$- (\tfrac{1}{4} \alpha_3 + \tfrac{1}{6} \alpha_2^2) a^3 \cos 3\phi \quad (7.147)$$

Eliminating the secular terms from u_3 demands that

$$A_2 = 0 \quad \Phi_2 = (\tfrac{3}{8} \alpha_3 - \tfrac{5}{12} \alpha_2^2) a^2 \quad (7.148)$$

Substitutin (7.144) and (7.148) into (7.130) and (7.131) yields

$$\dot{a} = 0 \quad (7.149)$$
$$\dot{\phi} = 1 + (\tfrac{3}{8} \alpha_3 - \tfrac{5}{12} \alpha_2^2) \epsilon^2 a^2 + \cdots \quad (7.150)$$

which are in agreement with (7.123) and (7.124) obtained by using the generalized method of averaging, and hence, with the solution obtained by using the method of multiple scales.

Exercises

7.1. Consider the equation

$$\ddot{x} - 2x - x^2 + x^3 = 0$$

Show that the equilibrium positions are $x = 0, -1$, and 2. Put $x = 2 + u$ and determine the equation governing u. Then, determine a second-order uniform expansion for small but finite amplitudes using

(a) the Lindstedt-Poincaré method,
(b) the method of multiple scales, and
(c) the generalized method of averaging.

7.2. Consider the equation

$$\ddot{u} - u + u^4 = 0$$

Show that $u = 1$ is an equilibrium point. Determine a second-order uniform expansion for small but finite motions around $u = 1$. *Hint:* put $u = 1 + x$, determine the equation describing x, and then use the Lindstedt-Poincaré method or the method of multiple scales or the generalized method of averaging.

7.3. Consider the equation

$$\ddot{u} - u + u^6 = 0$$

Show that $u = 1$ is an equilibrium position. Determine a second-order expansion for small but finite motions around $u = 1$.

7.4. Consider the equation

$$\ddot{x} + x - \frac{3}{16(1-x)} = 0$$

Show that the equilibrium points are $x = \frac{1}{4}$ and $\frac{3}{4}$. Examine the motion near these equilibrium points. Determine a second-order expansion around the stable equilibrium point (i.e., the one corresponding to sinusoidal motions).

7.5. Determine a second-order uniform expansion for

$$\ddot{u} + u + \epsilon^2 u^3 + \epsilon \dot{u}^2 = 0 \qquad \epsilon \ll 1.$$

7.6. Determine a second-order uniform expansion for

$$\ddot{u} + u + \epsilon u^2 + \epsilon \dot{u}^2 = 0 \qquad \epsilon \ll 1.$$

CHAPTER 8

General Weakly Nonlinear Systems

In this chapter, we consider systems having a single degree of freedom under the influence of general forces. Specifically, we consider the equation

$$\ddot{u} + u = \epsilon f(u, \dot{u}) \tag{8.1}$$

where ϵ is a small dimensionless parameter, the dot denotes the derivative with respect to the dimensionless time t, and u is a dimensionless dependent variable. The function f is general but piecewise continuous, so that the equations considered in the preceding four chapters are special cases of (8.1). In this chapter, we do not restrict f to be an analytic function of u and $\dot{u}$.

As before, we start by determining a first-order straightforward expansion and discuss its uniformity. In Section 8.2, we use the method of renormalization to render this straightforward expansion uniform. In Sections 8.3 and 8.4, respectively, we use the methods of multiple scales and averaging to determine first-order uniform expansions. Finally, in Section 8.5, we apply the results to nonanalytic functions f as well as to the cases treated in the preceding chapters.

8.1. The Straightforward Expansion

As before, we seek a first-order expansion in the form

$$u(t;\epsilon) = u_0(t) + \epsilon u_1(t) + \cdots \tag{8.2}$$

Substituting (8.2) into (8.1) gives

$$\ddot{u}_0 + \epsilon \ddot{u}_1 + \cdots + u_0 + \epsilon u_1 + \cdots = \epsilon f[u_0 + \epsilon u_1 + \cdots, \dot{u}_0 + \epsilon \dot{u}_1 + \cdots]$$

$$= \epsilon f(u_0, \dot{u}_0) + \cdots \tag{8.3}$$

Equating the coefficients of ϵ^0 and ϵ on both sides of (8.3), we obtain

$$\ddot{u}_0 + u_0 = 0 \tag{8.4}$$

$$\ddot{u}_1 + u_1 = f(u_0, \dot{u}_0) \tag{8.5}$$

178 GENERAL WEAKLY NONLINEAR SYSTEMS

The general solution of (8.4) can be written as

$$u_0 = a \cos(t + \beta) \tag{8.6}$$

where a and β are constants. Then, (8.5) becomes

$$\ddot{u}_1 + u_1 = f[a \cos(t + \beta), -a \sin(t + \beta)] \tag{8.7}$$

To determine a particular solution for the inhomogeneous equation (8.7), we find it convenient to express the inhomogeneous term in a Fourier series. To this end, we note that f is a periodic function of t having the period 2π. Hence, its Fourier expansion has the form

$$f[a \cos(t + \beta), -a \sin(t + \beta)] = f_0(a) + \sum_{n=1}^{\infty} f_n(a) \cos(nt + n\beta)$$

$$+ \sum_{n=1}^{\infty} g_n(a) \sin(nt + n\beta) \tag{8.8}$$

where

$$f_0(a) = \frac{1}{2\pi} \int_0^{2\pi} f(a \cos \phi, -a \sin \phi) \, d\phi \tag{8.9}$$

$$f_n(a) = \frac{1}{\pi} \int_0^{2\pi} f(a \cos \phi, -a \sin \phi) \cos n\phi \, d\phi \tag{8.10}$$

$$g_n(a) = \frac{1}{\pi} \int_0^{2\pi} f(a \cos \phi, -a \sin \phi) \sin n\phi \, d\phi \tag{8.11}$$

Using (8.8), we rewrite (8.7) as

$$\ddot{u}_1 + u_1 = f_0 + \sum_{n=1}^{\infty} f_n \cos(nt + n\beta) + \sum_{n=1}^{\infty} g_n \sin(nt + n\beta) \tag{8.12}$$

Since (8.12) is linear, one can use the principle of superposition and determine a particular solution as the sum of particular solutions, one corresponding to each inhomogeneous term. It follows from (B69), (B76), (B78), and (B82) that a particular solution for (8.12) is

$$u_1 = f_0 + \tfrac{1}{2} f_1 t \sin(t + \beta) - \tfrac{1}{2} g_1 t \cos(t + \beta) + \sum_{n=2}^{\infty} \frac{1}{1 - n^2} [f_n \cos(nt + n\beta)$$

$$+ g_n \sin(nt + n\beta)] \tag{8.13}$$

As before, we do not include the homogeneous solution of (8.12), so that (8.13)

THE METHOD OF RENORMALIZATION 179

gives its solution. Substituting (8.6) and (8.13) into (8.2) yields

$$u = a \cos(t + \beta) + \epsilon \{f_0 + \tfrac{1}{2} f_1 t \sin(t + \beta) - \tfrac{1}{2} g_1 t \cos(t + \beta)$$

$$+ \sum_{n=2}^{\infty} \frac{1}{1 - n^2} [f_n \cos(nt + n\beta) + g_n \sin(nt + n\beta)]\} + \cdots \quad (8.14)$$

We note that (8.14) is not valid for $t \geq O(\epsilon^{-1})$, owing to the presence of the mixed-secular terms. Next, we use the method of renormalization to render this straightforward expansion uniform.

8.2. The Method of Renormalization

To render (8.14) uniform, we introduce the transformation

$$\tau = \omega t \qquad \omega = 1 + \epsilon \omega_1 + \cdots \quad (8.15)$$

Then,

$$t = \omega^{-1} \tau = \tau (1 + \epsilon \omega_1 + \cdots)^{-1} = (1 - \epsilon \omega_1) \tau + \cdots \quad (8.16)$$

Substituting (8.16) into (8.14) gives

$$u = a \cos(\tau + \beta - \epsilon \omega_1 \tau + \cdots) + \epsilon \{f_0 + \tfrac{1}{2} f_1 (\tau - \epsilon \omega_1 \tau + \cdots) \sin(\tau + \beta$$

$$- \epsilon \omega_1 \tau + \cdots) - \tfrac{1}{2} g_1 (\tau - \epsilon \omega_1 \tau + \cdots) \cos(\tau + \beta - \epsilon \omega_1 \tau + \cdots)$$

$$+ \sum_{n=2}^{\infty} \frac{1}{1 - n^2} [f_n \cos(n\tau + n\beta - \epsilon \omega_1 n\tau + \cdots) + g_n \sin(n\tau + n\beta$$

$$- \epsilon \omega_1 n\tau + \cdots)]\} + \cdots \quad (8.17)$$

Using the expansions

$$\cos(n\tau + n\beta - \epsilon \omega_1 n\tau + \cdots) = \cos(n\tau + n\beta) + \epsilon \omega_1 n\tau \sin(n\tau + n\beta) + \cdots \quad (8.18)$$

$$\sin(n\tau + n\beta - \epsilon \omega_1 n\tau + \cdots) = \sin(n\tau + n\beta) - \epsilon \omega_1 n\tau \cos(n\tau + n\beta) + \cdots \quad (8.19)$$

we rewrite (8.17) as

$$u = a \cos(\tau + \beta) + \epsilon \{f_0 + (\tfrac{1}{2} f_1 + \omega_1 a) \tau \sin(\tau + \beta) - \tfrac{1}{2} g_1 \tau \cos(\tau + \beta)$$

$$+ \sum_{n=2}^{\infty} \frac{1}{1 - n^2} [f_n \cos(n\tau + n\beta) + g_n \sin(n\tau + n\beta)]\} + \cdots \quad (8.20)$$

Eliminating the secular terms from (8.20) demands that

$$\omega_1 a + \tfrac{1}{2} f_1(a) = 0 \quad (8.21)$$

180 GENERAL WEAKLY NONLINEAR SYSTEMS

$$g_1(a) = 0 \tag{8.22}$$

Equation (8.22) provides the values of a for which periodic solutions exist. Then, it follows from (8.15) and (8.21) that these periodic solutions have the frequencies

$$\omega = 1 - \frac{\epsilon}{2a} f_1(a) + \cdots \tag{8.23}$$

Using (8.10) and (8.11), we rewrite the conditions (8.22) and (8.23) for periodic solutions as

$$\int_0^{2\pi} f(a\cos\phi, -a\sin\phi) \sin\phi \, d\phi = 0 \tag{8.24}$$

$$\omega = 1 - \frac{\epsilon}{2\pi a} \int_0^{2\pi} f(a\cos\phi, -a\sin\phi) \cos\phi \, d\phi + \cdots \tag{8.25}$$

For the Duffing equation (4.7), $f = -u^3$ and (8.24) and (8.25) become

$$\int_0^{2\pi} a^3 \cos^3\phi \sin\phi \, d\phi = 0 \tag{8.26}$$

$$\omega = 1 + \frac{\epsilon}{2\pi a} \int_0^{2\pi} a^3 \cos^4\phi \, d\phi + \cdots = 1 + \tfrac{3}{8}\epsilon a^2 + \cdots \tag{8.27}$$

The integrations in (8.26) and (8.27) were carried out as in (A31) and (A37). Equation (8.26) is satisfied for all a, while (8.27) is in agreement with (4.80).

For the Rayleigh equation (6.4), $f = \dot{u} - \tfrac{1}{3}\dot{u}^3$ and (8.24) and (8.25) become

$$\int_0^{2\pi} (-a\sin\phi + \tfrac{1}{3}a^3 \sin^3\phi) \sin\phi \, d\phi = 0$$

$$\omega = 1 - \frac{\epsilon}{2\pi a} \int_0^{2\pi} (-a\sin\phi + \tfrac{1}{3}a^3 \sin^3\phi) \cos\phi \, d\phi + \cdots$$

After carrying out the integrations as in (A32), we have

$$-a + \tfrac{1}{4}a^3 = 0 \tag{8.28}$$

$$\omega = 1 + O(\epsilon^2) \tag{8.29}$$

in agreement with (6.27) and (6.28).

8.3. The Method of Multiple Scales

To determine a uniform first-order expansion by using the method of multiple scales, we introduce the two scales $T_0 = t$ and $T_1 = \epsilon t$. Then, the derivatives become

$$\frac{d}{dt} = D_0 + \epsilon D_1 + \cdots$$

$$\frac{d^2}{dt^2} = D_0^2 + 2\epsilon D_0 D_1 + \cdots$$

where $D_n = \partial/\partial T_n$. Hence, (8.1) becomes

$$D_0^2 u + 2\epsilon D_0 D_1 u + \cdots + u = \epsilon f[u, D_0 u + \epsilon D_1 u + \cdots] \qquad (8.30)$$

We seek an approximate solution to (8.30) in the form

$$u = u_0(T_0, T_1) + \epsilon u_1(T_0, T_1) + \cdots \qquad (8.31)$$

Substituting (8.31) into (8.30) gives

$$D_0^2 u_0 + \epsilon D_0^2 u_1 + 2\epsilon D_0 D_1 u_0 + \cdots + u_0 + \epsilon u_1 + \cdots = \epsilon f[u_0 + \epsilon u_1$$
$$+ \cdots, D_0 u_0 + \epsilon D_1 u_0 + \epsilon D_0 u_1 + \cdots] = \epsilon f(u_0, D_0 u_0) + \cdots$$

Equating coefficients of like powers of ϵ on both sides yields

$$D_0^2 u_0 + u_0 = 0 \qquad (8.32)$$

$$D_0^2 u_1 + u_1 = -2 D_0 D_1 u_0 + f(u_0, D_0 u_0) \qquad (8.33)$$

To be able to use directly the Fourier series expansion (8.8), we express the general solution of (8.32) in the following real rather than complex form as done before:

$$u_0 = a \cos(T_0 + \beta) \qquad (8.34)$$

Hence,

$$D_0 u_0 = -a \sin(T_0 + \beta)$$

$$D_1 D_0 u_0 = -a' \sin(T_0 + \beta) - a\beta' \cos(T_0 + \beta)$$

Then, (8.33) becomes

$$D_0^2 u_1 + u_1 = 2a' \sin(T_0 + \beta) + 2a\beta' \cos(T_0 + \beta) + f[a \cos(T_0 + \beta),$$
$$-a \sin(T_0 + \beta)] \qquad (8.35)$$

Using the Fourier-series expansion (8.8) for f, we rewrite (8.35) as

$$D_0^2 u_1 + u_1 = 2a' \sin(T_0 + \beta) + 2a\beta' \cos(T_0 + \beta) + f_0(a)$$

$$+ \sum_{n=1}^{\infty} f_n(a) \cos(nT_0 + n\beta) + \sum_{n=1}^{\infty} g_n(a) \sin(nT_0 + n\beta) \quad (8.36)$$

Eliminating the secular terms from u_1 demands that

$$2a' + g_1(a) = 0 \quad (8.37)$$

$$2a\beta' + f_1(a) = 0 \quad (8.38)$$

Substituting (8.10) and (8.11) into (8.37) and (8.38) yields

$$a' = -\frac{1}{2\pi} \int_0^{2\pi} f(a \cos \phi, -a \sin \phi) \sin \phi \, d\phi \quad (8.39)$$

$$a\beta' = -\frac{1}{2\pi} \int_0^{2\pi} f(a \cos \phi, -a \sin \phi) \cos \phi \, d\phi \quad (8.40)$$

Substituting (8.34) into (8.31) and setting $T_0 = t$, we find that to the first approximation

$$u = a \cos(t + \beta) + \cdots \quad (8.41)$$

where a and β are given by (8.39) and (8.40). We apply the present solution to a number of special cases in Section 8.5. Next, we derive (8.39) through (8.41) by using the method of averaging.

8.4. The Method of Averaging

First, we need to use the method of variation of parameters to transform the dependent variable from u to a and β where

$$u(t) = a(t) \cos[t + \beta(t)] \quad (8.42)$$

such that

$$\dot{u}(t) = -a(t) \sin[t + \beta(t)] \quad (8.43)$$

Thus, u and $\dot{u}$ have the same form as the unperturbed case for which $\epsilon = 0$. Differentiating (8.42) with respect to t yields

$$\dot{u} = -a \sin(t + \beta) + \dot{a} \cos(t + \beta) - a\dot{\beta} \sin(t + \beta) \quad (8.44)$$

Comparing (8.43) and (8.44), we conclude that

$$\dot{a} \cos(t + \beta) - a\dot{\beta} \sin(t + \beta) = 0 \quad (8.45)$$

Differentiating (8.43) with respect to t yields

$$\ddot{u} = -a \cos(t + \beta) - \dot{a} \sin(t + \beta) - a\dot{\beta} \cos(t + \beta) \quad (8.46)$$

THE METHOD OF AVERAGING   183

Substituting (8.42), (8.43), and (8.46) into (8.1) gives

$$\dot{a} \sin (t + \beta) + a\dot{\beta} \cos (t + \beta) = -\epsilon f [a \cos (t + \beta), -a \sin (t + \beta)] \tag{8.47}$$

Adding (8.45) times $\cos (t + \beta)$ to (8.47) times $\sin (t + \beta)$, we obtain

$$\dot{a} = -\epsilon \sin (t + \beta) f[a \cos (t + \beta), -a \sin (t + \beta)] \tag{8.48}$$

Substituting (8.48) into (8.45) and solving for $a\dot{\beta}$, we obtain

$$a\dot{\beta} = -\epsilon \cos (t + \beta) f[a \cos (t + \beta), -a \sin (t + \beta)] \tag{8.49}$$

Replacing f by its Fourier expansion (8.8), we rewrite (8.48) and (8.49) as

$$\dot{a} = -\epsilon \sin (t + \beta) \left[f_0(a) + \sum_{n=1}^{\infty} f_n(a) \cos (nt + n\beta) + \sum_{n=1}^{\infty} g_n(a) \sin (nt + n\beta) \right]$$

$$\tag{8.50}$$

$$a\dot{\beta} = -\epsilon \cos (t + \beta) \left[f_0(a) + \sum_{n=1}^{\infty} f_n(a) \cos (nt + n\beta) + \sum_{n=1}^{\infty} g_n(a) \sin (nt + n\beta) \right]$$

$$\tag{8.51}$$

Using trigonometric identities, we rewrite (8.50) and (8.51) as

$$\dot{a} = -\epsilon f_0(a) \sin (t + \beta) - \tfrac{1}{2} \epsilon \sum_{n=1}^{\infty} f_n(a) \{\sin [(n+1)t + (n+1)\beta] - \sin [(n-1)t$$

$$+ (n-1)\beta]\} - \tfrac{1}{2} \epsilon \sum_{n=1}^{\infty} g_n(a) \{\cos [(n-1)t + (n-1)\beta] - \cos [(n+1)t$$

$$+ (n+1)\beta]\} \tag{8.52}$$

$$a\dot{\beta} = -\epsilon f_0(a) \cos (t + \beta) - \tfrac{1}{2} \epsilon \sum_{n=1}^{\infty} f_n(a) \{\cos [(n+1)t + (n+1)\beta]$$

$$+ \cos [(n-1)t + (n-1)\beta]\} - \tfrac{1}{2} \epsilon \sum_{n=1}^{\infty} g_n(a) \{\sin [(n+1)t$$

$$+ (n+1)\beta] + \sin [(n-1)t + (n-1)\beta]\} \tag{8.53}$$

As before, to the first approximation, we keep only the slowly varying parts on the right-hand sides of (8.52) and (8.53). These parts are the terms that do not depend explicitly on t. Thus,

$$\dot{a} = -\tfrac{1}{2} \epsilon g_1(a) \tag{8.54}$$

$$a\dot{\beta} = -\tfrac{1}{2} \epsilon f_1(a) \tag{8.55}$$

184 GENERAL WEAKLY NONLINEAR SYSTEMS

which are in agreement with (8.39) and (8.40) obtained by using the method of multiple scales.

8.5. Applications

As a first application, we consider the Duffing equation (4.7). In this case $f = -u^3$ and (8.39) and (8.40) become

$$a' = \frac{1}{2\pi} \int_0^{2\pi} a^3 \cos^3 \phi \sin \phi \, d\phi = 0 \tag{8.56}$$

$$a\beta' = \frac{1}{2\pi} \int_0^{2\pi} a^3 \cos^4 \phi \, d\phi = \tfrac{3}{8} a^3 \tag{8.57}$$

in agreement with those obtained in Chapter 4. The integrations in (8.56) and (8.57) as well as in all following cases were performed as in Section A.3.

As a second application, we consider the linear damped oscillator (5.2). In this case, $f = -2\dot{u}$ and (8.39) and (8.40) become

$$a' = -\frac{1}{2\pi} \int_0^{2\pi} 2a \sin^2 \phi \, d\phi = -a \tag{8.58}$$

$$a\beta' = -\frac{1}{2\pi} \int_0^{2\pi} 2a \sin \phi \cos \phi \, d\phi = 0 \tag{8.59}$$

in agreement with those obtained in Chapter 5 to first-order.

As a third application, we consider the Rayleigh equation (6.4). In this case, $f = \dot{u} - \tfrac{1}{3}\dot{u}^3$ and (8.39) and (8.40) become

$$a' = -\frac{1}{2\pi} \int_0^{2\pi} (-a \sin \phi + \tfrac{1}{3} a^3 \sin^3 \phi) \sin \phi \, d\phi = \tfrac{1}{2} a - \tfrac{1}{8} a^3 \tag{8.60}$$

$$a\beta' = -\frac{1}{2\pi} \int_0^{2\pi} (-a \sin \phi + \tfrac{1}{3} a^3 \sin^3 \phi) \cos \phi \, d\phi = 0 \tag{8.61}$$

in agreement with those obtained in Chapter 6.

As a fourth application, we consider (7.9). Since there is no small parameter that appears explicitly in (7.9), we introduce one by setting $u = \epsilon v$ and obtain

$$\ddot{v} + v + \epsilon \alpha_2 v^2 + \epsilon^2 \alpha_3 v^3 = 0 \tag{8.62}$$

Hence, $f = -\alpha_2 v^2$ and (8.39) and (8.40) become

$$a' = \frac{1}{2\pi} \int_0^{2\pi} \alpha_2 a^2 \cos^2 \phi \sin \phi \, d\phi = 0 \tag{8.63}$$

$$a\beta' = \frac{1}{2\pi} \int_0^{2\pi} \alpha_2 a^2 \cos^3 \phi \, d\phi = 0 \tag{8.64}$$

in agreement with those obtained in Chapter 7. We note that the effect of nonlinearity on the amplitude and phase appears at second order.

In the remaining two applications, we consider nonanalytic functions f. In the first case, we consider

$$\ddot{u} + u = -\epsilon \dot{u}|\dot{u}| \tag{8.65}$$

Hence, $f = -\dot{u}|\dot{u}|$ and (8.39) and (8.40) become

$$a' = -\frac{1}{2\pi} \int_0^{2\pi} a^2 \sin^2 \phi \, |\sin \phi| \, d\phi \tag{8.66}$$

$$a\beta' = -\frac{1}{2\pi} \int_0^{2\pi} a^2 \sin \phi \cos \phi \, |\sin \phi| \, d\phi \tag{8.67}$$

To perform the integrations in (8.66) and (8.67), we note that $\sin \phi \geq 0$ for $0 \leq \phi \leq \pi$ and $\sin \phi \leq 0$ for $\pi \leq \phi \leq 2\pi$. Hence, $|\sin \phi| = \sin \phi$ in the first interval and $|\sin \phi| = -\sin \phi$ in the second interval. Consequently, we break the integration interval in (8.66) and (8.67) to the intervals $[0, \pi]$ and $[\pi, 2\pi]$, so that we replace $|\sin \phi|$ by $\sin \phi$ in the first interval and by $-\sin \phi$ in the second interval. Then, we rewrite (8.66) as

$$a' = -\frac{a^2}{2\pi} \int_0^\pi \sin^3 \phi \, d\phi + \frac{a^2}{2\pi} \int_\pi^{2\pi} \sin^3 \phi \, d\phi = -\frac{a^2}{8\pi} \int_0^\pi (3 \sin \phi - \sin 3\phi) \, d\phi$$

$$+ \frac{a^2}{8\pi} \int_\pi^{2\pi} (3 \sin \phi - \sin 3\phi) \, d\phi = \frac{a^2}{8\pi} (3 \cos \phi - \tfrac{1}{3} \cos 3\phi) \Big|_0^\pi$$

$$- \frac{a^2}{8\pi} (3 \cos \phi - \tfrac{1}{3} \cos 3\phi) \Big|_\pi^{2\pi}$$

or

$$a' = -\frac{4}{3\pi} a^2 \tag{8.68}$$

Also, we rewrite (8.67) as

186 GENERAL WEAKLY NONLINEAR SYSTEMS

$$a\beta' = -\frac{a^2}{2\pi}\int_0^\pi \sin^2\phi\cos\phi\, d\phi + \frac{a^2}{2\pi}\int_\pi^{2\pi}\sin^2\phi\cos\phi\, d\phi = -\frac{a^2}{6\pi}\sin^3\phi\Big|_0^\pi$$

$$+\frac{a^2}{6\pi}\sin^3\phi\Big|_\pi^{2\pi}$$

or

$$a\beta' = 0 \tag{8.69}$$

Equation (8.69) leads to $\beta = \beta_0 =$ constant, whereas (8.68) can be integrated by separation of variables. Thus,

$$-\frac{da}{a^2} = \frac{4}{3\pi}dT_1 \tag{8.70}$$

Integrating (8.70) gives

$$\frac{1}{a} + c = \frac{4}{3\pi}T_1 = \frac{4}{3\pi}\epsilon t \tag{8.71}$$

where c is a constant. If $a(0) = a_0$, then $c = -1/a_0$. Hence (8.71) becomes

$$\frac{1}{a} = \frac{1}{a_0} + \frac{4}{3\pi}\epsilon t$$

or

$$a = \frac{a_0}{1 + \frac{4}{3\pi}\epsilon t a_0} \tag{8.72}$$

Next, we consider

$$\ddot{u} + u = -\epsilon u|u| \tag{8.73}$$

In this case, $f = -u|u|$ and (8.39) and (8.40) become

$$a' = \frac{1}{2\pi}\int_0^{2\pi} a^2 \cos\phi\, |\cos\phi|\sin\phi\, d\phi \tag{8.74}$$

$$a\beta' = \frac{1}{2\pi}\int_0^{2\pi} a^2 \cos^2\phi\, |\cos\phi|\, d\phi \tag{8.75}$$

We note that $\cos\phi \geq 0$ for $-\frac{1}{2}\pi \leq \phi \leq \frac{1}{2}\pi$ and $\cos\phi \leq 0$ for $\frac{1}{2}\pi \leq \phi \leq \frac{3}{2}\pi$. Since the integrands in (8.74) and (8.75) are periodic with the period 2π, the values of

APPLICATIONS 187

the integrals are independent of the interval. Hence, we change the integration interval from $[0, 2\pi]$ to $[-\frac{1}{2}\pi, \frac{3}{2}\pi]$ and break it into the two intervals $[-\frac{1}{2}\pi, \frac{1}{2}\pi]$ and $[\frac{1}{2}\pi, \frac{3}{2}\pi]$, so that we can replace $|\cos\phi|$ by $\cos\phi$ in the first interval and by $-\cos\phi$ in the second interval. Consequently, we rewrite (8.74) as

$$a' = \frac{a^2}{2\pi} \int_{-(1/2)\pi}^{(1/2)\pi} \cos^2\phi \sin\phi \, d\phi - \frac{a^2}{2\pi} \int_{(1/2)\pi}^{(3/2)\pi} \cos^2\phi \sin\phi \, d\phi$$

$$= -\frac{a^2}{6\pi} \cos^3\phi \bigg|_{-(1/2)\pi}^{(1/2)\pi} + \frac{a^2}{6\pi} \cos^3\phi \bigg|_{(1/2)\pi}^{(3/2)\pi}$$

or

$$a' = 0 \tag{8.76}$$

Also, we rewrite (8.75) as

$$a\beta' = \frac{a^2}{2\pi} \int_{-(1/2)\pi}^{(1/2)\pi} \cos^3\phi \, d\phi - \frac{a^2}{2\pi} \int_{(1/2)\pi}^{(3/2)\pi} \cos^3\phi \, d\phi$$

$$= \frac{a^2}{8\pi} \int_{-(1/2)\pi}^{(1/2)\pi} (3\cos\phi + \cos 3\phi) \, d\phi - \frac{a^2}{8\pi} \int_{(1/2)\pi}^{(3/2)\pi} (3\cos\phi + \cos 3\phi) \, d\phi$$

$$= \frac{a^2}{8\pi} (3\sin\phi + \tfrac{1}{3}\sin 3\phi) \bigg|_{-(1/2)\pi}^{(1/2)\pi} - \frac{a^2}{8\pi} (3\sin\phi + \tfrac{1}{3}\sin 3\phi) \bigg|_{(1/2)\pi}^{(3/2)\pi}$$

or

$$a\beta' = \frac{4}{3\pi} a^2 \tag{8.77}$$

The solution of (8.76) is $a = a_0 =$ constant. If $a_0 \neq 0$, the solution of (8.77) is

$$\beta = \frac{4}{3\pi} a_0 T_1 + \beta_0 = \frac{4}{3\pi} \epsilon t a_0 + \beta_0 \tag{8.78}$$

where β_0 is a constant. Hence, it follows from (8.41) that to the first approximation

$$u = a_0 \cos\left[\left(1 + \frac{4}{3\pi}\epsilon a_0\right) t + \beta_0\right] + \cdots \tag{8.79}$$

These applications show that the method of multiple scales and method of averaging can be used effectively to determine first-order uniform expansions for weakly nonlinear oscillatory systems.

188 GENERAL WEAKLY NONLINEAR SYSTEMS

Exercises

8.1. For small ϵ, determine first-order uniform expansions for each of the following problems:

(a) $\ddot{u} + u + \epsilon u|u| = 0$
(b) $\ddot{u} + u + \epsilon(\text{sgn } \dot{u} + 2\mu_1 \dot{u}) = 0$
(c) $\ddot{u} + u + \epsilon(\text{sgn } \dot{u} + \mu_2 \dot{u}|\dot{u}|) = 0$

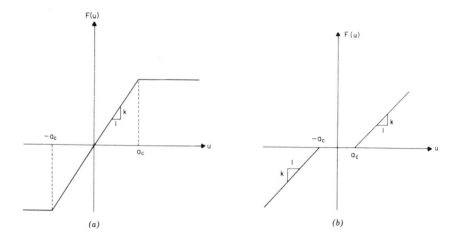

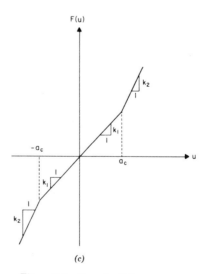

Figure 8-1. Exercise 8.2.

(d) $\ddot{u} + u + \epsilon(2\mu_1\dot{u} + \mu_2\dot{u}|\dot{u}|) = 0$
(e) $\ddot{u} + u + \epsilon(2\mu_1\dot{u} + \text{sgn } \dot{u} + \mu_2\dot{u}|\dot{u}|) = 0$

8.2. Consider the free oscillations of a system governed by

$$\ddot{u} + F(u) = 0$$

where $F(u)$ is defined in Figure 8-1 for three different cases. Show that in the first approximation

(a) $\omega_0^2 = \dfrac{2k}{\pi}\left[\sin^{-1}\left(\dfrac{a_c}{a}\right) + \dfrac{a_c}{a}\left(1 - \dfrac{a_c^2}{a^2}\right)^{1/2}\right]$

(b) $\omega_0^2 = k - \dfrac{2k}{\pi}\left[\sin^{-1}\left(\dfrac{a_c}{a}\right) + \dfrac{a_c}{a}\left(1 - \dfrac{a_c^2}{a^2}\right)^{1/2}\right]$

(c) $\omega_0^2 = k_2 - \dfrac{2}{\pi}(k_2 - k_1)\left[\sin^{-1}\left(\dfrac{a_c}{a}\right) + \dfrac{a_c}{a}\left(1 - \dfrac{a_c^2}{a^2}\right)^{1/2}\right]$

CHAPTER 9

Forced Oscillations of the Duffing Equation

In contrast with the five preceding chapters, which deal with free oscillations, this chapter and the next deal with forced oscillations. To compare the forced- and free-oscillation cases, we consider one of the systems discussed in the preceding chapter. We choose the Duffing equation (4.5) and consider in this chapter its response to a sinusoidal external excitation. That is, we consider

$$\frac{d^2 u^*}{dt^{*2}} + k_1 u^* + k_3 u^{*3} = F^* \cos \omega^* t^* \tag{9.1}$$

where F^* and ω^* are constants. As in Chapter 4, we introduce dimensionless quantities using a characteristic time T^* and a characteristic length U^*. Thus, we write

$$t = \frac{t^*}{T^*} \quad u = \frac{u^*}{U^*}$$

Then, (9.1) becomes

$$\frac{d^2 u}{dt^2} + k_1 T^{*2} u + k_3 T^{*2} U^{*2} u^3 = \frac{F^* T^{*2}}{U^*} \cos \omega^* T^* t \tag{9.2}$$

It is convenient to choose T^* so that

$$k_1 T^{*2} = 1 \quad \text{thus} \quad T^* = \frac{1}{\sqrt{k_1}} = \frac{1}{\omega_0^*}$$

where ω_0^* is the linear natural frequency of the system. Moreover, we let

$$\epsilon = k_3 T^{*2} U^{*2} \quad F = \frac{F^* T^{*2}}{U^*} \quad \omega = \omega^* T^* = \frac{\omega^*}{\omega_0^*}$$

Then, (9.2) becomes

$$\ddot{u} + u + \epsilon u^3 = F \cos \omega t \tag{9.3}$$

We note that ω is the ratio of the frequency of the excitation to the linear natural frequency of the system. Instead of treating (9.3), we treat the following slightly more general equation:

$$\ddot{u} + u + 2\epsilon\mu\dot{u} + \epsilon u^3 = F \cos \omega t \tag{9.4}$$

where μ is a positive constant. Thus, (9.4) includes the effect of small viscous damping.

In this chapter, we determine first-order solutions to (9.4) beginning with the straightforward expansion in the next section. We investigate this straightforward expansion and determine the conditions for its breakdown. This leads to the so-called resonant values of ω. In Sections 9.2 and 9.3, we use the methods of multiple scales and averaging, respectively, to determine first-order expansions for all resonant cases, including the effect of slight viscous damping.

9.1. The Straightforward Expansion

We seek a straightforward expansion for the solution of (9.4) in the form

$$u(t; \epsilon) = u_0(t) + \epsilon u_1(t) + \cdots \tag{9.5}$$

Substituting (9.5) into (9.4) yields

$$\ddot{u}_0 + \epsilon\ddot{u}_1 + \cdots + u_0 + \epsilon u_1 + \cdots + 2\epsilon\mu(\dot{u}_0 + \epsilon\dot{u}_1 + \cdots)$$
$$+ \epsilon(u_0 + \epsilon u_1 + \cdots)^3 = F \cos \omega t$$

or

$$\ddot{u}_0 + u_0 - F \cos \omega t + \epsilon(\ddot{u}_1 + u_1 + 2\mu\dot{u}_0 + u_0^3) + \cdots = 0 \tag{9.6}$$

Equating each of the coefficients of ϵ^0 and ϵ to zero, we have

$$\ddot{u}_0 + u_0 = F \cos \omega t \tag{9.7}$$

$$\ddot{u}_1 + u_1 = -2\mu\dot{u}_0 - u_0^3 \tag{9.8}$$

Since (9.7) is linear and inhomogeneous, its general solution can be obtained as the sum of a homogeneous solution and any particular solution. The homogeneous solution can be expressed as

$$u_{0_h} = a \cos(t + \beta) \tag{9.9}$$

where a and β are constants, whereas a particular solution is (B69)

$$u_{0_p} = \frac{F}{1 - \omega^2} \cos \omega t \tag{9.10}$$

Hence,

$$u_0 = a \cos(t + \beta) + 2\Lambda \cos \omega t \tag{9.11}$$

192 FORCED OSCILLATIONS OF THE DUFFING EQUATION

where
$$\Lambda = \tfrac{1}{2}(1 - \omega^2)^{-1} F \qquad (9.12)$$

Substituting (9.11) into (9.8) gives

$$\ddot{u}_1 + u_1 = 2\mu[a \sin(t + \beta) + 2\Lambda\omega \sin \omega t] - [a \cos(t + \beta) + 2\Lambda \cos \omega t]^3$$
$$= 2\mu a \sin(t + \beta) + 4\mu\Lambda\omega \sin \omega t - a^3 \cos^3(t + \beta)$$
$$- 6a^2\Lambda \cos^2(t + \beta) \cos \omega t - 12a\Lambda^2 \cos(t + \beta) \cos^2 \omega t$$
$$- 8\Lambda^3 \cos^3 \omega t \qquad (9.13)$$

Using trigonometric identities (Appendix A), we rewrite (9.13) as

$$\ddot{u}_1 + u_1 = 2\mu a \sin(t + \beta) + 4\mu\Lambda\omega \sin \omega t - \tfrac{1}{4}a^3 \cos(3t + 3\beta) - (\tfrac{3}{4}a^3$$
$$+ 6a\Lambda^2) \cos(t + \beta) - 2\Lambda^3 \cos 3\omega t - (6\Lambda^3 + 3a^2\Lambda) \cos \omega t$$
$$- \tfrac{3}{2}a^2\Lambda \cos[(2 + \omega)t + 2\beta] - \tfrac{3}{2}a^2\Lambda \cos[(2 - \omega)t + 2\beta]$$
$$- 3a\Lambda^2 \cos[(1 + 2\omega)t + \beta] - 3a\Lambda^2 \cos[(1 - 2\omega)t + \beta] \qquad (9.14)$$

As before, we do not include the homogeneous solution at any order except the first. Since (9.14) is linear, a particular solution can be obtained as the sum of particular solutions, each one corresponding to a different inhomogeneous term. It follows from Section B.4 that

$$u_1 = -\mu a t \cos(t + \beta) + \frac{4\mu\Lambda\omega}{1 - \omega^2} \sin \omega t + \tfrac{1}{32}a^3 \cos(3t + 3\beta)$$
$$- (\tfrac{3}{8}a^3 + 3a\Lambda^2)t \sin(t + \beta) - \frac{2\Lambda^3}{1 - 9\omega^2} \cos 3\omega t$$
$$- \frac{6\Lambda^3 + 3a^2\Lambda}{1 - \omega^2} \cos \omega t + \frac{3a^2\Lambda}{2(\omega^2 + 4\omega + 3)} \cos[(2 + \omega)t + 2\beta]$$
$$+ \frac{3a^2\Lambda}{2(\omega^2 - 4\omega + 3)} \cos[(2 - \omega)t + 2\beta] + \frac{3a\Lambda^2}{4(\omega^2 + \omega)}$$
$$\times \cos[(1 + 2\omega)t + \beta] + \frac{3a\Lambda^2}{4(\omega^2 - \omega)} \cos[(1 - 2\omega)t + \beta] \qquad (9.15)$$

Substituting (9.11) and (9.15) into (9.5) and using (9.12), we obtain

$$u = a \cos(t + \beta) + \frac{F}{1 - \omega^2} \cos \omega t + \epsilon \left\{ -\mu a t \cos(t + \beta) + \frac{2\mu\omega F}{(1 - \omega^2)^2} \sin \omega t \right.$$
$$\left. + \tfrac{1}{32}a^3 \cos(3t + 3\beta) - \tfrac{3}{4}a \left[\tfrac{1}{2}a^2 + \frac{F^2}{(1 - \omega^2)^2}\right] t \sin(t + \beta) \right.$$

$$-\frac{F^3}{4(1-\omega^2)^3(1-9\omega^2)}\cos 3\omega t - \frac{3F}{2(1-\omega^2)^2}\left[a^2 + \frac{F^2}{2(1-\omega^2)^2}\right]\cos \omega t$$

$$+\frac{3a^2 F}{4(1-\omega^2)(3+4\omega+\omega^2)}\cos[(2+\omega)t + 2\beta]$$

$$+\frac{3a^2 F}{4(1-\omega^2)(3-\omega)(1-\omega)}\cos[(2-\omega)t + 2\beta]$$

$$+\frac{3aF^2}{16(1-\omega^2)^2(\omega^2+\omega)}\cos[(1+2\omega)t + \beta]$$

$$+\frac{3aF^2}{16(1-\omega^2)^2(\omega^2-\omega)}\cos[(1-2\omega)t + \beta]\Bigg\} + \cdots \qquad (9.16)$$

In addition to the secular terms, we note that (9.16) contains terms whose denominators may be very small. Such terms are called *small-divisor terms*. If the frequencies are defined to be positive, small divisors occur when $\omega \approx 1$, $\omega \approx 0$, $\omega \approx \frac{1}{3}$, and $\omega \approx 3$. These special frequencies are called *resonant frequencies*. Thus, the straightforward expansion breaks down due to the presence of the small divisors as well as the secular terms.

When $\omega \approx 1$, small divisors first appear in the first term. Hence, when $\omega \approx 1$ we speak of a *primary or main resonance*. When $\omega \approx 0$, $\frac{1}{3}$, or 3, small divisors first appear in the second term. Hence, we speak of the resonances in these cases as *secondary resonances*. Carrying out the expansion to higher order, one finds that other resonances may occur. We note that the resonances that occur depend on the order of the nonlinearity. They can be easily identified by carrying out a straightforward expansion as done above.

In the next two sections, we use the methods of multiple scales and averaging to determine first-order uniform expansions for (9.4) that do not contain secular or small-divisor terms.

9.2. The Method of Multiple Scales

To determine an approximate solution to (9.4) free of secular and small-divisor terms, we need to distinguish between secondary and primary resonances. They are treated separately beginning with secondary resonances.

9.2.1. SECONDARY RESONANCES

In this case ω is away from 1 and small divisors first appear at $O(\epsilon)$. We introduce a slow scale $T_1 = \epsilon t$ in addition to the fast scale $T_0 = t$ so that the derivatives are transformed according to

$$\frac{d}{dt} = D_0 + \epsilon D_1 + \cdots$$

194 FORCED OSCILLATIONS OF THE DUFFING EQUATION

$$\frac{d^2}{dt^2} = D_0^2 + 2\epsilon D_0 D_1 + \cdots$$

where $D_n = \partial/\partial T_n$. Since t appears explicitly in the governing equation, the question arises as to whether it should be represented in terms of T_0 or T_1. To answer this question, we check whether the dependence on t is fast or slow. In this case, we check $\cos \omega t$. If ω is away from zero, $\cos \omega t$ is fast varying, and we write

$$\cos \omega t = \cos \omega T_0 \tag{9.17}$$

That is, t is represented in terms of T_0. On the other hand, if $\omega \approx 0$, $\cos \omega t$ is slowly varying. In this case, we write $\omega = \epsilon \sigma$, where $\sigma = O(1)$ to exhibit explicitly the smallness of ω. Then,

$$\cos \omega t = \cos \sigma \epsilon t = \cos \sigma T_1 \tag{9.18}$$

Thus, t is represented in terms of T_1. Consequently, the case $\omega \approx 0$ appears to require an independent treatment.

If ω is away from zero, (9.4) is transformed into

$$D_0^2 u + 2\epsilon D_0 D_1 u + 2\epsilon\mu D_0 u + \cdots + u + \epsilon u^3 = F \cos \omega T_0 \tag{9.19}$$

We seek an approximate solution to (9.19) in the form

$$u = u_0(T_0, T_1) + \epsilon u_1(T_0, T_1) + \cdots \tag{9.20}$$

Substituting (9.20) into (9.19) and equating coefficients of like powers of ϵ, we obtain

$$D_0^2 u_0 + u_0 = F \cos \omega T_0 \tag{9.21}$$

$$D_0^2 u_1 + u_1 = -2D_0 D_1 u_0 - 2\mu D_0 u_0 - u_0^3 \tag{9.22}$$

The general solution of (9.21) can be expressed as

$$u_0 = a(T_1) \cos [T_0 + \beta(T_1)] + 2\Lambda \cos \omega T_0 \tag{9.23}$$

or

$$u_0 = A(T_1)e^{iT_0} + \Lambda e^{i\omega T_0} + cc \tag{9.24}$$

where

$$A = \tfrac{1}{2} a e^{i\beta} \qquad \Lambda = \frac{F}{2(1 - \omega^2)} \tag{9.25}$$

Then, (9.22) becomes

$$D_0^2 u_1 + u_1 = -2i(A' + \mu A)e^{iT_0} + 2i(\bar{A}' + \mu \bar{A})e^{-iT_0} - 2i\mu\omega\Lambda e^{i\omega T_0}$$
$$+ 2i\mu\omega\Lambda e^{-i\omega T_0} - [Ae^{iT_0} + \Lambda e^{i\omega T_0} + \bar{A}e^{-iT_0} + \Lambda e^{-i\omega T_0}]^3$$

or

$$D_0^2 u_1 + u_1 = -[2i(A' + \mu A) + 3(A\bar{A} + 2\Lambda^2)A]e^{iT_0} - (2i\mu\omega + 6A\bar{A}$$
$$+ 3\Lambda^2)\Lambda e^{i\omega T_0} - A^3 e^{3iT_0} - \Lambda^3 e^{3i\omega T_0} - 3A^2 \Lambda e^{i(2+\omega)T_0}$$
$$- 3\bar{A}^2 \Lambda e^{i(\omega-2)T_0} - 3A\Lambda^2 e^{i(1+2\omega)T_0} - 3A\Lambda^2 e^{i(1-2\omega)T_0}$$
$$+ cc \tag{9.26}$$

As mentioned in the preceding section, the particular solution of (9.26) contains secular terms and small-divisor terms when $\omega \approx 3$, $\frac{1}{3}$, and 0. These cases need to be considered separately.

The Case $\omega \approx 3$. To express the nearness of ω to 3, we introduce a detuning parameter $\sigma = O(1)$ defined by

$$\omega = 3 + \epsilon\sigma \tag{9.27}$$

Substituting (9.27) into (9.26) gives

$$D_0^2 u_1 + u_1 = -[2iA' + 2i\mu A + 3A^2\bar{A} + 6A\Lambda^2]e^{iT_0} - A^3 e^{3iT_0} - (2i\mu\omega + 6A\bar{A}$$
$$+ 3\Lambda^2)\Lambda e^{3iT_0 + i\sigma\epsilon T_0} - \Lambda^3 e^{9iT_0 + 3i\sigma\epsilon T_0} - 3A^2 \Lambda e^{5iT_0 + i\sigma\epsilon T_0}$$
$$- 3\bar{A}^2 \Lambda e^{iT_0 + i\sigma\epsilon T_0} - 3A\Lambda^2 e^{7iT_0 + 2i\sigma\epsilon T_0}$$
$$- 3A\Lambda^2 e^{-5iT_0 - 2i\sigma\epsilon T_0} + cc \tag{9.28}$$

Although T_0 is a fast scale, the combination ϵT_0 is slow, and it should be expressed in terms of the slow scale. That is, $T_1 = \epsilon T_0$. Then, (9.28) can be written as

$$D_0^2 u_1 + u_1 = -[2iA' + 2i\mu A + 3A^2\bar{A} + 6A\Lambda^2]e^{iT_0} - [A^3 + (2i\mu\omega + 6A\bar{A}$$
$$+ 3\Lambda^2)\Lambda e^{i\sigma T_1}]e^{3iT_0} - \Lambda^3 e^{3i\sigma T_1} e^{9iT_0} - 3A^2 \Lambda e^{i\sigma T_1} e^{5iT_0}$$
$$- 3\bar{A}^2 \Lambda e^{i\sigma T_1} e^{iT_0} - 3A\Lambda^2 e^{2i\sigma T_1} e^{7iT_0} - 3A\Lambda^2 e^{-2i\sigma T_1} e^{-5iT_0} + cc$$
$$\tag{9.29}$$

We note that this introduction of the detuning parameter as in (9.27) resulted in the conversion of the term

$$-3\bar{A}^2 \Lambda \exp [i(\omega - 2)T_0]$$

which leads to a small-divisor term in the straightforward expansion to

$$-3\bar{A}^2 \Lambda \exp (i\sigma T_1) \exp (iT_0)$$

which leads to a secular term on the scale T_0. This is the approach we always use to deal with terms that lead to small-divisor terms. That is, we use detuning parameters to transform them into terms that lead to secular terms.

Eliminating the secular terms from u_1, we obtain from (9.29) that

$$2iA' + 2i\mu A + 3A^2\bar{A} + 6A\Lambda^2 + 3\bar{A}^2\Lambda e^{i\sigma T_1} = 0 \tag{9.30}$$

We should note that one does not need to write all the terms in (9.28) and (9.29). One needs only to write the terms that produce secular terms and the terms that produce small divisors. As before, we introduce the polar notation (9.25) at this stage. The result is

$$ia'e^{i\beta} - a\beta'e^{i\beta} + i\mu a e^{i\beta} + \tfrac{3}{8}a^3 e^{i\beta} + 3a\Lambda^2 e^{i\beta} + \tfrac{3}{4}a^2\Lambda e^{i(\sigma T_1 - 2\beta)} = 0 \tag{9.31}$$

Before separating (9.31) into real and imaginary parts, we multiply it by $\exp(-i\beta)$ so that a' and β' will not have an exponential multiplicative factor. This simplifies the resulting equations. Thus, multiplying (9.31) by $\exp(-i\beta)$ gives

$$ia' - a\beta' + i\mu a + \tfrac{3}{8}a^3 + 3a\Lambda^2 + \tfrac{3}{4}a^2\Lambda e^{i(\sigma T_1 - 3\beta)} = 0 \tag{9.32}$$

Consequently, there is only one term that contains an exponential factor. Since

$$e^{i\theta} = \cos\theta + i\sin\theta$$

we write (9.32) as

$$ia' - a\beta' + i\mu a + \tfrac{3}{8}a^3 + 3a\Lambda^2 + \tfrac{3}{4}a^2\Lambda[\cos(\sigma T_1 - 3\beta) + i\sin(\sigma T_1 - 3\beta)] = 0$$

or

$$i[a' + \mu a + \tfrac{3}{4}a^2\Lambda \sin(\sigma T_1 - 3\beta)] - a\beta' + \tfrac{3}{8}a^3 + 3a\Lambda^2 + \tfrac{3}{4}a^2\Lambda \cos(\sigma T_1 - 3\beta) = 0 \tag{9.33}$$

Since a complex number vanishes if and only if its real and imaginary parts vanish independently, (9.33) implies that

$$a' = -\mu a - \tfrac{3}{4}a^2\Lambda \sin(\sigma T_1 - 3\beta) \tag{9.34}$$

$$a\beta' = 3a\Lambda^2 + \tfrac{3}{8}a^3 + \tfrac{3}{4}a^2\Lambda \cos(\sigma T_1 - 3\beta) \tag{9.35}$$

These are the desired equations that describe the modulation of the amplitude and the phase of the free-oscillation term.

Substituting (9.23) into (9.20) and recalling that $T_0 = t$ gives

$$u = a\cos(t + \beta) + 2\Lambda\cos\omega t + O(\epsilon) \tag{9.36}$$

where a and β are given by (9.34) and (9.35). Since T_1 appears explicitly in (9.34) and (9.35), they are called a *nonautonomous system*. It is convenient to eliminate the explicit dependence on T_1, thereby transforming these equations into an *autonomous system*. This can be accomplished by introducing the new dependent variable γ defined by

$$\gamma = \sigma T_1 - 3\beta \tag{9.37}$$

Then,

THE METHOD OF MULTIPLE SCALES

$$\gamma' = \sigma - 3\beta' \tag{9.38}$$

Substituting (9.37) into (9.34) yields

$$a' = -\mu a - \tfrac{3}{4} a^2 \Lambda \sin \gamma \tag{9.39}$$

Substituting (9.35) into (9.38) and using (9.37) yields

$$a\gamma' = \sigma a - 9a\Lambda^2 - \tfrac{9}{8} a^3 - \tfrac{9}{4} a^2 \Lambda \cos \gamma \tag{9.40}$$

It follows from (9.37) that

$$\beta = \tfrac{1}{3} \sigma T_1 - \tfrac{1}{3}\gamma = \tfrac{1}{3}\epsilon\sigma t - \tfrac{1}{3}\gamma$$

Hence, (9.36) can be rewritten as

$$u = a \cos (t + \tfrac{1}{3}\epsilon\sigma t - \tfrac{1}{3}\gamma) + 2\Lambda \cos \omega t + O(\epsilon)$$

which in turn can be rewritten as

$$u = a \cos (\tfrac{1}{3}\omega t - \tfrac{1}{3}\gamma) + 2\Lambda \cos \omega t + O(\epsilon) \tag{9.41}$$

on account of (9.27). Thus to the first approximation, u is given by (9.41) where a and γ are given by the autonomous set of equations (9.39) and (9.40).

Figure 9-1 shows the variation of a and γ with T_1 as calculated by numerically integrating equations (9.39) and (9.40). Initially, a and γ oscillate with T_1, but as T_1 increases a and γ tend to constant values. There are two possibilities: either the steady-state value of a is zero or nonzero. These constant values are usually referred to as *stationary* or *steady-state values*. When the steady-state value of a is nonzero, the free-oscillation term is periodic. We note that in this case the frequency of the free-oscillation term is exactly $\tfrac{1}{3}\omega$, that is one third of the frequency of the external excitation. Consequently, we speak of such

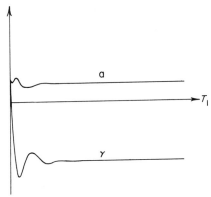

Figure 9-1. Variation of a and γ with t as numerically calculated from (9.39) and (9.40) for $\epsilon = 0.1$, $\Lambda = \sqrt{.08}$, $\mu = 0.1$, $\sigma = 1.0$, $a(0) = 1.0$, and $\gamma(0) = 1.0$.

198 FORCED OSCILLATIONS OF THE DUFFING EQUATION

resonances as *subharmonic resonances of one third*. In this case, u is periodic, that is, the steady-state response is periodic.

To determine the steady-state response, we need not integrate numerically equations (9.39) and (9.40) describing a and γ. Instead, we use the fact that a and γ are constants in the steady state, and hence set $a' = 0$ and $\gamma' = 0$ in (9.39) and (9.40). The result is

$$-3\mu a = \tfrac{9}{4} a^2 \Lambda \sin \gamma$$
$$\sigma a - 9a\Lambda^2 - \tfrac{9}{8} a^3 = \tfrac{9}{4} a^2 \Lambda \cos \gamma \qquad (9.42)$$

Squaring (9.42), adding the results, and noting that

$$\sin^2 \gamma + \cos^2 \gamma = 1 \qquad (9.43)$$

we obtain

$$9\mu^2 a^2 + (\sigma - 9\Lambda^2 - \tfrac{9}{8} a^2)^2 a^2 = \tfrac{81}{16} a^4 \Lambda^2 \qquad (9.44)$$

It follows from (9.44) that either $a = 0$ or

$$9\mu^2 + (\sigma - 9\Lambda^2 - \tfrac{9}{8} a^2)^2 = \tfrac{81}{16} a^2 \Lambda^2$$

Hence,

$$a^4 - \tfrac{16}{9}(\sigma - \tfrac{27}{4}\Lambda^2) a^2 + \tfrac{64}{81}[9\mu^2 + (\sigma - 9\Lambda^2)^2] = 0 \qquad (9.45)$$

Since (9.45) is quadratic in a^2, its solutions are

$$a^2 = \tfrac{8}{9}(\sigma - \tfrac{27}{4}\Lambda^2) \pm \tfrac{8}{9}[(\sigma - \tfrac{27}{4}\Lambda^2)^2 - 9\mu^2 - (\sigma - 9\Lambda^2)^2]^{1/2}$$

or

$$a^2 = \tfrac{8}{9}(\sigma - \tfrac{27}{4}\Lambda^2) \pm \tfrac{8}{3}[\tfrac{1}{4}\Lambda^2(2\sigma - \tfrac{63}{4}\Lambda^2) - \mu^2]^{1/2} \qquad (9.46)$$

Equation (9.46) is usually referred to as the *frequency-response equation*.

For nontrivial solutions, it follows from (9.46) that both the radical and the first term must be positive, that is,

$$\tfrac{1}{4}\Lambda^2(2\sigma - \tfrac{63}{4}\Lambda^2) \geq \mu^2 \qquad \sigma > \tfrac{27}{4}\Lambda^2$$

Thus, for a given σ, nontrivial solutions can exist only if

$$\frac{\sigma}{\mu} - \left(\frac{\sigma^2}{\mu^2} - 63\right)^{1/2} \leq \frac{63\Lambda^2}{4\mu} \leq \frac{\sigma}{\mu} + \left(\frac{\sigma^2}{\mu^2} - 63\right)^{1/2}$$

These conditions are represented graphically in Figure 9-2. Figure 9-3a shows several frequency-response curves, whereas Figure 9-3b shows the variation of the amplitude of the free-oscillation term with the amplitude of the excitation.

We note that although the frequency of the excitation is three times the natural frequency of the system, the response is quite large. For example, certain parts of an airplane can be violently excited by an engine running at an angular speed

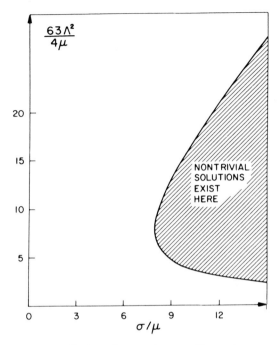

Figure 9-2. Regions where steady-state subharmonic responses exist.

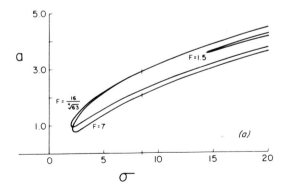

Figure 9.3. Steady-state subharmonic response for the Duffing equation; amplitude of the free-oscillation term versus (*a*) detuning and (*b*) amplitude of the excitation.

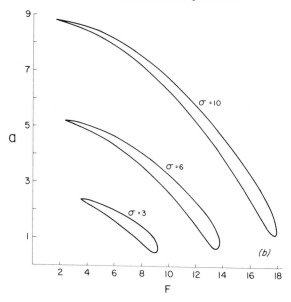

Figure 9-3. (cont.)

that is much larger than their natural frequencies. At one time, the propellers in a commercial airplane induced a subharmonic vibration of order $\frac{1}{2}$ in the wings, which in turn induced a subharmonic of order $\frac{1}{4}$ in the rudder. The oscillations were so violent that the airplane broke up.

The Case of $\omega \approx \frac{1}{3}$. To express the nearness of ω to $\frac{1}{3}$, we introduce a detuning parameter σ defined by

$$3\omega = 1 + \epsilon\sigma \tag{9.47}$$

As pointed earlier, we do not need to replace ω in all the terms in (9.26). We need only to replace the terms that produce small-divisor terms. These are the terms $-\Lambda^3 \exp(\pm 3i\omega T_0)$. To this end, we write

$$3\omega T_0 = (1 + \epsilon\sigma)T_0 = T_0 + \sigma\epsilon T_0 = T_0 + \sigma T_1 \tag{9.48}$$

Then, we rewrite (9.26) as

$$D_0^2 u_1 + u_1 = -[2iA' + 2i\mu A + 6\Lambda^2 A + 3A^2\overline{A}]e^{iT_0} - \Lambda^3 e^{i\sigma T_1}e^{iT_0} + cc + \text{NST} \tag{9.49}$$

where NST stands for the rest of the terms, which do not produce secular terms. Eliminating the secular terms from u_1 yields

$$2i(A' + \mu A) + 6\Lambda^2 A + 3A^2\overline{A} + \Lambda^3 e^{i\sigma T_1} = 0 \tag{9.50}$$

THE METHOD OF MULTIPLE SCALES 201

At this stage, we introduce the polar transformation (9.25) and rewrite (9.50) as

$$ia'e^{i\beta} - a\beta'e^{i\beta} + i\mu a e^{i\beta} + \tfrac{3}{8}a^3 e^{i\beta} + 3\Lambda^2 a e^{i\beta} + \Lambda^3 e^{i\sigma T_1} = 0 \quad (9.51)$$

Multiplying (9.51) by exp $(-i\beta)$ so that only one term contains an exponential term, we obtain

$$ia' - a\beta' + i\mu a + \tfrac{3}{8}a^3 + 3\Lambda^2 a + \Lambda^3 e^{i(\sigma T_1 - \beta)} = 0$$

or

$$ia' - a\beta' + i\mu a + \tfrac{3}{8}a^3 + 3\Lambda^2 a + \Lambda^3 \cos(\sigma T_1 - \beta) + i\Lambda^3 \sin(\sigma T_1 - \beta) = 0 \quad (9.52)$$

Separating real and imaginary parts in (9.52) gives

$$a' = -\mu a - \Lambda^3 \sin(\sigma T_1 - \beta) \quad (9.53)$$
$$a\beta' = 3\Lambda^2 a + \tfrac{3}{8}a^3 + \Lambda^3 \cos(\sigma T_1 - \beta) \quad (9.54)$$

As discussed earlier, we transform (9.53) and (9.54) into an autonomous system by introducing the transformation

$$\gamma = \sigma T_1 - \beta \quad (9.55)$$

Then,

$$\gamma' = \sigma - \beta' \quad (9.56)$$

Substituting (9.55) into (9.53) gives

$$a' = -\mu a - \Lambda^3 \sin \gamma \quad (9.57)$$

Substituting (9.54) into (9.56) and using (9.55) yields

$$a\gamma' = \sigma a - 3\Lambda^2 a - \tfrac{3}{8}a^3 - \Lambda^3 \cos \gamma \quad (9.58)$$

Eliminating β from (9.23) and (9.55) gives

$$u_0 = a \cos(T_0 + \sigma T_1 - \gamma) + 2\Lambda \cos \omega T_0$$

or

$$u_0 = a \cos(t + \epsilon \sigma t - \gamma) + 2\Lambda \cos \omega t \quad (9.59)$$

Substituting (9.59) into (9.20) and using (9.47), we obtain

$$u = a \cos(3\omega t - \gamma) + 2\Lambda \cos \omega t + O(\epsilon) \quad (9.60)$$

Thus, to the first approximation u is given by (9.60) where a and γ are given by (9.57) and (9.58).

Figure 9-4 shows the variation of a and γ with T_1 as calculated by numerically integrating equations (9.57) and (9.58). Initially, a and γ oscillate with T_1, but

202 FORCED OSCILLATIONS OF THE DUFFING EQUATION

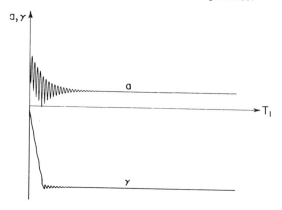

Figure 9-4. Variation of a and γ with T_1 as numerically calculated from (9.57) and (9.58) for $\epsilon = 0.05$, $\sigma = 0.1$, $\Lambda = 1.0$, $\mu = 0.1$, $a(0) = 1.1$, and $\gamma(0) = 2.5$.

as T_1 increases, a_1 and γ tend to constant values. As before, these constant values are usually referred to as stationary or steady-state values. Equation (9.60) shows that the stationary response is periodic. As before, to determine the steady-state response, we need not integrate numerically equations (9.57) and (9.58) describing a and γ. Instead, we use the fact that a and γ are constants in the steady state, set $a' = 0$ and $\gamma' = 0$ in (9.57) and (9.58), and obtain

$$-\mu a = \Lambda^3 \sin \gamma \qquad (9.61)$$

$$\sigma a - 3\Lambda^2 a - \tfrac{3}{8} a^3 = \Lambda^3 \cos \gamma \qquad (9.62)$$

Squaring (9.61) and (9.62), adding the results, and using (9.43), we obtain

$$\mu^2 a^2 + (\sigma - 3\Lambda^2 - \tfrac{3}{8} a^2)^2 a^2 = \Lambda^6 \qquad (9.63)$$

which is a cubic equation in a^2. This is usually referred to as the frequency-response equation. After a is known, γ can be obtained from either (9.61) or (9.62). With a and γ being constants, the frequency of the free-oscillation term is 3ω, which is exactly three times the frequency of the external excitation. Thus, we speak of such resonances as *superharmonic resonances of order three*.

Figure 9-5 shows a representative response curve. The bending of the response curve is due to the nonlinearity and it is responsible for a jump phenomenon. To explain this, we imagine that an experiment is performed in which the frequency of the excitation is kept fixed (i.e., σ is constant) whereas its amplitude (i.e., F or Λ) is slowly varied. If an experiment is started at a small value of Λ corresponding to point A, then as Λ is increased slowly, a slowly increases until point B is reached. As Λ is increased further, a jump from point B to point C takes place with an accompanying increase in a, after which a decreases slowly with increasing Λ until point E is reached. Further increases in Λ produce a

THE METHOD OF MULTIPLE SCALES   203

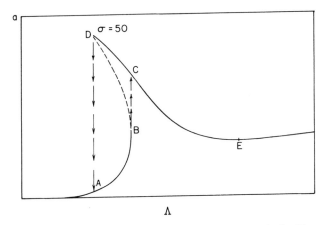

Figure 9-5. Jump phenomenon in the superharmonic response of the Duffing equation.

slow increase in a. If the experiment is started at point E and Λ is decreased, a increases slowly through point C until point D is reached. As Λ is decreased further, a jump from point D to point A takes place with an accompanying decrease in a, after which a decreases slowly with decreasing Λ. We note that the broken line cannot be reached by either increasing or decreasing Λ. Hence, this line corresponds to unstable motions.

The Case $\omega \approx 0$. As pointed out earlier, $\cos \omega t$ needs to be expressed in terms of T_1 as in (9.18). Then, (9.19) becomes

$$D_0^2 u + 2\epsilon D_0 D_1 u + 2\epsilon\mu D_0 u + \cdots + u + \epsilon u^3 = F \cos \sigma T_1 \quad (9.64)$$

Substituting (9.20) into (9.64) and equating coefficients of like powers of ϵ, we obtain

$$D_0^2 u_0 + u_0 = F \cos \sigma T_1 \quad (9.65)$$

$$D_0^2 u_1 + u_1 = -2D_0 D_1 u_0 - 2\mu D_0 u_0 - u_0^3 \quad (9.66)$$

The general solution of (9.65) can be expressed as

$$u_0 = A e^{iT_0} + \bar{A} e^{-iT_0} + F \cos \sigma T_1 \quad (9.67)$$

Then, (9.66) becomes

$$D_0^2 u_1 + u_1 = -2i(A' + \mu A)e^{iT_0} + 2i(\bar{A}' + \mu\bar{A})e^{-iT_0}$$
$$- [A e^{iT_0} + \bar{A} e^{-iT_0} + F \cos \sigma T_1]^3$$

or

$$D_0^2 u_1 + u_1 = -2i(A' + \mu A)e^{iT_0} - A^3 e^{3iT_0} - 3A^2 \bar{A} e^{iT_0} - 3F^2 \cos^2 \sigma T_1 A e^{iT_0}$$

$$- 3FA^2 \cos \sigma T_1 e^{2iT_0} - 3FA\bar{A} \cos \sigma T_1 - \tfrac{1}{2}F^3 \cos^3 \sigma T_1 + cc \quad (9.68)$$

Eliminating the secular terms from u_1, we have

$$2iA' + 2i\mu A + 3A^2\bar{A} + 3F^2 A \cos^2 \sigma T_1 = 0 \quad (9.69)$$

Expressing A in the polar form (9.25) gives

$$ia'e^{i\beta} - a\beta' e^{i\beta} + i\mu a e^{i\beta} + \tfrac{3}{8}a^3 e^{i\beta} + \tfrac{3}{2}F^2 a e^{i\beta} \cos^2 \sigma T_1 = 0$$

or

$$ia' - a\beta' + i\mu a + \tfrac{3}{8}a^3 + \tfrac{3}{2}F^2 a \cos^2 \sigma T_1 = 0 \quad (9.70)$$

Separating real and imaginary parts in (9.70) yields

$$a' = -\mu a \quad (9.71)$$

$$a\beta' = \tfrac{3}{8}a^3 + \tfrac{3}{2}F^2 a \cos^2 \sigma T_1 \quad (9.72)$$

It follows from (9.71) that

$$a = a_0 e^{-\mu T_1} \quad (9.73)$$

where a_0 is a constant. Then, it follows from (9.72) that

$$\beta' = \tfrac{3}{8}a_0^2 e^{-2\mu T_1} + \tfrac{3}{2}F^2 \cos^2 \sigma T_1$$

or

$$\beta' = \tfrac{3}{8}a_0^2 e^{-2\mu T_1} + \tfrac{3}{4}F^2 + \tfrac{3}{4}F^2 \cos 2\sigma T_1 \quad (9.74)$$

which on integration yields

$$\beta = -\frac{3}{16\mu} a_0^2 e^{-2\mu T_1} + \tfrac{3}{4}F^2 T_1 + \frac{3F^2}{8\sigma} \sin 2\sigma T_1 + \beta_0 \quad (9.75)$$

where β_0 is a constant. Substituting (9.67) into (9.20) and using the polar representation (9.25), we obtain

$$u = a \cos(T_0 + \beta) + F \cos \sigma T_1 + O(\epsilon) \quad (9.76)$$

Substituting (9.75) into (9.76), using the fact that $T_0 = t$, $T_1 = \epsilon t$, and $\omega = \epsilon \sigma$, we rewrite (9.76) as

$$u = a_0 e^{-\epsilon \mu t} \cos\left[(1 + \tfrac{3}{4}\epsilon F^2)t - \frac{3}{16\mu} a_0^2 e^{-2\epsilon \mu t} + \frac{3\epsilon F^2}{8\omega} \sin 2\omega t + \beta_0\right]$$

$$+ F \cos \omega t + O(\epsilon) \quad (9.77)$$

We should note that the above solution can be obtained as a special case of the general case ω away from 1. To this end, we replace ω in the last terms in (9.26) by $\epsilon \sigma$ and obtain

THE METHOD OF MULTIPLE SCALES 205

$$D_0^2 u_1 + u_1 = -[2iA' + 2i\mu A + 3A^2\bar{A} + 6\Lambda^2 A]e^{iT_0} - 3A\Lambda^2 e^{iT_0}$$
$$\times [e^{2iT_1\sigma} + e^{-2iT_1\sigma}] + cc + \text{NST} \qquad (9.78)$$

Eliminating the secular terms from (9.78) gives

$$2iA' + 2i\mu A + 3A^2\bar{A} + 6\Lambda^2 A + 6A\Lambda^2 \cos 2\sigma T_1 = 0 \qquad (9.79)$$

Since $\omega = \epsilon\sigma$, it follows from (9.25) that $\Lambda \approx \tfrac{1}{2} F$. Then, (9.79) becomes

$$2iA' + 2i\mu A + 3A^2\bar{A} + \tfrac{3}{2} F^2 A(1 + \cos 2\sigma T_1) = 0$$

or

$$2iA' + 2i\mu A + 3A^2\bar{A} + 3F^2 A \cos^2 \sigma T_1 = 0 \qquad (9.80)$$

in agreement with (9.69).

9.2.2. PRIMARY RESONANCE

In this case $\omega \approx 1$, and the small-divisor terms appear initially in the first term of the straightforward expansion (9.16). Thus, u_0 becomes very large as $\omega \to 1$. But then the nonlinear and damping terms become important, and the ordering of the terms in (9.6) is rendered invalid. Moreover, the homogeneous and particular solutions merge so that they are indistinguishable from one another. At this point, one has a choice. First, one can reorder the nonlinear and damping terms so that they will appear at $O(\epsilon^0)$, thereby balancing the effect of the primary-resonance excitation. However, this choice leads to

$$\ddot{u}_0 + u_0 + 2\hat{\mu}D_0 u_0 + u_0^3 - F \cos \omega T_0 = 0 \qquad (9.81)$$

which is the original equation. Second, one can reorder the excitation so that it appears at $O(\epsilon)$ where the nonlinear and damping terms first appear. This choice leads to a linear problem at $O(\epsilon^0)$. This is the approach used in this book, which is valid for weakly nonlinear systems. To this end, we let $F = \epsilon f$ and rewrite (9.4) as

$$\ddot{u} + u + 2\epsilon\mu\dot{u} + \epsilon u^3 = \epsilon f \cos \omega t \qquad (9.82)$$

As $\omega \to 1$, (9.16) shows that the first two terms have approximately the same frequency. It turns out that the free-oscillation term $a \cos(t + \beta)$ merges with the forced response $F(1 - \omega^2)^{-1} \cos \omega t$. This also justifies the reordering in (9.82).

To determine an approximate solution to (9.82), we introduce the scales $T_0 = t$ and $T_1 = \epsilon t$ and express $\cos \omega t$ as $\cos \omega T_0$. Then as before, (9.82) becomes

$$D_0^2 u + 2\epsilon D_0 D_1 u + 2\epsilon\mu D_0 u + u + \cdots + \epsilon u^3 = \epsilon f \cos \omega T_0 \qquad (9.83)$$

We seek the solution of (9.83) in the form

$$u = u_0(T_0, T_1) + \epsilon u_1(T_0, T_1) + \cdots \qquad (9.84)$$

206 FORCED OSCILLATIONS OF THE DUFFING EQUATION

Substituting (9.84) into (9.82) and equating coefficients of like powers of ϵ, we obtain

$$D_0^2 u_0 + u_0 = 0 \tag{9.85}$$

$$D_0^2 u_1 + u_1 = -2D_0 D_1 u_0 - 2\mu D_0 u_0 - u_0^3 + f \cos \omega T_0 \tag{9.86}$$

The general solution of (9.85) can be expressed as

$$u_0 = A(T_1)e^{iT_0} + \bar{A}(T_1)e^{-iT_0} \tag{9.87}$$

Then, (9.86) becomes

$$D_0^2 u_1 + u_1 = -2i(A' + \mu A)e^{iT_0} + 2i(\bar{A}' + \mu \bar{A})e^{-iT_0} - (Ae^{iT_0} + \bar{A}e^{-iT_0})^3$$
$$+ \tfrac{1}{2} f e^{i\omega T_0} + \tfrac{1}{2} f e^{-i\omega T_0}$$

or

$$D_0^2 u_1 + u_1 = -(2iA' + 2i\mu A + 3A^2 \bar{A})e^{iT_0} - A^3 e^{3iT_0} + \tfrac{1}{2} f e^{i\omega T_0} + cc \tag{9.88}$$

Since we are considering the case $\omega \approx 1$, we introduce a detuning parameter σ defined by

$$\omega = 1 + \epsilon \sigma \tag{9.89}$$

Then,

$$\omega T_0 = (1 + \epsilon \sigma) T_0 = T_0 + \sigma \epsilon T_0 = T_0 + \sigma T_1 \tag{9.90}$$

Substituting (9.90) into (9.88) gives

$$D_0^2 u_1 + u_1 = -(2iA' + 2i\mu A + 3A^2 \bar{A})e^{iT_0} - A^3 e^{3iT_0} + \tfrac{1}{2} f e^{i\sigma T_1} e^{iT_0} + cc \tag{9.91}$$

Eliminating the secular terms from (9.91) yields

$$2iA' + 2i\mu A + 3A^2 \bar{A} - \tfrac{1}{2} f e^{i\sigma T_1} = 0 \tag{9.92}$$

Expressing A in the polar form (9.25), we rewrite (9.92) as

$$ia'e^{i\beta} - a\beta'e^{i\beta} + i\mu a e^{i\beta} + \tfrac{3}{8} a^3 e^{i\beta} - \tfrac{1}{2} f e^{i\sigma T_1} = 0$$

or

$$ia' - a\beta' + i\mu a + \tfrac{3}{8} a^3 - \tfrac{1}{2} f \cos(\sigma T_1 - \beta) - \tfrac{1}{2} i f \sin(\sigma T_1 - \beta) = 0 \tag{9.93}$$

Separating real and imaginary parts in (9.93) yields

$$a' = -\mu a + \tfrac{1}{2} f \sin(\sigma T_1 - \beta) \tag{9.94}$$

$$a\beta' = \tfrac{3}{8} a^3 - \tfrac{1}{2} f \cos(\sigma T_1 - \beta) \tag{9.95}$$

As before, we transform (9.94) and (9.95) into an autonomous system by introducing the transformation

$$\gamma = \sigma T_1 - \beta \tag{9.96}$$

whence

$$\gamma' = \sigma - \beta' \tag{9.97}$$

Substituting (9.96) into (9.94) gives

$$a' = -\mu a + \tfrac{1}{2} f \sin \gamma \tag{9.98}$$

Substituting (9.95) into (9.97) and using (9.96), we obtain

$$a\gamma' = \sigma a - \tfrac{3}{8} a^3 + \tfrac{1}{2} f \cos \gamma \tag{9.99}$$

Substituting the polar form (9.25) into (9.87) yields

$$u_0 = a \cos (T_0 + \beta)$$

Then, it follows from (9.84) that

$$u = a \cos (T_0 + \beta) + O(\epsilon)$$

Since $\beta = \sigma T_1 - \gamma$ from (9.96) and $T_0 = t$ and $T_1 = \epsilon t$,

$$u = a \cos (\omega t - \gamma) + O(\epsilon) \tag{9.100}$$

on account of (9.89).

Figure 9-6 shows the variation of a and γ with T_1 as calculated by numerically integrating equations (9.98) and (9.99). Initially, a and γ oscillate, but as T_1 increases a and γ tend to constant values which are called stationary steady-state values. Then, it follows from (9.100) that the steady-state motion is periodic with the frequency ω. To determine the steady-state motion, we need not integrate numerically (9.98) and (9.99) for long times. Instead, we use the fact that a and γ are constants, set $a' = 0$ and $\gamma' = 0$, and find from (9.98) and (9.99) that

$$\mu a = \tfrac{1}{2} f \sin \gamma \tag{9.101}$$

$$-\sigma a + \tfrac{3}{8} a^3 = \tfrac{1}{2} f \cos \gamma \tag{9.102}$$

Squaring (9.101) and (9.102), adding the results, and using (9.43), we obtain

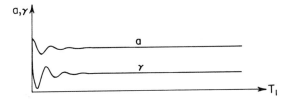

Figure 9-6. Variation of a and γ with T_1 as numerically calculated from (9.98) and (9.99) for $\epsilon = 0.5$, $\sigma = 0.05$, $f = 0.5$, $\mu = 0.1$, $a(0) = 1.1$, and $\gamma(0) = 0.5$.

$$\mu^2 a^2 + (\sigma - \tfrac{3}{8}a^2)^2 a^2 = \tfrac{1}{4}f^2 \qquad (9.103)$$

which is a cubic equation in a^2. It is usually called the *frequency-response equation*.

Figure 9-7 shows a representative curve, called a *frequency-response curve*, for the variation of a with σ. The bending of the frequency-response curve is responsible for a jump phenomenon. To explain this, we imagine that an experiment is performed in which the amplitude of the excitation is held fixed, the frequency of the excitation (i.e., σ) is very slowly varied up and down through the linear natural frequency, and the amplitude of the harmonic response is observed. The experiment is started at a frequency corresponding to point 1 on the curve in Figure 9-7. As the frequency is reduced, σ decreases and a slowly increases through point 2 until point 3 is reached. As σ is decreased further, a jump from point 3 to point 4 takes place with an accompanying increase in a, after which a decreases slowly with decreasing σ. If the experiment is started at point 5 and σ is increased, a increases slowly through point 4 until point 6 is reached. As σ is increased further, a jump from point 6 to point 2 takes place with an accompanying decrease in a, after which a decreases slowly with increasing σ. The maximum amplitude corresponding to point 6 is attainable only when approached from a lower frequency. The portion of the response curve between points 3 and 6 is unstable, and hence, cannot be produced experimentally.

If the experiment is performed with the frequency of the excitation ω held fixed while the amplitude of the excitation is varied slowly, a similar jump phenomenon can be observed. Suppose that the experiment is started at point 1 in Figure 9-8. As f is increased, a slowly increases through point 2 to point 3. As f is increased further, a jump takes place from point 3 to point 4, with an accompanying increase in a, after which a increases slowly with f. If the process is reversed, a decreases slowly as f decreases from point 5 to point 6. As f is decreased further, a jump from point 6 to point 2 takes place, with an accompanying decrease in a, after which a decreases slowly with decreasing f.

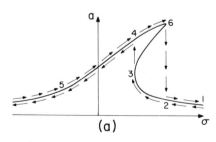

Figure 9-7. Jump phenomena for primary resonance of the Duffing equation.

THE METHOD OF AVERAGING 209

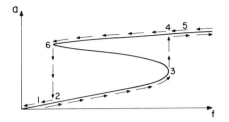

Figure 9-8. Jump phenomena for primary resonance of the Duffing equation.

9.3. The Method of Averaging

To apply the method of averaging to this problem, we need to treat secondary and primary resonances independently. We begin with secondary resonances.

9.3.1. SECONDARY RESONANCES

The first step in applying the method of averaging is the use of the method of variation of parameters to change the dependent variable u to the two dependent variables—the amplitude and phase of the free-oscillation term. To accomplish this, we note that when $\epsilon = 0$ the general solution of (9.4) is

$$u = a \cos(t + \beta) + 2\Lambda \cos \omega t \tag{9.104}$$

where a and β are constants and Λ is defined in (9.12). Differentiating (9.104) with respect to t yields

$$\dot{u} = -a \sin(t + \beta) - 2\Lambda\omega \sin \omega t \tag{9.105}$$

When $\epsilon \neq 0$, we still represent the solution in the form (9.104) subject to the constraint (9.105) but with time-varying rather than constant a and β. Differentiating (9.104) with respect to t gives

$$\dot{u} = -a \sin(t + \beta) + \dot{a} \cos(t + \beta) - a\dot{\beta} \sin(t + \beta) - 2\Lambda\omega \sin \omega t \tag{9.106}$$

Comparing (9.105) with (9.106), we find that

$$\dot{a} \cos(t + \beta) - a\dot{\beta} \sin(t + \beta) = 0 \tag{9.107}$$

Differentiating (9.105) with respect to t gives

$$\ddot{u} = -a \cos(t + \beta) - \dot{a} \sin(t + \beta) - a\dot{\beta} \cos(t + \beta) - 2\Lambda\omega^2 \cos \omega t \tag{9.108}$$

Substituting (9.104), (9.105), and (9.108) into (9.4) yields

$$-a \cos(t + \beta) - \dot{a} \sin(t + \beta) - a\dot{\beta} \cos(t + \beta) - 2\Lambda\omega^2 \cos \omega t$$

210 FORCED OSCILLATIONS OF THE DUFFING EQUATION

$$-2\epsilon\mu a \sin(t+\beta) - 4\epsilon\mu\omega\Lambda \sin\omega t + a\cos(t+\beta) + 2\Lambda\cos\omega t$$
$$+ \epsilon[a\cos(t+\beta) + 2\Lambda\cos\omega t]^3 = F\cos\omega t \qquad (9.109)$$

Since $2\Lambda = (1-\omega^2)^{-1} F$ according to (9.12), (9.109) reduces to

$$\dot{a}\sin(t+\beta) + a\dot{\beta}\cos(t+\beta) = -2\epsilon\mu[a\sin(t+\beta) + 2\omega\Lambda\sin\omega t]$$
$$+ \epsilon[a\cos(t+\beta) + 2\Lambda\cos\omega t]^3 \qquad (9.110)$$

Multiplying (9.107) by $\cos(t+\beta)$ and (9.110) by $\sin(t+\beta)$ and adding the results, we obtain

$$\dot{a} = -2\epsilon\mu\sin(t+\beta)[a\sin(t+\beta) + 2\omega\Lambda\sin\omega t] + \epsilon\sin(t+\beta)$$
$$\times [a\cos(t+\beta) + 2\Lambda\cos\omega t]^3 \qquad (9.111)$$

Substituting (9.111) into (9.107) and solving for $a\dot{\beta}$ yields

$$a\dot{\beta} = -2\epsilon\mu\cos(t+\beta)[a\sin(t+\beta) + 2\omega\Lambda\sin\omega t] + \epsilon\cos(t+\beta)$$
$$\times [a\cos(t+\beta) + 2\Lambda\cos\omega t]^3 \qquad (9.112)$$

Thus, the problem is transformed into solving (9.111) and (9.112) in place of (9.4).

Expanding the cubic term, we rewrite (9.111) and (9.112) as

$$\dot{a} = -2\epsilon\mu\sin(t+\beta)[a\sin(t+\beta) + 2\omega\Lambda\sin\omega t] + \epsilon\sin(t+\beta)$$
$$\times [a^3\cos^3(t+\beta) + 6a^2\Lambda\cos^2(t+\beta)\cos\omega t + 12a\Lambda^2\cos(t+\beta)$$
$$\times \cos^2\omega t + 8\Lambda^3\cos^3\omega t] \qquad (9.113)$$

$$a\dot{\beta} = -2\epsilon\mu\cos(t+\beta)[a\sin(t+\beta) + 2\omega\Lambda\sin\omega t] + \epsilon\cos(t+\beta)$$
$$\times [a^3\cos^3(t+\beta) + 6a^2\Lambda\cos^2(t+\beta)\cos\omega t + 12a\Lambda^2\cos(t+\beta)$$
$$\times \cos^2\omega t + 8\Lambda^3\cos^3\omega t] \qquad (9.114)$$

Using trigonometric identities (Appendix A), we rewrite (9.113) and (9.114) as

$$\dot{a} = \epsilon\{-\mu a + \mu a\cos(2t+2\beta) - 2\omega\mu\Lambda\cos[(1-\omega)t+\beta] + 2\omega\mu\Lambda$$
$$\times \cos[(1+\omega)t+\beta] + (\tfrac{1}{4}a^3 + 3a\Lambda^2)\sin(2t+2\beta) + \tfrac{1}{8}a^3\sin(4t+4\beta)$$
$$+ (\tfrac{3}{4}a^2\Lambda + 3\Lambda^3)\sin[(1+\omega)t+\beta] + (\tfrac{3}{4}a^2\Lambda + 3\Lambda^3)\sin[(1-\omega)t+\beta]$$
$$+ \tfrac{3}{4}a^2\Lambda\sin[(3+\omega)t+3\beta] + \tfrac{3}{4}a^2\Lambda\sin[(3-\omega)t+3\beta]$$
$$+ \tfrac{3}{2}a\Lambda^2\sin[(2+2\omega)t+2\beta] + \tfrac{3}{2}a\Lambda^2\sin[(2-2\omega)t+2\beta]$$
$$+ \Lambda^3\sin[(1+3\omega)t+\beta] + \Lambda^3\sin[(1-3\omega)t+\beta]\} \qquad (9.115)$$

$$a\dot{\beta} = \epsilon\{-\mu a\sin(2t+2\beta) - 2\omega\mu\Lambda\sin[(\omega+1)t+\beta] - 2\omega\mu\Lambda$$
$$\times \sin[(\omega-1)t-\beta] + \tfrac{3}{8}a^3 + 3a\Lambda^2 + (\tfrac{1}{2}a^3 + 3a\Lambda^2)\cos(2t+2\beta)$$

$$+ \tfrac{1}{8}a^3 \cos(4t + 4\beta) + (\tfrac{9}{4}a^2\Lambda + 3\Lambda^3)\cos[(1+\omega)t + \beta]$$
$$+ (\tfrac{9}{4}a^2\Lambda + 3\Lambda^3)\cos[(1-\omega)t + \beta] + \tfrac{3}{4}a^2\Lambda \cos[(3+\omega)t + 3\beta)$$
$$+ \tfrac{3}{4}a^2\Lambda \cos[(3-\omega)t + 3\beta] + \tfrac{3}{2}a\Lambda^2 \cos[(2+2\omega)t + 2\beta]$$
$$+ \tfrac{3}{2}a\Lambda^2 \cos[(2-2\omega)t + 2\beta] + 3a\Lambda^2 \cos 2\omega t$$
$$+ \Lambda^3 \cos[(1+3\omega)t + \beta] + \Lambda^3 \cos[(1-3\omega)t + \beta]\} \tag{9.116}$$

To the first approximation, we need to keep the slowly varying parts in (9.115) and (9.116). The terms $-\epsilon\mu a$ and $\epsilon(\tfrac{3}{8}a^3 + 3a\Lambda^2)$ do not depend on the value of ω, whereas others depend on the value of ω. A term is a slow varying term whenever the coefficient of t (i.e., the frequency) is small. Inspecting (9.115) and (9.116), we conclude that slowly varying terms occur when $\omega \approx 0, 1, 3$, and $\tfrac{1}{3}$. The case $\omega \approx 1$ should be excluded because u_0 has a small divisor. This case is treated in Section 9.3.2. Next, the other cases are treated separately.

The Case ω away from 0, 3, and $\tfrac{1}{3}$. In this case, the only slowly varying terms are the terms that are independent of t. Thus, it follows from (9.115) and (9.116) that

$$\dot{a} = -\epsilon\mu a \tag{9.117}$$

$$a\dot{\beta} = \tfrac{3}{8}\epsilon a^3 + 3\epsilon a\Lambda^2 \tag{9.118}$$

Hence

$$a = a_0 e^{-\epsilon\mu t} \tag{9.119}$$

$$\beta = -\frac{3}{16\mu}a_0^2 e^{-2\epsilon\mu t} + 3\epsilon\Lambda^2 t + \beta_0$$

where a_0 and β_0 are constants.

The Case $\omega \approx 3$. In this case, $\sin[(3-\omega)t + 3\beta]$ in (9.115) and $\cos[(3-\omega)t + 3\beta]$ in (9.116) are slowly varying terms. Hence,

$$\dot{a} = -\epsilon\mu a + \tfrac{3}{4}\epsilon a^2\Lambda \sin[(3-\omega)t + 3\beta] \tag{9.120}$$

$$a\dot{\beta} = \tfrac{3}{8}\epsilon a^3 + 3\epsilon a\Lambda^2 + \tfrac{3}{4}\epsilon a^2\Lambda \cos[(3-\omega)t + 3\beta] \tag{9.121}$$

which are in agreement with (9.34) and (9.35) obtained by using the method of multiple scales since $3 - \omega = -\epsilon\sigma$.

The Case $\omega \approx \tfrac{1}{3}$. In this case, $\sin[(1-3\omega)t + \beta]$ in (9.115) and $\cos[(1-3\omega)t + \beta]$ in (9.116) are slowly varying terms. Hence,

$$\dot{a} = -\epsilon\mu a + \epsilon\Lambda^3 \sin[(1-3\omega)t + \beta] \tag{9.122}$$

$$a\dot{\beta} = \tfrac{3}{8}\epsilon a^3 + 3\epsilon a\Lambda^2 + \epsilon\Lambda^3 \cos[(1-3\omega)t + \beta] \tag{9.123}$$

which are in agreement with (9.53) and (9.54) obtained by using the method of multiple scales since $3\omega - 1 = \epsilon\sigma$.

212 FORCED OSCILLATIONS OF THE DUFFING EQUATION

The Case $\omega \approx 0$. In this case, there are no slowly varying terms in (9.115), while $\cos 2\omega t$ in (9.116) is a slowly varying term. Hence,

$$\dot{a} = -\epsilon\mu a \qquad (9.124)$$

$$a\dot{\beta} = \tfrac{3}{8}\epsilon a^3 + 3\epsilon a\Lambda^2 + 3\epsilon a\Lambda^2 \cos 2\omega t \qquad (9.125)$$

which are in agreement with (9.71) and (9.72) obtained by using the method of multiple scales since $\omega = \epsilon\sigma$ and $\Lambda \approx \tfrac{1}{2}F$.

9.3.2. PRIMARY RESONANCE

When $\epsilon = 0$, the general solution of (9.82) can be expressed as

$$u = a \cos(t + \beta) \qquad (9.126)$$

where a and β are constants. Then,

$$\dot{u} = -a \sin(t + \beta) \qquad (9.127)$$

When $\epsilon \neq 0$, we still represent the solution by (9.126) subject to the constraint (9.127) but with time varying rather than constant a and β. Differentiating (9.126) with respect to t yields

$$\dot{u} = -a \sin(t + \beta) + \dot{a} \cos(t + \beta) - a\dot{\beta} \sin(t + \beta) \qquad (9.128)$$

It follows from (9.127) and (9.128) that

$$\dot{a} \cos(t + \beta) - a\dot{\beta} \sin(t + \beta) = 0 \qquad (9.129)$$

Differentiating (9.127) with respect to t yields

$$\ddot{u} = -a \cos(t + \beta) - \dot{a} \sin(t + \beta) - a\dot{\beta} \cos(t + \beta) \qquad (9.130)$$

Substituting (9.126), (9.127), and (9.130) into (9.82), we obtain

$$\dot{a} \sin(t + \beta) + a\dot{\beta} \cos(t + \beta) = -2\epsilon\mu a \sin(t + \beta) + \epsilon a^3 \cos^3(t + \beta)$$
$$- \epsilon f \cos \omega t \qquad (9.131)$$

Solving (9.129) and (9.131) for $\dot{a}$ and $a\dot{\beta}$ yields

$$\dot{a} = -2\epsilon\mu a \sin^2(t + \beta) + \epsilon a^3 \sin(t + \beta) \cos^3(t + \beta)$$
$$- \epsilon f \sin(t + \beta) \cos \omega t \qquad (9.132)$$

$$a\dot{\beta} = -2\epsilon\mu a \sin(t + \beta) \cos(t + \beta) + \epsilon a^3 \cos^4(t + \beta)$$
$$- \epsilon f \cos(t + \beta) \cos \omega t \qquad (9.133)$$

Using trigonometric identities (Appendix A), we rewrite (9.132) and (9.133) as

$$\dot{a} = -\epsilon\mu a + \epsilon\mu a \cos(2t + 2\beta) + \tfrac{1}{4}\epsilon a^3 \sin(2t + 2\beta) + \tfrac{1}{8}\epsilon a^3 \sin(4t + 4\beta)$$
$$- \tfrac{1}{2}\epsilon f \sin[(1 + \omega)t + \beta] - \tfrac{1}{2}\epsilon f \sin[(1 - \omega)t + \beta] \qquad (9.134)$$

$$a\dot\beta = -\epsilon\mu a \sin(2t + 2\beta) + \tfrac{3}{8}\epsilon a^3 + \tfrac{1}{2}\epsilon a^3 \cos(2t + 2\beta) + \tfrac{1}{8}\epsilon a^3 \cos(4t + 4\beta)$$
$$- \tfrac{1}{2}\epsilon f \cos[(1 + \omega)t + \beta] - \tfrac{1}{2}\epsilon f \cos[(1 - \omega)t + \beta] \tag{9.135}$$

To the first approximation, we need to keep only the slowly varying terms on the right-hand sides of (9.134) and (9.135). For a primary resonance, $\omega \approx 1$ and $\sin[(1 - \omega)t + \beta]$ in (9.134) and $\cos[(1 - \omega)t + \beta]$ in (9.135) are slowly varying terms. Hence,

$$\dot a = -\epsilon\mu a - \tfrac{1}{2}\epsilon f \sin[(1 - \omega)t + \beta] \tag{9.136}$$

$$a\dot\beta = \tfrac{3}{8}\epsilon a^3 - \tfrac{1}{2}\epsilon f \cos[(1 - \omega)t + \beta] \tag{9.137}$$

which are in agreement with (9.94) and (9.95) obtained by using the method of multiple scales since $1 - \omega = -\epsilon\sigma$.

Exercises

9.1. The response of a system with quadratic nonlinearities to a sinusoidal excitation is governed by

$$\ddot u + \omega_0^2 u = -2\epsilon\mu\dot u - \epsilon\alpha u^2 + K \cos\Omega t$$

(a) Use the method of multiple scales to show that

$$u_0 = A e^{i\omega_0 T_0} + \Lambda e^{i\Omega T_0} + c.c.$$

$$D_0^2 u_1 + \omega_0^2 u_1 = -2i\omega_0(A' + \mu A)e^{i\omega_0 T_0} - 2i\mu\Lambda\Omega e^{i\Omega T_0}$$
$$- \alpha\{A^2 e^{2i\omega_0 T_0} + \Lambda^2 e^{2i\Omega T_0} + A\bar A + \Lambda^2$$
$$+ 2\bar A \Lambda e^{i(\Omega - \omega_0)T_0} + 2A\Lambda e^{i(\Omega + \omega_0)T_0}\} + c.c.$$

What is Λ?

(b) When $2\Omega = \omega_0 + \epsilon\sigma$, show that

$$2i\omega_0(A' + \mu A) + \alpha\Lambda^2 e^{i\sigma T_1} = 0$$

Solve for A and then determine the steady-state solution.

(c) When $\Omega = 2\omega_0 + \epsilon\sigma$ show that

$$i\omega_0(A' + \mu A) + \alpha\bar A \Lambda e^{i\sigma T_1} = 0$$

Solve for A. Determine the conditions under which A is unbounded.

9.2. Consider

$$\ddot u + \omega_0^2 u = \epsilon(\dot u - \tfrac{1}{3}\dot u^3) + \epsilon k \cos\Omega t$$

When $\Omega = \omega_0 + \epsilon\sigma$, use the methods of multiple scales and averaging to show that

$$u = a \cos(\omega_0 t + \beta) + \cdots$$

where

214 FORCED OSCILLATIONS OF THE DUFFING EQUATION

$$\dot{a} = \tfrac{1}{2}\epsilon(1 - \tfrac{1}{4}\omega_0^2 a^2)a + \frac{\epsilon k}{2\omega_0}\sin(\epsilon\sigma t - \beta)$$

$$a\dot{\beta} = -\frac{\epsilon k}{2\omega_0}\cos(\epsilon\sigma t - \beta)$$

9.3. Consider

$$\ddot{u} + \omega_0^2 u = \epsilon(\dot{u} - \tfrac{1}{3}\dot{u}^3) + K\cos\Omega t$$

where Ω is away from $3\omega_0$, $\tfrac{1}{3}\omega_0$, and 0. Use the methods of multiple scales and averaging to show that

$$u = a\cos(\omega_0 t + \beta) - \frac{K}{\Omega^2 - \omega_0^2}\cos\Omega t + \cdots$$

where

$$\dot{a} = \tfrac{1}{2}\epsilon(\eta - \tfrac{1}{4}\omega_0^2 a^2)a \qquad \dot{\beta} = 0$$

Determine η.

9.4. Consider

$$\ddot{u} + \omega_0^2 u = \epsilon(\dot{u} - \tfrac{1}{3}\dot{u}^3) + K\cos\Omega t$$

when $3\Omega = \omega_0 + \epsilon\sigma$. Use the methods of multiple scales and averaging to show that

$$u = a\cos(\omega_0 t + \beta) - \frac{K}{\Omega^2 - \omega_0^2}\cos\Omega t + \cdots$$

where

$$\dot{a} = \tfrac{1}{2}\epsilon(\eta - \tfrac{1}{4}\omega_0^2 a^2)a + \epsilon\Gamma\cos(\epsilon\sigma t - \beta)$$
$$a\dot{\beta} = \epsilon\Gamma\sin(\epsilon\sigma t - \beta)$$

Determine η and Γ.

9.5. Consider

$$\ddot{u} + \omega_0^2 u = \epsilon(\dot{u} - \tfrac{1}{3}\dot{u}^3) + K\cos\Omega t$$

when $\Omega = 3\omega_0 + \epsilon\sigma$. Use the methods of multiple scales and averaging to show that

$$u = a\cos(\omega_0 t + \beta) - \frac{K}{\Omega^2 - \omega_0^2}\cos\Omega t + \cdots$$

where

$$\dot{a} = \tfrac{1}{2}\epsilon(\eta - \tfrac{1}{4}\omega_0^2 a^2)a + \tfrac{1}{4}\epsilon\omega_0\xi a^2\cos(\epsilon\sigma t - 3\beta)$$
$$a\dot{\beta} = \tfrac{1}{4}\epsilon\omega_0\xi a^2\sin(\epsilon\sigma t - 3\beta)$$

Determine η and ξ.

9.6. Consider

$$\ddot{u} + u = -\epsilon \dot{u}|\dot{u}| + 2\epsilon k \cos \Omega t$$

when $\Omega = 1 + \epsilon\sigma$. Use the methods of multiple scales and averaging to show that

$$u = a \cos(t + \beta) + \cdots$$

where

$$\dot{a} = -\frac{4\epsilon a^2}{3\pi} + \epsilon k \sin(\epsilon\sigma t - \beta)$$

$$a\dot{\beta} = -\epsilon k \cos(\epsilon\sigma t - \beta)$$

9.7. Consider

$$\ddot{u} + u + 2\epsilon u^2 \dot{u} = 2\epsilon k \cos \Omega t$$

when $\Omega = 1 + \epsilon\sigma$. Use the methods of multiple scales and averaging to show that

$$u = a \cos(t + \beta) + \cdots$$

where

$$\dot{a} = -\tfrac{1}{4}\epsilon a^3 + \epsilon k \sin(\epsilon\sigma t - \beta)$$

$$a\dot{\beta} = -\epsilon k \cos(\epsilon\sigma t - \beta)$$

9.8. Consider

$$\ddot{u} + \omega_0^2 u + \epsilon u^4 = 2K \cos \Omega t$$

Show that to first-order resonances exist when $\Omega \approx 4\omega_0$, $2\omega_0$, $\tfrac{1}{4}\omega_0$, $\tfrac{1}{2}\omega_0$, 0, $\tfrac{2}{3}\omega_0$, and $\tfrac{3}{2}\omega_0$. Use the methods of multiple scales and averaging to determine the equations describing the amplitude and phase for each case.

CHAPTER 10

Multifrequency Excitations

Whereas the preceding chapter deals with a single-frequency excitation, this chapter deals with multifrequency excitations. To minimize the algebraic manipulation, we consider a system with a quadratic nonlinearity under the influence of a two-frequency excitation. Thus, we consider the equation

$$\ddot{u} + u + 2\epsilon\mu\dot{u} + \epsilon u^2 = F_1 \cos \omega_1 t + F_2 \cos(\omega_2 t + \nu) \qquad (10.1)$$

where ν is a constant phase and the F_n and ω_n are constants. As before, we first determine a straightforward expansion and investigate its uniformity to exhibit the small divisors and hence the possible resonances. In Section 10.2, we use the method of multiple scales to determine uniform first-order expansions for some of these resonances. Finally in Section 10.3, we treat these resonances by using the first approximation of the method of averaging.

10.1. The Straightforward Expansion

We seek a first-order (two-term) expansion for u in powers of ϵ in the form

$$u(t;\epsilon) = u_0(t) + \epsilon u_1(t) + \cdots \qquad (10.2)$$

Substituting (10.2) into (10.1), we have

$$\ddot{u}_0 + \epsilon\ddot{u}_1 + \cdots + u_0 + \epsilon u_1 + \cdots + 2\epsilon\mu(\dot{u}_0 + \cdots) + \epsilon(u_0 + \cdots)^2$$
$$= F_1 \cos \omega_1 t + F_2 \cos(\omega_2 t + \nu)$$

or

$$\ddot{u}_0 + u_0 + \epsilon(\ddot{u}_1 + u_1 + 2\mu\dot{u}_0 + u_0^2) + \cdots = F_1 \cos \omega_1 t + F_2 \cos(\omega_2 t + \nu) \qquad (10.3)$$

Equating the coefficients of like powers of ϵ on both sides, we obtain

$$\ddot{u}_0 + u_0 = F_1 \cos \omega_1 t + F_2 \cos(\omega_2 t + \nu) \qquad (10.4)$$

$$\ddot{u}_1 + u_1 = -2\mu\dot{u}_0 - u_0^2 \qquad (10.5)$$

THE STRAIGHTFORWARD EXPANSION 217

The general solution of (10.4) can be obtained as in Section 9.1 or Appendix B to be

$$u_0 = a \cos(t + \beta) + 2\Lambda_1 \cos \omega_1 t + 2\Lambda_2 \cos(\omega_2 t + \nu) \tag{10.6}$$

where a and β are constants and

$$\Lambda_n = \tfrac{1}{2}(1 - \omega_n^2)^{-1} F_n$$

Then, (10.5) becomes

$$\ddot{u}_1 + u_1 = 2\mu[a \sin(t + \beta) + 2\Lambda_1 \omega_1 \sin \omega_1 t + 2\Lambda_2 \omega_2 \sin(\omega_2 t + \nu)]$$
$$- [a \cos(t + \beta) + 2\Lambda_1 \cos \omega_1 t + 2\Lambda_2 \cos(\omega_2 t + \nu)]^2$$

or

$$\ddot{u}_1 + u_1 = 2\mu[a \sin(t + \beta) + 2\Lambda_1 \omega_1 \sin \omega_1 t + 2\Lambda_2 \omega_2 \sin(\omega_2 t + \nu)]$$
$$- a^2 \cos^2(t + \beta) - 4\Lambda_1^2 \cos^2 \omega_1 t - 4\Lambda_2^2 \cos^2(\omega_2 t + \nu)$$
$$- 4a\Lambda_1 \cos(t + \beta) \cos \omega_1 t - 4a\Lambda_2 \cos(t + \beta) \cos(\omega_2 t + \nu)$$
$$- 8\Lambda_1 \Lambda_2 \cos \omega_1 t \cos(\omega_2 t + \nu) \tag{10.7}$$

Using trigonometric identities, we rewrite (10.7) as

$$\ddot{u}_1 + u_1 = 2\mu[a \sin(t + \beta) + 2\Lambda_1 \omega_1 \sin \omega_1 t + 2\Lambda_2 \omega_2 \sin(\omega_2 t + \nu)]$$
$$- (\tfrac{1}{2} a^2 + 2\Lambda_1^2 + 2\Lambda_2^2) - \tfrac{1}{2} a^2 \cos(2t + 2\beta) - 2\Lambda_1^2 \cos 2\omega_1 t$$
$$- 2\Lambda_2^2 \cos(2\omega_2 t + 2\nu) - 2a\Lambda_1 \cos[(1 + \omega_1)t + \beta]$$
$$- 2a\Lambda_1 \cos[(1 - \omega_1)t + \beta] - 2a\Lambda_2 \cos[(1 + \omega_2)t + \beta + \nu]$$
$$- 2a\Lambda_2 \cos[(1 - \omega_2)t + \beta - \nu] - 4\Lambda_1 \Lambda_2 \cos[(\omega_1 + \omega_2)t + \nu]$$
$$- 4\Lambda_1 \Lambda_2 \cos[(\omega_2 - \omega_1)t + \nu] \tag{10.8}$$

As discussed before, we do not include the solution of the homogeneous problem at any order except the first. Then a particular solution of (10.8) is (Section B.4)

$$u_1 = -\mu a t \cos(t + \beta) + \frac{4\mu \Lambda_1 \omega_1}{1 - \omega_1^2} \sin \omega_1 t + \frac{4\mu \Lambda_2 \omega_2}{1 - \omega_2^2} \sin(\omega_2 t + \nu)$$
$$- (\tfrac{1}{2} a^2 + 2\Lambda_1^2 + 2\Lambda_2^2) + \tfrac{1}{6} a^2 \cos(2t + 2\beta) - \frac{2\Lambda_1^2}{1 - 4\omega_1^2} \cos 2\omega_1 t$$
$$- \frac{2\Lambda_2^2}{1 - 4\omega_2^2} \cos(2\omega_2 t + 2\nu) + \frac{2a\Lambda_1}{2\omega_1 + \omega_1^2} \cos[(1 + \omega_1)t + \beta]$$
$$- \frac{2a\Lambda_1}{2\omega_1 - \omega_1^2} \cos[(1 - \omega_1)t + \beta] + \frac{2a\Lambda_2}{2\omega_2 + \omega_2^2} \cos[(1 + \omega_2)t + \beta + \nu]$$

218 MULTIFREQUENCY EXCITATIONS

$$-\frac{2a\Lambda_2}{2\omega_2 - \omega_2^2}\cos[(1-\omega_2)t + \beta - \nu]$$

$$-\frac{4\Lambda_1\Lambda_2}{1-(\omega_1+\omega_2)^2}\cos[(\omega_1+\omega_2)t + \nu]$$

$$-\frac{4\Lambda_1\Lambda_2}{1-(\omega_2-\omega_1)^2}\cos[(\omega_2-\omega_1)t + \nu] \tag{10.9}$$

Substituting (10.6) and (10.9) into (10.2) and using the definition of Λ_n, we have

$$u = a\cos(t + \beta) + \frac{F_1}{1-\omega_1^2}\cos\omega_1 t + \frac{F_2}{1-\omega_2^2}\cos(\omega_2 t + \nu)$$

$$+ \epsilon\left\{-\mu at\cos(t+\beta) + \frac{2\mu F_1 \omega_1}{(1-\omega_1^2)^2}\sin\omega_1 t + \frac{2\mu F_2 \omega_2}{(1-\omega_2^2)^2}\sin(\omega_2 t + \nu)\right.$$

$$-\tfrac{1}{2}a^2 - \frac{F_1^2}{2(1-\omega_1^2)^2} - \frac{F_2^2}{2(1-\omega_2^2)^2} + \tfrac{1}{6}a^2\cos(2t+2\beta)$$

$$-\frac{F_1^2}{2(1-\omega_1^2)^2(1-4\omega_1^2)}\cos 2\omega_1 t - \frac{F_2^2}{2(1-\omega_2^2)^2(1-4\omega_2^2)}\cos(2\omega_2 t + 2\nu)$$

$$+\frac{aF_1}{(1-\omega_1^2)(2\omega_1+\omega_1^2)}\cos[(1+\omega_1)t+\beta]$$

$$-\frac{aF_1}{(1-\omega_1^2)(2\omega_1-\omega_1^2)}\cos[(1-\omega_1)t+\beta] + \frac{aF_2}{(1-\omega_2^2)(2\omega_2+\omega_2^2)}$$

$$\times \cos[(1+\omega_2)t+\beta+\nu] - \frac{aF_2}{(1-\omega_2^2)(2\omega_2-\omega_2^2)}\cos[(1-\omega_2)t+\beta+\nu]$$

$$-\frac{F_1 F_2}{(1-\omega_1^2)(1-\omega_2^2)[1-(\omega_1+\omega_2)^2]}\cos[(\omega_1+\omega_2)t + \nu]$$

$$-\frac{F_1 F_2}{(1-\omega_1^2)(1-\omega_2^2)[1-(\omega_2-\omega_1)^2]}\cos[(\omega_2-\omega_1)t + \nu]\bigg\} + \cdots \tag{10.10}$$

Equation (10.10) shows that small divisors occur when $\omega_1 \approx 1$, $\omega_1 \approx \tfrac{1}{2}$, $\omega_1 \approx 2$, $\omega_2 \approx 1$, $\omega_2 \approx \tfrac{1}{2}$, $\omega_2 \approx 2$, $\omega_2 + \omega_1 \approx 1$, and $\omega_2 - \omega_1 \approx 1$. Since small divisors occur in the first term when $\omega_1 \approx 1$ or $\omega_2 \approx 1$, these are called primary or main resonances. All other resonances are called secondary resonances because their corresponding small divisors do not appear in the first term. The cases $\omega_1 \approx \tfrac{1}{2}$ and $\omega_2 \approx \tfrac{1}{2}$ are called superharmonic resonances of order 2 because they excite a free-oscillation term having the frequency 1 which is approximately $2\omega_1$ or $2\omega_2$. The cases $\omega_1 \approx 2$ and $\omega_2 \approx 2$ are called subharmonic resonances of order

THE METHOD OF MULTIPLE SCALES 219

$\frac{1}{2}$ because they excite a free-oscillation term having the frequency 1 which is approximately $\frac{1}{2}\omega_1$ or $\frac{1}{2}\omega_2$. The case $\omega_2 + \omega_1 \approx 1$ is called *a combination resonance of the summed type*, whereas the case $\omega_2 - \omega_1 \approx 1$ is called *a combination resonance of the difference type*. In the next section, we show how the method of multiple scales can be used to determine uniform expansions for secondary resonances. The main resonances can be treated as in the preceding chapter.

10.2. The Method of Multiple Scales

To first order, we need the slow scale $T_1 = \epsilon t$ in addition to the fast scale $T_0 = t$. Then,

$$\frac{d}{dt} = D_0 + \epsilon D_1 + \cdots$$

$$\frac{d^2}{dt^2} = D_0^2 + 2\epsilon D_0 D_1 + \cdots$$

where $D_n = \partial/\partial T_n$. Hence, (10.1) becomes

$$D_0^2 u + 2\epsilon D_0 D_1 u + \cdots + u + 2\epsilon\mu(D_0 u + \epsilon D_1 u + \cdots) + \epsilon u^2$$
$$= F_1 \cos \omega_1 T_0 + F_2 \cos(\omega_2 T_0 + \nu) \quad (10.11)$$

We seek an approximate solution to (10.11) in the form

$$u = u_0(T_0, T_1) + \epsilon u_1(T_0, T_1) + \cdots \quad (10.12)$$

Substituting (10.12) into (10.11) and equating coefficients of like powers of ϵ, we obtain

$$D_0^2 u_0 + u_0 = F_1 \cos \omega_1 T_0 + F_2 \cos(\omega_2 T_0 + \nu) \quad (10.13)$$

$$D_0^2 u_1 + u_1 = -2D_0 D_1 u_0 - 2\mu D_0 u_0 - u_0^2 \quad (10.14)$$

The general solution of (10.13) is written in the following complex form:

$$u_0 = A(T_1)e^{iT_0} + \overline{A}(T_1)e^{-iT_0} + \Lambda_1 e^{i\omega_1 T_0} + \overline{\Lambda}_1 e^{-i\omega_1 T_0} + \Lambda_2 e^{i\omega_2 T_0} + \overline{\Lambda}_2 e^{-i\omega_2 T_0}$$
$$(10.15)$$

where

$$\Lambda_1 = \frac{F_1}{2(1-\omega_1^2)} \qquad \Lambda_2 = \frac{F_2 e^{i\nu}}{2(1-\omega_2^2)} \quad (10.16)$$

Then, (10.14) becomes

$$D_0^2 u_1 + u_1 = -2i(A' + \mu A)e^{iT_0} - 2i\mu\omega_1 \Lambda_1 e^{i\omega_1 T_0} - 2i\mu\omega_2 \Lambda_2 e^{i\omega_2 T_0}$$

$$-A^2 e^{2iT_0} - \Lambda_1^2 e^{2i\omega_1 T_0} - \Lambda_2^2 e^{2i\omega_2 T_0} - A\bar{A} - \Lambda_1^2 - \Lambda_2\bar{\Lambda}_2$$
$$-2A\Lambda_1 e^{i(1+\omega_1)T_0} - 2A\Lambda_1 e^{i(1-\omega_1)T_0} - 2A\Lambda_2 e^{i(1+\omega_2)T_0}$$
$$-2A\bar{\Lambda}_2 e^{i(1-\omega_2)T_0} - 2\Lambda_1\Lambda_2 e^{i(\omega_2+\omega_1)T_0} - 2\Lambda_2\bar{\Lambda}_1 e^{i(\omega_2-\omega_1)T_0} + cc$$

(10.17)

As in the preceding section, the particular solutions of u_1 contain secular and small-divisor terms. For a uniform expansion, these secular and small-divisor terms must be eliminated by proper choices of A. These choices depend on the type of resonances present. Next, we treat the two cases: (a) $\omega_1 + \omega_2 \approx 1$, and (b) $\omega_2 - \omega_1 \approx 1$ and $\omega_1 \approx 2$.

10.2.1. THE CASE $\omega_2 + \omega_1 \approx 1$

In this case, we assume that $\omega_2 + \omega_1 \approx 1$ and no other resonances exist to first order. To describe quantitatively the nearness of $\omega_2 + \omega_1$ to 1, we introduce the detuning parameter σ defined by

$$\omega_2 + \omega_1 = 1 + \epsilon\sigma \tag{10.18}$$

In this case, the only terms in (10.17) that lead to small-divisor terms are $-2\Lambda_1\Lambda_2 \exp[i(\omega_2 + \omega_1)T_0]$ and its complex conjugate. Using (10.18), we rewrite this term as

$$-2\Lambda_1\Lambda_2 e^{i(\omega_1+\omega_2)T_0} = -2\Lambda_1\Lambda_2 e^{i(1+\epsilon\sigma)T_0} = -2\Lambda_1\Lambda_2 e^{iT_0 + i\sigma T_1}$$

Then, we rewrite (10.17) as

$$D_0^2 u_1 + u_1 = -2i(A' + \mu A)e^{iT_0} - 2\Lambda_1\Lambda_2 e^{i\sigma T_1}e^{iT_0} + cc + \text{NST} \tag{10.19}$$

where NST stands for terms that do not produce secular terms in u_1. Eliminating the secular terms from (10.19) demands that

$$2i(A' + \mu A) + 2\Lambda_1\Lambda_2 e^{i\sigma T_1} = 0$$

or

$$A' + \mu A = i\Lambda_1\Lambda_2 e^{i\sigma T_1} \tag{10.20}$$

In this case, we can find the exact solution of (10.20) because it is a linear inhomogeneous equation. Thus, we do not need to express A in its polar form. The homogeneous solution is

$$A = ce^{-\mu T_1} \tag{10.21}$$

where c is a complex constant. Since the inhomogeneous term in (10.20) is exponential, its corresponding particular solution can be sought in the form

$$A = be^{i\sigma T_1} \tag{10.22}$$

THE METHOD OF MULTIPLE SCALES 221

Substituting (10.22) into (10.20) gives

$$i\sigma b e^{i\sigma T_1} + \mu b e^{i\sigma T_1} = i\Lambda_1 \Lambda_2 e^{i\sigma T_1}$$

or

$$i\sigma b + \mu b = i\Lambda_1 \Lambda_2$$

Hence,

$$b = \frac{i\Lambda_1 \Lambda_2}{\mu + i\sigma}$$

Then, the general solution of (10.20) is

$$A = c e^{-\mu T_1} + \frac{i\Lambda_1 \Lambda_2}{\mu + i\sigma} e^{i\sigma T_1} \qquad (10.23)$$

Substituting for A into (10.15) and then substituting the result into (10.12), we obtain

$$u = \left[c e^{-\epsilon \mu t} + \frac{i\Lambda_1 \Lambda_2}{\mu + i\sigma} e^{i\epsilon\sigma t} \right] e^{it} + \Lambda_1 e^{i\omega_1 t} + \Lambda_2 e^{i\omega_2 t} + cc + O(\epsilon) \qquad (10.24)$$

We express c in the polar form $\tfrac{1}{2} a_0 \exp(i\beta_0)$ and substitute for the Λ_n from (10.16) into (10.24). The result is

$$u = \tfrac{1}{2} a_0 e^{-\epsilon \mu t} e^{i(t+\beta_0)} + \frac{iF_1 F_2}{4(\mu + i\sigma)(1 - \omega_1^2)(1 - \omega_2^2)} e^{i(1+\epsilon\sigma)t + i\nu}$$

$$+ \frac{F_1}{2(1 - \omega_1^2)} e^{i\omega_1 t} + \frac{F_2}{2(1 - \omega_2^2)} e^{i(\omega_2 t + \nu)} + cc + O(\epsilon) \qquad (10.25)$$

We note that $1 + \epsilon\sigma = \omega_1 + \omega_2$ according to (10.18) and that

$$\frac{1}{\mu + i\sigma} = \frac{1}{\sqrt{\mu^2 + \sigma^2}} e^{-i\,\tan^{-1}(\sigma/\mu)}$$

Hence, we rewrite (10.25) as

$$u = a_0 e^{-\epsilon\mu t} \cos(t + \beta_0) - \frac{F_1 F_2}{2(1 - \omega_1^2)(1 - \omega_2^2)\sqrt{\mu^2 + \sigma^2}} \sin\left[(\omega_1 + \omega_2)t\right.$$

$$\left. + \nu - \tan^{-1} \frac{\sigma}{\mu}\right] + \frac{F_1}{1 - \omega_1^2} \cos \omega_1 t + \frac{F_2}{1 - \omega_2^2} \cos(\omega_2 t + \nu) + O(\epsilon) \qquad (10.26)$$

As $t \to \infty$, the first term on the right-hand side of (10.26) tends to zero, and u tends to the following steady-state solution

$$u = -\frac{F_1 F_2}{2(1 - \omega_1^2)(1 - \omega_2^2)\sqrt{\mu^2 + \sigma^2}} \sin\left[(\omega_1 + \omega_2)t + \nu - \tan^{-1} \frac{\sigma}{\mu}\right]$$

$$+ \frac{F_1}{1 - \omega_1^2} \cos \omega_1 t + \frac{F_2}{1 - \omega_2^2} \cos (\omega_2 t + \nu) + O(\epsilon) \tag{10.27}$$

10.2.2. THE CASE $\omega_2 - \omega_1 \approx 1$ AND $\omega_1 \approx 2$

In this case, we assume that two resonances exist simultaneously, namely $\omega_2 - \omega_1 \approx 1$ and $\omega_1 \approx 2$. In other words, $\omega_1 \approx 2$ and $\omega_2 \approx 3$. To treat this case, we introduce the two detuning parameters σ_1 and σ_2 defined by

$$\omega_1 = 2 + \epsilon\sigma_1 \qquad \omega_2 = 3 + \epsilon\sigma_2 \tag{10.28}$$

Substituting (10.28) into (10.17), we have

$$D_0^2 u_1 + u_1 = -2i(A' + \mu A)e^{iT_0} - 2\Lambda_1 \overline{A} e^{iT_0 + i\epsilon\sigma_1 T_0}$$
$$- 2\Lambda_2 \Lambda_1 e^{iT_0 + i(\sigma_2 - \sigma_1)\epsilon T_0} + cc + \text{NST}$$

or

$$D_0^2 u_1 + u_1 = -2i(A' + \mu A)e^{iT_0} - 2\Lambda_1 \overline{A} e^{i\sigma_1 T_1} e^{iT_0}$$
$$- 2\Lambda_2 \Lambda_1 e^{i(\sigma_2 - \sigma_1)T_1} e^{iT_0} + cc + \text{NST} \tag{10.29}$$

Eliminating the secular terms from u_1, we obtain

$$i(A' + \mu A) + \Lambda_1 \overline{A} e^{i\sigma_1 T_1} + \Lambda_2 \Lambda_1 e^{i(\sigma_2 - \sigma_1)T_1} = 0 \tag{10.30}$$

Equation (10.30) is an inhomogeneous equation with variable coefficients. We first transform it into an inhomogeneous equation with constant coefficients. To accomplish this, we introduce the transformation

$$A = B e^{i\lambda T_1} \tag{10.31}$$

where λ is real and obtain

$$i(B' + i\lambda B + \mu B)e^{i\lambda T_1} + \Lambda_1 \overline{B} e^{i(\sigma_1 - \lambda)T_1} + \Lambda_2 \Lambda_1 e^{i(\sigma_2 - \sigma_1)T_1} = 0$$

or

$$i(B' + i\lambda B + \mu B) + \Lambda_1 \overline{B} e^{i(\sigma_1 - 2\lambda)T_1} + \Lambda_2 \Lambda_1 e^{i(\sigma_2 - \sigma_1 - \lambda)T_1} = 0 \tag{10.32}$$

We choose $\lambda = \frac{1}{2}\sigma_1$ so that the coefficients of B and $\overline{B}$ are independent of T_1. Then, (10.32) becomes

$$i(B' + \tfrac{1}{2} i\sigma_1 B + \mu B) + \Lambda_1 \overline{B} + \Lambda_2 \Lambda_1 e^{i(\sigma_2 - 3\sigma_1/2)T_1} = 0 \tag{10.33}$$

Instead of expressing B in polar form as in the preceding chapters, we express B in the form

$$B = B_r + iB_i \tag{10.34}$$

It turns out that the resulting equations are easier to solve in this form than those resulting from using the polar form. Substituting (10.34) into (10.33) and using (10.16) to express Λ_2 in polar form, we obtain

$$iB'_r - B'_i - \tfrac{1}{2}\sigma_1 B_r - \tfrac{1}{2}i\sigma_1 B_i + i\mu B_r - \mu B_i + \Lambda_1 B_r - i\Lambda_1 B_i$$
$$+ \Gamma \cos(\sigma T_1 + \nu) + i\Gamma \sin(\sigma T_1 + \nu) = 0 \quad (10.35)$$

where

$$\Gamma = \frac{\Lambda_1 F_2}{2(1 - \omega_2^2)} \qquad \sigma = \sigma_2 - \tfrac{3}{2}\sigma_1 \quad (10.36)$$

Separating real and imaginary parts in (10.35) yields

$$B'_r + \mu B_r - (\Lambda_1 + \tfrac{1}{2}\sigma_1)B_i = -\Gamma \sin(\sigma T_1 + \nu) \quad (10.37)$$

$$B'_i + \mu B_i - (\Lambda_1 - \tfrac{1}{2}\sigma_1)B_r = \Gamma \cos(\sigma T_1 + \nu) \quad (10.38)$$

Since (10.37) and (10.38) are linear coupled inhomogeneous equations for B_r and B_i, their general solutions can be obtained as the sum of any particular solution and a general homogeneous solution. The general homogeneous solution can be obtained by letting

$$B_r = b_r e^{\gamma T_1} \qquad B_i = b_i e^{\gamma T_1} \quad (10.39)$$

Substituting (10.39) into (10.37) and (10.38) and dropping the inhomogeneous terms, we obtain

$$(\gamma + \mu)b_r - (\Lambda_1 + \tfrac{1}{2}\sigma_1)b_i = 0 \quad (10.40)$$

$$-(\Lambda_1 - \tfrac{1}{2}\sigma_1)b_r + (\gamma + \mu)b_i = 0 \quad (10.41)$$

Thus, we ended up with a system of two linear coupled homogeneous equations. Since we are interested in nontrivial solutions for b_r and b_i, the determinant of the coefficient matrix

$$\begin{vmatrix} \gamma + \mu & -(\Lambda_1 + \tfrac{1}{2}\sigma_1) \\ -(\Lambda_1 - \tfrac{1}{2}\sigma_1) & \gamma + \mu \end{vmatrix}$$

must vanish. Then,

$$(\gamma + \mu)^2 - \Lambda_1^2 + \tfrac{1}{4}\sigma_1^2 = 0 \quad (10.42)$$

This condition could have been obtained by eliminating either b_r or b_i from (10.40) and (10.41). It follows from (10.42) that

$$\gamma + \mu = \pm \sqrt{\Lambda_1^2 - \tfrac{1}{4}\sigma_1^2} \quad (10.43)$$

Hence, there are two possible values γ_1 and γ_2 for γ. They are given by

$$\gamma_1 = -\mu - \sqrt{\Lambda_1^2 - \tfrac{1}{4}\sigma_1^2} \qquad \gamma_2 = -\mu + \sqrt{\Lambda_1^2 - \tfrac{1}{4}\sigma_1^2} \quad (10.44)$$

It follows from (10.40) that

$$b_i = \frac{\gamma + \mu}{\Lambda_1 + \tfrac{1}{2}\sigma_1} b_r \quad (10.45)$$

224 MULTIFREQUENCY EXCITATIONS

Hence, if
$$B_r = b_1 e^{\gamma_1 T_1} + b_2 e^{\gamma_2 T_1} \qquad (10.46)$$

it follows from (10.45) that

$$B_i = \frac{\gamma_1 + \mu}{\Lambda_1 + \frac{1}{2}\sigma_1} b_1 e^{\gamma_1 T_1} + \frac{\gamma_2 + \mu}{\Lambda_1 + \frac{1}{2}\sigma_1} b_2 e^{\gamma_2 T_1} \qquad (10.47)$$

where b_1 and b_2 are arbitrary constants.

A particular solution for (10.37) and (10.38) can be obtained by using the method of undetermined coefficients. Since the inhomogeneous terms consist of circular functions and the homogeneous problem has constant coefficients, we seek a particular solution in the form

$$B_r = c_1 \cos(\sigma T_1 + \nu) + c_2 \sin(\sigma T_1 + \nu) \qquad (10.48)$$

$$B_i = c_3 \cos(\sigma T_1 + \nu) + c_4 \sin(\sigma T_1 + \nu) \qquad (10.49)$$

Substituting (10.48) and (10.49) into (10.37) and (10.38) and equating the coefficients of $\cos(\sigma T_1 + \nu)$ and $\sin(\sigma T_1 + \nu)$ on both sides, we obtain

$$\mu c_1 + \sigma c_2 - (\Lambda_1 + \tfrac{1}{2}\sigma_1) c_3 = 0 \qquad (10.50)$$

$$-\sigma c_1 + \mu c_2 - (\Lambda_1 + \tfrac{1}{2}\sigma_1) c_4 = -\Gamma \qquad (10.51)$$

$$-(\Lambda_1 - \tfrac{1}{2}\sigma_1) c_1 + \mu c_3 + \sigma c_4 = \Gamma \qquad (10.52)$$

$$-(\Lambda_1 - \tfrac{1}{2}\sigma_1) c_2 - \sigma c_3 + \mu c_4 = 0 \qquad (10.53)$$

Using Cramer's rule, we find that the solution of (10.50) through (10.53) is

$$c_1 = \frac{1}{\Delta} \begin{vmatrix} 0 & \sigma & -(\Lambda_1 + \tfrac{1}{2}\sigma_1) & 0 \\ -\Gamma & \mu & 0 & -(\Lambda_1 + \tfrac{1}{2}\sigma_1) \\ \Gamma & 0 & \mu & \sigma \\ 0 & -(\Lambda_1 - \tfrac{1}{2}\sigma_1) & -\sigma & \mu \end{vmatrix}$$

$$(10.54)$$

$$c_2 = \frac{1}{\Delta} \begin{vmatrix} \mu & 0 & -(\Lambda_1 + \tfrac{1}{2}\sigma_1) & 0 \\ -\sigma & -\Gamma & 0 & -(\Lambda_1 + \tfrac{1}{2}\sigma_1) \\ -(\Lambda_1 - \tfrac{1}{2}\sigma_1) & \Gamma & \mu & \sigma \\ 0 & 0 & -\sigma & \mu \end{vmatrix}$$

$$(10.55)$$

$$c_3 = \frac{1}{\Delta} \begin{vmatrix} \mu & \sigma & 0 & 0 \\ -\sigma & \mu & -\Gamma & -(\Lambda_1 + \tfrac{1}{2}\sigma_1) \\ -(\Lambda_1 - \tfrac{1}{2}\sigma_1) & 0 & \Gamma & \sigma \\ 0 & -(\Lambda_1 - \tfrac{1}{2}\sigma_1) & 0 & \mu \end{vmatrix}$$

$$(10.56)$$

THE METHOD OF MULTIPLE SCALES 225

$$c_4 = \frac{1}{\Delta} \begin{vmatrix} \mu & \sigma & -(\Lambda_1 + \frac{1}{2}\sigma_1) & 0 \\ -\sigma & \mu & 0 & -\Gamma \\ -(\Lambda_1 - \frac{1}{2}\sigma_1) & 0 & \mu & \Gamma \\ 0 & -(\Lambda_1 - \frac{1}{2}\sigma_1) & -\sigma & 0 \end{vmatrix} \quad (10.57)$$

where

$$\Delta = \begin{vmatrix} \mu & \sigma & -(\Lambda_1 + \frac{1}{2}\sigma_1) & 0 \\ -\sigma & \mu & 0 & -(\Lambda_1 + \frac{1}{2}\sigma_1) \\ -(\Lambda_1 - \frac{1}{2}\sigma_1) & 0 & \mu & \sigma \\ 0 & -(\Lambda_1 - \frac{1}{2}\sigma_1) & -\sigma & \mu \end{vmatrix} \quad (10.58)$$

Adding the homogeneous solutions (10.46) and (10.47) to the particular solutions (10.48) and (10.49) yields the following general solutions for (10.37) and (10.38):

$$B_r = b_1 e^{\gamma_1 T_1} + b_2 e^{\gamma_2 T_1} + c_1 \cos(\sigma T_1 + \nu) + c_2 \sin(\sigma T_1 + \nu) \quad (10.59)$$

$$B_i = \frac{\gamma_1 + \mu}{\Lambda_1 + \frac{1}{2}\sigma_1} b_1 e^{\gamma_1 T_1} + \frac{\gamma_2 + \mu}{\Lambda_1 + \frac{1}{2}\sigma_1} b_2 e^{\gamma_2 T_1} + c_3 \cos(\sigma T_1 + \nu)$$

$$+ c_4 \sin(\sigma T_1 + \nu) \quad (10.60)$$

It follows from (10.31) and (10.34) and the fact that $\lambda = \frac{1}{2}\sigma_1$ that

$$A = (B_r + iB_i)e^{(1/2)i\sigma_1 T_1} \quad (10.61)$$

Substituting (10.61) into (10.15), then substituting the result into (10.12), and using (10.16), we have

$$u = (B_r + iB_i)e^{i[1 + (1/2)\epsilon\sigma_1]t} + (B_r - iB_i)e^{-i[1 + (1/2)\epsilon\sigma_1]t} + \frac{F_1}{1 - \omega_1^2}\cos\omega_1 t$$

$$+ \frac{F_2}{1 - \omega_2^2}\cos(\omega_2 t + \nu) + O(\epsilon)$$

But $\omega_1 = 2 + \epsilon\sigma_1$, hence

$$u = B_r[e^{(1/2)i\omega_1 t} + e^{-(1/2)i\omega_1 t}] + iB_i[e^{(1/2)i\omega_1 t} - e^{-(1/2)i\omega_1 t}]$$

$$+ \frac{F_1}{1 - \omega_1^2}\cos\omega_1 t + \frac{F_2}{1 - \omega_2^2}\cos(\omega_2 t + \nu) + O(\epsilon)$$

or

$$u = 2B_r \cos\tfrac{1}{2}\omega_1 t - 2B_i \sin\tfrac{1}{2}\omega_1 t + \frac{F_1}{1 - \omega_1^2}\cos\omega_1 t$$

226 MULTIFREQUENCY EXCITATIONS

$$+ \frac{F_2}{1 - \omega_2^2} \cos(\omega_2 t + \nu) + O(\epsilon) \tag{10.62}$$

Substituting (10.59) and (10.60) into (10.62) gives the following first approximation:

$$u = 2[b_1 e^{\epsilon \gamma_1 t} + b_2 e^{\epsilon \gamma_2 t} + c_1 \cos(\epsilon \sigma t + \nu) + c_2 \sin(\epsilon \sigma t + \nu)]$$

$$\times \cos \tfrac{1}{2} \omega_1 t - 2 \left[\frac{\gamma_1 + \mu}{\Lambda_1 + \tfrac{1}{2} \sigma_1} b_1 e^{\epsilon \gamma_1 t} + \frac{\gamma_2 + \mu}{\Lambda_1 + \tfrac{1}{2} \sigma_1} b_2 e^{\epsilon \gamma_2 t} \right.$$

$$\left. + c_3 \cos(\epsilon \sigma t + \nu) + c_4 \sin(\epsilon \sigma t + \nu) \right] \sin \tfrac{1}{2} \omega_1 t$$

$$+ \frac{F_1}{1 - \omega_1^2} \cos \omega_1 t + \frac{F_2}{1 - \omega_2^2} \cos(\omega_2 t + \nu) + O(\epsilon) \tag{10.63}$$

Equation (10.63) shows that u becomes unbounded in t if the real part of either γ_1 or γ_2 is positive. It follows from (10.44) that the real part of γ_1 is always negative, while the real part of γ_2 is positive if

$$\Lambda_1^2 > \tfrac{1}{4} \sigma_1^2 \quad \text{and} \quad \sqrt{\Lambda_1^2 - \tfrac{1}{4} \sigma_1^2} - \mu > 0$$

or

$$\Lambda_1^2 > \tfrac{1}{4} \sigma_1^2 + \mu^2$$

In this case, the free-oscillation terms proportional to $\cos \tfrac{1}{2} \omega_1 t$ and $\sin \tfrac{1}{2} \omega_1 t$ in (10.62) tend to infinity as $t \to \infty$. In reality, u does not tend to infinity but the free-oscillation term becomes large and we need to carry out the expansion to the next order.

10.3. The Method of Averaging

As before, the first step in the application of this technique is the use of the method of variation of parameters to transform the dependent variable from u to a and β, where a and β are the amplitude and phase of the free-oscillation term. To this end, we note that when $\epsilon = 0$ the solution of (10.1) is

$$u = a \cos(t + \beta) + \frac{F_1}{1 - \omega_1^2} \cos \omega_1 t + \frac{F_2}{1 - \omega_2^2} \cos(\omega_2 t + \nu) \tag{10.64}$$

where a and β are constants. Hence,

$$\dot{u} = -a \sin(t + \beta) - \frac{F_1 \omega_1}{1 - \omega_1^2} \sin \omega_1 t - \frac{F_2 \omega_2}{1 - \omega_2^2} \sin(\omega_2 t + \nu) \tag{10.65}$$

When $\epsilon \neq 0$, we still represent the solution by (10.64) subject to the constraint (10.65) but with time-varying a and β.

Differentiating (10.64) with respect to t yields

$$\dot{u} = -a \sin(t + \beta) + \dot{a} \cos(t + \beta) - a\dot{\beta} \sin(t + \beta)$$

$$- \frac{F_1 \omega_1}{1 - \omega_1^2} \sin \omega_1 t - \frac{F_2 \omega_2}{1 - \omega_2^2} \sin(\omega_2 t + \nu) \tag{10.66}$$

Comparing (10.65) and (10.66), we conclude that

$$\dot{a} \cos(t + \beta) - a\dot{\beta} \sin(t + \beta) = 0 \tag{10.67}$$

Differentiating (10.65) with respect to t yields

$$\ddot{u} = -a \cos(t + \beta) - \dot{a} \sin(t + \beta) - a\dot{\beta} \cos(t + \beta) - \frac{F_1 \omega_1^2}{1 - \omega_1^2} \cos \omega_1 t$$

$$- \frac{F_2 \omega_2^2}{1 - \omega_2^2} \cos(\omega_2 t + \nu) \tag{10.68}$$

Substituting (10.64), (10.65), and (10.68) into (10.1), we obtain

$$\dot{a} \sin(t + \beta) + a\dot{\beta} \cos(t + \beta) = \epsilon f \tag{10.69}$$

where

$$f = -2\mu \left[a \sin(t + \beta) + \frac{F_1 \omega_1}{1 - \omega_1^2} \sin \omega_1 t + \frac{F_2 \omega_2}{1 - \omega_2^2} \sin(\omega_2 t + \nu) \right]$$

$$+ a^2 \cos^2(t + \beta) + \frac{F_1^2}{(1 - \omega_1^2)^2} \cos^2 \omega_1 t + \frac{F_2^2}{(1 - \omega_2^2)^2} \cos^2(\omega_2 t + \nu)$$

$$+ \frac{2aF_1}{1 - \omega_1^2} \cos(t + \beta) \cos \omega_1 t + \frac{2aF_2}{1 - \omega_2^2} \cos(t + \beta) \cos(\omega_2 t + \nu)$$

$$+ \frac{2F_1 F_2}{(1 - \omega_1^2)(1 - \omega_2^2)} \cos \omega_1 t \cos(\omega_2 t + \nu) \tag{10.70}$$

Solving (10.67) and (10.69) for $\dot{a}$ and $a\dot{\beta}$, we have

$$\dot{a} = \epsilon f \sin(t + \beta) \tag{10.71}$$

$$a\dot{\beta} = \epsilon f \cos(t + \beta) \tag{10.72}$$

Substituting (10.70) into (10.71) and (10.72) and using trigonometric identities (Appendix A), we obtain

$$\dot{a} = -\epsilon \mu a + \epsilon \mu a \cos(2t + 2\beta) - \frac{\epsilon F_1 \omega_1 \mu}{1 - \omega_1^2} \{\cos[(\omega_1 - 1)t - \beta]$$

$$- \cos[(\omega_1 + 1)t + \beta]\} - \frac{\epsilon F_2 \omega_2 \mu}{1 - \omega_2^2} \{\cos[(\omega_2 - 1)t - \beta + \nu]$$

$$- \cos\left[(\omega_2 + 1)t + \beta + \nu\right]\} + \tfrac{1}{4}\epsilon a^2 \sin(t + \beta) + \tfrac{1}{4}\epsilon a^2 \sin(3t + 3\beta)$$

$$+ \frac{\epsilon F_1^2}{2(1 - \omega_1^2)^2}\{\sin(t + \beta) + \tfrac{1}{2}\sin[(1 + 2\omega_1)t + \beta] + \tfrac{1}{2}\sin[(1 - 2\omega_1)t + \beta]\}$$

$$+ \frac{\epsilon F_2^2}{2(1 - \omega_2^2)^2}\{\sin(t + \beta) + \tfrac{1}{2}\sin[(1 + 2\omega_2)t + \beta + 2\nu]$$

$$+ \tfrac{1}{2}\sin[(1 - 2\omega_2)t + \beta - 2\nu]\} + \frac{\epsilon a F_1}{2(1 - \omega_1^2)}\{\sin[(2 + \omega_1)t + 2\beta]$$

$$+ \sin[(2 - \omega_1)t + 2\beta]\} + \frac{\epsilon a F_2}{2(1 - \omega_2^2)}\{\sin[(2 + \omega_2)t + 2\beta + \nu]$$

$$+ \sin[(2 - \omega_2)t + 2\beta - \nu]\} + \frac{\epsilon F_1 F_2}{2(1 - \omega_1^2)(1 - \omega_2^2)}\{\sin[(1 + \omega_1 + \omega_2)t$$

$$+ \beta + \nu] + \sin[(1 - \omega_1 - \omega_2)t + \beta - \nu] + \sin[(1 + \omega_2 - \omega_1)t$$

$$+ \beta + \nu] + \sin[(1 - \omega_2 + \omega_1)t + \beta - \nu]\} \tag{10.73}$$

$$a\dot\beta = -\epsilon\mu a \sin(2t + 2\beta) - \frac{\epsilon F_1 \omega_1 \mu}{1 - \omega_1^2}\{\sin[(\omega_1 + 1)t + \beta]$$

$$+ \sin[(\omega_1 - 1)t - \beta]\} - \frac{\epsilon F_2 \omega_2 \mu}{1 - \omega_2^2}\{\sin[(\omega_2 + 1)t + \beta + \nu]$$

$$+ \sin[(\omega_2 - 1)t - \beta + \nu]\} + \tfrac{3}{4}a^2 \cos(t + \beta) + \tfrac{1}{4}a^2 \cos(3t + 3\beta)$$

$$+ \frac{\epsilon F_1^2}{2(1 - \omega_1^2)^2}\{\cos(t + \beta) + \tfrac{1}{2}\cos[(1 + 2\omega_1)t + \beta]$$

$$+ \tfrac{1}{2}\cos[(1 - 2\omega_1)t + \beta]\} + \frac{\epsilon F_2^2}{2(1 - \omega_2^2)^2}\{\cos(t + \beta)$$

$$+ \tfrac{1}{2}\cos[(1 + 2\omega_2)t + \beta + 2\nu] + \tfrac{1}{2}\cos[(1 - 2\omega_2)t + \beta - 2\nu]\}$$

$$+ \frac{\epsilon a F_1}{1 - \omega_1^2}\{\cos\omega_1 t + \tfrac{1}{2}\cos[(2 + \omega_1)t + 2\beta] + \tfrac{1}{2}\cos[(2 - \omega_1)t$$

$$+ 2\beta]\} + \frac{\epsilon a F_2}{1 - \omega_2^2}\{\cos(\omega_2 t + \nu) + \tfrac{1}{2}\cos[(2 + \omega_2)t + 2\beta + \nu]$$

$$+ \tfrac{1}{2}\cos[(2 - \omega_2)t + 2\beta - \nu]\}$$

$$+ \frac{\epsilon F_1 F_2}{2(1 - \omega_1^2)(1 - \omega_2^2)}\{\cos[(1 + \omega_1 + \omega_2)t + \beta + \nu]$$

$$+ \cos[(1 - \omega_1 - \omega_2)t + \beta - \nu] + \cos[(1 + \omega_2 - \omega_1)t + \beta + \nu]$$

$$+ \cos\left[(1 - \omega_2 + \omega_1)t + \beta - \nu\right]\} \qquad (10.74)$$

We note that (10.73) and (10.74) are lengthy. We could shorten them considerably if we express u and $\dot{u}$ in (10.64) and (10.65) in complex rather than real form. Thus, we let

$$u = Ae^{it} + \Lambda_1 e^{i\omega_1 t} + \Lambda_2 e^{i\omega_2 t} + cc \qquad (10.75)$$

$$\dot{u} = iAe^{it} + i\omega_1 \Lambda_1 e^{i\omega_1 t} + i\omega_2 \Lambda_2 e^{i\omega_2 t} + cc \qquad (10.76)$$

Differentiating u with respect to t and recalling that $A = A(t)$, we obtain

$$\dot{u} = iAe^{it} + \dot{A}e^{it} + i\omega_1 \Lambda_1 e^{i\omega_1 t} + i\omega_2 \Lambda_2 e^{i\omega_2 t} + cc \qquad (10.77)$$

Comparing (10.76) and (10.77), we conclude that

$$\dot{A}e^{it} + \dot{\overline{A}}e^{-it} = 0 \qquad (10.78)$$

Differentiating (10.76) with respect to t gives

$$\ddot{u} = -Ae^{it} + i\dot{A}e^{it} - \omega_1^2 \Lambda_1 e^{i\omega_1 t} - \omega_2^2 \Lambda_2 e^{i\omega_2 t} + cc \qquad (10.79)$$

Substituting (10.75), (10.76), and (10.79) into (10.1) and using (10.16), we obtain

$$i\dot{A}e^{it} - i\dot{\overline{A}}e^{-it} = -2i\epsilon\mu[Ae^{it} + \omega_1 \Lambda_1 e^{i\omega_1 t} + \omega_2 \Lambda_2 e^{i\omega_2 t}]$$

$$-\epsilon A^2 e^{2it} - \epsilon \Lambda_1^2 e^{2i\omega_1 t} - \epsilon \Lambda_2^2 e^{2i\omega_2 t} - 2\epsilon A \Lambda_1 e^{i(1+\omega_1)t}$$

$$- 2\epsilon A \Lambda_2 e^{i(1+\omega_2)t} - \epsilon A\overline{A} - \epsilon \Lambda_1^2 - \epsilon \Lambda_2 \overline{\Lambda}_2 - 2\epsilon \overline{A} \Lambda_1 e^{i(\omega_1 - 1)t}$$

$$- 2\epsilon \overline{A} \Lambda_2 e^{i(\omega_2 - 1)t} - 2\epsilon \Lambda_1 \Lambda_2 e^{i(\omega_2 + \omega_1)t} - 2\epsilon \Lambda_2 \Lambda_1 e^{i(\omega_2 - \omega_1)t}$$

$$+ cc \qquad (10.80)$$

It follows from (10.78) that $\dot{\overline{A}} \exp(-it) = -\dot{A}\exp(it)$. Then, it follows from (10.80) that

$$i\dot{A} = \{-i\epsilon\mu[Ae^{it} + \omega_1 \Lambda_1 e^{i\omega_1 t} + \omega_2 \Lambda_2 e^{i\omega_2 t}] - \tfrac{1}{2}\epsilon A^2 e^{2it}$$

$$- \tfrac{1}{2}\epsilon \Lambda_1^2 e^{i2\omega_1 t} - \tfrac{1}{2}\epsilon \Lambda_2^2 e^{i2\omega_2 t} - \epsilon A \Lambda_1 e^{i(\omega_1 + 1)t}$$

$$- \epsilon A \Lambda_2 e^{i(\omega_2 + 1)t} - \tfrac{1}{2}\epsilon(A\overline{A} + \Lambda_1^2 + \Lambda_2 \overline{\Lambda}_2) - \epsilon \overline{A} \Lambda_1 e^{i(\omega_1 - 1)t}$$

$$- \epsilon \overline{A} \Lambda_2 e^{i(\omega_2 - 1)t} - \epsilon \Lambda_1 \Lambda_2 e^{i(\omega_2 + \omega_1)t} - \epsilon \Lambda_2 \Lambda_1 e^{i(\omega_2 - \omega_1)t}$$

$$+ cc\} e^{-it} \qquad (10.81)$$

Expressing A in polar form and separating (10.81) into real and imaginary parts, we obtain (10.73) and (10.74).

Next, we consider the case $\omega_2 + \omega_1 = 1 + \epsilon\sigma$ discussed in Section (10.2.1) and the case $\omega_1 = 2 + \epsilon\sigma_1$ and $\omega_2 = 3 + \epsilon\sigma_2$ discussed in Section 10.2.2.

230 MULTIFREQUENCY EXCITATIONS

10.3.1. THE CASE $\omega_1 + \omega_2 \approx 1$

To the first approximation, we keep only the slowly varying terms on the right-hand sides of (10.73) and (10.74). The result is

$$\dot{a} = -\epsilon\mu a + \frac{\epsilon F_1 F_2}{2(1-\omega_1^2)(1-\omega_2^2)} \sin\left[(1-\omega_1-\omega_2)t + \beta - \nu\right] \quad (10.82)$$

$$a\dot{\beta} = \frac{\epsilon F_1 F_2}{2(1-\omega_1^2)(1-\omega_2^2)} \cos\left[(1-\omega_1-\omega_2)t + \beta - \nu\right] \quad (10.83)$$

which are equivalent to (10.20) obtained by using the method of multiple scales. We note that whereas (10.20) is linear, (10.82) and (10.83) are nonlinear. This is the reason we did not express (10.20) in polar form, but solved it in its complex form.

Keeping the slowly varying terms in (10.81) yields

$$\dot{A} = -\epsilon\mu A + i\epsilon\Lambda_1\Lambda_2 e^{i(\omega_1+\omega_2-1)t} \quad (10.84)$$

which is in full agreement with (10.20).

10.3.2. THE CASE $\omega_2 - \omega_1 \approx 1$ AND $\omega_1 \approx 2$

Keeping the slowly varying terms on the right-hand sides of (10.73) and (10.74), we obtain

$$\dot{a} = -\epsilon\mu a + \frac{\epsilon a F_1}{2(1-\omega_1^2)} \sin\left[(2-\omega_1)t + 2\beta\right] + \frac{\epsilon F_1 F_2}{2(1-\omega_1^2)(1-\omega_2^2)}$$
$$\times \sin\left[(1-\omega_2+\omega_1)t + \beta - \nu\right] \quad (10.85)$$

$$a\dot{\beta} = \frac{\epsilon a F_1}{2(1-\omega_1^2)} \cos\left[(2-\omega_1)t + 2\beta\right] + \frac{\epsilon F_1 F_2}{2(1-\omega_1^2)(1-\omega_2^2)}$$
$$\times \cos\left[(1-\omega_2+\omega_1)t + \beta - \nu\right] \quad (10.86)$$

which are equivalent to (10.30) obtained by using the method of multiple scales. Again, we note that the form (10.30) seems to be more convenient than (10.85) and (10.86).

Keeping the slowly varying terms in (10.81) yields

$$\dot{A} = -\epsilon\mu A + i\epsilon\overline{A}\Lambda_1 e^{i(\omega_1-2)t} + i\epsilon\Lambda_2\Lambda_1 e^{i(\omega_2-\omega_1-1)t} \quad (10.87)$$

which is in full agreement with (10.30). The development in (10.84) and (10.87) shows that the complex form has advantages over the real form of the solution.

Exercises

10.1. Use the methods of multiple scales and averaging to determine the equations governing the amplitudes and the phases to first order for a system governed by

$$\ddot{u} + \omega_0^2 u = \epsilon \dot{u}^2 + K_1 \cos \Omega_1 t + K_2 \cos \Omega_2 t \qquad \epsilon \ll 1$$

when $\Omega_2 \pm \Omega_1 \approx \omega_0$. Consider only the case when Ω_1 is away from zero.

10.2. Consider the equation

$$\ddot{u} + \omega_0^2 u + \epsilon u^3 = K_1 \cos(\Omega_1 t + \theta_1) + K_2 \cos(\Omega_2 t + \theta_2)$$

where ω_0, K_n, and θ_n are constants. When Ω_n is away from ω_0, use the method of multiple scales to show that

$$u_0 = A(T_1)e^{i\omega_0 T_0} + \Lambda_1 e^{i\Omega_1 T_0} + \Lambda_2 e^{i\Omega_2 T_0} + cc$$

$$\Lambda_n = \tfrac{1}{2} K_n (\omega_0^2 - \Omega_n^2)^{-1} e^{i\theta_n}$$

Then, show that

$$D_0^2 u_1 + \omega_0^2 u_1 = -[2i\omega_0 A' + 3(A\overline{A} + 2\Lambda_1 \overline{\Lambda}_1 + 2\Lambda_2 \overline{\Lambda}_2)A]e^{i\omega_0 T_0}$$
$$- 3(2A\overline{A} + \Lambda_1 \overline{\Lambda}_1 + 2\Lambda_2 \overline{\Lambda}_2)\Lambda_1 e^{i\Omega_1 T_0}$$
$$- 3(2A\overline{A} + 2\Lambda_1 \overline{\Lambda}_1 + \Lambda_2 \overline{\Lambda}_2)\Lambda_2 e^{i\Omega_2 T_0} - A^3 e^{3i\omega_0 T_0}$$
$$- \Lambda_1^3 e^{3i\Omega_1 T_0} - \Lambda_2^3 e^{3i\Omega_2 T_0} - 3A^2 \Lambda_1 e^{i(2\omega_0 + \Omega_1)T_0}$$
$$- 3A^2 \Lambda_2 e^{i(2\omega_0 + \Omega_2)T_0} - 3A^2 \overline{\Lambda}_1 e^{i(2\omega_0 - \Omega_1)T_0}$$
$$- 3A^2 \overline{\Lambda}_2 e^{i(2\omega_0 - \Omega_2)T_0} - 3A\Lambda_1^2 e^{i(\omega_0 + 2\Omega_1)T_0}$$
$$- 3A\Lambda_2^2 e^{i(\omega_0 + 2\Omega_2)T_0} - 3A\overline{\Lambda}_1^2 e^{i(\omega_0 - 2\Omega_1)T_0}$$
$$- 3A\overline{\Lambda}_2^2 e^{i(\omega_0 - 2\Omega_2)T_0} - 6A\Lambda_1 \Lambda_2 e^{i(\omega_0 + \Omega_1 + \Omega_2)T_0}$$
$$- 6A\overline{\Lambda}_1 \overline{\Lambda}_2 e^{i(\omega_0 - \Omega_1 - \Omega_2)T_0} - 6A\overline{\Lambda}_1 \Lambda_2 e^{i(\omega_0 - \Omega_1 + \Omega_2)T_0}$$
$$- 6A\Lambda_1 \overline{\Lambda}_2 e^{i(\omega_0 + \Omega_1 - \Omega_2)T_0} - 3\Lambda_1^2 \Lambda_2 e^{i(2\Omega_1 + \Omega_2)T_0}$$
$$- 3\Lambda_1^2 \overline{\Lambda}_2 e^{i(2\Omega_1 - \Omega_2)T_0} - 3\Lambda_1 \Lambda_2^2 e^{i(\Omega_1 + 2\Omega_2)T_0}$$
$$- 3\overline{\Lambda}_1 \Lambda_2^2 e^{i(2\Omega_2 - \Omega_1)T_0} + cc$$

(a) Show that, if $\Omega_2 > \Omega_1$, resonances occur whenever

$$\omega_0 \approx 3\Omega_1 \quad \text{or} \quad 3\Omega_2$$
$$\omega_0 \approx \tfrac{1}{3}\Omega_1 \quad \text{or} \quad \tfrac{1}{3}\Omega_2$$
$$\omega_0 \approx \Omega_2 \pm 2\Omega_1 \quad \text{or} \quad 2\Omega_1 - \Omega_2$$
$$\omega_0 \approx 2\Omega_2 \pm \Omega_1$$
$$\omega_0 \approx \tfrac{1}{2}(\Omega_2 \pm \Omega_1)$$

(b) Show that simultaneous resonances occur whenever

$$\Omega_2 \approx 9\Omega_1 \approx 3\omega_0$$
$$\Omega_2 \approx \Omega_1 \approx 3\omega_0$$
$$\Omega_2 \approx \Omega_1 \approx \tfrac{1}{3}\omega_0$$

232 MULTIFREQUENCY EXCITATIONS

$$\Omega_2 \approx 5\Omega_1 \approx \tfrac{5}{3}\omega_0$$

$$\Omega_2 \approx 7\Omega_1 \approx \tfrac{7}{3}\omega_0$$

$$\Omega_2 \approx 2\Omega_1 \approx \tfrac{2}{3}\omega_0$$

$$\Omega_2 \approx \tfrac{7}{3}\Omega_1 \approx 7\omega_0$$

$$\Omega_2 \approx \tfrac{5}{3}\Omega_1 \approx 5\omega_0$$

$$\Omega_2 \approx \tfrac{3}{2}\Omega_1 \approx 3\omega_0$$

$$\Omega_2 \approx 3\Omega_1 \approx \tfrac{3}{5}\omega_0$$

(c) When $\omega_0 = 2\Omega_1 + \Omega_2 + \epsilon\sigma$, show that

$$2i\omega_0 A' + (3A\overline{A} + 6\Lambda_1\overline{\Lambda}_1 + 6\Lambda_2\overline{\Lambda}_2)A + 3\Lambda_1^2\Lambda_2 e^{-i\sigma T_1} = 0$$

(d) When $3\Omega_1 = \omega_0 + \epsilon\sigma_1$ and $\Omega_2 = 3\omega_0 + \epsilon\sigma_2$, show that

$$2i\omega_0 A' + 3(A\overline{A} + 2\Lambda_1\overline{\Lambda}_1 + 2\Lambda_2\overline{\Lambda}_2)A + \Lambda_1^3 e^{i\sigma_1 T_1} + 3\Lambda_2 \overline{A}^2 e^{i\sigma_2 T_1} = 0$$

(e) When $3\Omega_1 = \omega_0 + \epsilon\sigma_1$ and $\Omega_2 + \Omega_1 = 2\omega_0 + \epsilon\sigma_2$, show that

$$\omega_0 A' + 3(A\overline{A} + 2\Lambda_2\overline{\Lambda}_2 + 2\Lambda_2\overline{\Lambda}_2)A + \Lambda_1^3 e^{i\sigma_1 T_1} + 6\overline{A}\Lambda_1\Lambda_2 e^{i\sigma_2 T_1} + 3\Lambda_2\overline{\Lambda}_1^2 e^{i(\sigma_2-\sigma_1)T_1} = 0$$

(f) When $3\Omega_1 = \omega_0 + \epsilon\sigma_1$ and $\Omega_2 - \Omega_1 = 2\omega_0 + \epsilon\sigma_2$, show that

$$2i\omega_0 A' + 3(A\overline{A} + 2\Lambda_1\overline{\Lambda}_1 + 2\Lambda_2\overline{\Lambda}_2)A + \Lambda_1^3 e^{i\sigma_1 T_1} + 6\overline{A}\Lambda_1\Lambda_2 e^{i\sigma_2 T_1} = 0$$

(g) When $3\Omega_1 = \omega_0 + \epsilon\sigma_1$ and $2\Omega_2 - \Omega_1 = \omega_0 + \epsilon\sigma_2$, show that

$$2i\omega_0 A' + 3(A\overline{A} + 2\Lambda_1\overline{\Lambda}_1 + 2\Lambda_2\overline{\Lambda}_2)A + \Lambda_1^3 e^{i\sigma_1 T_1} + 3\Lambda_2^2 \overline{\Lambda}_1 e^{i\sigma_2 T_1} = 0$$

(h) When $\Omega_1 = 3\omega_0 + \epsilon\sigma_1$ and $\Omega_2 - 2\Omega_1 = \omega_0 + \epsilon\sigma_2$, show that

$$2i\omega_0 A' + 3(A\overline{A} + 2\Lambda_1\overline{\Lambda}_1 + 2\Lambda_2\overline{\Lambda}_2)A + 3\overline{A}^2\Lambda_1 e^{i\sigma_1 T_1} + 3\Lambda_2 \overline{\Lambda}_1^2 e^{i\sigma_2 T_1} = 0$$

(i) When $\Omega_1 = 3\omega_0 + \epsilon\sigma_1$ and $\Omega_2 - \Omega_1 = 2\omega_0 + \epsilon\sigma_2$, show that

$$2i\omega_0 A' + 3(A\overline{A} + 2\Lambda_1\overline{\Lambda}_1 + 2\Lambda_2\overline{\Lambda}_2)A + 3\overline{A}^2\Lambda_1 e^{i\sigma_1 T_1} + 6\overline{A}\Lambda_1\Lambda_2 e^{i\sigma_2 T_1} + 3\Lambda_1^2\overline{\Lambda}_2 e^{i(\sigma_1-\sigma_2)T_1} = 0$$

10.3. Consider the problem in the preceding exercise. Let

$$u = A(t)e^{i\omega_0 t} + \Lambda_1 e^{i\Omega_1 t} + \Lambda_2 e^{i\Omega_2 t} + cc$$

Use the method of variation of parameters to show that

$$-2i\omega_0 A' = \{3(A\overline{A} + 2\Lambda_1\overline{\Lambda}_1 + 2\Lambda_2\overline{\Lambda}_2)A e^{i\omega_0 t} + 3(2A\overline{A} + \Lambda_1\overline{\Lambda}_1$$
$$+ 2\Lambda_2\overline{\Lambda}_2)\Lambda_1 e^{i\Omega_1 t} + 3(2A\overline{A} + 2\Lambda_1\overline{\Lambda}_1 + \Lambda_2\overline{\Lambda}_2)\Lambda_2 e^{i\Omega_2 t}$$
$$+ A^3 e^{3i\omega_0 t} + \Lambda_1^3 e^{3i\Omega_1 t} + \Lambda_2^3 e^{3i\Omega_2 t} + 3A^2\Lambda_1 e^{i(2\omega_0 + \Omega_1)t}$$
$$+ 3A^2\Lambda_2 e^{i(2\omega_0 + \Omega_2)t} + 3A^2\overline{\Lambda}_1 e^{i(2\omega_0 - \Omega_1)t} + 3A^2\overline{\Lambda}_2 e^{i(2\omega_0 - \Omega_2)t}$$
$$+ 3A\Lambda_1^2 e^{i(2\Omega_1 + \omega_0)t} + 3A\Lambda_2^2 e^{i(2\Omega_2 + \omega_0)t} + 3A\overline{\Lambda}_1^2 e^{i(\omega_0 - 2\Omega_1)t}$$
$$+ 3A\overline{\Lambda}_2^2 e^{i(\omega_0 - 2\Omega_2)t} + 6A\Lambda_1\Lambda_2 e^{i(\Omega_1 + \Omega_2 + \omega_0)t}$$
$$+ 6A\overline{\Lambda}_1\overline{\Lambda}_2 e^{-i(\Omega_1 + \Omega_2 - \omega_0)t} + 6A\overline{\Lambda}_1\Lambda_2 e^{i(\Omega_2 - \Omega_1 + \omega_0)t}$$
$$+ 6A\Lambda_1\overline{\Lambda}_2 e^{i(\Omega_1 - \Omega_2 + \omega_0)t} + 3\Lambda_1^2\Lambda_2 e^{i(2\Omega_1 + \Omega_2)t}$$

$$+ 3\Lambda_1^2 \overline{\Lambda}_2 e^{i(2\Omega_1 - \Omega_2)t} + 3\Lambda_1 \Lambda_2^2 e^{i(\Omega_1 + 2\Omega_2)t} + 3\overline{\Lambda}_1 \Lambda_2^2 e^{i(2\Omega_2 - \Omega_1)t}$$
$$+ cc \} e^{-i\omega_0 t}$$

Average this equation for the cases in the preceding exercise.

10.4. Use the methods of multiple scales and averaging to determine a first-order uniform expansion for

$$\ddot{u}_1 + \omega_1^2 u_1 = \alpha_1 u_1 u_2$$
$$\ddot{u}_2 + \omega_2^2 u_2 = \alpha_2 u_1^2$$

for small but finite amplitudes when $\omega_2 \approx 2\omega_1$.

10.5. Use the methods of multiple scales and averaging to determine the equations describing the amplitudes and phases of the system

$$\ddot{u}_1 + \omega_1^2 u_1 = \alpha_1 u_2 u_3$$
$$\ddot{u}_2 + \omega_2^2 u_2 = \alpha_2 u_1 u_3$$
$$\ddot{u}_3 + \omega_3^2 u_3 = \alpha_3 u_1 u_2$$

for small but finite amplitudes when $\omega_3 \approx \omega_1 + \omega_2$.

10.6. Use the methods of multiple scales and averaging to determine first-order uniform expansions for

$$\ddot{u}_1 + \omega_1^2 u_1 = \epsilon \alpha_1 u_1 u_2 + \epsilon k_1 \cos \Omega_1 t$$
$$\ddot{u}_2 + \omega_2^2 u_2 = \epsilon \alpha_2 u_1^2 + \epsilon k_2 \cos \Omega_2 t$$

when

(a) $\omega_2 \approx 2\omega_1$ and $\Omega_1 \approx \omega_1$.
(b) $\omega_2 \approx 2\omega_1$ and $\Omega_2 \approx \omega_2$.

CHAPTER 11

The Mathieu Equation

In contrast with the two preceding chapters, in which the excitations appear as inhomogeneities in the governing equations, in this chapter, we consider excitations that appear as coefficients in the governing equations. Such excitations are called *parametric excitations*. The simplest possible equation that describes the parametric excitations of a system having a single degree of freedom is the Mathieu equation

$$\frac{d^2 u^*}{dt^{*2}} + (\delta^* + \epsilon^* \cos \omega^* t^*) u^* = 0 \qquad (11.1)$$

As before, we introduce dimensionless quantities. We let

$$u = \frac{u^*}{u_0^*} \qquad t = \tfrac{1}{2} \omega^* t^*$$

where u_0^* is a representative value of u^*. Then, (11.1) can be put in the standard form

$$\ddot{u} + (\delta + 2\epsilon \cos 2t) u = 0 \qquad (11.2)$$

where

$$\delta = \frac{4\delta^*}{\omega^{*2}} \qquad \epsilon = \frac{4\epsilon^*}{\omega^{*2}}$$

In the next section, we determine a second-order straightforward expansion for small ϵ and investigate its uniformity. In Section 11.2, we describe the Floquet theory, which characterizes the exact solutions of (11.2). In Section 11.3, we describe the method of strained parameters to determine approximations to the periodic solutions, while in Section 11.4, we determine approximations to the exact solutions. In Sections 11.5 and 11.6, we show how the methods of multiple scales and averaging can be used to obtain uniform expansions for the solutions of (11.2).

11.1 The Straightforward Expansion

We seek a straightforward expansion for the solution of (11.2) in power series of ϵ in the form

$$u(t;\epsilon) = u_0(t) + \epsilon u_1(t) + \epsilon^2 u_2(t) + \cdots \quad (11.3)$$

Substituting (11.3) into (11.2) and equating coefficients of like powers of ϵ, we have

$$\ddot{u}_0 + \delta u_0 = 0 \quad (11.4)$$

$$\ddot{u}_1 + \delta u_1 = -2u_0 \cos 2t \quad (11.5)$$

$$\ddot{u}_2 + \delta u_2 = -2u_1 \cos 2t \quad (11.6)$$

The general solution of (11.4) can be expressed as

$$u_0 = a \cos(\omega t + \beta) \qquad \delta = \omega^2 \quad (11.7)$$

where a and β are constants. Then, (11.5) becomes

$$\ddot{u}_1 + \omega^2 u_1 = -2a \cos(\omega t + \beta) \cos 2t$$

or

$$\ddot{u}_1 + \omega^2 u_1 = -a \cos[(\omega + 2)t + \beta] - a \cos[(\omega - 2)t + \beta] \quad (11.8)$$

As before, disregarding the homogeneous solution, we write the solution of (11.8) as

$$u_1 = \frac{a \cos[(\omega + 2)t + \beta]}{4(1 + \omega)} + \frac{a \cos[(\omega - 2)t + \beta]}{4(1 - \omega)} \quad (11.9)$$

Then, (11.6) becomes

$$\ddot{u}_2 + \omega^2 u_2 = -\frac{a}{2(1+\omega)} \cos[(\omega+2)t + \beta] \cos 2t$$

$$- \frac{a}{2(1-\omega)} \cos[(\omega-2)t + \beta] \cos 2t$$

or

$$\ddot{u}_2 + \omega^2 u_2 = -\frac{a}{4(1+\omega)} \cos[(\omega+4)t+\beta] - \frac{a}{4(1-\omega)} \cos[(\omega-4)t+\beta]$$

$$- \left[\frac{a}{4(1+\omega)} + \frac{a}{4(1-\omega)}\right] \cos(\omega t + \beta) \quad (11.10)$$

Disregarding the homogeneous solution, we write the solution of (11.10) as

$$u_2 = -\frac{a}{4\omega(1-\omega^2)} t \sin(\omega t + \beta) + \frac{a \cos[(\omega+4)t+\beta]}{32(1+\omega)(2+\omega)}$$

$$+ \frac{a \cos [(\omega - 4)t + \beta]}{32(1 - \omega)(2 - \omega)} \qquad (11.11)$$

Substituting (11.7), (11.9), and (11.11) into (11.3), we obtain

$$u = a \cos (\omega t + \beta) + \tfrac{1}{4}\epsilon a \left\{ \frac{\cos [(\omega + 2)t + \beta]}{1 + \omega} + \frac{\cos [(\omega - 2)t + \beta]}{1 - \omega} \right\}$$

$$+ \tfrac{1}{32}\epsilon^2 a \left\{ -\frac{8t \sin (\omega t + \beta)}{\omega(1 - \omega^2)} + \frac{\cos [(\omega + 4)t + \beta]}{(1 + \omega)(2 + \omega)} \right.$$

$$\left. + \frac{\cos [(\omega - 4)t + \beta]}{(1 - \omega)(2 - \omega)} \right\} + \cdots \qquad (11.12)$$

The straightforward expansion breaks down for large t because of the presence of the mixed secular term. It also breaks down when $\omega \approx 0, 1$, and 2 because of the presence of small-divisor terms. Carrying out the expansion to higher order, one finds that small-divisor terms occur when $\omega \approx n$, where $n = 0, 1, 2, 3, \cdots$. Next, we discuss the Floquet theory, which characterizes the exact solutions of (11.2).

11.2. The Floquet Theory

In this section, we determine general properties of the solutions of (11.2) without actually solving for them. These properties are then used in the subsequent sections to determine uniform approximations to these solutions.

Since (11.2) is a second-order linear homogeneous equation, it possesses two linearly independent solutions $u_1(t)$ and $u_2(t)$ satisfying the initial conditions

$$\begin{aligned} u_1(0) &= 1 & \dot{u}_1(0) &= 0 \\ u_2(0) &= 0 & \dot{u}_2(0) &= 1 \end{aligned} \qquad (11.13)$$

because the determinant of the Wronskian matrix (see Section B.1.) is different from zero. We show next that, if $u_1(t)$ is a solution of (11.2), then $u_1(t + \pi)$ is also a solution of (11.2). To this end, we change the independent variable from t to $z = t + \pi$. Then, (11.2) becomes

$$\frac{d^2 u}{dz^2} + [\delta + 2\epsilon \cos (2z - 2\pi)] u = 0$$

or

$$\frac{d^2 u}{dz^2} + (\delta + 2\epsilon \cos 2z) u = 0 \qquad (11.14)$$

since $\cos (2z - 2\pi) = \cos 2z$. But (11.14) has the same form as (11.2); therefore, if $u_1(t)$ is a solution of (11.2), then $u_1(z) = u_1(t + \pi)$ is also a solution of (11.14), that is, (11.2).

From the preceding discussion, it follows that, if $u_1(t)$ and $u_2(t)$ are any two solutions of (11.2), then $u_1(t + \pi)$ and $u_2(t + \pi)$ are solutions of (11.2). Moreover, if $u_1(t)$ and $u_2(t)$ are linearly independent, then $u_1(t + \pi)$ must be linearly dependent on $u_1(t)$ and $u_2(t)$, because a second-order equation has only two linearly independent solutions. Hence, there exist two constants a_{11} and a_{12}; both do not vanish simultaneously, such that

$$u_1(t + \pi) = a_{11} u_1(t) + a_{12} u_2(t) \tag{11.15}$$

Similarly, there exist two constants a_{21} and a_{22}; both do not vanish simultaneously, such that

$$u_2(t + \pi) = a_{21} u_1(t) + a_{22} u_2(t) \tag{11.16}$$

because $u_2(t + \pi)$ must be linearly dependent on $u_1(t)$ and $u_2(t)$. For the initial conditions (11.13), we find from (11.15) and (11.16) that

$$a_{11} = u_1(\pi) \qquad a_{21} = u_2(\pi) \tag{11.17}$$

Differentiating (11.15) and (11.16) with respect to t gives

$$\begin{aligned} \dot{u}_1(t + \pi) &= a_{11} \dot{u}_1(t) + a_{12} \dot{u}_2(t) \\ \dot{u}_2(t + \pi) &= a_{21} \dot{u}_1(t) + a_{22} \dot{u}_2(t) \end{aligned} \tag{11.18}$$

Substituting (11.13) into (11.18), we have

$$a_{12} = \dot{u}_1(\pi) \qquad a_{22} = \dot{u}_2(\pi) \tag{11.19}$$

Thus, once $u_1(t)$ and $u_2(t)$ are known, the coefficients a_{ij} in (11.15) and (11.16) can be uniquely determined as in (11.17) and (11.19).

We return to (11.15) and (11.16) and write them in matrix notation as

$$\mathbf{u}(t + \pi) = A\mathbf{u}(t) \tag{11.20}$$

where

$$A = \begin{bmatrix} a_{11} & a_{12} \\ a_{21} & a_{22} \end{bmatrix} \qquad \mathbf{u} = \begin{bmatrix} u_1 \\ u_2 \end{bmatrix} \tag{11.21}$$

Next, we investigate the effect of linearly transforming $\mathbf{u}(t)$ to $\mathbf{v}(t)$. Thus, we let

$$\mathbf{v}(t) = P\mathbf{u}(t) \tag{11.22}$$

where P is a 2×2 constant nonsingular matrix. In scalar notation, (11.22) can be rewritten as

$$\begin{aligned} v_1 &= p_{11} u_1 + p_{12} u_2 \\ v_2 &= p_{21} u_1 + p_{22} u_2 \end{aligned} \tag{11.23}$$

238 THE MATHIEU EQUATION

It follows from (11.22) that

$$\mathbf{u}(t) = P^{-1}\mathbf{v}(t) \tag{11.24}$$

where P^{-1} is the inverse of P; that is,

$$PP^{-1} = I \tag{11.25}$$

where I is the identity matrix

$$I = \begin{bmatrix} 1 & 0 \\ 0 & 1 \end{bmatrix}$$

Substituting (11.24) into (11.20) gives

$$P^{-1}\mathbf{v}(t + \pi) = AP^{-1}\mathbf{v}(t) \tag{11.26}$$

Multiplying (11.26) from the left by the matrix P and using (11.25), we obtain

$$\mathbf{v}(t + \pi) = PAP^{-1}\mathbf{v}(t) \tag{11.27}$$

or

$$\mathbf{v}(t + \pi) = B\mathbf{v}(t) \tag{11.28}$$

where

$$B = PAP^{-1} \tag{11.29}$$

The matrices A and B are usually called *similar matrices* because they have the same eigenvalues. To show this, we note that

$$|B - \lambda I| = |PAP^{-1} - \lambda PP^{-1}| = |P(A - \lambda I)P^{-1}|$$
$$= |P| |A - \lambda I| |P^{-1}| = |A - \lambda I| \tag{11.30}$$

because $|P| |P^{-1}| = 1$.

It follows from any book on linear algebra that a nonsingular constant matrix P can be chosen so that B will have its simplest possible form, the so-called *Jordan canonical form*. This form depends on the eigenvalues and eigenvectors of A. The eigenvalues of A and hence B are given by

$$\begin{vmatrix} a_{11} - \lambda & a_{12} \\ a_{21} & a_{22} - \lambda \end{vmatrix} = 0$$

or

$$(a_{11} - \lambda)(a_{22} - \lambda) - a_{12}a_{21} = 0$$

Hence,

$$\lambda^2 - (a_{11} + a_{22})\lambda + a_{11}a_{22} - a_{12}a_{21} = 0 \tag{11.31}$$

But

$$a_{11}a_{22} - a_{12}a_{21} = u_1(\pi)\dot{u}_2(\pi) - u_2(\pi)\dot{u}_1(\pi) = 1$$

according to (11.17) and (11.19) and the fact that the Wronskian is unity. Then, (11.31) becomes

$$\lambda^2 - 2\alpha\lambda + 1 = 0 \tag{11.32}$$

where

$$\alpha = \tfrac{1}{2}(a_{11} + a_{22}) = \tfrac{1}{2}[u_1(\pi) + \dot{u}_2(\pi)]$$

The solutions of (11.32) are

$$\lambda = \alpha \pm \sqrt{\alpha^2 - 1} \tag{11.33}$$

If $\alpha \neq 1$, (11.33) yields the two distinct eigenvalues

$$\lambda_1 = \alpha + \sqrt{\alpha^2 - 1} \qquad \lambda_2 = \alpha - \sqrt{\alpha^2 - 1} \tag{11.34}$$

and B has the diagonal form

$$B = \begin{bmatrix} \lambda_1 & 0 \\ 0 & \lambda_2 \end{bmatrix} \tag{11.35}$$

If $\alpha = \pm 1$, (11.33) yields only one eigenvalue, namely $\lambda = \alpha = \pm 1$, and B has either the form

$$B = \begin{bmatrix} \pm 1 & 0 \\ 0 & \pm 1 \end{bmatrix} \tag{11.36}$$

or the form

$$B = \begin{bmatrix} \pm 1 & 0 \\ 1 & \pm 1 \end{bmatrix} \tag{11.37}$$

When B has either the form (11.35) or (11.36), (11.28) can be rewritten in scalar form as

$$\begin{aligned} v_1(t + \pi) &= \lambda_1 v_1(t) \\ v_2(t + \pi) &= \lambda_2 v_2(t) \end{aligned} \tag{11.38}$$

where $\lambda_1 = \lambda_2 = \pm 1$ when B has the form (11.36). It follows from (11.38) that

$$\begin{aligned} v_1(t + 2\pi) &= \lambda_1 v_1(t + \pi) = \lambda_1^2 v_1(t) \\ v_1(t + 3\pi) &= \lambda_1 v_1(t + 2\pi) = \lambda_1^3 v_1(t) \\ v_1(t + 4\pi) &= \lambda_1 v_1(t + 3\pi) = \lambda_1^4 v_1(t) \\ &\cdots\cdots\cdots\cdots\cdots\cdots \\ v_1(t + n\pi) &= \lambda_1^n v_1(t) \end{aligned} \tag{11.39}$$

where n is an integer. Similarly,

240 THE MATHIEU EQUATION

$$v_2(t + n\pi) = \lambda_2^n v_2(t) \tag{11.40}$$

Consequently, as $t \to \infty$ (i.e., $n \to \infty$)

$$v_i(t) \to \begin{cases} 0 & \text{if } |\lambda_i| < 1 \\ \infty & \text{if } |\lambda_i| > 1 \end{cases} \tag{11.41}$$

and the solution becomes unbounded with time if the absolute value of any of the λ_i is larger than unity. When $\lambda_1 = \lambda_2 = 1$, (11.38) shows that v_i is periodic with period π. When $\lambda_1 = \lambda_2 = -1$,

$$\begin{aligned} v_i(t + \pi) &= -v_i(t) \\ v_i(t + 2\pi) &= -v_i(t + \pi) = v_i(t) \end{aligned} \tag{11.42}$$

so that v_i is periodic with period 2π. Thus, the cases $\lambda_1 = \lambda_2 = \pm 1$ separate stable from unstable solutions and they are usually referred to as *transition values*.

Equations (11.38) can be used to express the $v_i(t)$ in the so-called *normal or Floquet form*. To accomplish this, we multiply the first of (11.38) by $\exp \cdot [-\gamma_1(t + \pi)]$, where γ_1 is specified later, and obtain

$$e^{-\gamma_1(t+\pi)}v_1(t+\pi) = \lambda_1 e^{-\gamma_1 \pi} e^{-\gamma_1 t} v_1(t) \tag{11.43}$$

If we let

$$\lambda_1 e^{-\gamma_1 \pi} = 1 \quad \text{then} \quad e^{\gamma_1 \pi} = \lambda_1 \quad \text{and} \quad \gamma_1 = \frac{1}{\pi} \ln \lambda_1 \tag{11.44}$$

Consequently, (11.43) becomes

$$e^{-\gamma_1(t+\pi)}v_1(t+\pi) = e^{-\gamma_1 t}v_1(t) \tag{11.45}$$

Hence, $\exp(-\gamma_1 t)v_1(t)$ is periodic with the period π so that it can be expressed as

$$e^{-\gamma_1 t}v_1(t) = \phi_1(t) \tag{11.46}$$

where $\phi_1(t + \pi) = \phi_1(t)$. Hence, $v_1(t)$ can be expressed in the normal form

$$v_1(t) = e^{\gamma_1 t}\phi_1(t) \tag{11.47}$$

where $\gamma_1 = (1/\pi)\ln \lambda_1$ is called the *characteristic exponent*. Similarly, $v_2(t)$ can be expressed in the normal form

$$v_2(t) = e^{\gamma_2 t}\phi_2(t) \tag{11.48}$$

where

$$\phi_2(t + \pi) = \phi_2(t) \quad \text{and} \quad \gamma_2 = \frac{1}{\pi} \ln \lambda_2$$

When B has the form (11.37), (11.28) can be rewritten as

$$v_1(t + \pi) = \lambda v_1(t) \tag{11.49a}$$

$$v_2(t + \pi) = \lambda v_2(t) + v_1(t) \tag{11.49b}$$

where $\lambda = \pm 1$. Using a reasoning similar to the above, we can express $v_1(t)$ in the normal form

$$v_1(t) = e^{\gamma t}\phi_1(t)$$
$$\phi_1(t + \pi) = \phi_1(t) \quad \text{and} \quad \gamma = \frac{1}{\pi}\ln\lambda \tag{11.50}$$

Then, the second equation in (11.49) becomes

$$v_2(t + \pi) = \lambda v_2(t) + e^{\gamma t}\phi_1(t) \tag{11.51}$$

Multiplying, as before, (11.51) by $\exp[-\gamma(t + \pi)]$ gives

$$e^{-\gamma(t+\pi)}v_2(t+\pi) = \lambda e^{-\gamma \pi}e^{-\gamma t}v_2(t) + e^{-\gamma \pi}\phi_1(t)$$

which can be rewritten as

$$e^{-\gamma(t+\pi)}v_2(t+\pi) = e^{-\gamma t}v_2(t) + \frac{1}{\lambda}\phi_1(t) \tag{11.51}$$

In this case, $v_2(t)$ does not have the form (11.48) due to the presence of the term $\lambda^{-1}\phi_1(t)$. Instead, one can easily verify that

$$v_2(t) = e^{\gamma t}[\phi_2(t) + \frac{t}{\pi\lambda}\phi_1(t)], \quad \phi_2(t + \pi) = \phi_2(t) \tag{11.52}$$

When $|\alpha| > 1$, the absolute value of one of the λ_i is larger than unity, whereas that of the other is less than unity because $\lambda_1\lambda_2 = 1$ according to (11.32). Since $\gamma_i = (1/\pi)\ln\lambda_i$, the real part of one of the γ_i is positive and the other is negative. Hence, it follows from (11.41) or (11.47), (11.48), and (11.52) that one of the solutions is unbounded and the other is bounded with time. Figure 11-1 shows two possible types of unbounded solutions. The first type is oscillatory but with an amplitude that increases exponentially with time, whereas the second type is nonoscillatory and also increases exponentially with time.

When $|\alpha| < 1$, the λ_i are complex conjugates and have unit moduli so that the real parts of the γ_i are zero. Consequently, the normal solutions are bounded.

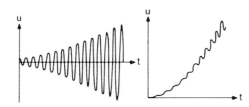

Figure 11-1. Unbounded solutions of the Mathieu equation.

242 THE MATHIEU EQUATION

The bounded solutions are aperiodic varying with two frequencies—the imaginary part of γ and the frequency of the excitation 2. Depending on the ratio of these two frequencies, the solution may exhibit many shapes besides the transition periodic shapes. Three of these shapes are shown in Figure 11-2. The transition from stability to instability occurs for $|\alpha| = 1$, which corresponds to the repeated roots $\lambda_1 = \lambda_2 = 1$ and $\lambda_1 = \lambda_2 = -1$ so that $\gamma_1 = \gamma_2 = 0$ or i. As discussed before, the former case corresponds to the existence of a periodic normal solution with period π, whereas the second case corresponds to the existence of a periodic normal solution with period 2π. These ideas are the basis of the method of *strained parameters* for determining the values of δ and ϵ, which correspond to $|\alpha| = 1$, and hence, the transition from stability to instability. The locus of transition values separates the $\epsilon\delta$ plane into regions of stability and instability, as shown in Figure 11-3. Along these curves at least one of the normal solutions is periodic, with the period π or 2π.

The characteristic exponents for (11.2) can be obtained by numerically calculating two linearly independent solutions using the initial conditions (11.13) during the first period of oscillation. Using the values and first derivatives of these solutions at $t = \pi$, one can calculate $\alpha = \frac{1}{2}[u_1(\pi) + \dot{u}_2(\pi)]$, then determine the λ_i from (11.34), and use them in turn to calculate $\gamma_i = (1/\pi)\ln \lambda_i$. Thus, for each pair of values δ and ϵ, one needs to repeat the above procedure, which is costly and time consuming. When the γ_i are small, we describe in Section 11.4

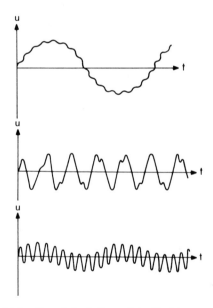

Figure 11-2. Bounded solutions of the Mathieu equation.

THE METHOD OF STRAINED PARAMETERS 243

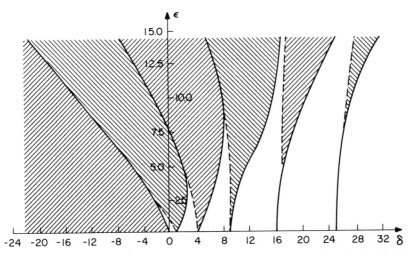

Figure 11-3. Stable and unstable (shaded) regions in the parameter plane for the Mathieu equation.

a scheme called Whittaker's method for determining approximations to the characteristic exponents.

The above discussion shows that one needs a qualitative form of the solutions, namely the normal or Floquet form, to apply either the method of strained parameters or Whittaker's method. In Sections 11.5 and 11.6, we show that one need not know the normal form of the solutions to apply either the method of multiple scales or the method of averaging.

11.3. The Method of Strained Parameters

In Section 11.1, we found that the straightforward expansion fails when $\delta \approx 0, 1, 4, \cdots, n^2$. This suggests expanding δ around these values in powers of ϵ in addition to expanding $u(t; \epsilon)$. Thus, we seek a uniform expansion in the form

$$u(t; \epsilon) = u_0(t) + \epsilon u_1(t) + \epsilon^2 u_2(t) + \cdots \qquad (11.53)$$

$$\delta = n^2 + \epsilon \delta_1 + \epsilon^2 \delta_2 + \cdots \qquad (11.54)$$

and determine the δ_n so that the resulting expansion is periodic. Consequently, the resulting expressions for δ define the transition curves separating stability from instability, as discussed in the preceding section.

Substituting (11.53) and (11.54) into (11.2) gives

$$\ddot{u}_0 + \epsilon \ddot{u}_1 + \epsilon^2 \ddot{u}_2 + \cdots + (n^2 + \epsilon \delta_1 + \epsilon^2 \delta_2 + \cdots)(u_0 + \epsilon u_1 + \epsilon^2 u_2 + \cdots)$$
$$+ 2\epsilon \cos 2t (u_0 + \epsilon u_1 + \epsilon^2 u_2 + \cdots) = 0$$

244 THE MATHIEU EQUATION

Then,

$$\ddot{u}_0 + n^2 u_0 + \epsilon(\ddot{u}_1 + n^2 u_1 + \delta_1 u_0 + 2u_0 \cos 2t)$$
$$+ \epsilon^2(\ddot{u}_2 + n^2 u_2 + \delta_1 u_1 + \delta_2 u_0 + 2u_1 \cos 2t) + \cdots = 0$$

which upon equating the coefficient of each power of ϵ to zero yields

$$\ddot{u}_0 + n^2 u_0 = 0 \tag{11.55}$$

$$\ddot{u}_1 + n^2 u_1 = -\delta_1 u_0 - 2u_0 \cos 2t \tag{11.56}$$

$$\ddot{u}_2 + n^2 u_2 = -\delta_1 u_1 - \delta_2 u_0 - 2u_1 \cos 2t \tag{11.57}$$

The general solution of (11.55) is

$$u_0 = a \cos nt + b \sin nt \tag{11.58}$$

which shows that u_0 is periodic with period π if n is even and period 2π if n is odd. In what follows, we consider the three cases $n = 0, 1$, and 2, beginning with the first.

THE CASE $n = 0$

In this case, the bounded solution of (11.55) is $u_0 = a$. Then, (11.56) becomes

$$\ddot{u}_1 = -\delta_1 a - 2a \cos 2t \tag{11.59}$$

Eliminating the term that produces a secular term in u_1 (i.e., imposing the condition that u_1 be periodic) demands that

$$\delta_1 = 0 \tag{11.60}$$

so that (11.59) reduces to

$$\ddot{u}_1 = -2a \cos 2t \tag{11.61}$$

The solution of (11.61), disregarding the homogeneous solution, is

$$u_1 = \tfrac{1}{2} a \cos 2t \tag{11.62}$$

Then, (11.57) becomes

$$\ddot{u}_2 = -\delta_2 a - a \cos 2t \cos 2t$$

which upon using trigonometric identities becomes

$$\ddot{u}_2 = -\delta_2 a - \tfrac{1}{2} a - \tfrac{1}{2} a \cos 4t \tag{11.63}$$

In order that u_2 be periodic

$$\delta_2 a + \tfrac{1}{2} a = 0$$

which when $a \neq 0$ (nontrivial solution) yields

$$\delta_2 = -\tfrac{1}{2} \tag{11.64}$$

THE METHOD OF STRAINED PARAMETERS 245

Using the above results, we find that to the second approximation

$$u = a + \tfrac{1}{2}\epsilon a \cos 2t + \cdots \qquad (11.65)$$

$$\delta = -\tfrac{1}{2}\epsilon^2 + \cdots \qquad (11.66)$$

Thus, the transition curve separating stability from instability and emanating from $\delta = 0$ is given by (11.66). Equation (11.65) shows that along this curve u is periodic with period π.

THE CASE $n = 1$

Substituting (11.58) into (11.56) and setting $n = 1$, we obtain

$$\ddot{u}_1 + u_1 = -\delta_1(a \cos t + b \sin t) - 2(a \cos t + b \sin t) \cos 2t$$

which upon using trigonometric identities becomes

$$\ddot{u}_1 + u_1 = -(\delta_1 + 1)a \cos t - (\delta_1 - 1)b \sin t$$
$$- a \cos 3t - b \sin 3t \qquad (11.67)$$

Eliminating the terms that produce secular terms in u_1 demands that

$$(\delta_1 + 1)a = 0 \qquad (11.68)$$

$$(\delta_1 - 1)b = 0 \qquad (11.69)$$

so that (11.67) becomes

$$\ddot{u}_1 + u_1 = -a \cos 3t - b \sin 3t \qquad (11.70)$$

It follows from (11.68) that either $\delta_1 = -1$ or $a = 0$, while it follows from (11.69) that either $\delta_1 = 1$ or $b = 0$. When $\delta_1 = -1$, (11.69) demands that $b = 0$, while when $\delta_1 = 1$, (11.68) demands that $a = 0$. Thus, we have the two possibilities

$$\delta_1 = -1 \quad b = 0 \qquad (11.71)$$

$$\delta_1 = 1 \quad a = 0 \qquad (11.72)$$

Then, disregarding the homogeneous solution we find that the solution of (11.70) is either

$$u_1 = \tfrac{1}{8}a \cos 3t \qquad (11.73)$$

or

$$u_1 = \tfrac{1}{8}b \sin 3t \qquad (11.74)$$

Substituting (11.71) and (11.73) into (11.57) gives

$$\ddot{u}_2 + u_2 = \tfrac{1}{8}a \cos 3t - \delta_2 a \cos t - \tfrac{1}{4}a \cos 3t \cos 2t$$

or

$$\ddot{u}_2 + u_2 = -(\delta_2 + \tfrac{1}{8})a \cos t + \tfrac{1}{8}a \cos 3t - \tfrac{1}{8}a \cos 5t \qquad (11.75)$$

246 THE MATHIEU EQUATION

Eliminating the term that produces a secular term in u_2 demands that

$$\delta_2 = -\tfrac{1}{8} \tag{11.76}$$

Therefore, to the second approximation

$$u = a \cos t + \tfrac{1}{8} \epsilon a \cos 3t + \cdots \tag{11.77}$$

$$\delta = 1 - \epsilon - \tfrac{1}{8}\epsilon^2 + \cdots \tag{11.78}$$

Equation (11.78) defines one of the transition curves emanating from $\delta = 1$, and (11.77) shows that along this curve u is periodic with period 2π.

Substituting (11.72) and (11.74) into (11.57) gives

$$\ddot{u}_2 + u_2 = -\tfrac{1}{8}b \sin 3t - \delta_2 b \sin t - \tfrac{1}{4}b \sin 3t \cos 2t$$

or

$$\ddot{u}_2 + u_2 = -(\delta_2 + \tfrac{1}{8})b \sin t - \tfrac{1}{8}b \sin 3t - \tfrac{1}{8}b \sin 5t \tag{11.79}$$

Eliminating the term that produces a secular term in u_2 demands that

$$\delta_2 = -\tfrac{1}{8} \tag{11.80}$$

Therefore, to the second approximation,

$$u = b \sin t + \tfrac{1}{8}\epsilon b \sin 3t + \cdots \tag{11.81}$$

$$\delta = 1 + \epsilon - \tfrac{1}{8}\epsilon^2 + \cdots \tag{11.82}$$

Equation (11.82) defines the second transition curve emanating from $\delta = 1$, and (11.81) shows that along this curve u is periodic with period 2π.

THE CASE $n = 2$

Substituting (11.58) into (11.56) and setting $n = 2$, we obtain

$$\ddot{u}_1 + 4u_1 = -\delta_1(a \cos 2t + b \sin 2t) - 2(a \cos 2t + b \sin 2t) \cos 2t$$

or

$$\ddot{u}_1 + 4u_1 = -\delta_1 a \cos 2t - \delta_1 b \sin 2t - a - a \cos 4t - b \sin 4t \tag{11.83}$$

Eliminating the secular terms from u_1 demands that $\delta_1 = 0$. Then, disregarding the homogeneous solution of (11.83), we have

$$u_1 = -\tfrac{1}{4}a + \tfrac{1}{12}a \cos 4t + \tfrac{1}{12}b \sin 4t \tag{11.84}$$

Then, (11.57) becomes

$$\ddot{u}_2 + 4u_2 = -\delta_2(a \cos 2t + b \sin 2t) - 2(-\tfrac{1}{4}a + \tfrac{1}{12}a \cos 4t + \tfrac{1}{12}b \sin 4t) \cos 2t$$

or

$$\ddot{u}_2 + 4u_2 = -(\delta_2 - \tfrac{5}{12})a \cos 2t - (\delta_2 + \tfrac{1}{12})b \sin 2t + \text{NST} \quad (11.85)$$

Eliminating the secular terms from u_2 demands that

$$(\delta_2 - \tfrac{5}{12})a = 0 \quad \text{and} \quad (\delta_2 + \tfrac{1}{12})b = 0 \quad (11.86)$$

Hence, either

$$\delta_2 = \tfrac{5}{12} \quad \text{and} \quad b = 0 \quad (11.87)$$

or

$$\delta_2 = -\tfrac{1}{12} \quad \text{and} \quad a = 0 \quad (11.88)$$

Therefore, to the second approximation, either

$$u = a \cos 2t - \tfrac{1}{4}\epsilon a(1 - \tfrac{1}{3}\cos 4t) + \cdots \quad (11.89)$$

$$\delta = 4 + \tfrac{5}{12}\epsilon^2 + \cdots \quad (11.90)$$

or

$$u = b \sin 2t + \tfrac{1}{12}\epsilon b \sin 4t + \cdots \quad (11.91)$$

$$\delta = 4 - \tfrac{1}{12}\epsilon^2 + \cdots \quad (11.92)$$

Equations (11.90) and (11.92) define the transition curves emanating from $\delta = 4$ along which (11.89) and (11.91) show that u is periodic with period π.

11.4. Whittaker's Method

In the preceding section, we found that the method of strained parameters yielded the transition curves and the periodic solutions along them. If we are interested in the solutions in the neighborhoods of the transition curves, we need to use a different method. In this section, we use the normal or Floquet forms of the solution and put

$$u(t) = e^{\gamma t}\phi(t) \quad (11.93)$$

where $\phi(t + \pi) = \phi(t)$ according to Floquet theory. Differentiating (11.93) with respect to t yields

$$\dot{u} = e^{\gamma t}\dot{\phi} + \gamma e^{\gamma t}\phi \quad (11.94)$$

which upon differentiation with repect to t yields

$$\ddot{u} = e^{\gamma t}\ddot{\phi} + 2\gamma e^{\gamma t}\dot{\phi} + \gamma^2 e^{\gamma t}\phi \quad (11.95)$$

Substituting (11.93) and (11.95) into (11.2), we have

$$\ddot{\phi} + 2\gamma\dot{\phi} + (\delta + \gamma^2 + 2\epsilon \cos 2t)\phi = 0 \quad (11.96)$$

Thus, the problem is transformed into one of determining γ and the periodic solutions of (11.96) for a given δ. Near the transition curves, γ is small, and hence, we seek an expansion in the form

$$\phi(t;\epsilon) = \phi_0(t) + \epsilon\phi_1(t) + \epsilon^2\phi_2(t) + \cdots \quad (11.97)$$

$$\delta = \delta_0 + \epsilon\delta_1 + \epsilon^2\delta_2 + \cdots \quad (11.98)$$

$$\gamma = \epsilon\gamma_1 + \epsilon^2\gamma_2 + \cdots \quad (11.99)$$

To describe the method, we limit ourselves to $O(\epsilon)$. Thus, substituting (11.97) through (11.99) into (11.96) yields

$$\ddot{\phi}_0 + \epsilon\ddot{\phi}_1 + 2\epsilon\gamma_1(\dot{\phi}_0 + \epsilon\dot{\phi}_1) + (\delta_0 + \epsilon\delta_1 + \epsilon^2\gamma_1^2 + 2\epsilon\cos 2t)(\phi_0 + \epsilon\phi_1) + \cdots = 0$$

or

$$\ddot{\phi}_0 + \delta_0\phi_0 + \epsilon(\ddot{\phi}_1 + \delta_0\phi_1 + 2\gamma_1\dot{\phi}_0 + \delta_1\phi_0 + 2\phi_0\cos 2t) + \cdots = 0 \quad (11.100)$$

Equating each of the coefficients of ϵ^0 and ϵ to zero in (11.100), we obtain

$$\ddot{\phi}_0 + \delta_0\phi_0 = 0 \quad (11.101)$$

$$\ddot{\phi}_1 + \delta_0\phi_1 = -2\gamma_1\dot{\phi}_0 - \delta_1\phi_0 - 2\phi_0\cos 2t \quad (11.102)$$

The general solution of (11.101) can be expressed as

$$\phi_0 = a\cos\sqrt{\delta_0}\,t + b\sin\sqrt{\delta_0}\,t \quad (11.103)$$

Since ϕ is periodic with period π according to Floquet theory, $\sqrt{\delta_0} = n$ so that $\delta_0 = n^2$, where n is an integer. Then, (11.102) becomes

$$\ddot{\phi}_1 + n^2\phi_1 = -2\gamma_1(-an\sin nt + bn\cos nt) - \delta_1(a\cos nt + b\sin nt)$$
$$- 2(a\cos nt + b\sin nt)\cos 2t \quad (11.104)$$

At this order, secular terms in ϕ_1 are produced only from the terms proportional to γ_1 and δ_1 unless $n = 1$. Then, (11.104) can be rewritten as

$$\ddot{\phi}_1 + \phi_1 = (2\gamma_1 a - \delta_1 b + b)\sin t - (2\gamma_1 b + \delta_1 a + a)\cos t + \text{NST} \quad (11.105)$$

Eliminating the terms that produce secular terms in ϕ_1, we have

$$2\gamma_1 a + (1 - \delta_1)b = 0$$
$$(1 + \delta_1)a + 2\gamma_1 b = 0 \quad (11.106)$$

For a nontrivial solution, the determinant of the coefficient matrix in (11.106) must vanish. The result is

$$4\gamma_1^2 - (1 - \delta_1^2) = 0 \quad \text{or} \quad \gamma_1^2 = \tfrac{1}{4}(1 - \delta_1^2)$$

Hence,

$$\gamma_1 = \pm\tfrac{1}{2}\sqrt{1 - \delta_1^2} \quad (11.107)$$

It follows from the first equation in (11.106) that

$$b = -\frac{2\gamma_1}{1-\delta_1} a = \mp \left(\frac{1+\delta_1}{1-\delta_1}\right)^{1/2} a, \quad \delta_1 \neq 1 \qquad (11.108)$$

on account of (11.107). Therefore, to the first approximation

$$u = a_1 e^{(1/2)\epsilon t \sqrt{1-\delta_1^2}} \left[\cos t - \left(\frac{1+\delta_1}{1-\delta_1}\right)^{1/2} \sin t\right]$$

$$+ a_2 e^{-(1/2)\epsilon t \sqrt{1-\delta_1^2}} \left[\cos t + \left(\frac{1+\delta_1}{1-\delta_1}\right)^{1/2} \sin t\right] + \cdots \qquad (11.109)$$

where a_1 and a_2 are arbitrary constants to be determined from the initial conditions. When $\delta_1 \approx 1$, a should be expressed in terms of b.

Equation (11.109) provides a first approximation to u on and near the transition curves. The characteristic exponents are $\pm\frac{1}{2}\epsilon\sqrt{1-\delta_1^2}$. Consequently, the motion is unbounded when $1 > \delta_1^2$ and bounded when $1 \leq \delta_1^2$. Then, $\delta_1^2 = 1$ or $\delta_1 = \pm 1$ corresponds to the transition from stability to instability. Therefore, to the first approximation, the transition curves emanating from $\delta = 1$ are given by

$$\delta = 1 \pm \epsilon + \cdots \qquad (11.110)$$

11.5. The Method of Multiple Scales

Although Whittaker's method yielded a uniform approximation to the solutions of (11.2) on and near the transition curves, it is valid only for linear problems for which one can appeal to Floquet theory to determine the forms of the solutions. For nonlinear problems, Floquet theory does not apply and one may not know the form of the solutions a priori. Consequently, one may not be able to use Whittaker's method for such problems. However, these problems can be treated effectively by using the method of multiple scales, because one need not know the form of the solution a priori and because it is not limited by the nonlinearity.

To apply the method of multiple scales to (11.2), we seek a uniform expansion in the form

$$u(t; \epsilon) = u(T_0, T_1, T_2, \cdots, T_n; \epsilon) = u_0 + \epsilon u_1 + \epsilon^2 u_2 + \cdots \qquad (11.111)$$

where $T_n = \epsilon^n t$. In this section, we obtain only a first-order expansion and leave the second-order expansion for the exercises. Thus, we stop at $O(\epsilon)$ and use only T_0 and T_1. Substituting (11.111) into (11.2) and using (5.45) and (5.46), we obtain

$$(D_0^2 + 2\epsilon D_0 D_1)(u_0 + \epsilon u_1) + (\delta + 2\epsilon \cos 2T_0)(u_0 + \epsilon u_1) + \cdots = 0$$

where $\cos 2t$ is expressed in terms of the fast scale. Equating each of the coefficients of ϵ^0 and ϵ to zero, we have

250 THE MATHIEU EQUATION

$$D_0^2 u_0 + \delta u_0 = 0 \tag{11.113}$$

$$D_0^2 u_1 + \delta u_1 = -2D_0 D_1 u_0 - 2u_0 \cos 2T_0 \tag{11.114}$$

The general solution of (11.113) can be expressed in the complex form

$$u_0 = A(T_1)e^{i\omega T_0} + \bar{A}(T_1)e^{-i\omega T_0} \tag{11.115}$$

where

$$\omega = \sqrt{\delta} \quad \text{or} \quad \omega^2 = \delta \tag{11.116}$$

Then, (11.114) becomes

$$D_0^2 u_1 + \omega^2 u_1 = -2i\omega A' e^{i\omega T_0} + 2i\omega \bar{A}' e^{-i\omega T_0}$$
$$- (e^{2iT_0} + e^{-2iT_0})(A e^{i\omega T_0} + \bar{A} e^{-i\omega T_0})$$

or

$$D_0^2 u_1 + \omega^2 u_1 = -2i\omega A' e^{i\omega T_0} - \bar{A} e^{i(2-\omega)T_0} - \bar{A} e^{-i(2+\omega)T_0} + cc \tag{11.117}$$

There are two cases to be considered: ω is away from 1 and $\omega \approx 1$.

THE CASE ω AWAY FROM 1
In this case, eliminating the terms that lead to secular terms in u_1 demands that

$$A' = 0 \quad \text{or} \quad A = A_0 = \text{constant} \tag{11.118}$$

Therefore, to the first approximation,

$$u = A_0 e^{i\omega T_0} + \bar{A}_0 e^{-i\omega T_0} + \cdots$$

If we let $A_0 = \tfrac{1}{2} a_0 \exp(i\beta_0)$, where a_0 and β_0 are real constants,

$$u = a_0 \cos(\omega t + \beta_0) + \cdots \tag{11.119}$$

Thus, the motion is bounded.

THE CASE $\omega \approx 1$
In this case, we introduce the detuning parameter ω_1 defined by

$$\omega = 1 + \epsilon \omega_1 \quad \text{or} \quad 1 = \omega - \epsilon \omega_1 \tag{11.120}$$

and put

$$(2 - \omega)T_0 = \omega T_0 - 2\epsilon \omega_1 T_0 = \omega T_0 - 2\omega_1 T_1 \tag{11.121}$$

It follows from (11.117) that the secular terms in u_1 will be eliminated if

$$2i\omega A' + \bar{A} e^{-2i\omega_1 T_1} = 0 \tag{11.122}$$

To solve (11.122), we have two choices—either express A in polar form or separate A into real and imaginary parts. With the first choice

THE METHOD OF MULTIPLE SCALES 251

$$A = \tfrac{1}{2} a e^{i\beta} \qquad (11.123)$$

so that (11.122) becomes

$$i\omega(a' + ia\beta')e^{i\beta} + \tfrac{1}{2} a e^{-i(\beta + 2\omega_1 T_1)} = 0$$

or

$$i\omega(a' + ia\beta') + \tfrac{1}{2} a e^{-i\chi} = i\omega(a' + ia\beta') + \tfrac{1}{2} a \cos \chi - \tfrac{1}{2} ia \sin \chi = 0 \qquad (11.124)$$

where

$$\chi = 2\beta + 2\omega_1 T_1 \qquad (11.125)$$

Separating real and imaginary parts leads to

$$\omega a' = \tfrac{1}{2} a \sin \chi \qquad (11.126)$$

$$\omega a \beta' = \tfrac{1}{2} a \cos \chi \qquad (11.127)$$

Eliminating β from (11.125) and (11.127) yields

$$\omega a \chi' = 2\omega \omega_1 a + a \cos \chi \qquad (11.128)$$

It follows from (11.126) and (11.128) that when $a \neq 0$

$$\frac{\omega a'}{\omega a \chi'} = \frac{\tfrac{1}{2} a \sin \chi}{2\omega \omega_1 a + a \cos \chi}$$

Hence,

$$\frac{da}{a} = \frac{1}{2} \frac{\sin \chi \, d\chi}{2\omega \omega_1 + \cos \chi} = -\frac{1}{2} \frac{d(\cos \chi)}{2\omega \omega_1 + \cos \chi}$$

which upon integration yields

$$\ln a = -\tfrac{1}{2} \ln (2\omega \omega_1 + \cos \chi) + \ln c$$

where c is a constant. Consequently

$$a = c[2\omega \omega_1 + \cos \chi]^{-1/2} \qquad (11.129)$$

Equation (11.129) provides a relation between a and χ and one needs to solve either (11.126) or (11.128) to determine a and hence χ as functions of T_1.

Since χ takes on all possible values, $\cos \chi$ ranges from -1 to 1. Hence, the bracketed term in (11.129) vanishes and a becomes infinite (i.e., the motion is unbounded) if $2\omega \omega_1 \leq 1$ or $2\omega \omega_1 \geq -1$; otherwise, the bracketed term never vanishes and a is always bounded. Thus, the transition from stability to instability corresponds to

$$2\omega \omega_1 = 1 \quad \text{and} \quad 2\omega \omega_1 = -1$$

Since $\omega \approx 1$, these conditions yield $\omega_1 \approx \tfrac{1}{2}$ or $-\tfrac{1}{2}$. Hence, the transition curves are given by

$$\omega = 1 \pm \tfrac{1}{2}\epsilon + \cdots$$

But $\delta = \omega^2$; therefore, the transition curves emanating from $\delta = 1$ are given by

$$\delta = (1 \pm \tfrac{1}{2}\epsilon + \cdots)^2$$

or

$$\delta = 1 \pm \epsilon + \cdots \tag{11.130}$$

With the second choice, we first introduce the transformation

$$A = Be^{-i\omega_1 T_1} \tag{11.131}$$

in (11.122) and obtain

$$2i\omega B' + 2\omega\omega_1 B + \bar{B} = 0 \tag{11.132}$$

Thus, the above transformation transformed (11.122) into the constant-coefficient equation (11.132). To solve (11.132), we express B as

$$B = B_r + iB_i \tag{11.133}$$

and obtain

$$2i\omega(B'_r + iB'_i) + 2\omega\omega_1(B_r + iB_i) + B_r - iB_i = 0 \tag{11.134}$$

Separating real and imaginary parts in (11.134) yields

$$\begin{aligned} 2\omega B'_r + (2\omega\omega_1 - 1)B_i &= 0 \\ 2\omega B'_i - (2\omega\omega_1 + 1)B_r &= 0 \end{aligned} \tag{11.135}$$

Equations (11.135) have constant coefficients so that their general solutions can be sought in the form

$$B_r = b_r e^{\gamma_1 T_1} \quad B_i = b_i e^{\gamma_1 T_1} \tag{11.136}$$

which when substituted into (11.135) gives

$$\begin{aligned} 2\omega\gamma_1 b_r + (2\omega\omega_1 - 1)b_i &= 0 \\ -(2\omega\omega_1 + 1)b_r + 2\omega\gamma_1 b_i &= 0 \end{aligned} \tag{11.137}$$

For a nontrivial solution, the determinant of the coefficient matrix in (11.137) must vanish. The result is

$$4\omega^2\gamma_1^2 + (4\omega^2\omega_1^2 - 1) = 0 \quad \text{or} \quad \gamma_1^2 = \tfrac{1}{4}(\omega^{-2} - 4\omega_1^2)$$

or

$$\gamma_1 = \pm\tfrac{1}{2}\sqrt{\omega^{-2} - 4\omega_1^2} \tag{11.138}$$

Since $\omega \approx 1$,

$$\gamma_1 \approx \pm\tfrac{1}{2}\sqrt{1 - 4\omega_1^2} \tag{11.139}$$

Then, it follows from (11.137) that

$$b_i = \frac{2\omega\gamma_1}{1 - 2\omega\omega_1} b_r \approx \pm \left(\frac{1 + 2\omega_1}{1 - 2\omega_1}\right)^{1/2} b_r, \quad \omega_1 \neq \tfrac{1}{2} \quad (11.140)$$

Substituting (11.131) and (11.133) into (11.115) and substituting the result into (11.111), we have

$$u = (B_r + iB_i)e^{i(\omega T_0 - \omega_1 T_1)} + (B_r - iB_i)e^{-i(\omega T_0 - \omega_1^* T_1)} + \cdots$$

$$= (B_r + iB_i)e^{it} + (B_r - iB_i)e^{-it} + \cdots$$

$$= B_r(e^{it} + e^{-it}) + iB_i(e^{it} - e^{-it})$$

on account of (11.120). Hence,

$$u = 2B_r \cos t - 2B_i \sin t + \cdots \quad (11.141)$$

Using (11.136) and (11.140) in (11.141), we obtain

$$u = a_1 e^{(1/2)\epsilon t \sqrt{1 - 4\omega_1^2}} \left[\cos t - \left(\frac{1 + 2\omega_1}{1 - 2\omega_1}\right)^{1/2} \sin t\right]$$

$$+ a_2 e^{-(1/2)\epsilon t \sqrt{1 - 4\omega_1^2}} \left[\cos t + \left(\frac{1 + 2\omega_1}{1 - 2\omega_1}\right)^{1/2} \sin t\right] + \cdots \quad (11.142)$$

in agreement with (11.109) obtained by using Whittaker's method because $2\omega_1 = \delta_1$. When $\omega_1 \approx \tfrac{1}{2}$, b_r should be expressed in terms of b_i.

11.6. The Method of Averaging

To apply the method of averaging to (11.2), we first need to change the dependent variable from $u(t)$ to $a(t)$ and $\beta(t)$, where

$$u(t) = a(t) \cos [\omega t + \beta(t)] \quad (11.143)$$

and as before $\delta = \omega^2$. Differentiating (11.143) with respect to t yields

$$\dot{u} = -\omega a \sin (\omega t + \beta) + \dot{a} \cos (\omega t + \beta) - a\dot{\beta} \sin (\omega t + \beta) \quad (11.144)$$

As before, imposing the condition that

$$\dot{u} = -\omega a \sin (\omega t + \beta) \quad (11.145)$$

we obtain from (11.144) that

$$\dot{a} \cos (\omega t + \beta) - a\dot{\beta} \sin (\omega t + \beta) = 0 \quad (11.146)$$

Differentiating (11.145) with respect to t yields

$$\ddot{u} = -\omega^2 a \cos (\omega t + \beta) - \omega\dot{a} \sin (\omega t + \beta) - \omega a \dot{\beta} \cos (\omega t + \beta) \quad (11.147)$$

Substituting (11.143) and (11.147) into (11.2), we have

$$-\omega^2 a \cos(\omega t + \beta) - \omega \dot{a} \sin(\omega t + \beta) - \omega a \dot{\beta} \cos(\omega t + \beta)$$
$$+ (\delta + 2\epsilon \cos 2t) a \cos(\omega t + \beta) = 0$$

which can be simplified to

$$\omega \dot{a} \sin(\omega t + \beta) + \omega a \dot{\beta} \cos(\omega t + \beta) = 2\epsilon a \cos 2t \cos(\omega t + \beta) \quad (11.148)$$

because $\delta = \omega^2$. Solving (11.146) and (11.148) for $\dot{a}$ and $\dot{\beta}$ yields

$$\omega \dot{a} = 2\epsilon a \cos 2t \sin(\omega t + \beta) \cos(\omega t + \beta)$$
$$\omega a \dot{\beta} = 2\epsilon a \cos 2t \cos^2(\omega t + \beta) \quad (11.149)$$

Using trigonometric identities, we rewrite (11.149) as

$$\omega \dot{a} = \tfrac{1}{2} \epsilon a \{\sin[2(\omega + 1)t + 2\beta] + \sin[2(\omega - 1)t + 2\beta]\} \quad (11.150)$$
$$\omega a \dot{\beta} = \tfrac{1}{2} \epsilon a \{2 \cos 2t + \cos[2(\omega + 1)t + 2\beta] + \cos[2(\omega - 1)t + 2\beta]\} \quad (11.151)$$

Again, there are two possibilities—either ω is away from 1 or $\omega \approx 1$. When ω is away from 1, all the terms on the right-hand sides of (11.150) and (11.151) are fast varying so that to the first approximation

$$\dot{a} = 0 \quad \text{and} \quad \dot{\beta} = 0$$

in agreement with (11.118) obtained by using the method of multiple scales. When $\omega \approx 1$, $(\omega - 1)t + \beta$ is a slowly varying function of t so that, to the first approximation, it follows from (11.150) and (11.151) that

$$\omega \dot{a} = \tfrac{1}{2} \epsilon a \sin[2(\omega - 1)t + 2\beta]$$
$$\omega a \dot{\beta} = \tfrac{1}{2} \epsilon a \cos[2(\omega - 1)t + 2\beta] \quad (11.152)$$

in agreement with (11.125) through (11.127) obtained by using the method of multiple scales.

Exercises

11.1. Consider the Mathieu equation

$$\ddot{u} + (\delta + 2\epsilon \cos 2t)u = 0$$

Determine a second-order uniform expansion by using the method of multiple scales when $\delta \approx 0$ and $\delta \approx 4$.

11.2. Consider the Mathieu equation

$$\ddot{u} + (\delta + 2\epsilon \cos 2t)u = 0$$

Use Whittaker's method to determine second-order uniform expansions when $\delta \approx 0$ and $\delta \approx 4$.

11.3. Consider the equation

$$\ddot{u} + \frac{\delta u}{1 + \epsilon \cos 2t} = 0$$

(a) Determine second-order expansions for the transition curves near $\delta = 0$, 1, and 4.
(b) Use Whittaker's technique to determine second-order expansions for u near these curves.

11.4. Consider the equation

$$\ddot{u} + \frac{\delta - \epsilon \cos^2 t}{1 - \epsilon \cos^2 t} u = 0$$

(a) Determine second-order expansions of the first three transition curves (i.e., near $\delta = 0$, 1, and 4).
(b) Use Whittaker's technique to determine u near these curves.

11.5. Consider the equation

$$\ddot{u} + (\delta + \epsilon \cos^3 t)u = 0$$

Determine second-order expansions for the first three transition curves using both the method of strained parameters and Whittaker's technique.

11.6. Consider the equation

$$\ddot{u} + (\delta + \epsilon \cos 2t - \tfrac{1}{2}\epsilon^2 \alpha \sin 2t + \tfrac{1}{8}\epsilon^2 \cos 4t)u = 0$$

Determine three terms for the transition curves when $\delta \approx 1$ and $\delta \approx 4$.

11.7. Consider the equation

$$\ddot{u} + \tfrac{1}{4}(1 - \epsilon \cos t)^{-2} a^{-2}[2(1 - \epsilon \cos t)(2 - \epsilon a^2 \cos t) + \epsilon^2 a^2 \sin^2 t]u = 0$$

Show that three of the transition curves are given by

$$a = 2 \pm \tfrac{1}{2}\epsilon + \cdots$$

$$a = 1 + \tfrac{1}{6}\epsilon^2 + \cdots$$

11.8. Consider the problem

$$\ddot{u} + (\omega^2 + 2\epsilon \cos 3t)u = 0, \quad \epsilon \ll 1$$

(a) Use the method of variation of parameters to determine the equations describing the amplitude and the phase.
(b) Use the method of averaging to determine the equations describing the slow variations in the amplitude and the phase. Consider all cases.

11.9. Use the methods of multiple scales and averaging to determine the equations describing the amplitudes and the phases to first order for a system governed by

$$\ddot{u} + \omega_0^2 u = \epsilon u^2 \cos \Omega t$$

where Ω is away from zero when

(a) ω_0 is away from Ω and $\frac{1}{3}\Omega$,
(b) $\omega_0 \approx \Omega$,
(c) $3\omega_0 \approx \Omega$.

11.10. Consider the equation

$$\ddot{u} + \omega_0^2 u + 2\epsilon u^3 \cos 2t = 0 \qquad \epsilon \ll 1$$

Use the methods of multiple scales and averaging to determine the equations describing the amplitude and the phase to first order when

(a) ω_0 is away from 1 and $\frac{1}{2}$,
(b) $\omega_0 \approx 1$,
(c) $\omega_0 \approx \frac{1}{2}$.

11.11. The parametric excitation of a two-degree-of-freedom system is governed by

$$\ddot{u}_1 + \omega_1^2 u_1 + \epsilon \cos \Omega t (f_{11} u_1 + f_{12} u_2) = 0$$

$$\ddot{u}_2 + \omega_2^2 u_2 + \epsilon \cos \Omega t (f_{21} u_2 + f_{22} u_2) = 0$$

Use the methods of multiple scales and averaging to determine the equations describing the amplitudes and the phases when $\Omega \approx \omega_2 \mp \omega_1$.

CHAPTER 12

Boundary-Layer Problems

In Chapters 4-11, the effect of the perturbations is small but cumulative over a long period of time. Consequently, the amplitudes and phases are slowly varying functions of time and can be handled by slow time scales or mild strainings of the independent variable. In this chapter, we consider problems in which the perturbations are operative over very narrow regions across which the dependent variables undergo very rapid changes. These narrow regions frequently adjoin the boundaries of the domain of interest, owing to the fact that the small parameter multiplies the highest derivative. Consequently, they are usually referred to as *boundary layers* in fluid mechanics, *edge layers* in solid mechanics, and *skin layers* in electrical applications. There are many physical situations in which the sharp changes occur inside the domain of interest, and the narrow regions across which these changes take place are usually referred to as *shock layers* in fluid and solid mechanics, *transition points* in quantum mechanics, and *Stokes lines and surfaces* in mathematics. These rapid changes cannot be handled by slow scales, but they can be handled by fast or magnified or stretched scales.

There are a number of methods available for treating boundary-layer problems including the method of matched asymptotic expansions, the method of composite expansions, the method of multiple scales, the WKBJ method, and the Langer transformation. The latter two methods are applicable to linear problems with a large parameter and they are discussed in Chapter 14. In this chapter, we concentrate our discussion on the method of matched asymptotic expansions, briefly introduce the method of composite expansions, and apply the method of multiple scales to one example.

We begin with a simple example whose exact solution is available for comparison and motivation of the methods to be used. Then, we consider linear and nonlinear problems whose exact solutions are not available.

12.1. A Simple Example

We consider the simple boundary-value problem

$$\epsilon y'' + (1 + \epsilon^2)y' + (1 - \epsilon^2)y = 0 \qquad (12.1)$$

258 BOUNDARY-LAYER PROBLEMS

$$y(0) = \alpha \quad y(1) = \beta \tag{12.2}$$

where ϵ is a small dimensionless positive number. It is assumed that the equation and boundary conditions have been made dimensionless.

As a start, we seek a straightforward expansion in the form

$$y(x; \epsilon) = y_0(x) + \epsilon y_1(x) + \cdots \tag{12.3}$$

Substituting (12.3) into (12.1) and (12.2) gives

$$\epsilon(y_0'' + \epsilon y_1'' + \cdots) + (1 + \epsilon^2)(y_0' + \epsilon y_1' + \cdots) + (1 - \epsilon^2)(y_0 + \epsilon y_1 + \cdots) = 0$$

$$y_0(0) + \epsilon y_1(0) + \cdots = \alpha$$

$$y_0(1) + \epsilon y_1(1) + \cdots = \beta$$

Equating coefficients of like powers of ϵ, we have

Order ϵ^0

$$y_0' + y_0 = 0 \tag{12.4}$$

$$y_0(0) = \alpha \quad y_0(1) = \beta \tag{12.5}$$

Order ϵ

$$y_1' + y_1 = -y_0'' \tag{12.6}$$

$$y_1(0) = 0 \quad y_1(1) = 0 \tag{12.7}$$

The general solution of (12.4) is

$$y_0 = c_0 e^{-x} \tag{12.8}$$

where c_0 is an arbitrary constant.

We note from (12.5) that there are two boundary conditions on y_0, whereas the general solution (12.8) of y_0 contains only one arbitrary constant. Thus, y_0 cannot (except by coincidence) satisfy both boundary conditions. For example, if we impose the condition $y_0(0) = \alpha$, we obtain from (12.8) that

$$\alpha = c_0 \quad \text{so that} \quad y_0 = \alpha e^{-x} \tag{12.9}$$

Then, imposing the boundary condition $y_0(1) = \beta$, we obtain from (12.8) that

$$\beta = c_0 e^{-1} \quad \text{or} \quad c_0 = \beta e \tag{12.10}$$

so that

$$y_0 = \beta e^{1-x} \tag{12.11}$$

Comparing (12.9) and (12.10), we find that the boundary conditions demand two different values for c_0, namely $c_0 = \alpha$ and $c_0 = \beta e$, which are inconsistent unless it happens by coincidence that $\alpha = \beta e$.

Comparing (12.4) and (12.1), we note that the order of the differential equation is reduced from second order, which can cope with two boundary conditions, to first order, which can cope with only one boundary condition. Hence, one of the boundary conditions cannot be satisfied and must be dropped. Consequently, the resulting expansion is not expected to be valid at or near the end point where the boundary condition has been dropped. The question arises as to which of the boundary conditions must be dropped. This question can be answered using either physical or mathematical arguments as discussed below. As shown below, when the coefficient of y' in (12.1) is positive, the boundary condition at the left end must be dropped.

Dropping the boundary condition $y(0) = \alpha$, we conclude that $c_0 = \beta e$ and that y_0 is given by (12.11). Then, (12.6) becomes

$$y'_1 + y_1 = -\beta e^{1-x} \qquad (12.12)$$

whose general solution is

$$y_1 = c_1 e^{-x} - \beta x e^{1-x} \qquad (12.13)$$

Again, (12.12) is first order. Hence, y_1 contains only one arbitrary constant and cannot cope with the two boundary conditions (12.7). Hence, no relief arose at first-order. In fact, no relief can be expected at any order, because the differential equation at any order is first order. Again, we drop the boundary condition at $x = 0$, use the boundary condition $y_1(1) = 0$, and find from (12.13) that $c_1 = \beta e$. Hence,

$$y_1 = \beta e^{1-x} - \beta x e^{1-x} = \beta(1-x)e^{1-x} \qquad (12.14)$$

Substituting (12.11) and (12.14) into (12.3), we obtain

$$y = \beta e^{1-x} + \epsilon\beta(1-x)e^{1-x} + \cdots \qquad (12.15)$$

At the origin, $y = \beta e(1 + \epsilon)$, which is in general different from the α in (12.2).

To determine the source of the nonuniformity and how it can be circumvented, we next investigate the exact solution.

EXACT SOLUTION

Since (12.1) is linear having constant coefficients, its solution can be sought by putting

$$y = e^{sx}$$

and obtaining

$$\epsilon s^2 + (1 + \epsilon^2)s + 1 - \epsilon^2 = 0$$

or

$$(\epsilon s + 1 - \epsilon)(s + 1 + \epsilon) = 0$$

260 BOUNDARY-LAYER PROBLEMS

Hence,

$$s = -(1 + \epsilon) \quad \text{or} \quad -\frac{1}{\epsilon} + 1$$

and the general solution of (12.1) is

$$y = a_1 e^{-(1+\epsilon)x} + a_2 e^{-[(1/\epsilon) - 1]x} \qquad (12.16)$$

Using the boundary conditions (12.2) in (12.16), we have

$$\alpha = a_1 + a_2 \qquad \beta = a_1 e^{-(1+\epsilon)} + a_2 e^{-[(1/\epsilon) - 1]}$$

whose solution is

$$a_1 = \frac{\beta - \alpha e^{-[(1/\epsilon) - 1]}}{e^{-(1+\epsilon)} - e^{-[(1/\epsilon) - 1]}} \qquad a_2 = \frac{\alpha e^{-(1+\epsilon)} - \beta}{e^{-(1+\epsilon)} - e^{-[(1/\epsilon) - 1]}}$$

Therefore, the exact solution of (12.1) is

$$y = \frac{[\beta - \alpha e^{-[(1/\epsilon) - 1]}] e^{-(1+\epsilon)x} + [\alpha e^{-(1+\epsilon)} - \beta] e^{-[(1/\epsilon) - 1]x}}{e^{-(1+\epsilon)} - e^{-[(1/\epsilon) - 1]}} \qquad (12.17)$$

To understand the nature of the nonuniformity at the origin in the straightforward expansion, we expand the exact solution (12.17) for small ϵ. To this end, we note that $\exp(-1/\epsilon)$ is smaller than any power of ϵ as $\epsilon \to 0$. Hence, we can rewrite (12.17) as

$$y = \beta e^{(1+\epsilon)(1-x)} + [\alpha - \beta e^{1+\epsilon}] e^{-(x/\epsilon) + x} + \text{EST} \qquad (12.18)$$

where EST stands for exponentially small terms. In deriving the straightforward expansion (12.15), we assumed that x is fixed at a value different from zero and then expanded y for small ϵ. If we keep x fixed and positive, then $\exp(-x/\epsilon)$ is exponentially small and (12.18) can be rewritten as

$$y = \beta e^{(1+\epsilon)(1-x)} + \text{EST} \qquad (12.19)$$

Expanding (12.19) for small ϵ, we have

$$y = \beta e^{1-x} e^{\epsilon(1-x)} + \text{EST}$$

$$= \beta e^{1-x} [1 + \epsilon(1-x) + \frac{1}{2!} \epsilon^2 (1-x)^2 + \cdots]$$

Hence,

$$y = \beta e^{1-x} + \epsilon \beta (1-x) e^{1-x} + \cdots \qquad (12.20)$$

in agreement with the straightforward expansion.

As in the case of the straightforward expansion, (12.20) is not valid at or near the origin because $y(0) = \beta e(1 + \epsilon)$, which is in general different from the α in the boundary condition in (12.2). Therefore, in the process of expanding (12.17) for small ϵ, we must have performed one or more illegitimate operations that caused the nonuniformity. To trace such illegitimate operations, we review our process of expanding the exact solution. We first assumed that $\exp(-1/\epsilon)$ is exponentially small and arrived at (12.18), which is uniform because at $x = 0$ it yields $y(0) = \alpha$, the imposed boundary condition in (12.2). Next, we fixed x at a positive value that is different from zero, concluded that $\exp(-x/\epsilon)$ is exponentially small, and obtained (12.19). Putting $x = 0$ in (12.19), we find that $y(0) = \beta \exp(1 + \epsilon)$, which is in general different from the α in (12.2). Hence, this step is the one responsible for the nonuniformity. Looking at this step closely, we find that $\exp(-x/\epsilon)$ is exponentially small as $\epsilon \to 0$ only when x is positive and away from zero. Therefore, it is not surprising that any expansion based on this assumption is not valid when $x \approx 0$. In fact, at $x = 0$, $\exp(-x/\epsilon) = 1$, which is much bigger than any ϵ^m, where $m > 0$. When $x = \epsilon$, $\exp(-x/\epsilon) = e^{-1}$, which is $O(1)$ and not an exponentially small term.

Then, there arises the question of how good the straightforward expansion (12.15) is when we know that it is not valid near $x = 0$. The answer to this question can be seen in Figure 12-1, which compares the value of y calculated from (12.15) and denoted by y^0, with the value of y calculated from the exact solution (12.17) and denoted by y^e for $\epsilon = 0.01$. It can be seen that, for this small ϵ, y^0 agrees with y^e except in a small interval near the origin called the *boundary layer*, where y^e changes quickly in order to satisfy the boundary condition there. Figure 12-2 shows that as ϵ decreases the boundary layer becomes thinner; thus, $y(x; \epsilon)$ is continuous for $\epsilon > 0$ but discontinuous for $\epsilon = 0$. Hence, the limits of y as $x \to 0$ and $\epsilon \to 0$ are not interchangeable. In fact,

$$\lim_{x \to 0} y(x; \epsilon) = \alpha$$

according to (12.17), and hence,

$$\lim_{\epsilon \to 0} \lim_{x \to 0} y(x; \epsilon) = \alpha$$

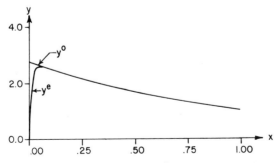

Figure 12-1. Comparison of the outer expansion y^0 with the exact solution y^e for $\epsilon = 0.01$, $\beta = 1.0$, and $\alpha = 0.0$.

262 BOUNDARY-LAYER PROBLEMS

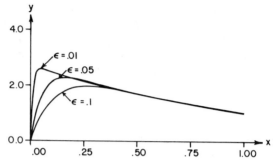

Figure 12-2. Effect of ϵ on the thickness of the boundary layer for $\beta = 1.0$ and $\alpha = 0.0$.

On the other hand,

$$\lim_{\epsilon \to 0} y(x; \epsilon) = \beta e^{1-x}$$

according to (12.17), and hence,

$$\lim_{x \to 0} \lim_{\epsilon \to 0} y(x; \epsilon) = \beta e$$

Consequently,

$$\lim_{\epsilon \to 0} \lim_{x \to 0} y(x; \epsilon) \neq \lim_{x \to 0} \lim_{\epsilon \to 0} y(x; \epsilon)$$

thereby showing the nonuniform convergence of the exact solution $y(x; \epsilon)$ to the straightforward expansion y^0. Therefore, problems of this kind are usually referred to as *singular-perturbation problems*. Inspection of the exact solution (12.17) shows that y depends on x and ϵ in the combinations (scales) x/ϵ and ϵx as well as x alone. There arises the question of whether one can determine a uniform expansion by using another scale that is a function of x and ϵ instead of x. To answer this question, we investigate next the effect of scales on the resulting expansion.

EFFECT OF SCALES ON EXPANSION

To devise a method for determining uniform expansions for singular-perturbation problems, we magnify or stretch the boundary layer and investigate the behavior of the solution in the boundary layer as a function of the magnified scale. To facilitate this step, we investigate the effect of changing the scale on the resulting expansion.

For example, let us change the scale from x to the magnified scale $\xi = x/\epsilon$. Then, (12.18), which does not contain the exponentially small terms, becomes

$$y = \beta e^{(1 + \epsilon)(1 - \epsilon \xi)} + [\alpha - \beta e^{1 + \epsilon}] e^{-\xi + \epsilon \xi} + \text{EST}$$

Expanding for small ϵ with ξ being kept fixed yields

$$y = \beta e + (\alpha - \beta e) e^{-\xi} + \epsilon \{\beta e(1 - \xi) + [-\beta e + (\alpha - \beta e)\xi] e^{-\xi}\} + \cdots \qquad (12.21)$$

In terms of x, (12.21) can be rewritten as

$$y = \beta e(1 - x) + (\alpha - \beta e)(1 + x)e^{-x/\epsilon} + \epsilon\beta e[1 - e^{-x/\epsilon}] + \cdots \quad (12.22)$$

At the origin, $y(0) = \alpha$, while at the right end $y(1) = \epsilon\beta e$. Hence, (12.22) is not valid near $x = 1$ although it seems to be valid at the origin. Figure 12-3 shows that the expansion (12.22) agrees with the exact solution in a small neighborhood of the origin but deviates very much from the exact solution away from the origin.

As a second example, we consider the more magnified scale $\zeta = x/\epsilon^2$. Then, (12.18) becomes

$$y = \beta e^{(1+\epsilon)(1-\epsilon^2\zeta)} + [\alpha - \beta e^{1+\epsilon}]e^{-\epsilon\zeta + \epsilon^2\zeta} + \text{EST}$$

which, when expanded for small ϵ with ζ being kept fixed, yields

$$y = \alpha - \epsilon(\alpha - \beta e)\zeta + \cdots \quad (12.23a)$$

or

$$y = \alpha - \frac{x}{\epsilon}(\alpha - \beta e) + \cdots \quad (12.23b)$$

Again, at the origin $y(0) = \alpha$, whereas, at the right end, $y(1) = \alpha - (\alpha - \beta e)/\epsilon$, which is different from the β in (12.2). Hence, (12.23) is not valid near $x = 1$. In fact, Figure 12-4 shows that it is valid only in a very small neighborhood of the origin.

As a third example, we consider the moderately magnified scale $\eta = x/\epsilon^{1/2}$. Then, (12.18) becomes

$$y = \beta e^{(1+\epsilon)(1-\epsilon^{1/2}\eta)} + [\alpha - \beta e^{1+\epsilon}]e^{-\eta\epsilon^{-1/2} + \epsilon^{1/2}\eta} + \text{EST}$$

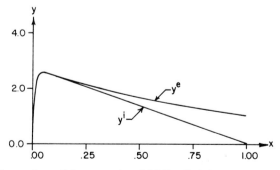

Figure 12-3. Comparison of the expansion (12.22) called the inner expansion and denoted by y^i with the exact solution y^e for $\epsilon = 0.01$, $\beta = 1.0$, and $\alpha = 0.0$.

264 BOUNDARY-LAYER PROBLEMS

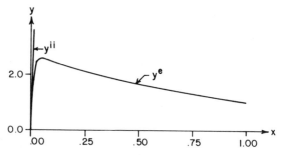

Figure 12-4. Comparison of the expansion (12.23) called the inner inner expansion and denoted by y^{ii} with the exact solution y^e for $\epsilon = 0.01$, $\beta = 1.0$, and $\alpha = 0.0$.

which, when expanded for small ϵ with η being kept fixed, yields

$$y = \beta e(1 + \epsilon + \cdots)(1 - \epsilon^{1/2}\eta + \tfrac{1}{2}\epsilon\eta^2 + \cdots)$$
$$= \beta e[1 - \epsilon^{1/2}\eta + \epsilon(1 + \tfrac{1}{2}\eta^2) + \cdots] \qquad (12.24a)$$

or

$$y = \beta e[1 - x + \tfrac{1}{2}x^2 + \epsilon + \cdots] \qquad (12.24b)$$

At the origin, $y(0) = \beta e(1 + \epsilon) \neq \alpha$, whereas at the right end $y(1) = \beta e(\tfrac{1}{2} + \epsilon) \neq \beta$. Hence, (12.24) is not valid near either of the end points. However, Figure 12-5 shows that it agrees with the exact solution in a small interior interval.

The above discussion shows that an expansion of a function that depends on an independent variable and a small parameter, such as $y(x; \epsilon)$, depends strongly on the scale being used (i.e., the independent variable being kept fixed). In the present example, when x was kept fixed, we obtained (12.20), which is valid everywhere except near the origin. When $\xi = x/\epsilon$ was kept fixed, we obtained (12.22), which is valid only in a small neighborhood of the origin. When $\zeta = x/\epsilon^2$

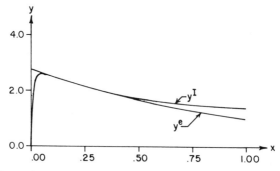

Figure 12-5. Comparison of the expansion (12.24) called the intermediate expansion and denoted by y^I with the exact solution y^e for $\epsilon = 0.01$, $\beta = 1.0$, and $\alpha = 0.0$.

was kept fixed, we obtained (12.23), which is valid only in a very small neighborhood of the origin. When $\eta = x/\epsilon^{1/2}$ was kept fixed, we obtained (12.24), which is not valid near any of the end points but valid in a small interior interval. Thus, it appears that a uniform expansion of the solution of a singular-perturbation problem cannot be expressed in terms of a single scale (i.e., a single combination of x and ϵ), such as x or x/ϵ or x/ϵ^2 or $x/\epsilon^{1/2}$, making it an ideal problem for application of the method of multiple scales. However, for nonlinear problems, especially those governed by nonlinear partial-differential equations, application of the method of multiple scales may not be straightforward, and an alternative method, the method of matched asymptotic expansions, is frequently used. In this chapter, we stress this method.

The basic idea underlying the method of matched asymptotic expansions is that an approximate solution to a given problem is sought not as a single expansion in terms of a single scale but as two or more separate expansions in terms of two or more scales each of which is valid in part of the domain. The scales are chosen so that (a) the expansions as a whole cover the whole domain of interest and (b) the domains of validity of neighboring expansions overlap. Because the domains overlap, the neighboring expansions can be *matched (blended)*, and hence, connected. Next, we discuss the matching or blending process.

MATCHING

As an example, to illustrate the basic idea, we attempt to match the expansion (12.15), which is valid everywhere except in a small neighborhood of the origin, and the expansion (12.21), which is valid in a small neighborhood of the origin. If these two expansions have overlapping domains of validity, then we should be able to match them. Expansion (12.15) was obtained by expanding $y(x; \epsilon)$ with x being kept fixed, while (12.21) was obtained by expanding $y(x; \epsilon)$ with $\xi = x/\epsilon$ being kept fixed. There arises the question of what the effect would be of changing the scales in these two expansions.

Before answering this question, let us denote the expansion obtained by keeping x fixed with the superscript o, and hence, replace y in (12.15) with y^o. Moreover, let us denote the expansion obtained by keeping $\xi = x/\epsilon$ fixed with the superscript i, and hence, replace y in (12.21) with y^i. Next, let us change the scale in (12.15) from x to ξ and obtain

$$y^o = \beta e^{1 - \epsilon\xi} + \epsilon\beta(1 - \epsilon\xi)e^{1 - \epsilon\xi} + \cdots \quad (12.25)$$

Expanding (12.25) for small ϵ and keeping ξ fixed, we obtain

$$(y^o)^i = \beta e + \epsilon\beta e(1 - \xi) + \cdots \quad (12.26)$$

where the superscript i is used in (12.26) to indicate that y^o has been expanded by keeping ξ fixed. We express (12.21) in terms of x, recall that y should be replaced with y^i, and obtain

266 BOUNDARY-LAYER PROBLEMS

$$y^i = \beta e(1 - x) + (\alpha - \beta e)(1 + x)e^{-x/\epsilon} + \epsilon\beta e[1 - e^{-x/\epsilon}] + \cdots \quad (12.27)$$

Expanding (12.27) for small ϵ with x being kept fixed, we have

$$(y^i)^o = \beta e(1 - x) + \epsilon\beta e + \cdots \quad (12.28)$$

where the superscript o is used in (12.28) to indicate that y^i has been expanded by keeping x fixed. If we replace ξ with x/ϵ in (12.26), we have

$$(y^o)^i = \beta e(1 - x) + \epsilon\beta e + \cdots \quad (12.29)$$

Comparing (12.28) and (12.29), we conclude that

$$(y^i)^o = (y^o)^i \quad (12.30a)$$

In other words,

the outer expansion of (the inner expansion)

$$= \text{the inner expansion of the outer expansion} \quad (12.30b)$$

It turns out that (12.30) holds whenever two neighboring expansions have overlapping domains. It is usually referred to as the *matching principle* and it serves to connect neighboring expansions. Hence, the union of the domains of validity of (12.15) and (12.21) covers the whole domain of interest, namely the interval $[0, 1]$. Consequently, they can be used to represent y over $[0, 1]$.

Next, let us change the scales in (12.15) and (12.23a). Again, let us denote by the superscript o the expansion obtained by keeping x fixed. Also, let us denote by the superscript ii the expansion obtained by keeping $\zeta = x/\epsilon^2$ fixed. We refer to this expansion as an *inner inner expansion*. Replacing x in (12.15) with ζ, we have

$$y^o = \beta e^{1 - \epsilon^2 \zeta} + \epsilon\beta(1 - \epsilon^2 \zeta)e^{1 - \epsilon^2 \zeta} + \cdots$$

which, when expanded for small ϵ with ζ being kept fixed, yields

$$(y^o)^{ii} = \beta e + \epsilon\beta e + \cdots \quad (12.31)$$

In terms of x, the expansion (12.23a) is given by (12.23b), which, when expanded for small ϵ with x being kept fixed, yields

$$(y^{ii})^o = -(\alpha - \beta e)\frac{x}{\epsilon} + \alpha \quad (12.32)$$

Comparing (12.31) and (12.32), we find that

$$(y^o)^{ii} \neq (y^{ii})^o \quad (12.33)$$

indicating that (12.15) and (12.23a) do not have overlapping domains. This is not surprising because it is clear from Figure 12-4 that (12.23) is valid only in a very small neighborhood of the origin, and this is the reason we referred to it as an inner inner expansion. We note that there are many inner inner expansions characterized by the scales $\zeta = x/\epsilon^\nu$, where $\nu > 1$. Hence, the union of the domains of validity of (12.15) and (12.23a) does not cover the whole domain of interest, namely the interval $[0, 1]$.

As a third case, let us consider interchanging the scales in (12.15) and (12.24a). Let us denote by the superscript I the expansion obtained by keeping $\eta = x/\epsilon^{1/2}$ fixed. We refer to this expansion as an *intermediate expansion*. Changing the scale from x to η in (12.15), we have

$$y = \beta e^{1 - \epsilon^{1/2}\eta} + \epsilon\beta(1 - \epsilon^{1/2}\eta)e^{1 - \epsilon^{1/2}\eta} + \cdots \quad (12.34)$$

which, when expanded for small ϵ with η being kept fixed, yields

$$(y^0)^I = \beta e - \epsilon^{1/2}\beta e\eta + \epsilon(\beta e + \tfrac{1}{2}\eta^2) \quad (12.35)$$

Replacing η with x in (12.24a) yields (12.24b), which, when expanded for small ϵ with x being kept fixed, yields

$$(y^I)^0 = \beta e(1 - x + \tfrac{1}{2}x^2) + \epsilon\beta e + \cdots \quad (12.36)$$

Comparing (12.35) and (12.36) and recalling that $\eta = x/\epsilon^{1/2}$, we conclude that they are identical, and hence, write

$$(y^0)^I = (y^I)^0 \quad (12.37)$$

Hence, the expansions (12.15) and (12.24a) have overlapping domains, making the expansions obtained by interchanging the scales identical. It is clear from Figure 12-5 that it is not surprising that (12.15) and (12.24a) have overlapping domains. Although (12.15) and (12.24a) have overlapping domains, the union of their domains of validity does not cover the whole interval of interest because neither of them is valid near the origin.

The above discussion shows that neighboring expansions obtained by using different scales need not have overlapping domains. Moreover for neighboring expansions, the union of their overlapping domains need not cover the whole domain of interest. Thus, the objective of the method of matched asymptotic expansions is to determine expansions that cover the whole domain of interest such that neighboring expansions have overlapping domains. Neighboring expansions are matched or blended using the matching principle (12.30).

In this case, one can use (12.15) and (12.21) because they cover the whole domain of interest and yet have overlapping domains. However, one cannot use (12.15) and (12.23a) because they do not have overlapping domains. Moreover, one cannot use (12.15) and (12.24a) because the union of their domains does not cover the whole domain of interest. We should note that the domain of validity of (12.24a) overlaps with that of (12.21). To see this, we express (12.21) in terms of $\eta = x/\epsilon^{1/2} = \epsilon^{1/2}\xi$ and obtain

268 BOUNDARY-LAYER PROBLEMS

$$y^i = \beta e + (\alpha - \beta e)e^{-\eta/\epsilon^{1/2}} + \epsilon \left\{ \beta e \left(1 - \frac{\eta}{\epsilon^{1/2}}\right) + \left[-\beta e + (\alpha - \beta e)\frac{\eta}{\epsilon^{1/2}}\right] e^{-\eta/\epsilon^{1/2}} \right\}$$
$$+ \cdots$$

which, when expanded for small ϵ with η being kept fixed, yields

$$(y^i)^I = \beta e - \beta e \epsilon^{1/2}\eta + \epsilon\beta e + \cdots \tag{12.38}$$

Replacing η in (12.24a) with ξ, we have

$$y^I = \beta e[1 - \epsilon\xi + \epsilon(1 + \tfrac{1}{2}\epsilon\xi^2) + \cdots]$$

which, when expanded for small ϵ with ξ being kept fixed, yields

$$(y^I)^i = \beta e + \epsilon\beta e(1 - \xi) + \cdots \tag{12.39}$$

Comparing (12.38) and (12.39) and recalling that $\eta = \epsilon^{1/2}\xi$, we conclude that they are identical, and hence, write

$$(y^i)^I = (y^I)^i \tag{12.40}$$

The preceding discussion shows that the domains of validity of (12.15) and (12.21) overlap and their union covers the whole domain of interest. Expansion (12.15) is usually referred to as an *outer expansion* and denoted by the superscript o. Expansion (12.21) is usually referred to as *inner expansion* and denoted by the superscript i. The variable x is called the *outer variable*, while the variable $\xi = x/\epsilon$ is called the *inner variable*. Moreover, the domain of (12.24a) overlaps that of (12.15) on the one hand and that of (12.21) on the other hand. Consequently, expansion (12.24a) is usually referred to as an *intermediate expansion*, and the variable $\eta = x/\epsilon^{1/2}$ is called an *intermediate variable* because the magnification provided by η falls in between x and $\xi = x/\epsilon$. Thus, any variable $\eta = x/\epsilon^\nu$, where $0 < \nu < 1$, is an intermediate variable. Instead of matching the outer and inner expansions directly, one can match them by equating their respective intermediate expansions. We demonstrate this process in Section 12.3. In the next section, we show how the method of multiple scales can be used to determine a uniform expansion for (12.1).

12.2. The Method of Multiple Scales

As discussed in the preceding section, $y(x; \epsilon)$ depends on x and ϵ in the combinations x/ϵ and ϵx in addition to x and ϵ alone. This makes this problem ideal for application of the method of multiple scales. Since the domain is finite, ϵx stays small, and hence, nonuniformities will not arise from the presence of the secular terms ϵx, $\epsilon^2 x^2$, $\epsilon^3 x^3$, $\cdots$, in contrast with the cases of infinite domains discussed in Chapters 4 through 11. Thus, it is sufficient to introduce the stretched scale $\xi = x/\epsilon$, which in this case is the same as the inner variable, and $x_0 = x$, which in this case is the outer variable. In terms of these scales

THE METHOD OF MULTIPLE SCALES 269

$$\frac{d}{dx} = \frac{1}{\epsilon}\frac{\partial}{\partial \xi} + \frac{\partial}{\partial x_0}$$

$$\frac{d^2}{dx^2} = \frac{1}{\epsilon^2}\frac{\partial^2}{\partial \xi^2} + \frac{2}{\epsilon}\frac{\partial^2}{\partial \xi \partial x_0} + \frac{\partial^2}{\partial x_0^2}$$
(12.41)

Then, (12.1) becomes

$$\frac{1}{\epsilon}\frac{\partial^2 y}{\partial \xi^2} + \frac{2\partial^2 y}{\partial \xi \partial x_0} + \epsilon \frac{\partial^2 y}{\partial x_0^2} + (1+\epsilon^2)\left(\frac{1}{\epsilon}\frac{\partial y}{\partial \xi} + \frac{\partial y}{\partial x_0}\right) + (1-\epsilon^2)y = 0 \quad (12.42)$$

We seek a first-order uniform expansion for y in the form

$$y = y_0(\xi, x_0) + \epsilon y_1(\xi, x_0) + \cdots \quad (12.43)$$

As in nonlinear oscillation problems, we need to investigate the term $O(\epsilon)$ to determine the arbitrary functions that appear in y_0. Substituting (12.43) into (12.42) and equating coefficients of like powers of ϵ, we have

$$\frac{\partial^2 y_0}{\partial \xi^2} + \frac{\partial y_0}{\partial \xi} = 0 \quad (12.44)$$

$$\frac{\partial^2 y_1}{\partial \xi^2} + \frac{\partial y_1}{\partial \xi} = -\frac{2\partial^2 y_0}{\partial \xi \partial x_0} - \frac{\partial y_0}{\partial x_0} - y_0 \quad (12.45)$$

The general solution of (12.44) is

$$y_0 = A(x_0) + B(x_0)e^{-\xi} \quad (12.46)$$

where A and B are undetermined at this level of approximation; they are determined at the next level of approximation by imposing the solvability conditions. Putting y_0 in (12.45) gives

$$\frac{\partial^2 y_1}{\partial \xi^2} + \frac{\partial y_1}{\partial \xi} = 2B'e^{-\xi} - A' - B'e^{-\xi} - A - Be^{-\xi}$$

or

$$\frac{\partial^2 y_1}{\partial \xi^2} + \frac{\partial y_1}{\partial \xi} = -(A' + A) + (B' - B)e^{-\xi} \quad (12.47)$$

A particular solution of (12.47) is

$$y_{1p} = -(A' + A)\xi - (B' - B)\xi e^{-\xi} \quad (12.48)$$

which makes ϵy_1 much bigger than y_0 as $\xi \to \infty$. Hence, for a uniform expansion, the coefficients of ξ and $\xi \exp(-\xi)$ in (12.48) must vanish independently. The result is

$$A' + A = 0$$
$$B' - B = 0$$
(12.49)

270 BOUNDARY-LAYER PROBLEMS

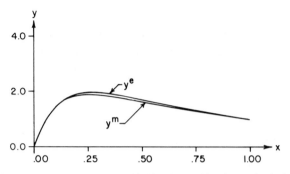

Figure 12-6. Comparison of the solution (12.52) obtained by the method of multiple scales and denoted by y^m with the exact solution y^e for $\epsilon = 0.1$, $\beta = 1.0$, and $\alpha = 0.0$.

The solutions of (12.49) are

$$A = ae^{-x_0} \qquad B = be^{x_0} \qquad (12.50)$$

where a and b are arbitrary constants. Then, (12.46) becomes

$$y_0 = ae^{-x_0} + be^{-\xi + x_0}$$

or, in terms of the original variable,

$$y_0 = ae^{-x} + be^{-(x/\epsilon) + x}$$

Substituting for y_0 in (12.43) gives

$$y = ae^{-x} + be^{-(x/\epsilon) + x} + \cdots \qquad (12.51)$$

Imposing the boundary conditions (12.2) yields

$$\alpha = a + b \qquad \beta = ae^{-1} + be^{-(1/\epsilon) + 1}$$

Neglecting the exponentially small term $\exp(-1/\epsilon)$ and solving for a and b, we obtain

$$a = \beta e \qquad b = \alpha - \beta e$$

Hence, it follows from (12.51) that, to the first approximation,

$$y = \beta e^{1-x} + (\alpha - \beta e)e^{-(x/\epsilon) + x} + \cdots \qquad (12.52)$$

Figure 12-6 shows that (12.52) is everywhere in close agreement with the exact solution (12.17). We took $\epsilon = 0.1$ in Figure 12-6, so that one can distinguish the two solutions.

12.3. The Method of Matched Asymptotic Expansions

As discussed in Section 12.1, the basic idea underlying the method of matched asymptotic expansions is the representation of the solution by more than one expansion, each of which is valid in part of the domain, and neighboring expansions overlap so that they can be matched. We describe this method by its application to a number of examples, beginning in this section with the simple example (12.1) and (12.2).

As $\epsilon \to 0$, (12.1) reduces to

$$y' + y = 0$$

which is of first order, and hence, cannot cope with the two boundary conditions. Consequently, one of them must be dropped and a boundary layer must be introduced. In Section 12.1, we used the exact solution to conclude that the boundary layer is at the origin. Of course, in the normal application of the present technique, the exact solution is not available; otherwise, there may not be any need to carry out an approximate solution. Consequently, we need to determine the locations of the boundary layers. In many situations, one can use physical arguments; in this Chapter, we describe how one can decide analytically the locations of the boundary layers. Moreover, in the preceding two sections, we were guided by the exact solution to decide that the scale $\xi = x/\epsilon$ is the inner variable. In this section, we also describe the selection of the inner variable.

To determine the location of the boundary layer, we assume that it exists at one of the ends. Then, we carry out one-term expansions. If neighboring expansions can be matched, our assumption is correct; otherwise, the boundary layer exists at the other end.

Let us assume that the boundary layer is at the right end, and hence, $y(1) = \beta$ must be dropped. Then, we seek an outer expansion in the form

$$y^o(x; \epsilon) = y_0(x) + \cdots$$

which, when substituted into (12.1) and $y(0) = \alpha$, yields

$$y'_0 + y_0 = 0 \qquad y_0(0) = \alpha$$

Hence,

$$y_0 = c_0 e^{-x} \qquad \text{so that} \qquad \alpha = c_0$$

and

$$y^o = \alpha e^{-x} + \cdots \tag{12.53}$$

To analyze the behavior of the solution in the assumed boundary layer, we need to magnify the neighborhood of $x = 1$. Since the interval of interest is $[0, 1]$, we magnify a small interval $1 - x$ near 1, where $x < 1$. Thus, we let

$$\xi = \frac{1 - x}{\epsilon^\nu} \tag{12.54}$$

where ν must be greater than zero, in order that ξ be a magnified or stretched scale. The value of ν is not known, in general, a priori and must be determined from the analysis. We note that by design ξ is positive.

272 BOUNDARY-LAYER PROBLEMS

It follows from (12.54) that

$$x = 1 - e^\nu \xi$$

and that

$$\frac{d}{dx} = \frac{d}{d\xi}\frac{d\xi}{dx} = -\frac{1}{e^\nu}\frac{d}{d\xi}$$

$$\frac{d^2}{dx^2} = \frac{1}{e^{2\nu}}\frac{d^2}{d\xi^2}$$

In terms of ξ, (12.1) becomes

$$e^{1-2\nu}\frac{d^2y}{d\xi^2} - e^{-\nu}(1+e^2)\frac{dy}{d\xi} + (1-e^2)y = 0 \tag{12.55}$$

As $e \to 0$ with ξ being kept fixed, the dominant terms in (12.55) are

$$e^{1-2\nu}\frac{d^2y}{d\xi^2} - e^{-\nu}\frac{dy}{d\xi} + y + \cdots = 0 \tag{12.56}$$

The limiting form of (12.56) as $e \to 0$ depends on the value of ν. There are three possibilities: $\nu > 1, \nu < 1$, and $\nu = 1$.

When $\nu > 1$, the limiting form of (12.56) is

$$\frac{d^2 y^i}{d\xi^2} = 0$$

whose general solution is

$$y^i = a_0 + b_0 \xi \tag{12.57}$$

Since the boundary layer is assumed to be at $x = 1$, it must satisfy the boundary condition $y = \beta$ at $x = 1$. But $x = 1$ corresponds to $\xi = 0$; therefore, $y^i = \beta$ at $\xi = 0$. Then, it follows from (12.57) that

$$\beta = a_0$$

and hence,

$$y^i = \beta + b_0 \xi \tag{12.58}$$

To match (12.53) with (12.58), we need $(y^o)^i$ and $(y^i)^o$. Thus, we express (12.53) in terms of ξ and obtain

$$y^o = \alpha e^{-1 + e^\nu \xi}$$

which, when expanded for small e with ξ being kept fixed, yields

$$(y^o)^i = \alpha e^{-1} \tag{12.59}$$

Also, we express (12.58) in terms of x and obtain

$$y^i = \beta + \frac{b_0(1-x)}{\epsilon^\nu}$$

which, when expanded for small ϵ with x being kept fixed, yields

$$(y^i)^o = \begin{cases} \beta & \text{if } b_0 = 0 \\ \dfrac{b_0(1-x)}{\epsilon^\nu} & \text{if } b_0 \neq 0 \end{cases} \quad (12.60)$$

Equating (12.59) and (12.60) according to the matching principle demands that $b_0 = 0$ and $\alpha e^{-1} = \beta$, which is not true, in general. Hence, the case $\nu > 1$ must be discarded.

When $\nu < 1$, the limiting form of (12.56) is

$$\frac{dy^i}{d\xi} = 0$$

whose solution is

$$y^i = a_0$$

Again, y^i should satisfy the boundary condition $y(1) = \beta$ or $y^i = \beta$ at $\xi = 0$. Hence, $a_0 = \beta$ and

$$y^i = \beta \quad (12.61)$$

Since y^i is constant,

$$(y^i)^o = \beta \quad (12.62)$$

As in the preceding case, $(y^o)^i = \alpha e^{-1}$. Hence, the matching principle

$$(y^o)^i = (y^i)^o$$

demands that $\beta = \alpha e^{-1}$, which is not true, in general. Therefore, the case $\nu < 1$ must also be discarded.

When $\nu = 1$, the limiting form of (12.56) is

$$\frac{d^2 y^i}{d\xi^2} - \frac{dy^i}{d\xi} = 0$$

whose general solution is

$$y^i = a_0 + b_0 e^\xi$$

Putting $y^i = \beta$ when $\xi = 0$ gives

$$a_0 + b_0 = \beta \quad \text{or} \quad a_0 = \beta - b_0$$

Hence,

$$y^i = \beta - b_0 + b_0 e^\xi$$

274 BOUNDARY-LAYER PROBLEMS

To perform the matching, we express y^i in terms of x and obtain

$$y^i = \beta - b_0 + b_0 \exp\left(\frac{1-x}{\epsilon}\right)$$

which, when expanded for small ϵ with x being kept fixed, yields

$$(y^i)^o = \begin{cases} \beta & \text{if } b_0 = 0 \\ b_0 \exp\left(\dfrac{1-x}{\epsilon}\right) & \text{if } b_0 \neq 0 \end{cases} \qquad (12.63)$$

Again, $(y^o)^i$ is given by (12.59). Then, equating (12.59) and (12.63) according to the matching principle demands that $b_0 = 0$ and hence $\beta = \alpha e^{-1}$, which is not true, in general. Hence, this last case must also be discarded and the boundary layer does not exist at $x = 1$. It must exist at the origin. This is checked next.

To check whether the boundary layer exists at the origin, we introduce the stretching transformation

$$\xi = \frac{x}{\epsilon^\nu} \quad \text{or} \quad x = \epsilon^\nu \xi \qquad (12.64)$$

where ν must be greater than zero. Then,

$$\frac{d}{dx} = \frac{d}{d\xi}\frac{d\xi}{dx} = \frac{1}{\epsilon^\nu}\frac{d}{d\xi}$$

$$\frac{d^2}{dx^2} = \frac{1}{\epsilon^{2\nu}}\frac{d^2}{d\xi^2}$$

and (12.1) becomes

$$\epsilon^{1-2\nu}\frac{d^2 y}{d\xi^2} + \epsilon^{-\nu}(1+\epsilon^2)\frac{dy}{d\xi} + (1-\epsilon^2)y = 0$$

whose dominant part as $\epsilon \to 0$ is

$$\epsilon^{1-2\nu}\frac{d^2 y}{d\xi^2} + \epsilon^{-\nu}\frac{dy}{d\xi} + y = 0 \qquad (12.65)$$

As before, there are three possibilities: $\nu > 1$, $\nu < 1$, and $\nu = 1$.

When $\nu > 1$, the limiting form of (12.65) as $\epsilon \to 0$ is

$$\frac{d^2 y^i}{d\xi^2} = 0 \qquad (12.66a)$$

whose general solution is

$$y^i = a_0 + b_0 \xi$$

Since the boundary layer is assumed to be at the origin, it must satisfy the

THE METHOD OF MATCHED ASYMPTOTIC EXPANSIONS 275

boundary condition $y(0) = \alpha$. But $x = 0$ corresponds to $\xi = 0$ according to (12.64); hence, $y^i = \alpha$ at $\xi = 0$. Then, $\alpha = a_0$ and

$$y^i = \alpha + b_0 \xi \tag{12.66b}$$

Since the boundary layer is assumed to be at the origin, the outer expansion must be valid at $x = 1$. Thus, we seek an outer expansion in the form

$$y^o(x; \epsilon) = y_0(x) + \cdots$$

which, when substituted into (12.1) and $y(1) = \beta$, yields

$$y_0' + y_0 = 0 \qquad y_0(1) = \beta$$

Hence,

$$y_0 = c_0 e^{-x} \quad \text{so that} \quad c_0 = \beta e$$

and

$$y^o = \beta e^{1-x} + \cdots \tag{12.67}$$

To match (12.66b) and (12.67), we express the former in terms of x and obtain

$$y^i = \alpha + \frac{b_0 x}{\epsilon^\nu}$$

which, when expanded for small ϵ with x being kept fixed, yields

$$(y^i)^o = \begin{cases} \alpha & \text{if} \quad b_0 = 0 \\ \dfrac{b_0 x}{\epsilon^\nu} & \text{if} \quad b_0 \neq 0 \end{cases} \tag{12.68}$$

Next, we express (12.67) in terms of ξ and obtain

$$y^o = \beta e^{1 - \epsilon^\nu \xi}$$

which, when expanded for small ϵ with ξ being kept fixed, yields

$$(y^o)^i = \beta e + \cdots \tag{12.69}$$

Equating (12.68) and (12.69) according to the matching principle demands that $b_0 = 0$ and $\alpha = \beta e$, which is not true, in general. Hence, the case $\nu > 1$ must be discarded.

When $\nu < 1$, the limiting form of (12.65) as $\epsilon \to 0$ is

$$\frac{dy^i}{d\xi} = 0 \tag{12.70a}$$

whose solution is

$$y^i = a_0$$

Putting $y^i = \alpha$ at $\xi = 0$ yields $a_0 = \alpha$, and hence,

$$y^i = \alpha \tag{12.70b}$$

To match (12.67) with (12.70b), we note that $(y^o)^i$ is still given by (12.69) and that

$$(y^i)^o = \alpha$$

because y^i is a constant. Again, the matching principle demands that $\alpha = \beta e$, which is not true, in general. Hence, the case $\nu < 1$ must also be discarded.

When $\nu = 1$, the limiting form of (12.65) as $\epsilon \to 0$ is

$$\frac{d^2 y^i}{d\xi^2} + \frac{dy^i}{d\xi} = 0 \tag{12.71}$$

whose general solution is

$$y^i = a_0 + b_0 e^{-\xi}$$

Putting $y^i = \alpha$ when $\xi = 0$ gives

$$a_0 + b_0 = \alpha \quad \text{or} \quad a_0 = \alpha - b_0$$

Hence,

$$y^i = \alpha - b_0 + b_0 e^{-\xi} \tag{12.72}$$

To match (12.67) with (12.72), we note that $(y^o)^i$ is still given by (12.69). Expressing (12.72) in terms of x, we have

$$y^i = \alpha - b_0 + b_0 e^{-x/\epsilon}$$

which, when expanded for small ϵ with x being kept fixed, yields

$$(y^i)^o = \alpha - b_0 \tag{12.73}$$

Equating (12.69) and (12.73) according to the matching principle demands that

$$\alpha - b_0 = \beta e \quad \text{or} \quad b_0 = \alpha - \beta e$$

Hence, to the first approximation

$$\begin{aligned} y^o &= \beta e^{1-x} + \cdots \\ y^i &= \beta e + (\alpha - \beta e) e^{-x/\epsilon} + \cdots \end{aligned} \tag{12.74}$$

in agreement with those obtained in Section 12.1 by expanding the exact solution.

Since matching has been achieved, our assumption about the location of the boundary layer is correct. Moreover, the choice $\xi = x/\epsilon$ for the inner variable yielded an inner expansion that overlaps the outer expansion, and hence, the stretching provided with $\nu = 1$ is the proper one. We note that the stretchings

THE METHOD OF MATCHED ASYMPTOTIC EXPANSIONS 277

of the boundary layer provided by $\nu < 1$ and $\nu > 1$ demanded that $\alpha = \beta e$, which is not true, in general. The conditions $\nu < 1$ and $\nu > 1$ are indefinite because each of them can be satisfied by an infinite number of values of ν. On the other hand, the condition $\nu = 1$ is definite, and it is usually referred to as the *distinguished limit*. It turns out that the *proper stretchings* are always provided by the *distinguished limits*. Moreover, the limiting forms (12.66a) and (12.70a) of the differential equation when $\nu > 1$ and $\nu < 1$ are special cases of the limiting form (12.71) when $\nu = 1$. Thus, one speaks of (12.71) as the *least-degenerate form* of the limiting equation in the boundary layer. Also, it turns out that the proper scale provided by the distinguished limit always yields the least-degenerate limiting form of the equation in the boundary layer. Therefore, in subsequent sections, *we will always determine the proper stretchings by choosing the distinguished limits*.

Equations (12.74) provide two separate expansions, y^o valid everywhere except in a small interval order $O(\epsilon)$ near the origin and y^i valid only in a small interval $O(\epsilon)$ near the origin. Although y^o and y^i have overlapping domains, one needs to switch from one expansion to the other if a numerical solution is desired over the whole interval. Moreover, the switching location is not known precisely. To circumvent this difficulty of switching from one expansion to another, one usually combines both expansions into a so-called *composite expansion* denoted by the superscript c and defined by

$$y^c = y^o + y^i - (y^o)^i = y^o + y^i - (y^i)^o \qquad (12.75)$$

The two alternatives in (12.75) are equivalent because $(y^o)^i = (y^i)^o$ according to the matching principle. This expression shows that the composite expansion is formed by adding the inner and outer expansions and subtracting their common part $(y^o)^i$ or $(y^i)^o$ from the result. Expansion (12.75) agrees with the outer and inner expansions in their respective domains of validity because

$$(y^c)^o = (y^o)^o + (y^i)^o - [(y^i)^o]^o$$

But $(f^o)^o = f^o$, hence

$$(y^o)^o = y^o \qquad [(y^i)^o]^o = (y^i)^o$$

and

$$(y^c)^o = y^o$$

Also,

$$(y^c)^i = (y^o)^i + (y^i)^i - [(y^o)^i]^i$$

But $(f^i)^i = f^i$, hence

$$(y^i)^i = y^i \qquad [(y^o)^i]^i = (y^o)^i$$

and

$$(y^c)^i = y^i \tag{12.77}$$

Since y^c reproduces the outer expansion in the outer domain and the inner expansion in the inner domain, we postulate that it is valid everywhere.

In the present example, the common part between y^o and y^i is given by either (12.69) or (12.73), that is,

$$(y^o)^i = (y^i)^o = \beta e$$

Hence,

$$y^c = \beta e^{1-x} + (\alpha - \beta e)e^{-x/\epsilon} + \cdots \tag{12.78}$$

Letting $\epsilon \to 0$ in (12.78) with x being kept fixed yields

$$(y^c)^o = \beta e^{1-x} + \cdots = y^o$$

Expressing (12.78) in terms of ξ, we have

$$y^c = \beta e^{1-\epsilon\xi} + (\alpha - \beta e)e^{-\xi} + \cdots$$

which, when expanded for small ϵ with ξ being kept fixed, yields

$$(y^c)^i = \beta e + (\alpha - \beta e)e^{-\xi} + \cdots = y^i$$

Thus, as expected, y^c reproduces the inner and outer expansions in their respective domains of validity. Hence, it is not surprising that (12.78) is in close agreement with the exact solution everywhere, as shown in Figure 12-7. We took $\epsilon = 0.1$ so that the two solutions can be distinguished from each other.

Before concluding this section, we compare (12.78) with (12.52) obtained by using the method of multiple scales. Whereas the method of matched asymptotic expansions yields a composite expansion that is separable in the outer and inner scales, the method of multiple scales yields a nonseparable expansion. Moreover, in the case of equations with variable coefficients, the inner variable may be a

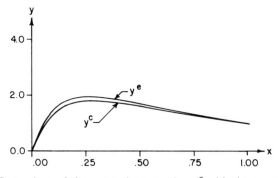

Figure 12-7. Comparison of the composite expansion y^c with the exact solution y^e for $\epsilon = 0.1$, $\beta = 1.0$, and $\alpha = 0.0$.

HIGHER APPROXIMATIONS

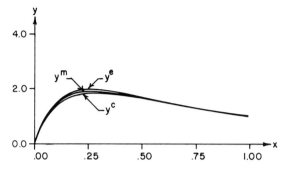

Figure 12-8. Comparison of the method-of-multiple-scales solution y^m with the composite expansion y^c and the exact solution y^e for $\epsilon = 0.1$, $\beta = 1.0$, and $\alpha = 0.0$.

nonlinear rather than a linear function of x. Letting $\epsilon \to 0$ with x being kept fixed, we find that (12.52) and (12.78) yield the same outer expansion. Putting $x = \epsilon\xi$ and letting $\epsilon \to 0$ with ξ being fixed, we find that (12.52) and (12.78) also yield the same inner expansion. Thus, (12.52) and (12.78) agree in the outer and inner domains. However, the presence of the extra factor $\exp(x)$ in (12.52) makes it agree more closely with the exact solution, as shown in Figure 12-8 for larger values of ϵ. This does not mean that the method of multiple scales is superior to the method of matched asymptotic expansions because the application of the method of multiple scales to nonlinear differential equations, especially partial-differential equations such as the Navier-Stokes equations, is not straightforward.

12.4. Higher Approximations

To determine a higher-order asymptotic expansion of (12.1) and (12.2), we need to determine an outer expansion and an inner expansion, match them, and then form a composite expansion.

OUTER EXPANSION

We seek an outer expansion in the form

$$y^o = y_0(x) + \epsilon y_1(x) + \cdots \qquad (12.79)$$

This expansion is expected to satisfy the boundary condition $y(1) = \beta$ because the boundary layer is at the origin. Hence, substituting (12.79) into (12.1) and the condition $y(1) = \beta$ and equating coefficients of like powers of ϵ, we have

$$y_0' + y_0 = 0 \qquad y_0(1) = \beta \qquad (12.80)$$

$$y_1' + y_1 = -y_0'' \qquad y_1(1) = 0 \qquad (12.81)$$

280 BOUNDARY-LAYER PROBLEMS

As before, the solution of y_0 is

$$y_0 = \beta e^{1-x}$$

Then, (12.81) becomes

$$y_1' + y_1 = -\beta e^{1-x}$$

whose general solution is

$$y_1 = c_1 e^{-x} - \beta x e^{1-x}$$

But $y_1(1) = 0$, hence $c_1 = \beta e$. Therefore,

$$y^o = \beta e^{1-x} + \epsilon\beta(1-x)e^{1-x} + \cdots \qquad (12.82)$$

INNER EXPANSION

To determine an inner expansion, we first change the independent variable from x to $\xi = x/\epsilon$ in (12.1), add a superscript i to y, and obtain

$$\frac{d^2 y^i}{d\xi^2} + (1 + \epsilon^2)\frac{dy^i}{d\xi} + \epsilon(1 - \epsilon^2)y^i = 0 \qquad (12.83)$$

Since the boundary layer is at the origin, the inner expansion must satisfy the boundary condition $y = \alpha$ at $x = 0$. But $x = 0$ corresponds to $\xi = 0$, hence

$$y^i(0) = \alpha \qquad (12.84)$$

Now, we seek an inner expansion in the form

$$y^i = Y_0(\xi) + \epsilon Y_1(\xi) + \cdots \qquad (12.85)$$

Substituting (12.85) into (12.83) and (12.84) and equating coefficients of like powers of ϵ, we have

$$Y_0'' + Y_0' = 0 \qquad Y_0(0) = \alpha \qquad (12.86)$$

$$Y_1'' + Y_1' = -Y_0 \qquad Y_1(0) = 0 \qquad (12.87)$$

The general solution of Y_0 is

$$Y_0 = a_0 + b_0 e^{-\xi}$$

Putting $Y_0(0) = \alpha$ gives

$$\alpha = a_0 + b_0 \quad \text{or} \quad a_0 = \alpha - b_0$$

and hence,

$$Y_0 = \alpha - b_0 + b_0 e^{-\xi}$$

Then, (12.87) becomes

$$Y_1'' + Y_1' = -(\alpha - b_0) - b_0 e^{-\xi}$$

whose general solution is
$$Y_1 = a_1 + b_1 e^{-\xi} - (\alpha - b_0)\xi + b_0 \xi e^{-\xi}$$
Putting $Y_1(0) = 0$ gives
$$a_1 + b_1 = 0 \quad \text{or} \quad a_1 = -b_1$$
and hence,
$$Y_1 = -b_1 + b_1 e^{-\xi} - (\alpha - b_0)\xi + b_0 \xi e^{-\xi}$$
Therefore,
$$y^i = \alpha - b_0 + b_0 e^{-\xi} + \epsilon[-b_1 + b_1 e^{-\xi} - (\alpha - b_0)\xi + b_0 \xi e^{-\xi}] + \cdots \quad (12.88)$$
where the remaining constants b_0 and b_1 need to be determined from the matching condition.

MATCHING

To match the outer expansion (12.82) and the inner expansion (12.88), we use the matching condition $(y^o)^i = (y^i)^o$. Since we have two terms in each of them, we need to determine the first two terms in each of $(y^o)^i$ and $(y^i)^o$. Expressing (12.82) in terms of $\xi = x/\epsilon$, we have
$$y^o = \beta e^{1-\epsilon\xi} + \epsilon\beta(1 - \epsilon\xi)e^{1-\epsilon\xi} + \cdots$$
which, when expanded for small ϵ with ξ being kept fixed, yields
$$(y^o)^i = \beta e + \epsilon\beta e(1 - \xi) + \cdots \quad (12.89)$$
Expressing (12.88) in terms of x, we have
$$y^i = \alpha - b_0 + b_0 e^{-x/\epsilon} + \epsilon\left[-b_1 + b_1 e^{-x/\epsilon} - (\alpha - b_0)\frac{x}{\epsilon} + \frac{b_0 x}{\epsilon} e^{-x/\epsilon}\right] + \cdots$$
which, when expanded for small ϵ with x being kept fixed, yields
$$(y^i)^o = \alpha - b_0 - (\alpha - b_0)x - \epsilon b_1 + \cdots \quad (12.90)$$
Equating (12.89) and (12.90) according to the matching principle, we have
$$\beta e + \epsilon\beta e(1 - \xi) = \alpha - b_0 - (\alpha - b_0)x - \epsilon b_1$$
which, since $\xi = x/\epsilon$, can be rewritten as
$$\beta e - \beta e x + \epsilon\beta e = \alpha - b_0 - (\alpha - b_0)x - \epsilon b_1 \quad (12.91)$$
Equating coefficients of like powers of ϵ in (12.91) yields
$$\beta e - \beta e x = \alpha - b_0 - (\alpha - b_0)x \quad (12.92)$$
$$\beta e = -b_1 \quad (12.93)$$

282 BOUNDARY-LAYER PROBLEMS

If the boundary layer is where it was assumed to be, the stretching transformation is correct, and there are no algebraic errors, the matching conditions will be consistent, and hence, they can be solved for b_0 and b_1. Equating the coefficients of like powers of x in (12.92) gives

$$\beta e = \alpha - b_0 \qquad \beta e = \alpha - b_0$$

which are consistent. Hence, $b_0 = \alpha - \beta e$. It follows from (12.93) that $b_1 = -\beta e$. Hence, (12.88) becomes

$$y^i = \beta e + (\alpha - \beta e)e^{-\xi} + \epsilon[\beta e - \beta e e^{-\xi} - \beta e \xi + (\alpha - \beta e)\xi e^{-\xi}] + \cdots \quad (12.94)$$

As mentioned in Section 12.1, one can alternatively perform the matching by using intermediate expansions. For example, let us consider the intermediate variable $\eta = x/\epsilon^{1/2}$, which yields a stretching that is intermediate between x and $\xi = x/\epsilon$. Expressing x in terms of η, we rewrite (12.82) as

$$y^o = \beta e^{1-\epsilon^{1/2}\eta} + \epsilon\beta(1 - \epsilon^{1/2}\eta)e^{1-\epsilon^{1/2}\eta} + \cdots \quad (12.95a)$$

which, when expanded for small ϵ with η being kept fixed, yields

$$(y^o)^I = \beta e - \epsilon^{1/2}\beta e \eta + \epsilon\beta e(1 + \tfrac{1}{2}\eta^2) + \cdots \quad (12.95b)$$

Putting $\xi = \eta/\epsilon^{1/2}$ in (12.88), we have

$$y^i = \alpha - b_0 + b_0 e^{-\eta/\epsilon^{1/2}} + \epsilon\left[-b_1 + b_1 e^{-\eta/\epsilon^{1/2}} - \frac{(\alpha - b_0)\eta}{\epsilon^{1/2}}\right.$$

$$\left. + \frac{b_0 \eta}{\epsilon^{1/2}} e^{-\eta/\epsilon^{1/2}}\right] + \cdots \quad (12.96a)$$

which, when expanded for small ϵ with η being kept fixed, yields

$$(y^i)^I = \alpha - b_0 - \epsilon^{1/2}(\alpha - b_0)\eta - \epsilon b_1 + \cdots \quad (12.96b)$$

Equating (12.95b) and (12.96b) according to the intermediate matching principle, we find that they match up to $O(\epsilon^{1/2})$. Thus, one can only determine b_0 to this order. To match the expansions to $O(\epsilon)$, one needs to carry out the inner expansion to second order. Alternatively, the intermediate matching can be performed by subtracting (12.96a) from (12.95a) and then letting $\epsilon \to 0$ with η being kept fixed. This process will determine b_0 and b_1. Comparing the intermediate matching with the straightforward matching, we conclude that the intermediate matching is an unnecessary complication, and hence, we will not use it further.

An attractive alternative to the straightforward matching is Van Dyke's matching principle, which states that

HIGHER APPROXIMATIONS   283

The m-term inner expansion of (the n-term outer expansion) equals the n-term outer expansion of (the m-term inner expansion)   (12.97)

where m and n may be any two integers that need not be equal. To determine the m-term inner expansion of the (n-term outer expansion), we rewrite the first n-terms of the outer expansion in terms of the inner variable, expand it for small ϵ with the inner variable being kept fixed, and truncate the resulting expansion after m terms, and conversely for the right-hand side of (12.97).

To show the application of van Dyke's matching principle, we use it to match the two-term outer expansion (12.82) with the two-term inner expansion (12.88). We proceed systematically as follows:

Two-term outer expansion: $\quad y \sim \beta e^{1-x} + \epsilon\beta(1-x)e^{1-x}$

Rewritten in inner variable: $\quad = \beta e^{1-\epsilon\xi} + \epsilon\beta(1-\epsilon\xi)e^{1-\epsilon\xi}$

Expanded for small ϵ: $\quad = \beta e(1 - \epsilon\xi + \frac{1}{2}\epsilon^2\xi^2 + \cdots)$
$\quad\quad + \epsilon\beta e(1 - \epsilon\xi)(1 - \epsilon\xi + \cdots)$

Two-term inner expansion: $\quad = \beta e + \epsilon\beta e(1 - \xi)$ (12.98)

Two-term inner expansion: $\quad y \sim \alpha - b_0 + b_0 e^{-\xi} + \epsilon[-b_1 + b_1 e^{-\xi}$
$\quad\quad - (\alpha - b_0)\xi + b_0 \xi e^{-\xi}]$

Rewritten in outer variable: $\quad = \alpha - b_0 + b_0 e^{-x/\epsilon} + \epsilon\left[-b_1 + b_1 e^{-x/\epsilon}\right.$
$\quad\quad \left. - (\alpha - b_0)\frac{x}{\epsilon} + \frac{b_0 x}{\epsilon}e^{-x/\epsilon}\right]$

Expanded for small ϵ: $\quad = \alpha - b_0 - (\alpha - b_0)x - \epsilon b_1 + \text{EST}$

Two-term outer expansion: $\quad = \alpha - b_0 - (\alpha - b_0)x - \epsilon b_1$ (12.99)

Equating (12.98) and (12.99) according to the matching principle (12.97) and expressing ξ in terms of x, we obtain exactly (12.91).

It should be noted that there are some special examples for which Van Dyke's principle does not apply. Nonetheless, it is systematic and hence widely used.

In the above example, we can determine as many terms as we want in the inner or outer expansions to be sure, with arbitrary constants that must be determined from matching. This is not always possible. For example, in determining an asymptotic solution for viscous flow past an arbitrary body, one must determine a one-term outer expansion, which is used to determine a one-term inner expansion. Then, one proceeds to determine the second term in the outer expansion and uses it in turn to determine the second term in the inner expansion, and so on.

284 BOUNDARY-LAYER PROBLEMS

COMPOSITE EXPANSION

Once the inner and outer expansions have been determined and matched, and hence, all the constants have been determined, we can form a composite expansion that is uniform everywhere. To this end, we substitute for y^o, y^i, and $(y^o)^i$ from (12.82), (12.94), and (12.89), respectively, into (12.75) and obtain

$$y^c = \beta e^{1-x} + (\alpha - \beta e)e^{-\xi} + \epsilon[\beta(1-x)e^{1-x} - \beta e^{1-\xi} + (\alpha - \beta e)\xi e^{-\xi}] + \cdots \tag{12.100}$$

as a single uniform expansion.

12.5. Equations With Variable Coefficients

In this section, we use the method of matched asymptotic expansions to determine first-order asymptotic solutions of

$$\epsilon y'' + p_1(x)y' + p_0(x)y = 0 \quad \epsilon \ll 1 \tag{12.101}$$

$$y(0) = \alpha \quad y(1) = \beta \tag{12.102}$$

for special functions p_1 and p_0. This problem is treated in Chapter 14 by using the WKB approximation and the Langer transformation. Here, we treat two examples in which $p_1(x) \neq 0$ in [0, 1] and we treat two examples in which $p_1(x)$ has zeros in [0,1].

EXAMPLE 1
We consider

$$\epsilon y'' - (2x+1)y' + 2y = 0 \tag{12.103}$$

As $\epsilon \to 0$, the order of the differential equation reduces to one, and hence, one of the boundary conditions must be dropped. It turns out that, if $p_1(x) > 0$ in [0, 1], the boundary layer is at the left end and, if $p_1(x) < 0$ in [0, 1], the boundary layer is at the right end. Hence, in this case the boundary layer is at the right end, and the outer expansion satisfies the boundary condition $y(0) = \alpha$. Anyhow the existence of the boundary layer at the right end will be verified if the resulting expansions are matchable and the results are mathematically consistent.

We seek an outer expansion in the form

$$y^o = y_0(x) + \epsilon y_1(x) + \cdots \tag{12.104}$$

Substituting (12.104) into (12.103) and $y(0) = \alpha$ and equating the coefficients of ϵ^o on both sides, we obtain

$$-(2x+1)y_0' + 2y_0 = 0 \quad y_0(0) = \alpha \tag{12.105}$$

Separating variables, we have

EQUATIONS WITH VARIABLE COEFFICIENTS 285

$$\frac{dy_0}{y_0} = \frac{2dx}{2x+1}$$

Hence,

$$\ln y_0 = \ln(2x+1) + \ln c_0$$

where c_0 is a constant. Then,

$$y_0 = c_0(2x+1)$$

But $y_0(0) = \alpha$, hence $c_0 = \alpha$. Therefore,

$$y_0 = \alpha(2x+1)$$

and

$$y^o = \alpha(2x+1) + \cdots \tag{12.106}$$

To determine an expansion valid in the boundary layer, we need to stretch the neighborhood of $x = 1$. Thus, we let

$$\xi = \frac{1-x}{\epsilon^\nu} \quad \text{or} \quad x = 1 - \epsilon^\nu \xi \tag{12.107}$$

where ν must be greater than zero and it is determined in the course of analysis. Putting (12.107) in (12.103) and denoting the inner expansion by the superscript i, we have

$$\epsilon^{1-2\nu} \frac{d^2 y^i}{d\xi^2} + \epsilon^{-\nu}(3 - 2\epsilon^\nu \xi)\frac{dy^i}{d\xi} + 2y^i = 0 \tag{12.108}$$

As $\epsilon \to 0$, the limiting form of (12.108) depends on the value of ν. As discussed before, the value of ν corresponding to the distinguished limit must be selected. Thus, we put $\nu = 1$ and rewrite (12.108) as

$$\frac{d^2 y^i}{d\xi^2} + (3 - 2\epsilon\xi)\frac{dy^i}{d\xi} + 2\epsilon y^i = 0 \tag{12.109}$$

Since the boundary layer is assumed to be at $x = 1$, it must satisfy the boundary condition $y(1) = \beta$. But $x = 1$ corresponds to $\xi = 0$, hence

$$y^i(0) = \beta \tag{12.110}$$

We seek an inner expansion in the form

$$y^i = Y_0(\xi) + \epsilon Y_1(\xi) + \cdots \tag{12.111}$$

Substituting (12.111) into (12.109) and (12.110) and equating the coefficients of ϵ^o on both sides, we obtain

$$Y_0'' + 3Y_0' = 0 \quad Y_0(0) = \beta \tag{12.112}$$

286 BOUNDARY-LAYER PROBLEMS

Hence,

$$Y_0 = a_0 + b_0 e^{-3\xi} \qquad \beta = a_0 + b_0$$

and

$$y^i = \beta - b_0 + b_0 e^{-3\xi} + \cdots \qquad (12.113)$$

where b_0 needs to be determined from matching the inner and outer expansions.

Using Van Dyke's matching principle, we proceed as follows:

One-term outer expansion: $\qquad y \sim \alpha(2x + 1)$

Rewritten in inner variable: $\qquad = \alpha(3 - 2\epsilon\xi)$

Expanded for small ϵ: $\qquad = 3\alpha - 2\epsilon\alpha\xi$

One-term inner expansion: $\qquad = 3\alpha \qquad (12.114)$

One-term inner expansion: $\qquad y \sim \beta - b_0 + b_0 e^{-3\xi}$

Rewritten in outer variable: $\qquad = \beta - b_0 + b_0 e^{-3(1-x)/\epsilon}$

Expanded for small ϵ: $\qquad = \beta - b_0 + \text{EST}$

One-term outer expansion: $\qquad = \beta - b_0 \qquad (12.115)$

Equating (12.114) and (12.115), we have

$$3\alpha = \beta - b_0 \quad \text{or} \quad b_0 = \beta - 3\alpha$$

Hence,

$$y^i = 3\alpha + (\beta - 3\alpha) e^{-3\xi} + \cdots \qquad (12.116)$$

and

$$y^c = \alpha(2x + 1) + 3\alpha + (\beta - 3\alpha) e^{-3\xi} - 3\alpha + \cdots$$

or

$$y^c = \alpha(2x + 1) + (\beta - 3\alpha) e^{-3(1-x)/\epsilon} + \cdots \qquad (12.117)$$

Figure 12-9 shows that y^c is close to the solution obtained by numerically integrating (12.103) and (12.102). Since the inner and outer expansions are matchable and the results are mathematically consistent, our assumption about the location of the boundary layer is correct.

EXAMPLE 2

As a second example, we consider (12.101) with $p_1(x)$ being either positive or negative everywhere in $[0, 1]$.

When $p_1(x) > 0$ in $[0, 1]$, the boundary layer is assumed to be at the left end because only this assumption leads to matchable expansions and mathematically

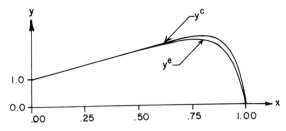

Figure 12-9. Comparison of composite expansion y^c with the exact solution y^e obtained by numerically integrating (12.103) and (12.102) for $\epsilon = 0.2$, $\alpha = 1.0$, and $\beta = 0.0$.

consistent results. Then, the outer expansion must satisfy the boundary condition $y(1) = \beta$, whereas the inner expansion must satisfy the boundary condition $y(0) = \alpha$. We seek an outer expansion in the form

$$y^o = y_0(x) + \epsilon y_1(x) + \cdots \qquad (12.118)$$

Substituting (12.118) into (12.101) and $y(1) = \beta$ and equating the coefficients of ϵ^o on both sides, we obtain

$$p_1 y_0' + p_0 y_0 = 0 \qquad y_0(1) = \beta \qquad (12.119)$$

which is a first-order linear differential equation, and hence, it is solvable. Separating variables, we rewrite the equation governing y_0 as

$$\frac{dy_0}{y_0} = -\frac{p_0}{p_1} dx$$

which upon integration yields

$$\ln y_0 = -\int_1^x \frac{p_0(\tau)}{p_1(\tau)} d\tau + \ln c_0$$

Hence,

$$y_0 = c_0 \exp\left[-\int_1^x \frac{p_0(\tau)}{p_1(\tau)} d\tau\right]$$

where the lower limit of integration was taken to be 1 to facilitate satisfaction of the boundary condition. Putting $y_0(1) = \beta$ gives $c_0 = \beta$, and hence,

$$y_0 = \beta \exp\left[-\int_1^x \frac{p_0}{p_1} d\tau\right] = \beta \exp\left[\int_x^1 \frac{p_0}{p_1} d\tau\right]$$

and

288 BOUNDARY-LAYER PROBLEMS

$$y^0 = \beta \exp\left[\int_x^1 \frac{p_0}{p_1} d\tau\right] + \cdots \tag{12.120}$$

To determine an inner expansion valid near the origin, we introduce the stretching transformation

$$\xi = \frac{x}{\epsilon^\nu} \quad \text{or} \quad x = \epsilon^\nu \xi \quad \nu > 0 \tag{12.121}$$

Then, (12.101) becomes

$$\epsilon^{1-2\nu} \frac{d^2 y^i}{d\xi^2} + \epsilon^{-\nu} p_1(\epsilon^\nu \xi) \frac{dy^i}{d\xi} + p_0(\epsilon^\nu \xi) y^i = 0 \tag{12.122}$$

As $\epsilon \to 0$, $p_1(\epsilon^\nu \xi) \to p_1(0)$ and $p_0(\epsilon^\nu \xi) \to p_0(0)$, where p_0 is assumed to be regular at the origin. Then, (12.122) tends to

$$\epsilon^{1-2\nu} \frac{d^2 y^i}{d\xi^2} + \epsilon^{-\nu} p_1(0) \frac{dy^i}{d\xi} + p_0(0) y^i + \cdots = 0 \tag{12.123}$$

whose limiting form as $\epsilon \to 0$ depends on the value of ν. Choosing $\nu = 1$, corresponding to the distinguished limit, we find that the limiting form of (12.123) is

$$\frac{d^2 y^i}{d\xi^2} + p_1(0) \frac{dy^i}{d\xi} = 0 \tag{12.124}$$

whose general solution is

$$y^i = a_0 + b_0 e^{-p_1(0)\xi} \tag{12.125}$$

It follows from (12.121) that $x = 0$ corresponds to $\xi = 0$, and hence, the boundary condition $y(0) = \alpha$ transforms into $y^i(0) = \alpha$. Then, it follows from (12.125) that

$$\alpha = a_0 + b_0 \quad \text{or} \quad a_0 = \alpha - b_0$$

and hence,

$$y^i = \alpha - b_0 + b_0 e^{-p_1(0)\xi} \tag{12.126}$$

Next, we match the one-term outer expansion (12.120) with the one-term inner expansion (12.126). We proceed formally as follows:

One-term outer expansion: $\quad y \sim \beta \exp\left[\int_x^1 \frac{p_0}{p_1} d\tau\right]$

Rewritten in inner variable: $\quad = \beta \exp\left[\int_{\epsilon\xi}^1 \frac{p_0}{p_1} d\tau\right]$

Expanded for small ϵ: $= \beta \exp\left[\int_0^1 \frac{p_0}{p_1} d\tau\right] + \cdots$

One-term inner expansion: $= \beta \exp\left[\int_0^1 \frac{p_0}{p_1} d\tau\right]$ (12.127)

One-term inner expansion: $y \sim \alpha - b_0 + b_0 e^{-p_1(0)\xi}$

Rewritten in outer variable: $= \alpha - b_0 + b_0 e^{-p_1(0)x/\epsilon}$

Expanded for small ϵ: $= \alpha - b_0 + \text{EST}$

One-term outer expansion: $= \alpha - b_0$ (12.128)

We note that $\exp[-p_1(0)x/\epsilon]$ is exponentially small as $\epsilon \to 0$ because $p_1(0) > 0$. If $p_1(0)$ were negative, $\exp[-p_1(0)x/\epsilon]$ would have been exponentially large as $\epsilon \to 0$, and hence, the inner expansion could not be matched with the outer expansion. Consequently, the boundary layer would be at the right end. Thus, the absence of the exponential growth is essential for matching.

Equating (12.127) and (12.128), we have

$$\alpha - b_0 = \beta \exp\left[\int_0^1 \frac{p_0}{p_1} d\tau\right]$$

or

$$b_0 = \alpha - \beta \exp\left[\int_0^1 \frac{p_0}{p_1} d\tau\right]$$

Hence,

$$y^i = \beta \exp\left[\int_0^1 \frac{p_0}{p_1} d\tau\right] + \left\{\alpha - \beta \exp\left[\int_0^1 \frac{p_0}{p_1} d\tau\right]\right\} e^{-p_1(0)\xi} + \cdots \quad (12.129)$$

Adding the outer expansion (12.120) to the inner expansion (12.129) and subtracting from the result their common part (12.127), we obtain the composite expansion

$$y^c = \beta \exp\left[\int_x^1 \frac{p_0}{p_1} d\tau\right] + \left\{\alpha - \beta \exp\left[\int_0^1 \frac{p_0}{p_1} d\tau\right]\right\} e^{-p_1(0)\xi} + \cdots \quad (12.130)$$

EXAMPLE 3

As a third example, we consider a case in which $p_1(x)$ vanishes at the origin. Thus, we consider the equation

290 BOUNDARY-LAYER PROBLEMS

$$\epsilon y'' + xy' - xy = 0 \qquad (12.131)$$

subject to the boundary conditions in (12.102). Since the coefficient of y' is positive, we expect the boundary layer to be at the left end even if x vanishes there. If this is not the way it is, the resulting expansions cannot be matched and the results will not be mathematically consistent.

Seeking an outer expansion in the form

$$y^0(x) = y_0(x) + \epsilon y_1(x) + \cdots$$

we find from (12.131) that

$$xy_0' - xy_0 = 0$$

whose general solution is

$$y_0 = c_0 e^x$$

Since the boundary layer is assumed to be at the origin, the outer expansion must satisfy $y(1) = \beta$ or $y^0(1) = \beta$. Hence, $y_0(1) = \beta$ and

$$\beta = c_0 e \quad \text{or} \quad c_0 = \beta e^{-1}$$

Then,

$$y_0 = \beta e^{x-1}$$

and

$$y^0 = \beta e^{x-1} + \cdots \qquad (12.132)$$

To investigate the boundary layer at the origin, we introduce the stretching transformation

$$\xi = \frac{x}{\epsilon^\nu} \quad \text{or} \quad x = \epsilon^\nu \xi \quad \nu > 0 \qquad (12.133)$$

in (12.131) and obtain

$$\epsilon^{1-2\nu} \frac{d^2 y^i}{d\xi^2} + \xi \frac{dy^i}{d\xi} - \epsilon^\nu \xi y^i = 0 \qquad (12.134)$$

As $\epsilon \to 0$, the limiting form of (12.134) depends on the value of ν. As before, only the distinguished limit is chosen. In this case, the distinguished limit is

$$\frac{d^2 y^i}{d\xi^2} + \xi \frac{dy^i}{d\xi} = 0 \qquad (12.135)$$

corresponding to $\nu = \frac{1}{2}$. Equation (12.135) is a first-order differential equation in $dy^i/d\xi$, and hence, it is solvable. Putting

$$v = \frac{dy^i}{d\xi}$$

in (12.135), we have

$$\frac{dv}{d\xi} + \xi v = 0$$

whose general solution is

$$v = a_0 e^{-(1/2)\xi^2}$$

Hence,

$$\frac{dy^i}{d\xi} = a_0 e^{-(1/2)\xi^2}$$

which upon integration yields

$$y^i = a_0 \int_0^\xi e^{-(1/2)\tau^2} \, d\tau + b_0$$

where the lower limit in the integral was taken to be zero to facilitate satisfaction of the boundary condition. Since $x = 0$ corresponds to $\xi = 0$ according to (12.133), imposing the boundary condition $y(0) = \alpha$ or $y^i(0) = \alpha$ leads to

$$\alpha = b_0$$

Hence,

$$y^i = a_0 \int_0^\xi e^{-(1/2)\tau^2} \, d\tau + \alpha \tag{12.136}$$

where the constant a_0 needs to be determined by matching the inner and outer expansions.

To match the one-term outer expansion (12.132) with the one-term inner expansion (12.136), we proceed as follows:

One-term outer expansion: $\quad y \sim \beta e^{x-1}$

Rewritten in inner variable: $\quad = \beta e^{\epsilon^{1/2}\xi - 1}$

Expanded for small ϵ: $\quad = \beta e^{-1}(1 + \epsilon^{1/2}\xi + \cdots)$

One-term inner expansion: $\quad = \beta e^{-1} \tag{12.137}$

One-term inner expansion: $\quad y \sim a_0 \int_0^\xi e^{-(1/2)\tau^2} \, d\tau + \alpha$

Rewritten in outer variable: $\quad = a_0 \int_0^{x/\epsilon^{1/2}} e^{-(1/2)\tau^2} \, d\tau + \alpha$

Expanded for small ϵ: $\quad = a_0 \int_0^\infty e^{-(1/2)\tau^2} d\tau + \alpha + \cdots$

One-term outer expansion: $\quad = \dfrac{a_0 \sqrt{\pi}}{\sqrt{2}} + \alpha \qquad\qquad (12.138)$

where the integral is evaluated as follows. Putting $t = \tau/\sqrt{2}$, we have

$$\int_0^\infty e^{-(1/2)\tau^2} d\tau = \sqrt{2} \int_0^\infty e^{-t^2} dt = \sqrt{2} \dfrac{\sqrt{\pi}}{2} = \dfrac{\sqrt{\pi}}{\sqrt{2}}$$

according to (3.25). Equating (12.137) and (12.138) yields

$$\beta e^{-1} = \dfrac{a_0 \sqrt{\pi}}{\sqrt{2}} + \alpha \quad \text{or} \quad a_0 = \dfrac{\sqrt{2}}{\sqrt{\pi}} (\beta e^{-1} - \alpha)$$

Hence, the outer and inner expansions are matchable and the results are mathematically consistent, thereby justifying our assumption that the boundary layer is at the origin. Substituting for a_0 in (12.136), we have

$$y^i = \alpha + \dfrac{\sqrt{2}}{\sqrt{\pi}} (\beta e^{-1} - \alpha) \int_0^\xi e^{-(1/2)\tau^2} d\tau + \cdots \qquad (12.139)$$

Finally, we determine a single composite uniform expansion by adding the outer expansion (12.132) to the inner expansion (12.139) and subtracting from the result their common part (12.137). Thus, we have

$$y^c = \beta e^{x-1} + \alpha + \dfrac{\sqrt{2}}{\sqrt{\pi}} (\beta e^{-1} - \alpha) \int_0^\xi e^{-(1/2)\tau^2} d\tau - \beta e^{-1}$$

or

$$y^c = \beta e^{x-1} - (\beta e^{-1} - \alpha) \left(1 - \dfrac{\sqrt{2}}{\sqrt{\pi}} \int_0^\xi e^{-(1/2)\tau^2} d\tau\right) + \cdots \qquad (12.140)$$

Figure 12.10 shows that the composite expansion y^c is very close to the exact solution y^e even for $\epsilon = 0.2$. When $\epsilon = 0.1$, the composite expansion is indistinguishable from the exact solution.

EXAMPLE 4

As a fourth example, we consider a case in which $p_1(x)$ has a simple zero at a point inside the interval $[0, 1]$, thereby leading to an interior "boundary layer." Thus, we consider the differential equation

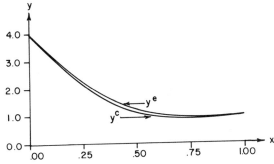

Figure 12-10. Comparison of the composite expansion y^c with the exact solution y^e obtained by numerically integrating (12.131) and (12.102) for $\epsilon = 0.2$, $\alpha = 4.0$, $\beta = 1.0$.

$$\epsilon y'' + (x - \tfrac{1}{2})y' - (x - \tfrac{1}{2})y = 0 \qquad (12.141)$$

subject to the boundary conditions (12.102). In this case, $p_1(x) = x - \tfrac{1}{2}$ is negative for $x < \tfrac{1}{2}$ and positive for $x > \tfrac{1}{2}$. Consequently, our expectation about the location of the boundary layer needs to be examined more closely. It turns out that there are no boundary layers at the end points. Instead, there is a boundary layer at $x = \tfrac{1}{2}$. This statement is checked a posteriori.

Seeking an outer expansion in the form

$$y^o = y_0(x) + \epsilon y_1(x) + \cdots$$

we obtain from (12.141) that

$$(x - \tfrac{1}{2})y'_0 - (x - \tfrac{1}{2})y_0 = 0$$

Its general solution is

$$y_0 = c_0 e^x$$

which is expected to be valid everywhere except in a small neighborhood surrounding $x = \tfrac{1}{2}$. In the interval $x > \tfrac{1}{2}$, y^o must satisfy the boundary condition $y^o(1) = \beta$. Hence, $c_0 e = \beta$ and

$$y^o_r = \beta e^{x-1} + \cdots \qquad (12.142)$$

where the subscript r indicates the right part of the interval. In the interval $x < \tfrac{1}{2}$, y^o must satisfy the boundary condition $y^o(0) = \alpha$. Hence, $c_0 = \alpha$ and

$$y^o_l = \alpha e^x + \cdots \qquad (12.143)$$

where the subscript l indicates the left part of the interval.

Next, we investigate the neighborhood of $x = \tfrac{1}{2}$ by introducing the stretching transformation

294 BOUNDARY-LAYER PROBLEMS

$$\xi = \frac{x - \frac{1}{2}}{\epsilon^\nu} \quad \text{or} \quad x = \tfrac{1}{2} + \epsilon^\nu \xi \quad \nu > 0 \tag{12.144}$$

Then, (12.141) becomes

$$\epsilon^{1-2\nu} \frac{d^2 y^i}{d\xi^2} + \xi \frac{dy^i}{d\xi} - \epsilon^\nu \xi y^i = 0 \tag{12.145}$$

As $\epsilon \to 0$, the limiting form of (12.145) depends on the value of ν. As before, we choose the distinguished limit

$$\frac{d^2 y^i}{d\xi^2} + \xi \frac{dy^i}{d\xi} = 0 \tag{12.146}$$

corresponding to $\nu = \tfrac{1}{2}$. As in the preceding section, the general solution of (12.146) is

$$y^i = b_0 + a_0 \int_0^\xi e^{-(1/2)\tau^2} d\tau \tag{12.147}$$

where the lower limit is taken to be $\xi = 0$ corresponding to the assumed location $x = \tfrac{1}{2}$ of the boundary layer. The constants a_0 and b_0 need to be determined by matching inner and outer expansions.

To match the one-term outer expansion (12.142) with the one-term inner expansion (12.147), we proceed as follows:

One-term outer expansion: $\quad y \sim \beta e^{x-1}$

Rewritten in inner variable: $\quad = \beta e^{\epsilon^{1/2}\xi - (1/2)}$

Expanded for small ϵ: $\quad = \beta e^{-1/2}(1 + \epsilon^{1/2}\xi + \cdots)$

One-term inner expansion: $\quad = \beta e^{-1/2} \tag{12.148}$

One-term inner expansion: $\quad y \sim b_0 + a_0 \int_0^\xi e^{-(1/2)\tau^2} d\tau$

Rewritten in outer variable: $\quad = b_0 + a_0 \int_0^{[x-(1/2)]/\epsilon^{1/2}} e^{-(1/2)\tau^2} d\tau$

Expanded for small ϵ: $\quad = b_0 + a_0 \int_0^\infty e^{-(1/2)\tau^2} d\tau + \cdots$

One-term outer expansion: $\quad = b_0 + \dfrac{a_0 \sqrt{\pi}}{\sqrt{2}} \tag{12.149}$

where the integral was evaluated in the preceding example. Equating (12.148) and (12.149), we have

$$b_0 + \frac{a_0 \sqrt{\pi}}{\sqrt{2}} = \beta e^{-1/2} \qquad (12.150)$$

which is an equation governing a_0 and b_0.

To determine a second equation for a_0 and b_0, we match the inner expansion (12.147) with the outer expansion (12.143). Following the asymptotic matching procedure, we find that

$$(y_i^o)^i = \alpha e^{1/2}$$

$$(y^i)^o = b_0 + a_0 \int_0^{-\infty} e^{-(1/2)\tau^2} d\tau = b_0 - a_0 \int_0^{\infty} e^{-(1/2)\tau^2} d\tau$$

Hence,

$$b_0 - \frac{a_0 \sqrt{\pi}}{\sqrt{2}} = \alpha e^{1/2} \qquad (12.151)$$

Solving (12.150) and (12.151) yields

$$b_0 = \tfrac{1}{2}(\beta e^{-1/2} + \alpha e^{1/2}) \qquad a_0 = \frac{1}{\sqrt{2\pi}}(\beta e^{-1/2} - \alpha e^{1/2})$$

and hence,

$$y^i = \tfrac{1}{2}(\beta e^{-1/2} + \alpha e^{1/2}) + \frac{1}{\sqrt{2\pi}}(\beta e^{-1/2} - \alpha e^{1/2}) \int_0^{\xi} e^{-(1/2)\tau^2} d\tau + \cdots$$

$$(12.152)$$

In this case, we cannot form a single composite expansion that is uniformly valid over the whole interval. Instead, we form two composite expansions, one valid in $[0, \tfrac{1}{2}]$ and the other valid in $[\tfrac{1}{2}, 1]$. Thus, we put

$$y_l^c = y_l^o + y^i - (y_l^o)^i$$

$$= \alpha e^x + \tfrac{1}{2}(\beta e^{-1/2} + \alpha e^{1/2})$$

$$+ \frac{1}{\sqrt{2\pi}}(\beta e^{-1/2} - \alpha e^{1/2}) \int_0^{\xi} e^{-(1/2)\tau^2} d\tau - \alpha e^{1/2} + \cdots$$

or

$$y_l^c = \alpha e^x + (\beta e^{-1/2} - \alpha e^{1/2})\left(\tfrac{1}{2} + \frac{1}{\sqrt{2\pi}} \int_0^{\xi} e^{-(1/2)\tau^2} d\tau\right) + \cdots \qquad (12.153)$$

296 BOUNDARY-LAYER PROBLEMS

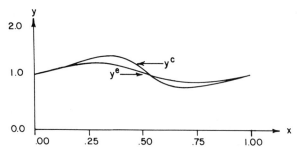

Figure 12-11. Comparison of the composite expansion y^c with the exact solution y^e obtained by numerically integrating (12.141) and (12.102) for $\epsilon = 0.1$, $\alpha = 1.0$, and $\beta = 1.0$.

Similarly, we form

$$y_r^c = \beta e^{x-1} + (\alpha e^{1/2} - \beta e^{-1/2})\left(\frac{1}{2} - \frac{1}{\sqrt{2\pi}}\int_0^\xi e^{-(1/2)\tau^2}\, d\tau\right) + \cdots \quad (12.154)$$

Figure 12-11 shows that y^c is very close to the exact solution y^e for $\epsilon = 0.1$.

In this example, $p_1(x)$ is monotonically increasing, and there is only an interior boundary layer. When $p_1(x)$ is monotonically decreasing, the situation is more complicated and boundary layers may exist at the boundaries in addition to an interior point. Currently, this problem is being vigorously investigated by many researchers.

12.6. Problems with Two Boundary Layers

In the preceding sections, we considered problems having one boundary layer. In this section, we consider a problem with two boundary layers; specifically, we consider

$$\epsilon^2 y^{iv} - (1+x)^2 y'' = 1 \quad (12.155)$$

$$y(0) = \alpha \quad y'(0) = \beta \quad y(1) = \gamma \quad y'(1) = \delta \quad (12.156)$$

In this case, the order of the highest derivative is greater than the order of the second term by two, suggesting that there are two boundary layers, one at each end. This suggestion will be verified a posteriori by checking the mathematical consistency of the results.

OUTER EXPANSION

We seek a two-term outer expansion in the form

$$y^o = y_0(x) + \epsilon y_1(x) + \cdots \quad (12.157)$$

Substituting (12.157) into (12.155) and equating coefficients of like powers of ϵ, we obtain

$$-(1+x)^2 y_0'' = 1 \tag{12.158}$$
$$-(1+x)^2 y_1'' = 0 \tag{12.159}$$

Equation (12.158) can be rewritten as

$$y_0'' = -\frac{1}{(1+x)^2}$$

which can be integrated to give

$$y_0' = \frac{1}{1+x} + A_0$$

and hence,

$$y_0 = \ln(1+x) + A_0 x + B_0 \tag{12.160}$$

where A_0 and B_0 are arbitrary constants. The general solution of (12.159) is

$$y_1 = A_1 x + B_1 \tag{12.161}$$

where A_1 and B_1 are arbitrary constants. Therefore,

$$y^o = \ln(1+x) + A_0 x + B_0 + \epsilon(A_1 x + B_1) + \cdots \tag{12.162}$$

where this outer expansion is not expected to satisfy any of the boundary conditions. Hence, it needs to be matched with two boundary-layer expansions, one valid near $x = 0$ and the other valid near $x = 1$.

INNER EXPANSION NEAR $x = 0$
In this case, we introduce the stretching transformation

$$\xi = \frac{x}{\epsilon^\nu} \quad \text{or} \quad x = \epsilon^\nu \xi \quad \nu > 0 \tag{12.163}$$

in (12.155), denote the inner expansion near $x = 0$ by the superscript i, and obtain

$$\epsilon^{2-4\nu} \frac{d^4 y^i}{d\xi^4} - \epsilon^{-2\nu}(1+\epsilon^\nu \xi)^2 \frac{d^2 y^i}{d\xi^2} = 1$$

As $\epsilon \to 0$, the distinguished limit corresponds to $\nu = 1$. Then, y^i is governed by

$$\frac{d^4 y^i}{d\xi^4} - (1 + 2\epsilon\xi + \epsilon^2 \xi^2) \frac{d^2 y^i}{d\xi^2} = \epsilon^2 \tag{12.164}$$

We seek a two-term inner expansion in the form

$$y^i = Y_0(\xi) + \epsilon Y_1(\xi) + \cdots \tag{12.165}$$

Substituting (12.165) into (12.164) and equating like powers of ϵ, we have

298 BOUNDARY-LAYER PROBLEMS

$$Y_0^{iv} - Y_0'' = 0 \tag{12.166}$$

$$Y_1^{iv} - Y_1'' = 2\xi Y_0'' \tag{12.167}$$

This inner expansion must satisfy the boundary conditions at $x = 0$, corresponding to $\xi = 0$. Then, $y(0) = \alpha$ transforms into

$$y^i(0) = \alpha \tag{12.168}$$

Since

$$y' = \frac{dy}{dx} = \frac{dy}{d\xi}\frac{d\xi}{dx} = \frac{1}{\epsilon}\frac{dy}{d\xi}$$

the boundary condition $y'(0) = \beta$ transforms into

$$\frac{dy^i}{d\xi}(0) = \epsilon\beta \tag{12.169}$$

Substituting (12.165) into (12.168) and (12.169) and equating coefficients of like powers of ϵ, we have

$$Y_0(0) = \alpha \quad Y_0'(0) = 0 \tag{12.170}$$

$$Y_1(0) = 0 \quad Y_1'(0) = \beta \tag{12.171}$$

We note that the effect of β appears at first order and this is the reason we are determining two terms in the expansion.

The general solution of (12.166) is

$$Y_0 = a_0 + b_0\xi + c_0 e^{-\xi} + d_0 e^{\xi}$$

where the constant d_0 must be zero; otherwise, Y_0 would grow exponentially with ξ, making it unmatchable with the outer expansion. Then, the boundary conditions (12.170) demand that

$$a_0 + c_0 = \alpha \quad b_0 - c_0 = 0$$

Hence, $b_0 = c_0$ and $a_0 = \alpha - c_0$, so that

$$Y_0 = \alpha + c_0(e^{-\xi} + \xi - 1) \tag{12.172}$$

In this case, it is advantageous to perform the matching at this stage because c_0 turns out to be zero. To match a one-term outer expansion with a one-term inner expansion, we take $m = n = 1$ in (12.97) and proceed as follows:

One-term outer expansion: $y \sim \ln(1 + x) + A_0 x + B_0$

Rewritten in inner variable: $= \ln(1 + \epsilon\xi) + \epsilon A_0 \xi + B_0$

Expanded for small ϵ: $= \epsilon\xi + \epsilon A_0 \xi + B_0 + \cdots$

One-term inner expansion: $= B_0 \tag{12.173}$

PROBLEMS WITH TWO BOUNDARY LAYERS   299

One-term inner expansion: $\quad y \sim \alpha + c_0(e^{-\xi} + \xi - 1)$

Rewritten in outer variable: $\quad = \alpha + c_0(e^{-x/\epsilon} + \dfrac{x}{\epsilon} - 1)$

Expanded for small ϵ: $\quad = \alpha - c_0 + \dfrac{c_0 x}{\epsilon} + \text{EST}$

One-term outer expansion: $\quad = \begin{cases} \alpha & \text{if } c_0 = 0 \\ \dfrac{c_0 x}{\epsilon} & \text{if } c_0 \neq 0 \end{cases}$ (12.174)

Equating (12.173) and (12.174) demands that

$$c_0 = 0 \quad B_0 = \alpha \tag{12.175}$$

and

$$Y_0 = \alpha \tag{12.176}$$

Substituting for Y_0 in (12.167) gives

$$Y_1^{iv} - Y_1'' = 0$$

whose general solution is

$$Y_1 = a_1 + b_1\xi + c_1 e^{-\xi} + d_1 e^{\xi}$$

Again, d_1 must be zero; otherwise, Y_1 would grow exponentially with ξ, making it unmatchable with the outer expansion. The boundary conditions (12.171) demand that

$$a_1 + c_1 = 0 \quad b_1 - c_1 = \beta$$

Hence, $a_1 = -c_1$ and $b_1 = c_1 + \beta$ so that

$$Y_1 = \beta\xi + c_1(e^{-\xi} + \xi - 1) \tag{12.177}$$

Therefore,

$$y^i = \alpha + \epsilon[\beta\xi + c_1(e^{-\xi} + \xi - 1)] + \cdots \tag{12.178}$$

Next, we match the two-term outer expansion with the two-term inner expansion and proceed as follows:

Two-term outer expansion: $\quad y \sim \ln(1+x) + A_0 x + \alpha + \epsilon(A_1 x + B_1)$

Rewritten in inner variable: $\quad = \ln(1 + \epsilon\xi) + \epsilon A_0 \xi + \alpha + \epsilon(\epsilon A_1 \xi + B_1)$

Expanded for small ϵ: $\quad = \epsilon\xi + \epsilon A_0 \xi + \alpha + \epsilon^2 A_1 \xi + \epsilon B_1 + \cdots$

Two-term inner expansion: $\quad = \alpha + \epsilon(\xi + A_0 \xi + B_1) \tag{12.179}$

Two-term inner expansion: $\quad y \sim \alpha + \epsilon[\beta\xi + c_1(e^{-\xi} + \xi - 1)]$

300 BOUNDARY-LAYER PROBLEMS

Rewritten in outer variable: $= \alpha + \epsilon \left[\dfrac{\beta x}{\epsilon} + c_1 \left(e^{-x/\epsilon} + \dfrac{x}{\epsilon} - 1 \right) \right]$

Expanded for small ϵ: $= \alpha + \beta x + c_1 x - \epsilon c_1 + \text{EST}$

Two-term outer expansion: $= \alpha + \beta x + c_1 x - \epsilon c_1$ (12.180)

Expressing (12.179) in terms of x and equating it to (12.180), we have

$$\alpha + x + A_0 x + \epsilon B_1 = \alpha + \beta x + c_1 x - \epsilon c_1$$

Hence,

$$1 + A_0 = \beta + c_1 \quad \text{and} \quad B_1 = -c_1 \qquad (12.181)$$

INNER EXPANSION NEAR $x = 1$

In this case, we introduce the stretching transformation

$$\zeta = \dfrac{1-x}{\epsilon^\nu} \quad \text{or} \quad x = 1 - \epsilon^\nu \zeta \quad \nu > 0 \qquad (12.182)$$

in (12.155), denote the inner expansion near $x = 1$ by the superscript I, and obtain

$$\epsilon^{2-4\nu} \dfrac{d^4 y^I}{d\zeta^4} - \epsilon^{-2\nu}(2 - \epsilon^\nu \zeta)^2 \dfrac{d^2 y^I}{d\zeta^2} = 1$$

As $\epsilon \to 0$, the distinguished limit corresponds to $\nu = 1$. Then, y^I is governed by

$$\dfrac{d^4 y^I}{d\zeta^4} - (4 - 4\epsilon\zeta + \epsilon^2 \zeta^2) \dfrac{d^2 y^I}{d\zeta^2} = \epsilon^2 \qquad (12.183)$$

We seek a two-term expansion for y^I in the form

$$y^I = \tilde{Y}_0(\zeta) + \epsilon \tilde{Y}_1(\zeta) + \cdots \qquad (12.184)$$

Substituting (12.184) into (12.183) and equating coefficients of like powers of ϵ, we have

$$\tilde{Y}_0^{iv} - 4\tilde{Y}_0'' = 0 \qquad (12.185)$$

$$\tilde{Y}_1^{iv} - 4\tilde{Y}_1'' = -4\zeta \tilde{Y}_0'' \qquad (12.186)$$

The present inner expansion must satisfy the boundary conditions at $x = 1$, corresponding to $\zeta = 0$. Then, $y(1) = \gamma$ transforms into

$$y^I(0) = \gamma \qquad (12.187)$$

Since

$$y' = \dfrac{dy}{dx} = \dfrac{dy}{d\zeta} \dfrac{d\zeta}{dx} = -\dfrac{1}{\epsilon} \dfrac{dy}{d\zeta}$$

the boundary condition $y'(1) = \delta$ transforms into

$$\frac{dy^I}{d\xi}(0) = -\epsilon\delta \qquad (12.188)$$

Substituting (12.184) into (12.187) and (12.188) and equating coefficients of like powers of ϵ, we obtain

$$\tilde{Y}_0(0) = \gamma \qquad \tilde{Y}'_0(0) = 0 \qquad (12.189)$$

$$\tilde{Y}_1(0) = 0 \qquad \tilde{Y}'_1(0) = -\delta \qquad (12.190)$$

The general solution of (12.185) is

$$\tilde{Y}_0 = \tilde{a}_0 + \tilde{b}_0\zeta + \tilde{c}_0 e^{-2\zeta} + \tilde{d}_0 e^{2\zeta}$$

where $\tilde{d}_0$ must be zero; otherwise, $\tilde{Y}_0$ would grow exponentially with ζ and could not be matched with the outer expansion. Then, the boundary conditions (12.189) demand that

$$\tilde{a}_0 + \tilde{c}_0 = \gamma \qquad \tilde{b}_0 - 2\tilde{c}_0 = 0$$

Hence, $\tilde{b}_0 = 2\tilde{c}_0$ and $\tilde{a}_0 = \gamma - \tilde{c}_0$, so that

$$\tilde{Y}_0 = \gamma + \tilde{c}_0(e^{-2\zeta} + 2\zeta - 1) \qquad (12.191)$$

Again, we match a one-term outer expansion with a one-term inner expansion and proceed as follows:

One-term outer expansion:	$y \sim \ln(1 + x) + A_0 x + \alpha$
Rewritten in inner variable:	$= \ln(2 - \epsilon\zeta) + A_0(1 - \epsilon\zeta) + \alpha$
Expanded for small ϵ:	$= \ln 2 - \frac{1}{2}\epsilon\zeta + A_0 - \epsilon A_0\zeta + \alpha + \cdots$
One-term inner expansion:	$= A_0 + \alpha + \ln 2 \qquad (12.192)$
One-term inner expansion:	$y \sim \gamma + \tilde{c}_0(e^{-2\zeta} + 2\zeta - 1)$
Rewritten in outer variable:	$= \gamma + \tilde{c}_0\left(e^{-2\epsilon^{-1}(1-x)} + \frac{2(1-x)}{\epsilon} - 1\right)$
Expanded for small ϵ:	$= \gamma + \frac{2\tilde{c}_0(1-x)}{\epsilon} - \tilde{c}_0 + \text{EST}$
One-term outer expansion:	$= \begin{cases} \gamma & \text{if } \tilde{c}_0 = 0 \\ 2\frac{\tilde{c}_0(1-x)}{\epsilon} & \text{if } \tilde{c}_0 \neq 0 \end{cases} \qquad (12.193)$

Equating (12.192) and (12.193) demands that $\tilde{c}_0 = 0$ and

$$A_0 + \alpha + \ln 2 = \gamma \quad \text{or} \quad A_0 = \gamma - \alpha - \ln 2 \qquad (12.194)$$

302 BOUNDARY-LAYER PROBLEMS

Hence,
$$Y_0 = \gamma \tag{12.195}$$

Substituting for $\tilde{Y}_0$ in (12.186), we have
$$\tilde{Y}_1^{iv} - 4\tilde{Y}_1'' = 0$$

whose general solution is
$$\tilde{Y}_1 = \tilde{a}_1 + \tilde{b}_1 \zeta + \tilde{c}_1 e^{-2\zeta} + \tilde{d}_1 e^{2\zeta}$$

Again, $\tilde{d}_1$ must be zero in order that the inner expansion be matchable with the outer expansion. Then, the boundary conditions (12.190) demand that
$$\tilde{a}_1 + \tilde{c}_1 = 0 \qquad \tilde{b}_1 - 2\tilde{c}_1 = -\delta$$

Hence, $\tilde{a}_1 = -\tilde{c}_1$ and $\tilde{b}_1 = 2\tilde{c}_1 - \delta$, so that
$$\tilde{Y}_1 = -\delta\zeta + \tilde{c}_1(e^{-2\zeta} + 2\zeta - 1) \tag{12.196}$$

Next, we match the two-term outer expansion with the two-term inner expansion and proceed as follows:

Two-term outer expansion: $\quad y \sim \ln(1+x) + A_0 x + \alpha + \epsilon(A_1 x + B_1)$

Rewritten in inner variable: $\quad = \ln(2 - \epsilon\zeta) + A_0(1 - \epsilon\zeta) + \alpha$
$$+ \epsilon(A_1 - \epsilon A_1\zeta + B_1)$$

Expanded for small ϵ: $\quad = \ln 2 - \tfrac{1}{2}\epsilon\zeta + A_0 - \epsilon A_0\zeta + \alpha$
$$+ \epsilon A_1 - \epsilon^2 A_1\zeta + \epsilon B_1 + \cdots$$

Two-term inner expansion: $\quad = A_0 + \alpha + \ln 2 + \epsilon(A_1 + B_1 - \tfrac{1}{2}\zeta - A_0\zeta)$
$$\tag{12.197}$$

Two-term inner expansion: $\quad y \sim \gamma + \epsilon[-\delta\zeta + \tilde{c}_1(e^{-2\zeta} + 2\zeta - 1)]$

Rewritten in outer variable: $\quad = \gamma + \epsilon\left\{-\dfrac{\delta(1-x)}{\epsilon} + \tilde{c}_1\left[e^{-2\epsilon^{-1}(1-x)}\right.\right.$
$$\left.\left. + \dfrac{2(1-x)}{\epsilon} - 1\right]\right\}$$

Expanded for small ϵ: $\quad = \gamma - \delta(1-x) + 2\tilde{c}_1(1-x) - \epsilon\tilde{c}_1 + \text{EST}$

Two-term outer expansion: $\quad = \gamma + (2\tilde{c}_1 - \delta)(1-x) - \epsilon\tilde{c}_1 \tag{12.198}$

Expressing (12.197) in terms of x and equating it to (12.198), we have
$$A_0 + \alpha + \ln 2 - (\tfrac{1}{2} + A_0)(1-x) + \epsilon(A_1 + B_1) = \gamma + (2\tilde{c}_1 - \delta)(1-x) - \epsilon\tilde{c}_1$$

which yields (12.194) and

PROBLEMS WITH TWO BOUNDARY LAYERS

$$-\tfrac{1}{2} - A_0 = 2\tilde{c}_1 - \delta \qquad A_1 + B_1 = -\tilde{c}_1 \tag{12.199}$$

Substituting for A_0 from (12.194) into (12.181), we obtain

$$B_1 = -c_1 = -\gamma + \alpha + \beta - 1 + \ln 2 \tag{12.200}$$

whereas substituting for A_0 and B_1 from (12.194) and (12.200) into (12.199), we obtain

$$\tilde{c}_1 = \tfrac{1}{2}(\alpha + \delta - \gamma - \tfrac{1}{2} + \ln 2) \tag{12.201}$$

$$A_1 = -\tfrac{1}{2}(3\alpha + 2\beta - 3\gamma + \delta - \tfrac{5}{2} + 3\ln 2) \tag{12.202}$$

Therefore, all the arbitrary constants in the outer and inner expansions have been expressed in terms of the boundary values.

Finally, a composite expansion can be obtained as follows:

$$y^c = y^o + y^i + y^I - (y^i)^o - (y^I)^o$$

where $(y^i)^o$ and $(y^I)^o$ are given by (12.180) and (12.198). The result is

$$y^c = \ln(1+x) + A_0 x + \alpha + \epsilon(A_1 x + B_1) + \alpha + \epsilon[\beta\xi - B_1(e^{-\xi} + \xi - 1)]$$
$$+ \gamma + \epsilon[-\delta\zeta + \tilde{c}_1(e^{-2\zeta} + 2\zeta - 1)] - \alpha - \beta x + B_1 x - \epsilon B_1 - \gamma$$
$$- (2\tilde{c}_1 - \delta)(1-x) + \epsilon\tilde{c}_1$$

or

$$y^c = \alpha + \ln(1+x) + (\gamma - \alpha - \ln 2)x + \epsilon[-\tfrac{1}{2}(3\alpha + 2\beta - 3\gamma + \delta - \tfrac{5}{2}$$
$$+ 3\ln 2)x - (\gamma - \alpha - \beta + 1 - \ln 2)(1 - e^{-x/\epsilon})$$
$$+ \tfrac{1}{2}(\alpha + \delta - \gamma - \tfrac{1}{2} + \ln 2)e^{-2(1-x)/\epsilon}] + \cdots \tag{12.203}$$

Figure 12-12 shows that y^c is close to the exact solution.

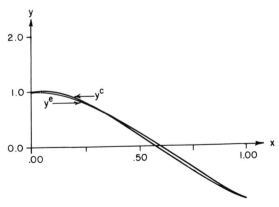

Figure 12-12. Comparison of the composite expansion y^c with the exact solution y^e obtained by numerically integrating (12.155) and (12.156) for $\epsilon = 0.2$, $\alpha = \beta = 1.0$, and $\gamma = \delta = -1.0$.

12.7. Multiple Decks

In all preceding examples, only one distinguished limit exists in a given boundary layer. In this section, we consider a problem in which more than one distinguished limit exist in a given boundary layer. When two distinguished limits exist, the resulting expansion consists of two inner expansions in addition to the outer expansion. The domains of validity of each of them are often called *decks,* and hence, the problem is usually referred to as a *triple-deck problem.* Thus, when two or more distinguished limits exist in a given boundary layer, one speaks of a multiple-deck problem. Here, we consider a case with two distinguished limits, that is, a triple-deck problem.

We consider the problem

$$\epsilon^3 y'' + x^3 y' + (x^3 - \epsilon)y = 0 \qquad (12.204)$$

$$y(0) = \alpha \qquad y(1) = \beta \qquad (12.205)$$

Since the coefficient of y' is positive, the boundary layer is expected to be at the origin. Hence, the outer expansion must satisfy the boundary condition $y(1) = \beta$ and it is not expected to satisfy the boundary condition $y(0) = \alpha$.

We seek an outer expansion in the form

$$y^o = y_0(x) + \epsilon y_1(x) + \cdots$$

Substituting y^o into (12.204) and $y(1) = \beta$ and equating the coefficient of ϵ^0 on both sides, we obtain

$$x^3 y'_0 + x^3 y_0 = 0 \qquad y_0(1) = \beta \qquad (12.206)$$

The general solution for y_0 is

$$y_0 = c_0 e^{-x}$$

It follows from $y_0(1) = \beta$ that $c_0 = \beta e$. Hence,

$$y_0 = \beta e^{1-x}$$

and

$$y^o = \beta e^{1-x} + \cdots \qquad (12.207)$$

which does not satisfy the boundary condition at the origin.

As before, to investigate the neighborhood of the origin, we introduce the stretching transformation

$$\xi = \frac{x}{\epsilon^\nu} \qquad \text{or} \qquad x = \epsilon^\nu \xi \qquad \nu > 0 \qquad (12.208)$$

into (12.204) and obtain

$$\epsilon^{3-2\nu} \frac{d^2 y}{d\xi^2} + \epsilon^{2\nu} \xi^3 \frac{dy}{d\xi} + (\epsilon^{3\nu} \xi^3 - \epsilon)y = 0 \qquad (12.209)$$

As $\epsilon \to 0$, the term $O(\epsilon^{3\nu})$ is small compared with $\epsilon^{2\nu}$ since $\nu > 0$. Hence, the dominant part of (12.209) is

$$\epsilon^{3-2\nu} \frac{d^2y}{d\xi^2} + \epsilon^{2\nu}\xi^3 \frac{dy}{d\xi} - \epsilon y + \cdots = 0 \qquad (12.210)$$

The distinguished limits can be obtained by balancing any two terms in (12.210). Balancing the first and second terms demands that $3 - 2\nu = 2\nu$ or $\nu = \frac{3}{4}$. Then, (12.209) becomes

$$\epsilon^{3/2} \frac{d^2y}{d\xi^2} + \epsilon^{3/2}\xi^3 \frac{dy}{d\xi} + (\epsilon^{9/4}\xi^3 - \epsilon)y = 0$$

whose dominant part is a trivial case $y = 0$. Hence, this case must be discarded. Balancing the first and third terms in (12.210) demands that $3 - 2\nu = 1$ or $\nu = 1$. Then, (12.209) becomes

$$\epsilon \frac{d^2y}{d\xi^2} + \epsilon^2 \xi^3 \frac{dy}{d\xi} + (\epsilon^3 \xi^3 - \epsilon)y = 0 \qquad (12.211a)$$

whose dominant part is nontrivial, and hence, this case must be included. Balancing the second and third terms in (12.210) demands that $2\nu = 1$ or $\nu = \frac{1}{2}$. To distinguish this case from the preceding one, we use the variable ζ instead of ξ so that $\xi = x/\epsilon$ and $\zeta = x/\epsilon^{1/2}$. Then, (12.209) becomes

$$\epsilon^2 \frac{d^2y}{d\zeta^2} + \epsilon \zeta^3 \frac{dy}{d\zeta} + (\epsilon^{3/2}\zeta^3 - \epsilon)y = 0 \qquad (12.211b)$$

whose dominant part is nontrivial, and hence, this case must also be included. The stretching $\xi = x/\epsilon$ corresponding to $\nu = 1$ describes a deck (layer) close to the origin, which we refer to as the *left deck*. The stretching $\zeta = x/\epsilon^{1/2}$, corresponding to $\nu = \frac{1}{2}$, describes a deck that is between the left deck and the outer expansion *(right deck)*. Consequently, we refer to this deck as the *middle deck*. In the case of viscous-inviscid interactions, these decks are usually referred to as the lower, middle, and upper decks, respectively.

It follows from (12.211a) that the leading term Y_0 in the left-deck expansion y^l is governed by

$$\frac{d^2 Y_0}{d\xi^2} - Y_0 = 0$$

whose general solution is

$$Y_0 = a_0 e^{-\xi} + b_0 e^{\xi}$$

where b_0 must be zero; otherwise, Y_0 would grow exponentially with ξ, making it unmatchable with either the middle or right deck. Since this deck is valid at the origin, it must satisfy the boundary condition $y(0) = \alpha$. But $x = 0$ corresponds to $\xi = 0$, hence, $y^l(0) = \alpha$. Then, it follows that $a_0 = \alpha$, so that

$$Y_0 = \alpha e^{-\xi}$$

and

$$y^l = \alpha e^{-\xi} + \cdots \qquad (12.212)$$

It follows from (12.211b) that the leading term $\tilde{Y}_0$ in the middle-deck expansion y^m is governed by

$$\zeta^3 \tilde{Y}_0' - \tilde{Y}_0 = 0$$

Separating variables, we have

$$\frac{d\tilde{Y}_0}{\tilde{Y}_0} = \frac{d\zeta}{\zeta^3}$$

which upon integration gives

$$\ln \tilde{Y}_0 = -\frac{1}{2\zeta^2} + \ln d_0$$

Hence,

$$\tilde{Y}_0 = d_0 e^{-1/2\zeta^2}$$

and

$$y^m = d_0 e^{-1/2\zeta^2} + \cdots \qquad (12.213)$$

Since $x = 0$ corresponds to $\zeta = 0$, it follows from (12.213) that $y^m \to 0$ as $\zeta \to 0$, and hence, it cannot satisfy the boundary condition $y(0) = \alpha$. Therefore, the constant d_0 needs to be determined from matching y^m with either y^l or y^o.

We note that y^l cannot be matched directly with y^o because

$$(y^l)^o = 0 \qquad (y^o)^l = \beta e$$

Hence, y^m must be used to bridge the gap between y^l and y^o. To match y^m with y^o, we note that

$$(y^m)^o = d_0 \quad \text{and} \quad (y^o)^m = \beta e$$

Hence, $d_0 = \beta e$ and

$$y^m = \beta e^{1 - 1/2\zeta^2} + \cdots \qquad (12.214)$$

To match y^m with y^l, we note that

$$(y^m)^l = 0 \quad \text{and} \quad (y^l)^m = 0$$

and hence, y^m and y^l are matchable.

Finally, we form a composite expansion as follows:

NONLINEAR PROBLEMS 307

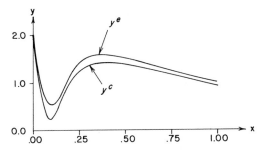

Figure 12-13. Comparison of composite expansion y^c with the exact solution y^e obtained by numerically integrating (12.204) and (12.205) for $\epsilon = 0.05$, $\alpha = 2.0$, and $\beta = 1.0$.

$$y^c = y^o + y^m + y^I - (y^o)^m - (y^I)^m$$

Hence,

$$y^c = \beta e^{1-x} + \beta e^{1-(1/2\xi^2)} + \alpha e^{-\xi} - \beta e + \cdots \qquad (12.215)$$

It follows from (12.215) that

$$(y^c)^o = \beta e^{1-x} = y^o$$
$$(y^c)^m = \beta e^{1-1/2\xi^2} = y^m$$
$$(y^c)^I = \alpha e^{-\xi}$$

and therefore, we postulate that y^c is valid everywhere in $[0, 1]$. Figure 12-13 shows that y^c is close to the exact solution.

12.8. Nonlinear Problems

To conclude this chapter, we consider two nonlinear problems.

PROBLEM 1

As a first problem, we consider the case in which a *contraction* rather than a stretching *transformation* is needed. Specifically, we consider

$$y'' + \frac{2}{x} y' + \epsilon y y' = 0 \qquad 0 < \epsilon \ll 1 \qquad (12.216)$$

$$y(1) = 0 \quad \text{and} \quad y(\infty) = 1 \qquad (12.217)$$

It turns out that the zeroth-order expansion is uniform and that the nonuniformity appears at first order. We note that the small parameter does not multiply the highest derivative. In this case, as in nonlinear-oscillation problems, the domain is infinite, thereby causing the nonuniformity.

We seek a two-term straightforward expansion in the form

$$y = y_0(x) + \epsilon y_1(x) + \cdots \qquad (12.218)$$

Substituting (12.218) into (12.216) and (12.217) and equating coefficients of like powers of ϵ, we have

Order ϵ^0

$$y_0'' + \frac{2}{x} y_0' = 0 \qquad (12.219)$$

$$y_0(1) = 0 \qquad y_0(\infty) = 1 \qquad (12.220)$$

Order ϵ

$$y_1'' + \frac{2}{x} y_1' = -y_0 y_0' \qquad (12.221)$$

$$y_1(1) = 0 \qquad y_1(\infty) = 0 \qquad (12.222)$$

Multiplying (12.219) by x^2, we rewrite it as

$$x^2 y_0'' + 2x y_0' = \frac{d}{dx}\left(x^2 \frac{dy_0}{dx}\right) = 0$$

which can be integrated once to yield

$$x^2 \frac{dy_0}{dx} = a_0$$

Then,

$$\frac{dy_0}{dx} = \frac{a_0}{x^2}$$

which upon integration gives

$$y_0 = -\frac{a_0}{x} + b_0 \qquad (12.223)$$

Imposing the boundary conditions (12.220), we have

$$-a_0 + b_0 = 0 \quad \text{and} \quad b_0 = 1$$

Hence, $a_0 = b_0 = 1$ so that

$$y_0 = 1 - \frac{1}{x} \qquad (12.224)$$

which is uniform because it satisfies the reduced differential equation and boundary conditions.

Substituting for y_0 in (12.221), we obtain

$$y_1'' + \frac{2}{x} y_1' = -\left(1 - \frac{1}{x}\right) \cdot \frac{1}{x^2} = -\frac{1}{x^2} + \frac{1}{x^3}$$

which upon multiplication by x^2 becomes

$$x^2 y_1'' + 2x y_1' = -1 + \frac{1}{x}$$

or

$$\frac{d}{dx}\left(x^2 \frac{dy_1}{dx}\right) = -1 + \frac{1}{x}$$

Integrating once yields

$$x^2 \frac{dy_1}{dx} = -x + \ln x + a_1$$

which can be rewritten as

$$\frac{dy_1}{dx} = -\frac{1}{x} + \frac{\ln x}{x^2} + \frac{a_1}{x^2}$$

Integrating again gives

$$y_1 = -\ln x - \frac{\ln x}{x} - \frac{1}{x} - \frac{a_1}{x} + b_1 \quad (12.225)$$

Using the boundary condition $y_1(1) = 0$, we have $b_1 = a_1 + 1$ so that

$$y_1 = -\ln x - \frac{\ln x}{x} + b_1 \left(1 - \frac{1}{x}\right) \quad (12.226)$$

As $x \to \infty$, $y_1 \to -\ln x$, and hence, it cannot satisfy the second boundary condition $y_1(\infty) = 0$ in (12.222). Consequently, the straightforward expansion, which we will call the outer expansion and denote by the superscript o, is

$$y^o = 1 - \frac{1}{x} + \epsilon\left[-\left(1 + \frac{1}{x}\right) \ln x + b_1 \left(1 - \frac{1}{x}\right)\right] + \cdots \quad (12.227)$$

is not valid for large x.

To investigate the behavior of the solution for large x, we introduce the contraction transformation

$$\xi = x \epsilon^\nu \quad \nu > 0 \quad (12.228)$$

so that

$$\frac{d}{dx} = \frac{d}{d\xi}\frac{d\xi}{dx} = \epsilon^\nu \frac{d}{d\xi}$$

$$\frac{d^2}{dx^2} = \epsilon^{2\nu}\frac{d^2}{d\xi^2}$$

Then, (12.216) becomes

$$\epsilon^{2\nu}\frac{d^2 y^i}{d\xi^2} + \frac{2\epsilon^{2\nu}}{\xi}\frac{dy^i}{d\xi} + \epsilon^{1+\nu} y^i \frac{dy^i}{d\xi} = 0$$

whose distinguished limit is

$$\frac{d^2 y^i}{d\xi^2} + \frac{2}{\xi}\frac{dy^i}{d\xi} + y^i \frac{dy^i}{d\xi} = 0 \qquad (12.229)$$

corresponding to $\nu = 1$. The resulting expansion we call the inner expansion and we have indicated it by the superscript i.

Instead of assuming the form of the expansion for y^i, finding and solving the equations for the different orders, and then matching with the outer expansion, one can use the outer expansion as a guide for determining the form of the inner expansion. This is very convenient in the case of partial-differential and differential-integral equations, especially nonlinear equations. Moreover, in this case, the first term given by (12.224) is valid everywhere, and hence, it is valid in the inner region. Thus, expressing it in terms of the inner variable, we have

$$y_0 = 1 - \frac{\epsilon}{\xi} \qquad (12.230)$$

Then, letting $\epsilon \to 0$ with ξ being kept fixed yields $y_0 = 1$, which is the first term in the inner expansion. Moreover, the form of y_0 in (12.230) suggests that y^i goes in powers of ϵ. Hence, we seek an inner expansion in the form

$$y^i = 1 + \epsilon Y_1(\xi) + \cdots \qquad (12.231)$$

Substituting (12.231) into (12.229) and equating the coefficient of ϵ to zero, we obtain

$$Y_1'' + \left(\frac{2}{\xi} + 1\right) Y_1' = 0 \qquad (12.232)$$

Using the integrating factor

$$\exp\left[\int \left(\frac{2}{\xi} + 1\right) d\xi\right] = \exp(2 \ln \xi + \xi) = \xi^2 e^\xi$$

we rewrite (12.232) as

NONLINEAR PROBLEMS 311

$$\frac{d}{d\xi}\left(\xi^2 e^\xi \frac{dY_1}{d\xi}\right) = 0$$

Hence

$$\xi^2 e^\xi \frac{dY_1}{d\xi} = c_1$$

or

$$\frac{dY_1}{d\xi} = c_1 \frac{e^{-\xi}}{\xi^2}$$

which upon integration gives

$$Y_1 = c_1 \int_\infty^\xi \frac{e^{-\tau}}{\tau^2} d\tau + d_1 \qquad (12.233)$$

where c_1 and d_1 are arbitrary constants. The inner expansion must satisfy the boundary condition at infinity, and this is the reason we set the lower limit equal to ∞ because $x = \infty$ corresponds to $\xi = \infty$. It follows from (12.217) and (12.231) that $Y_1(\infty) = 0$, and hence, it follows from (12.233) that $d_1 = 0$. Then, (12.233) can be rewritten as

$$Y_1 = c_1 \int_\infty^\xi \frac{e^{-\tau}}{\tau^2} d\tau$$

so that (12.231) becomes

$$y^i = 1 - \epsilon c_1 \int_\xi^\infty \frac{e^{-\tau}}{\tau^2} d\tau + \cdots \qquad (12.234)$$

where the limits of integration have been interchanged so that the resulting integral is positive. The constant c_1 needs to be determined by matching the inner and outer expansions.

We match the two-term outer expansion (12.227) with the two-term inner expansion (12.234) and proceed as follows:

Two-term outer expansion: $\quad y \sim 1 - \dfrac{1}{x} + \epsilon\left[-\left(1 + \dfrac{1}{x}\right)\ln x + b_1\left(1 - \dfrac{1}{x}\right)\right]$

Rewritten in inner variable: $\quad = 1 - \dfrac{\epsilon}{\xi} + \epsilon\left[-\left(1 + \dfrac{\epsilon}{\xi}\right)\ln\left(\dfrac{\xi}{\epsilon}\right) + b_1\left(1 - \dfrac{\epsilon}{\xi}\right)\right]$

Expanded for small ϵ: $\quad = 1 - \dfrac{\epsilon}{\xi} + \epsilon \ln \epsilon - \epsilon \ln \xi + \epsilon b_1 + \cdots$

312 BOUNDARY-LAYER PROBLEMS

Two-term inner expansion: $\quad = 1 - \epsilon \left(\dfrac{1}{\xi} + \ln \xi - \ln \epsilon - b_1\right)$ (12.235)

We note the presence of the logarithmic term in (12.235).

Two-term inner expansion: $\quad y \sim 1 - \epsilon c_1 \displaystyle\int_\xi^\infty \dfrac{e^{-\tau}}{\tau^2}\, d\tau$

Rewritten in outer variable: $\quad = 1 - \epsilon c_1 \displaystyle\int_{\epsilon x}^\infty \dfrac{e^{-\tau}}{\tau^2}\, d\tau$

Expanded for small ϵ: $\quad = 1 - \epsilon c_1 \left[\dfrac{1 - \epsilon x}{\epsilon x} + \gamma + \ln \epsilon x\right] + \cdots$

Two-term outer expansion: $\quad = 1 - \dfrac{c_1}{x} - \epsilon c_1(-1 + \gamma + \ln x + \ln \epsilon)$ (12.236)

where γ is Euler constant. In expanding the integral, we used integration by parts as follows:

$$\int_{\epsilon x}^\infty \dfrac{e^{-\tau}}{\tau^2}\, d\tau = -\dfrac{e^{-\tau}}{\tau}\Big|_{\epsilon x}^\infty - \int_{\epsilon x}^\infty \dfrac{e^{-\tau}}{\tau}\, d\tau$$

$$= -\left[\dfrac{e^{-\tau}}{\tau} + e^{-\tau}\ln \tau\right]_{\epsilon x}^\infty - \int_0^\infty \ln \tau\, e^{-\tau}\, d\tau + \int_0^{\epsilon x} \ln \tau\, e^{-\tau}\, d\tau$$

$$= \dfrac{e^{-\epsilon x}}{\epsilon x} + e^{-\epsilon x}\ln \epsilon x + \gamma + O(\epsilon \ln \epsilon)$$

Expressing ξ in (12.235) in terms of x and equating the result to (12.236), we have

$$1 - \dfrac{1}{x} - \epsilon \ln x + \epsilon b_1 = 1 - \dfrac{c_1}{x} - \epsilon c_1(-1 + \gamma + \ln x + \ln \epsilon) \quad (12.237)$$

Therefore, $c_1 = 1$ and $b_1 = 1 - \gamma - \ln \epsilon$ so that

$$y^o = 1 - \dfrac{1}{x} + \epsilon\left[-\left(1 + \dfrac{1}{x}\right)\ln x + (1 - \gamma - \ln \epsilon)\left(1 - \dfrac{1}{x}\right)\right] + \cdots \quad (12.238)$$

$$y^i = 1 - \epsilon \int_\xi^\infty \dfrac{e^{-\tau}}{\tau^2}\, d\tau + \cdots \quad (12.239)$$

Finally, we form a composite expansion by adding the outer expansion (12.238) and the inner expansion (12.239) and subtracting from the result their common part (12.236) with $c_1 = 1$. The result is

$$y^c = 1 - \frac{1}{x} + \epsilon \left[-\left(1 + \frac{1}{x}\right) \ln x + (1 - \gamma - \ln \epsilon)\left(1 - \frac{1}{x}\right)\right] + 1$$

$$- \epsilon \int_\xi^\infty \frac{e^{-\tau}}{\tau^2} d\tau - 1 + \frac{1}{x} + \epsilon(-1 + \gamma + \ln x + \ln \epsilon) + \cdots$$

or

$$y^c = 1 - \epsilon \int_\xi^\infty \frac{e^{-\tau}}{\tau^2} d\tau - \epsilon \left[\frac{\ln x}{x} + \frac{1 - \gamma - \ln \epsilon}{x}\right] + \cdots \quad (12.240)$$

PROBLEM 2

As a second problem, we consider after Cole (1968)

$$\epsilon y'' + yy' - y = 0 \quad 0 \leq x \leq 1 \quad (12.241)$$

$$y(0) = \alpha \quad y(1) = \beta \quad (12.242)$$

where $0 < \epsilon \ll 1$ and α and β are independent of ϵ. In this case, the small parameter multiplies the highest derivative, and hence, a boundary layer is expected. However, the location of the boundary layer depends on the sign of the coefficient y of y'. But the value of y is a function of its values α and β at the boundaries. Hence, the location of the boundary layer depends on the values of α and β, as verified below.

We seek an outer expansion in the form

$$y^o = y_0(x) + \epsilon y_1(x) + \cdots$$

Substituting for y^o in (12.241) and (12.242) and equating the coefficients of ϵ^o to zero, we obtain

$$y_0 y_0' - y_0 = 0 \quad (12.243)$$

$$y_0(0) = \alpha \quad y_0(1) = \beta \quad (12.244)$$

Equation (12.243) provides two branches for the outer expansion, namely

$$y_0 = 0 \quad (12.245)$$

and

$$y_0 = x + c_0 \quad (12.246)$$

The first branch must be discarded because it cannot satisfy general boundary conditions. Then, the second branch yields the two special outer solutions

$$y^r = x + \beta - 1 \quad (12.247)$$

$$y^l = x + \alpha \quad (12.248)$$

where y^r satisfies the right boundary condition and y^l satisfies the left boundary condition. It follows from (12.247) and (12.248) that

$$y^r(0) = \beta - 1 \quad \text{and} \quad y^l(1) = \alpha + 1$$

Hence, if $\alpha \neq \beta - 1$, y^r is not valid near $x = 0$ and y^l is not valid near $x = 1$. Hence, a boundary layer is needed. If the boundary layer is at the left end, y^l is discarded; if the boundary layer is at the right end, y^r is discarded; and if the boundary layer (shock layer) is in the interior of the interval, both y^r and y^l are needed. If $\alpha = \beta - 1$, then $y^r = y^l = x + \beta - 1$ satisfies the differential equation and boundary conditions, and hence, it is the exact solution.

When $\alpha \neq \beta - 1$, a boundary layer develops somewhere in [0, 1]. To investigate the behavior of y in the boundary layer, we introduce the stretching transformation

$$\xi = \frac{x - x_b}{\epsilon^\nu} \quad \text{or} \quad x = x_b + \epsilon^\nu \xi \quad \nu > 0 \tag{12.249}$$

where x_b is the location of the boundary layer, which is not known a priori. For linear problems, it is unnecessary to scale the dependent variable because the scale does not affect the solution. However, for a nonlinear problem, the dependent variable may need to be scaled. Thus, we put

$$y^i = Y(\xi; \epsilon) + \cdots = \frac{y(x; \epsilon)}{\epsilon^\lambda} \tag{12.250}$$

where λ is determined in the course of analysis. Substituting (12.249) and (12.250) into (12.241), we have

$$\epsilon^{1-2\nu+\lambda} \frac{d^2 y^i}{d\xi^2} + \epsilon^{2\lambda-\nu} y^i \frac{dy^i}{d\xi} - \epsilon^\lambda y^i = 0$$

or

$$\frac{d^2 y^i}{d\xi^2} + \epsilon^{\lambda+\nu-1} y^i \frac{dy^i}{d\xi} - \epsilon^{2\nu-1} y^i = 0$$

If $\lambda = 0$, the distinguished limit is

$$\frac{d^2 y^i}{d\xi^2} + y^i \frac{dy^i}{d\xi} = 0 \tag{12.251}$$

corresponding to $\nu = 1$. If $\lambda \neq 0$, the distinguished limit is

$$\frac{d^2 y^i}{d\xi^2} + y^i \frac{dy^i}{d\xi} - y^i = 0 \tag{12.252}$$

corresponding to

$$\lambda + \nu - 1 = 0 \quad \text{and} \quad 2\nu - 1 = 0$$

or

$$\nu = \tfrac{1}{2} \quad \text{and} \quad \lambda = \tfrac{1}{2}$$

It follows from (12.250) that, in the first case (i.e., $\lambda = 0$), $y = O(1)$, whereas, in the second case (i.e., $\lambda = \tfrac{1}{2}$), $y = O(\epsilon^{1/2})$. In the second case, the distinguished limit (12.252) is the same as the original differential equation. Thus, it needs to be integrated numerically and no simplification is achieved by carrying out the expansions. However, Cole (1968) pointed out that the boundary conditions are canonical so that the numerical integration can be done once for all problems. In this book, we consider only the first case in which $y = O(1)$, $\nu = 1$, and the distinguished inner limit (12.251) is simpler than the original differential equation.

To the first approximation, y^i can be replaced by Y in (12.251), which upon integration gives

$$Y' + \tfrac{1}{2} Y^2 = \tfrac{1}{2} b \quad \text{or} \quad Y' = \tfrac{1}{2}(b - Y^2) \qquad (12.253)$$

where b is a constant of integration. It must be positive; otherwise, as $\xi \to \infty$, $Y \to -\infty$, whereas as $\xi \to -\infty$, $Y \to \infty$, making y^i unmatchable with the outer expansion(s). Then, separating variables in (12.253), we have

$$\frac{2dY}{k^2 - Y^2} = d\xi \qquad (12.254)$$

where b is replaced with k^2, since it is positive. In integrating (12.254), we consider two cases: $Y^2 \leq k^2$ and $Y^2 \geq k^2$. In the first case, we assume that

$$Y = k \tanh \theta \quad \text{so that} \quad dY = k \operatorname{sech}^2 \theta \, d\theta$$

and obtain from (12.254) that

$$\frac{2k \operatorname{sech}^2 \theta \, d\theta}{k^2 - k^2 \tanh^2 \theta} = d\xi = \frac{2}{k} d\theta$$

Hence,

$$\theta = \tfrac{1}{2} k(\xi + d)$$

and

$$y^i = Y + \cdots = k \tanh \left[\tfrac{1}{2} k(\xi + d) \right] + \cdots \qquad (12.255)$$

where d is a constant of integration. In the second case, we assume that

$$Y = k \coth \theta \quad \text{so that} \quad dY = -k \operatorname{cosech}^2 \theta \, d\theta$$

and obtain from (12.254) that

$$-\frac{2k \operatorname{cosech}^2 \theta \, d\theta}{k^2 - k^2 \coth^2 \theta} = d\xi = \frac{2}{k} d\theta$$

Hence,
$$\theta = \tfrac{1}{2} k(\xi + d)$$
and
$$y^i = Y + \cdots = k \coth\left[\tfrac{1}{2} k(\xi + d)\right] + \cdots \qquad (12.256)$$

We note that k may be taken to be positive because tanh and coth are odd. The constants of integration k and d in either form of the inner expansion need to be determined from matching the inner and outer expansions. We consider the three possible locations of the boundary layer: at the left end, at the right end, and at an interior point.

When the boundary layer is at the left end, $x_b = 0$, the outer solution (12.248) must be discarded, and the proper outer solution is (12.247). Next, this outer solution is matched with one of the forms of the inner expansion and conditions are obtained for the boundary layer to be at the left end. Expressing y^r in terms of ξ and expanding the result for small ϵ, we have

$$(y^o)^i = \beta - 1 \qquad (12.257)$$

Expressing y^i in terms of x, expanding for small ϵ, and noting that x is positive, we obtain from either (12.255) or (12.256) that

$$(y^i)^o = k > 0 \qquad (12.258)$$

Hence, $k = \beta - 1$ and

$$Y = \begin{cases} (\beta - 1) \tanh\left[\tfrac{1}{2} (\beta - 1)(\xi + d)\right] & \text{if } Y \leq \beta - 1 \\ (\beta - 1) \coth\left[\tfrac{1}{2} (\beta - 1)(\xi + d)\right] & \text{if } Y \geq \beta - 1 \end{cases} \qquad (12.259)$$

Since the boundary layer is assumed to be at $x = 0$ corresponding to $\xi = 0$, it must satisfy $y(0) = \alpha$ or $Y(0) = \alpha$. Hence, it follows from (12.259) that either

$$\alpha = (\beta - 1) \tanh\left[\tfrac{1}{2} (\beta - 1)d\right] \qquad (12.260)$$

or

$$\alpha = (\beta - 1) \coth\left[\tfrac{1}{2} (\beta - 1)d\right] \qquad (12.261)$$

It follows from (12.258) and the behavior of tanh and coth in Figure 12-14 that β must be greater than 1 so that the inner solution either descends or ascends to $\beta - 1 > 0$. Moreover, if $\alpha > \beta - 1$, the inner solution is given by a coth that descends from α to $\beta - 1$. If α is less than $\beta - 1$, it must also be greater than $-(\beta - 1)$ so that the inner solution will be given by a tanh that ascends from α to $\beta - 1$. These solutions are shown in Figure 12-15.

Next, we investigate the case in which the boundary layer is at the right end. In this case, $x_b = 1$, the outer solution (12.247) must be discarded, and the proper outer solution is (12.248). To match this outer solution with one of the

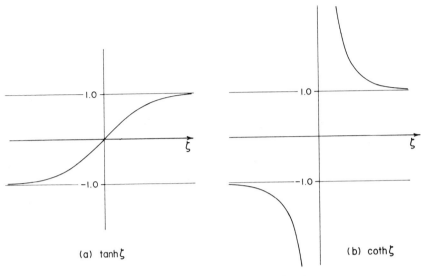

Figure 12-14. Behavior of $\tanh\zeta$ and $\coth\zeta$.

forms of the inner expansion, we express y^I in terms of $\xi = (x - 1)/\epsilon$, expand the result for small ϵ, and obtain

$$(y^o)^i = \alpha + 1 \tag{12.262}$$

Moreover, we express y^i in (12.255) and (12.256) in terms of x and obtain

$$Y = k \tanh\left[\tfrac{1}{2} k\left(\frac{x-1}{\epsilon} + d\right)\right]$$

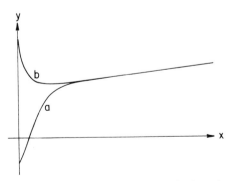

Figure 12-15. Behavior of the composite solution when the boundary layer is at $x = 0$ for $\epsilon = 0.1$, $\beta = 3$ and: (a) $\alpha = -0.5$; (b) $\alpha = 4$.

318 BOUNDARY-LAYER PROBLEMS

$$Y = k \coth \left[\tfrac{1}{2} k \left(\frac{x-1}{\epsilon} + d \right) \right]$$

Letting $\epsilon \to 0$ and noting that $x < 1$, we find that the argument tends to $-\infty$, and hence,

$$(y^i)^o = -k \qquad (12.263)$$

Equating (12.262) and (12.263) according to the matching principle, we have $k = -(\alpha + 1)$. Hence,

$$Y = \begin{cases} -(\alpha+1) \tanh\left[-\tfrac{1}{2}(\alpha+1)(\xi+d)\right] & Y \geqslant \alpha+1 \\ -(\alpha+1) \coth\left[-\tfrac{1}{2}(\alpha+1)(\xi+d)\right] & Y \leqslant \alpha+1 \end{cases} \qquad (12.264)$$

which must satisfy the boundary condition $y(1) = \beta$ or $Y(0) = \beta$ because $x = 1$ corresponds to $\xi = 0$. Then, either

$$\beta = -(\alpha+1) \tanh\left[-\tfrac{1}{2}(\alpha+1)d\right] \qquad (12.265)$$

or

$$\beta = -(\alpha+1) \coth\left[-\tfrac{1}{2}(\alpha+1)d\right] \qquad (12.266)$$

Since k is assumed to be positive, $\alpha + 1$ must be negative. When $\beta < \alpha + 1 < 0$, the inner solution is given by a coth so that it rises from β to $\alpha + 1$. When $|\beta| < |\alpha + 1|$ and $\alpha + 1 < 0$, the inner solution is given by a tanh, which descends from β to $\alpha + 1$. These solutions are shown in Figure 12-16.

Finally, we investigate the case in which the boundary layer is at an interior point. In this case, both outer solutions are needed and they must be matched with the inner solution. To match y^r of (12.247) with y^i, we express it in terms of $\xi = (x - x_b)/\epsilon$, expand the result for small ϵ, and obtain

$$(y^r)^i = x_b + \beta - 1 \qquad (12.267)$$

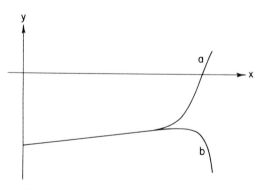

Figure 12-16. Behavior of the composite solution when the boundary layer is at $x = 1$ for $\epsilon = 0.1$, $\alpha = -3$ and: (a) $\beta = 1$; (b) $\beta = -4$.

NONLINEAR PROBLEMS 319

To match y^l of (12.248) with y^i, we express it in terms of ξ, expand the result for small ϵ, and obtain

$$(y^l)^i = x_b + \alpha \tag{12.268}$$

Since $\xi = (x - x_b)/\epsilon$ tends to $-\infty$ to the left of x_b and to $+\infty$ to the right of x_b, the inner solution must rise from $x_b + \alpha$ to $x_b + \beta - 1$. Figure 12-14 shows that only tanh rises from $-k$ at $\xi = -\infty$ to $+k$ at $\xi = \infty$. Thus, the inner solution must be given by (12.255). Letting $\xi \to \infty$, we find that

$$(y^i)^r = k \tag{12.269}$$

whereas letting $\xi \to -\infty$, we find that

$$(y^i)^l = -k \tag{12.270}$$

where the superscript r and l refer to the outer limits to the right and left of x_b, respectively. Equating (12.267) and (12.269), we have

$$x_b + \beta - 1 = k \tag{12.271}$$

whereas equating (12.268) and (12.270), we have

$$x_b + \alpha = -k \tag{12.272}$$

It follows from (12.271) and (12.272) that

$$x_b = \tfrac{1}{2}(1 - \alpha - \beta) \qquad k = \tfrac{1}{2}(\beta - 1 - \alpha) \tag{12.273}$$

and the inner solution is

$$y^i = \tfrac{1}{2}(\beta - 1 - \alpha) \tanh\left[\tfrac{1}{4}(\beta - 1 - \alpha)\xi\right] + \cdots \tag{12.274}$$

Thus, the inner solution rises from $-\tfrac{1}{2}(\beta - 1 - \alpha)$ to $\tfrac{1}{2}(\beta - 1 - \alpha)$, as shown in Figure 12-17, where $d = 0$ because the shock is assumed to be centered at $x = x_b$ or $\xi = 0$. Since $0 < x_b < 1$ and k is assumed to be positive, it follows from (12.273) that the presence of a shock layer (i.e., an interior boundary layer) demands that

$$0 < \tfrac{1}{2}(1 - \alpha - \beta) < 1 \qquad \text{and} \qquad \beta - 1 - \alpha > 0$$

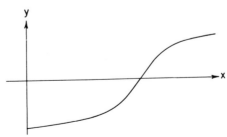

Figure 12-17. Behavior of the composite expansion when the boundary layer is at an interior point for $\alpha = -2$, $\beta = 1.8$, and $\epsilon = 0.1$.

320 BOUNDARY-LAYER PROBLEMS

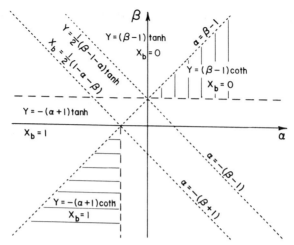

Figure 12-18. A diagram that summarizes the location of the boundary layer in the $\beta\alpha$-plane.

or

$$-1 < \alpha + \beta < 1 \quad \text{and} \quad \beta > 1 + \alpha \qquad (12.275)$$

Figure 12-18 summarizes the locations of the boundary layer in the $\beta\alpha$-plane. The unshaded area corresponds to the case in which $y = O(\epsilon^{1/2})$ and the distinguished limit is given by (12.252). Hence, the solution needs to be determined numerically. This example shows that in a nonlinear boundary-layer problem, the location of the boundary layer is strongly dependent on the end values.

Exercises

12.1. Consider the problem

$$\epsilon y'' + y' + y = 0$$
$$y(0) = \alpha \qquad y(1) = \beta$$

(a) Determine the exact solution.

(b) Use the method of matched asymptotic expansions to determine a first-order uniform expansion. Compare your answer with the exact solution.

(c) Use the method of multiple scales to determine a first-order uniform expansion. Compare your answer with those in (a) and (b).

12.2. Consider the problem

$$\epsilon y'' - y' + y = 0$$
$$y(0) = \alpha \qquad y(1) = \beta$$

(a) Determine the exact solution.

(b) Use the method of matched asymptotic expansions to determine a first-order uniform expansion. Compare your answer with the exact solution.

(c) Use the method of multiple scales to determine a first-order uniform expansion. Compare your answer with those in (a) and (b).

12.3. Consider the problem
$$\epsilon y'' - y' = 1$$
$$y(0) = \alpha \quad y(1) = \beta$$

(a) Determine the exact solution.

(b) Use the method of matched asymptotic expansions to determine a first-order uniform expansion. Compare your answer with the exact solution.

(c) Use the method of multiple scales to determine a first-order uniform expansion. Compare your answer with those in (a) and (b).

12.4. Consider the problem
$$\epsilon y'' + y' = 1$$
$$y(0) = \alpha \quad y(1) = \beta$$

(a) Determine the exact solution.

(b) Use the method of matched asymptotic expansions to determine a first-order uniform expansion. Compare your answer with the exact solution.

(c) Use the method of multiple scales to determine a first-order uniform expansion. Compare your answer with those in (a) and (b).

12.5. Determine first-order (one-term) uniform expansions for
$$\epsilon y'' \pm (3x + 1)y' = 1$$
$$y(0) = \alpha \quad y(1) = \beta$$

12.6. Determine first-order uniform expansions for
$$\epsilon y'' \pm y' = 2x$$
$$y(0) = \alpha \quad y(1) = \beta$$

In each case, compare your answer with the exact solution.

12.7. Determine first-order uniform expansions for
$$\epsilon y'' \pm (2x^2 + x + 1)y' = 4x + 1$$
$$y(0) = \alpha \quad y(1) = \beta$$

12.8. Determine first-order uniform expansions for

(a) $\epsilon y'' + xy' + xy = 0$
$y(0) = \alpha \quad y(1) = \beta$

(b) $\epsilon y'' - (1 - x)y' - (1 - x)y = 0$
$y(0) = \alpha \quad y(1) = \beta$

12.9. Determine a first-order uniform expansions for

$$\epsilon y'' + x^2 y' - x^3 y = 0$$
$$y(0) = \alpha \qquad y(1) = \beta$$

12.10. Consider the problem

$$\epsilon y'' + x^n y' - x^m y = 0$$
$$y(0) = \alpha \qquad y(1) = \beta$$

Under what conditions will there be a boundary layer at the origin? Then, determine a first-order uniform expansion when a boundary layer exists there.

12.11. Determine a first-order uniform expansion for

$$\epsilon y'' + xy' - xy = 0$$
$$y(-1) = \alpha \qquad y(1) = \beta$$

12.12. Consider the problem

$$\epsilon y''' - y' = 1$$
$$y(0) = \alpha \qquad y'(0) = \beta \qquad y(1) = \gamma$$

(a) Determine the exact solution and use it to show that there is in general a boundary layer at each end.

(b) Determine a two-term uniform expansion and compare your answer with the exact solution.

12.13. Consider the problem

$$\epsilon y''' - y' + y = 0$$
$$y(0) = \alpha \qquad y'(0) = \beta \qquad y(1) = \gamma$$

(a) Determine the exact solution and use it to show that there is in general a boundary layer at each end.

(b) Determine a two-term uniform expansion and compare your answer with the exact solution.

12.14. Consider the problem

$$\epsilon y''' - (2x + 1)y' = 1$$
$$y(0) = \alpha \qquad y'(0) = \beta \qquad y(1) = \gamma$$

Determine a two-term uniform expansion.

12.15. Consider the problem

$$\epsilon y^{iv} - y'' = 1$$
$$y(0) = \alpha \qquad y'(0) = \beta \qquad y(1) = \gamma \qquad y'(1) = \delta$$

(a) Determine the exact solution and use it to show that there is in general a boundary layer at each end.

(b) Determine a two-term uniform expansion and compare your answer with the exact solution.

12.16. Consider the problem

$$\epsilon y^{iv} - (2x + 1)y'' = 1$$

$$y(0) = \alpha \quad y'(0) = \beta \quad y(1) = \gamma \quad y'(1) = \delta$$

Determine a two-term uniform expansion.

12.17. Consider the problem

$$\epsilon^2 y'' + x^2 y' - (x^2 + \epsilon^{1/2})y = 0$$

$$y(0) = \alpha \quad y(1) = \beta$$

Show that there are two distinguished limits. Then, develop a first-order triple-deck uniform solution.

12.18. Consider the problem

$$u'' + \frac{3}{r} u' + \epsilon u u' = 0$$

$$u(1) = 0 \quad u(\infty) = 1$$

Determine a two-term uniform expansion.

12.19. Consider the problem

$$u'' + \frac{1}{r} u' + \epsilon u u' = 0$$

$$u(1) = 0 \quad u(\infty) = 1$$

Determine a one-term uniform expansion. Note that the nonuniformity appears in the first term. Answer: $u^o = [\ln(1/\epsilon)]^{-1} \ln r + \cdots$,

$$u^i = 1 + \left(\ln \frac{1}{\epsilon}\right)^{-1} \int_{\epsilon r}^{\infty} \frac{e^{-\tau}}{\tau} d\tau + \cdots, \quad u^c = u^i$$

12.20. Consider the problem

$$\tfrac{1}{2}\dot{x}^2 = \frac{1-\epsilon}{x} + \frac{\epsilon}{1-x} \quad t(0) = 0$$

Show that

$$\sqrt{2}\, t^c = \tfrac{2}{3} x^{3/2} + \epsilon[\tfrac{1}{2} - \ln 2 + \tfrac{1}{2} \ln \epsilon + \tfrac{2}{3} x^{3/2} + x^{1/2} - \ln(1 + x^{1/2})$$
$$+ \xi - \sqrt{\xi(\xi + 1)} + \sinh^{-1} \sqrt{\xi}] + \cdots$$

where $\xi = (1 - x)/\epsilon$.

12.21. Consider the problem

$$\epsilon y'' - yy' - y = 0$$
$$y(0) = \alpha \qquad y(1) = \beta$$

Determine a first-order uniform solution for the case $y = 0(1)$.

12.22. Consider the problem
$$\epsilon y'' + yy' - xy = 0$$
$$y(0) = \alpha \qquad y(1) = \beta$$

Determine a first-order uniform expansion for the case $y = 0(1)$.

12.23. Consider the problem
$$\epsilon y'' - y^2 = 0$$
$$y(0) = \alpha \qquad y(1) = \beta$$

Determine a first-order uniform expansion.

12.24. Consider the problem
$$\epsilon y'' \pm (2x + 1)y' + y^2 = 0$$
$$y(0) = \alpha \qquad y(1) = \beta$$

Determine a first-order uniform expansion.

CHAPTER 13

Linear Equations with Variable Coefficients

In this chapter, we consider linear differential equations with variable coefficients having the form

$$\mathbf{y}'(x) = F(x)\,\mathbf{y}(x) + \mathbf{h}(x) \tag{13.1}$$

where $\mathbf{y}$ and $\mathbf{h}$ are column vectors with n components and $F(x)$ is an $n \times n$ variable-coefficient matrix. We concentrate on determining the solutions of the homogeneous problem

$$\mathbf{y}'(x) = F(x)\,\mathbf{y}(x) \tag{13.2}$$

because, once these are known, we can use the method of variation of parameters to determine a particular solution and then the general solution. If $\mathbf{y}_1(x)$, $\mathbf{y}_2(x)$, $\cdots$, $\mathbf{y}_n(x)$ are n linearly independent vector solutions of (13.2), we use them to express the solution of the inhomogeneous problem (13.1) as

$$\mathbf{y}(x) = c_1(x)\,\mathbf{y}_1(x) + c_2(x)\,\mathbf{y}_2(x) + \cdots + c_n(x)\,\mathbf{y}_n(x) \tag{13.3}$$

where the $c_n(x)$ are scalar functions of x to be determined. Differentiating (13.3) with respect to x yields

$$\mathbf{y}'(x) = c_1'\mathbf{y}_1 + c_2'\mathbf{y}_2 + \cdots + c_n'\mathbf{y}_n + c_1\mathbf{y}_1' + c_2\mathbf{y}_2' + \cdots + c_n\mathbf{y}_n' \tag{13.4}$$

Substituting (13.4) into (13.1) gives

$$c_1'\mathbf{y}_1 + c_2'\mathbf{y}_2 + \cdots + c_n'\mathbf{y}_n + c_1\mathbf{y}_1' + c_2\mathbf{y}_2' + \cdots + c_n\mathbf{y}_n'$$
$$= c_1 F\mathbf{y}_1 + c_2 F\mathbf{y}_2 + \cdots + c_n F\mathbf{y}_n + \mathbf{h} \tag{13.5}$$

Since $\mathbf{y}_n' = F\mathbf{y}_n$, (13.5) simplifies to

$$c_1'\mathbf{y}_1 + c_2'\mathbf{y}_2 + \cdots + c_n'\mathbf{y}_n = \mathbf{h} \tag{13.6}$$

which is a system of n linear equations for the n scalars c_m'. If the components of $\mathbf{y}_m$ are y_{ms}, then (13.6) can be rewritten as

326 LINEAR EQUATIONS WITH VARIABLE COEFFICIENTS

$$Yc' = h \tag{13.7}$$

where

$$Y = \begin{bmatrix} y_{11} & y_{12} & \cdots & y_{1n} \\ y_{21} & y_{22} & \cdots & y_{2n} \\ \vdots & \vdots & \cdots & \vdots \\ y_{n1} & y_{n2} & \cdots & y_{nn} \end{bmatrix} \quad c' = \begin{bmatrix} c'_1 \\ c'_2 \\ \vdots \\ c'_n \end{bmatrix} \tag{13.8}$$

Since $y_1, y_2, \cdots, y_n$ are linearly independent, the matrix Y is nonsingular and has an inverse, which we denote by Y^{-1}. Multiplying (13.7) from the left by Y^{-1}, we have

$$c' = Y^{-1}h \tag{13.8}$$

which yields a system of n uncoupled equations for the c_m. Since y and h are known functions, the solution of (13.8) can be obtained in quadratures as

$$c = \int Y^{-1}(x)\, h(x)\, dx \tag{13.9}$$

Therefore, once the solutions of the homogeneous equation in (13.1) are known, a particular solution of (13.1) can be determined from (13.9). Consequently, we consider only the homogeneous solutions of (13.2). We begin with first-order scalar equations, then we take up second-order equations.

13.1. First-Order Scalar Equations

We begin with the case in which $y(x)$ and $F(x)$ are scalar functions because the exact solution of (13.2) can be expressed as

$$y = ce^{\int F(x) dx} \tag{13.10}$$

where c is a constant of integration that can be determined from the initial condition.

If $F(x)$ has a Taylor series representation convergent for $|x - x_0| < R$, we say that x_0 is an *ordinary point* of the differential equation; otherwise, we say that x_0 is a *singular point* of the differential equation. Moreover, if x is considered as a complex variable, then $F(x)$ is an *analytic function, regular* in the neighborhood $|x - x_0| < R$ of x_0.

The point x_0 if finite can always be transferred to the origin by letting

$$\xi = x - x_0 \tag{13.11}$$

so that (13.2) becomes

$$\frac{dy}{d\xi} = F(\xi + x_0)y \tag{13.12}$$

where $F(\xi + x_0)$ has a Taylor series representation convergent for $|\xi| < R$. Therefore, without loss of generality, we consider the case in which the point under consideration is the origin.

If

$$F(x) = \sum_{n=0}^{\infty} F_n x^n \qquad (13.13)$$

where the F_n are constants, is a convergent series in $|x| < R$, then (13.10) becomes

$$y = c \exp\left[\int \left(\sum_{n=0}^{\infty} F_n x^n\right) dx\right] = c \exp\left[\sum_{n=0}^{\infty} F_n \int x^n \, dx\right]$$

or

$$y = c \exp\left[\sum_{n=0}^{\infty} \frac{F_n x^{n+1}}{n+1}\right] \qquad (13.14)$$

We note that the series in the exponent is convergent because

$$\lim_{n \to \infty}\left|\frac{n\text{th term}}{(n-1)\text{th term}}\right| = \lim_{n \to \infty}\left|\frac{F_n x^{n+1} n}{(n+1)F_{n-1} x^n}\right| = x \lim_{n \to \infty}\left|\frac{F_n}{F_{n-1}}\right| = \frac{x}{R} \qquad (13.15)$$

on account of (13.13) being a convergent series of radius R. Since $\exp(z)$ can be represented by a Taylor series convergent for all z, (13.14) can be represented by a Taylor series that converges in $|x| < R$.

If $F(x)$ has an isolated singularity that is a pole of order N at $x = 0$, then it can be represented by the Laurent series

$$F(x) = \frac{1}{x^N} \sum_{n=0}^{\infty} F_n x^n \qquad (13.16)$$

where the F_n are constants and $F_0 \neq 0$. Substituting (13.16) into (13.10) yields

$$y = c \exp\left[\int \left(\sum_{n=0}^{\infty} F_n x^{n-N}\right) dx\right] = c \exp\left[\sum_{n=0}^{\infty} F_n \int x^{n-N} \, dx\right]$$

or

$$y = c \exp\left[-\frac{F_0}{(N-1)x^{N-1}} - \frac{F_1}{(N-2)x^{N-2}} - \cdots - \frac{F_{N-2}}{x} + F_{N-1} \ln x \right.$$
$$\left. + F_N x + \tfrac{1}{2} F_{N+1} x^2 + \tfrac{1}{3} F_{N+2} x^3 + \cdots \right] \qquad (13.17)$$

When $N = 1$, (13.17) becomes

$$y = c \exp\left[F_0 \ln x + F_1 x + \tfrac{1}{2} F_2 x^2 + \tfrac{1}{3} F_3 x^3 + \cdots\right]$$

which can be rewritten as

$$y = cx^{F_0} \exp [F_1 x + \tfrac{1}{2} F_2 x^2 + \tfrac{1}{3} F_3 x^3 + \cdots] \quad (13.18)$$

The exponential term can be represented in a power series in x, so that (13.18) can be expressed as

$$y = cx^{F_0} \sum_{n=0}^{\infty} a_n x^n \quad (13.19)$$

where the a_n are independent of x. When $N = 2$, (13.17) can be rewritten as

$$y = c \exp \left[-\frac{F_0}{x} + F_1 \ln x + F_2 x + \tfrac{1}{2} F_3 x^2 + \tfrac{1}{3} F_4 x^3 + \cdots \right]$$

which can be rewritten as

$$y = cx^{F_1} \exp \left(-\frac{F_0}{x} \right) \exp [F_2 x + \tfrac{1}{2} F_3 x^2 + \tfrac{1}{3} F_4 x^3 + \cdots] \quad (13.20)$$

The term $\exp(-F_0/x)$ cannot be expressed as a power series in x or x^{-1} because it tends to zero faster than any power of x as $x \to 0$ if $F_0 > 0$ and it tends to infinity faster than any power of x^{-1} as $x \to 0$ if $F_0 < 0$. The last exponential in (13.20) can be expressed as a power series in x, so that (13.20) can be rewritten as

$$y = cx^{F_1} e^{-F_0 x^{-1}} \sum_{n=0}^{\infty} a_n x^n \quad (13.21)$$

where the a_n are independent of x. Similarly, for $N \geq 2$, (13.17) can be expressed as

$$y = cx^{F_{N-1}} \exp \left[-\frac{F_0}{(N-1)x^{N-1}} - \frac{F_1}{(N-2)x^{N-2}} - \cdots - \frac{F_{N-2}}{x} \right]$$

$$\cdot \sum_{n=0}^{\infty} a_n x^n \quad (13.22)$$

where the a_n are independent of x.

Equations (13.19), (13.21), and (13.22) show that one can distinguish the pole of order one from those with order greater than one. When the pole is of order one, the form of the solution (13.19) differs from that in the case of an ordinary point by the factor x^{F_0}. When the pole is of order greater than one, the form of the solution differs from that in the case of an ordinary point by an exponential factor that cannot be expanded in powers of x in addition to the factor x^σ, where σ is a constant. When $N = 1$, the origin is called a *regular singular point*, while when $N \geq 2$, the origin is called an *irregular singular point*.

SECOND-ORDER EQUATIONS

It turns out that the above forms of the solution do not change when y is a column vector and F is an $n \times n$ matrix. Next, we consider second-order equations.

13.2. Second-Order Equations

We consider the solutions of

$$y'' + p(x)y' + q(x)y = 0 \qquad (13.23)$$

in the neighborhood of the origin. If $p(x)$ and $q(x)$ are analytic functions, regular in the neighborhood of the origin, the origin is called an *ordinary point* of the differential equation. In this case, $p(x)$ and $q(x)$ can be expanded in convergent power series in x with nonvanishing radii of convergence and y possesses convergent power-series solutions in x. When either $p(x)$ or $q(x)$ or both are singular at the origin, the origin is called a *singularity* of the differential equation. If $p(x)$ has at most a pole of order one and $q(x)$ has at most a pole of order 2 at the origin, the origin is called a *regular singular point*. In this case, at least one of the solutions of (13.23) has the form (13.19). If $p(x)$ has a pole of order higher than one or $q(x)$ has a pole of order higher than 2 or both at the origin, the origin is called an *irregular singular point* and at least one of the solutions of (13.23) has the form (13.22). Next, we show how to determine the solutions of (13.23). To minimize the algebra, we consider special cases that show the characteristics of the solutions. In this section, we consider the case when the origin is an ordinary point of the differential equation.

Let us consider

$$y'' + xy' + 2y = 0 \qquad (13.24)$$

Here, $p(x) = x$ and $q(x) = 2$, so that the origin is an ordinary point of (13.24) and y can be expanded in a Taylor series as

$$y(x) = \sum_{n=0}^{\infty} a_n x^n \qquad (13.25)$$

Differentiating (13.25) with respect to x yields

$$y' = \sum_{n=0}^{\infty} n a_n x^{n-1} \qquad (13.26)$$

Differentiating (13.26) with respect to x yields

$$y''(x) = \sum_{n=2}^{\infty} n(n-1) a_n x^{n-2} \qquad (13.27)$$

330 LINEAR EQUATIONS WITH VARIABLE COEFFICIENTS

where the terms proportional to a_0 and a_1 have disappeared. Substituting (13.25) through (13.27) into (13.24), we have

$$\sum_{n=2}^{\infty} n(n-1)a_n x^{n-2} + \sum_{n=0}^{\infty} na_n x^n + 2 \sum_{n=0}^{\infty} a_n x^n = 0 \qquad (13.28)$$

The next step involves equating coefficients of like powers of x. To accomplish it, we change the dummy summation index n so that the powers of x under the summation are the same. Thus, we let $m = n - 2$ in the first summation and $m = n$ in the second and third summations. The result is

$$\sum_{m=0}^{\infty} (m+2)(m+1)a_{m+2}x^m + \sum_{m=0}^{\infty} ma_m x^m + 2 \sum_{m=0}^{\infty} a_m x^m = 0$$

or

$$\sum_{m=0}^{\infty} [(m+1)(m+2)a_{m+2} + (m+2)a_m]x^m = 0 \qquad (13.29)$$

Equating each of the coefficients of x^m to zero, we obtain the recurrence relation

$$a_{m+2} = -\frac{a_m}{m+1} \qquad \text{for} \qquad m = 0, 1, 2, \cdots \qquad (13.30)$$

It follows from (13.30) that

$$a_2 = -a_0 \qquad a_3 = -\frac{a_1}{2} \qquad a_4 = -\frac{a_2}{3} = \frac{a_0}{3}$$

$$a_5 = -\frac{a_3}{4} = \frac{a_1}{2 \cdot 4} \qquad a_6 = -\frac{a_4}{5} = -\frac{a_0}{3 \cdot 5}$$

Thus, the odd and even coefficients are uncoupled; the even coefficients can be expressed in terms of a_0, whereas the odd coefficients can be expressed in terms of a_1. The result is two linearly independent solutions, one multiplied by a_0 and the other multiplied by a_1. Therefore,

$$y(x) = a_0 y_1(x) + a_1 y_2(x) \qquad (13.31)$$

where

$$y_1(x) = 1 - x^2 + \frac{x^4}{3} - \frac{x^6}{3 \cdot 5} + \frac{x^8}{3 \cdot 5 \cdot 7} + \cdots$$

$$= \sum_{m=0}^{\infty} \frac{(-2)^m m! x^{2m}}{(2m)!} \qquad (13.32)$$

$$y_2(x) = x - \frac{x^3}{2} + \frac{x^5}{2 \cdot 4} - \frac{x^7}{2 \cdot 4 \cdot 6} + \cdots$$

SOLUTIONS NEAR REGULAR SINGULAR POINTS

$$= x \sum_{m=0}^{\infty} \frac{(-1)^m x^{2m}}{2^m m!} = xe^{-(1/2)x^2} \qquad (13.33)$$

It is clear that the series in (13.33) converges because we are able to find its sum in closed form. Using the ratio test on (13.32), we have

$$\lim_{m \to \infty} \frac{m\text{th term}}{(m-1)\text{th term}} = \lim_{m \to \infty} \frac{(-2)^m m! x^{2m} (2m-2)!}{(2m)!(-2)^{m-1}(m-1)! x^{2m-2}}$$

$$= \lim_{m \to \infty} \frac{-2mx^2}{(2m)(2m-1)}$$

$$= -x^2 \lim_{m \to \infty} \frac{1}{2m-1} = 0$$

and the series in (13.32) converges for all values of x. Therefore, the general solution of (13.24) is given by (13.31) for all values of x.

13.3. Solutions Near Regular Singular Points

In this case, $p(x)$ or $q(x)$ or both are singular at $x = 0$ but $xp(x)$ and $x^2 q(x)$ can be represented by Taylor series in powers of x. The simplest possible equation of this kind occurs when

$$p(x) = p_0 x^{-1} \quad \text{and} \quad q(x) = q_0 x^{-2}$$

where p_0 and q_0 are constants, so that (13.23) becomes

$$y'' + \frac{p_0}{x} y' + \frac{q_0}{x^2} y = 0 \qquad (13.34)$$

It is called *Euler's equation* and its exact solution can be found easily, as shown next.

THE EULER EQUATION

Equation (13.34) can be rewritten as

$$x^2 y'' + p_0 x y' + q_0 y = 0 \qquad (13.35)$$

which belongs to the class of differential equations where in each term the power of x is the same as the order of the derivative of y. Such equations have solutions of the form

$$y = x^{\sigma} \qquad (13.36)$$

where σ is a constant called the *index* to be determined from the equation. Substituting (13.36) into (13.35) gives

332 LINEAR EQUATIONS WITH VARIABLE COEFFICIENTS

$$\sigma(\sigma - 1)x^\sigma + p_0 \sigma x^\sigma + q_0 x^\sigma = 0$$

or

$$\sigma^2 + (p_0 - 1)\sigma + q_0 = 0 \qquad (13.37)$$

which is usually called the *indicial equation*. The roots of the quadratic equation (13.37) can be easily found to be

$$\sigma_1, \sigma_2 = \tfrac{1}{2}(1 - p_0) \pm [\tfrac{1}{4}(1 - p_0)^2 - q_0]^{1/2} \qquad (13.38)$$

If σ_1 and σ_2 are different, then x^{σ_1} and x^{σ_2} are two linearly independent solutions and the general solution of (13.35) can be expressed as

$$y = c_1 x^{\sigma_1} + c_2 x^{\sigma_2} \qquad (13.39)$$

where c_1 and c_2 are arbitrary constants. If $\sigma_1 = \sigma_2 = \tfrac{1}{2}(1 - p_0)$, then $(1 - p_0)^2 = 4q_0$, and the above procedure yields only one of the two possible linearly independent solutions, namely

$$y_1(x) = x^{\sigma_1}$$

The second linearly independent solution can be determined by letting

$$y_2(x) = u(x) y_1(x)$$

where the function $u(x)$ needs to be determined. Substituting y_2 into (13.35) gives

$$x^2 y_1 u'' + 2x^2 y_1' u' + x^2 y_1'' u + p_0 x u' y_1 + p_0 x u y_1' + q_0 u y_1 = 0$$

which simplifies to

$$x^2 y_1 u'' + (2x^2 y_1' + p_0 x y_1) u' = 0 \qquad (13.40)$$

because y_1 is a solution of (13.35). Equation (13.40) is a first-order equation in u'. Thus, this method is general and can be used to determine a second linearly independent solution of a second-order equation once one of the solutions is known. Substituting for y_1 in (13.40) gives

$$x^{\sigma_1 + 2} u'' + (2\sigma_1 + p_0) x^{\sigma_1 + 1} u' = 0$$

or

$$xu'' + u' = 0$$

Hence,

$$\frac{u''}{u'} + \frac{1}{x} = 0$$

or

SOLUTIONS NEAR REGULAR SINGULAR POINTS   333

$$\ln u' + \ln x = 0$$

Thus,

$$u' = \frac{1}{x} \quad \text{then} \quad u = \ln x$$

Therefore,

$$y_2(x) = x^{\sigma_1} \ln x \tag{13.41}$$

The solution (13.41) can be obtained by using the following alternate method. Putting $y = x^\sigma$ in (13.35) leads to

$$x^2 y'' + p_0 x y' + q_0 y = [\sigma^2 + (p_0 - 1)\sigma + q_0] x^\sigma = (\sigma - \sigma_1)^2 x^\sigma \tag{13.42}$$

when $\sigma_1 = \sigma_2$. Differentiating (13.42) with respect to σ gives

$$x^2 \frac{d^2}{dx^2}\left(\frac{\partial y}{\partial \sigma}\right) + p_0 x \frac{d}{dx}\left(\frac{\partial y}{\partial \sigma}\right) + q_0 \frac{\partial y}{\partial \sigma} = 2(\sigma - \sigma_1) x^\sigma + (\sigma - \sigma_1)^2 x^\sigma \ln x \tag{13.43}$$

Setting $\sigma = \sigma_1$ in (13.42) and (13.43) shows that y and $\partial y/\partial \sigma$ satisfy the differential equation when $\sigma = \sigma_1$. Thus, the occurrence of the factor $(\sigma - \sigma_1)^2$ makes y and $\partial y/\partial \sigma$ satisfy the differential equation when $\sigma = \sigma_1$. Therefore, one of the solutions is $y_1 = x^{\sigma_1}$, whereas the second solution is

$$y_2 = \left.\frac{\partial y}{\partial \sigma}\right|_{\sigma = \sigma_1} = \left.\frac{\partial}{\partial \sigma}(x^\sigma)\right|_{\sigma = \sigma_1} = \left.\frac{\partial}{\partial \sigma}(e^{\sigma \ln x})\right|_{\sigma = \sigma_1}$$

$$= \left. e^{\sigma \ln x} \ln x \right|_{\sigma = \sigma_1} = x^{\sigma_1} \ln x$$

in agreement with (13.41).

The above example shows that the case of equal indices needs special treatment. As shown below, even the case of different indices may need special treatment if $\sigma_2 - \sigma_1$ is an integer. Next, we consider equations more general than the Euler equation beginning with the case of unequal indices differing by a noninteger.

THE CASE $\sigma_2 - \sigma_1 \neq$ INTEGER

We consider the equation

$$4xy'' + 2y' + y = 0 \tag{13.44}$$

Here, $p(x) = \frac{1}{2}x^{-1}$ and $q(x) = \frac{1}{4}x^{-1}$ so that $x = 0$ is a regular singular point. Hence, (13.44) possesses solutions of the so-called *Frobenius* (after Frobenius) *form*

334 LINEAR EQUATIONS WITH VARIABLE COEFFICIENTS

$$y = x^\sigma \sum_{n=0}^{\infty} a_n x^n = \sum_{n=0}^{\infty} a_n x^{\sigma+n} \tag{13.45}$$

Substituting y into (13.44) gives

$$4 \sum_{n=0}^{\infty} (\sigma+n)(\sigma+n-1) a_n x^{\sigma+n-1} + 2 \sum_{n=0}^{\infty} (\sigma+n) a_n x^{\sigma+n-1}$$

$$+ \sum_{n=0}^{\infty} a_n x^{\sigma+n} = 0 \tag{13.46}$$

The leading term in each of the first two summations in (13.46) is proportional to $x^{\sigma-1}$, whereas the leading term in the last summation is proportional to x^σ. Hence, the dominant term that yields the indices can be obtained by setting the coefficient of $x^{\sigma-1}$ to zero, that is,

$$4\sigma(\sigma - 1)a_0 + 2\sigma a_0 = 0 \tag{13.47}$$

Then, (13.46) becomes

$$\sum_{n=1}^{\infty} 2(\sigma+n)(2\sigma+2n-1) a_n x^{\sigma+n-1} + \sum_{n=0}^{\infty} a_n x^{\sigma+n} = 0$$

The power of x can be made the same in the two summations by putting $n - 1 = m$ in the first summation and $n = m$ in the second summation. The result is

$$\sum_{m=0}^{\infty} [2(\sigma+m+1)(2\sigma+2m+1) a_{m+1} + a_m] x^{\sigma+m} = 0$$

which, upon equating each of the coefficients of $x^{\sigma+m}$ to zero, yields the recurrence relation

$$a_{m+1} = -\frac{a_m}{2(\sigma+m+1)(2\sigma+2m+1)} \tag{13.48}$$

It follows from (13.47) that either $a_0 = 0$ (trivial solution) or

$$4\sigma^2 - 4\sigma + 2\sigma = 0$$

Thus,

$$4\sigma^2 - 2\sigma = 0 \quad \text{then} \quad \sigma = 0 \quad \text{or} \quad \tfrac{1}{2}$$

In this case, the indices are unequal and differ by $\tfrac{1}{2}$, which is not an integer. Hence, we expect to obtain two linearly independent solutions, one corresponding to each index.

Putting $\sigma = 0$ in (13.48), we have

$$a_{m+1} = -\frac{a_m}{2(m+1)(2m+1)}$$

Hence,

$$a_1 = -\frac{a_0}{2} \qquad a_2 = -\frac{a_1}{2 \cdot 2 \cdot 3} = \frac{a_0}{4!}$$

$$a_3 = -\frac{a_2}{2 \cdot 3 \cdot 5} = -\frac{a_0}{6!}$$

Therefore, when $a_0 = 1$

$$y_1(x) = 1 - \frac{x}{2!} + \frac{x^2}{4!} - \frac{x^3}{6!} + \frac{x^4}{8!} - \cdots$$

$$= \sum_{m=0}^{\infty} \frac{(-x)^m}{(2m)!} \qquad (13.49)$$

This series converges for all x because

$$\lim_{m \to \infty} \frac{m\text{th term}}{(m-1)\text{th term}} = \lim_{m \to \infty} \frac{(-x)^m (2m-2)!}{(2m)! \, (-x)^{m-1}} = 0$$

Putting $\sigma = \frac{1}{2}$ in (13.48), we have

$$a_{m+1} = -\frac{a_m}{2(m+1)(2m+3)}$$

Hence,

$$a_1 = -\frac{a_0}{2 \cdot 3} \qquad a_2 = -\frac{a_1}{2 \cdot 2 \cdot 5} = \frac{a_0}{5!}$$

$$a_3 = -\frac{a_2}{2 \cdot 3 \cdot 7} = -\frac{a_0}{7!}$$

Therefore, when $a_0 = 1$

$$y_2(x) = x^{1/2} \left(1 - \frac{x}{3!} + \frac{x^2}{5!} - \frac{x^3}{7!} + \frac{x^4}{9!} - \cdots \right)$$

$$= x^{1/2} \sum_{m=0}^{\infty} \frac{(-x)^m}{(2m+1)!} \qquad (13.50)$$

which can be shown to converge for all values of x. Thus, in this case as in the case of the Euler equation, two linearly independent solutions are obtained, one corresponding to each index.

THE CASE $\sigma_2 - \sigma_1$ = INTEGER

We consider two equations. The solutions of one of them can be obtained using the above procedure, whereas the solutions of the second are obtained by using a modification of the above procedure. We begin with the equation

$$(1 - x^2)y'' + 2xy' + y = 0 \tag{13.51}$$

Proceeding as before, we assume that $y(x)$ has the form in (13.45) and find from (13.51) that

$$(1 - x^2) \sum_{n=0}^{\infty} (\sigma + n)(\sigma + n - 1)a_n x^{\sigma+n-2}$$

$$+ 2 \sum_{n=0}^{\infty} (\sigma + n)a_n x^{\sigma+n} + \sum_{n=0}^{\infty} a_n x^{\sigma+n} = 0$$

or

$$\sum_{n=0}^{\infty} (\sigma + n)(\sigma + n - 1)a_n x^{\sigma+n-2} + \sum_{n=0}^{\infty} [1 + 3(\sigma + n) - (\sigma + n)^2]a_n x^{\sigma+n} = 0$$

$$\tag{13.52}$$

The dominant terms in (13.52) are the coefficients of $x^{\sigma-2}$ and $x^{\sigma-1}$, corresponding to $n = 0$ and 1 in the first summation. Setting these coefficients equal to zero yields

$$\sigma(\sigma - 1)a_0 = 0 \tag{13.53}$$

$$(\sigma + 1)\sigma a_1 = 0 \tag{13.54}$$

Then, (13.52) becomes

$$\sum_{n=2}^{\infty} (\sigma + n)(\sigma + n - 1)a_n x^{\sigma+n-2} + \sum_{n=0}^{\infty} [1 + 3(\sigma + n) - (\sigma + n)^2]a_n x^{\sigma+n} = 0$$

As before, we make the powers of x the same in these summations. To this end, we put $n - 2 = m$ in the first summation and $n = m$ in the second summation and obtain

$$\sum_{m=0}^{\infty} \{(\sigma + m + 2)(\sigma + m + 1)a_{m+2} + [1 + 3(\sigma + m) - (\sigma + m)^2]a_m\}x^{\sigma+m} = 0$$

Then, equating the coefficient of each power of x to zero, we obtain

$$a_{m+2} = -\frac{1 + 3(\sigma + m) - (\sigma + m)^2}{(\sigma + m + 2)(\sigma + m + 1)} a_m \tag{13.55}$$

It follows from (13.53) that either $\sigma = 0$ or 1 if $a_0 \neq 0$. When $\sigma = 0$, (13.54) is automatically satisfied and a_1 is arbitrary. When $\sigma = 1$, (13.54) demands that $a_1 = 0$.

When $\sigma = 0$, (13.55) reduces to

$$a_{m+2} = -\frac{1 + 3m - m^2}{(m+1)(m+2)} a_m$$

Then,

$$a_2 = -\frac{a_0}{2} \quad a_4 = -\frac{a_2}{4} = \frac{a_0}{8} \quad a_6 = \frac{a_4}{10} = \frac{a_0}{80}$$

$$a_3 = -\frac{a_1}{2} \quad a_5 = -\frac{a_3}{20} = \frac{a_1}{40} \quad a_7 = \frac{3a_5}{14} = \frac{3a_1}{560}$$

Hence,

$$y(x) = a_0 \left[1 - \frac{x^2}{2} + \frac{x^4}{8} + \frac{x^6}{80} + \cdots \right] + a_1 \left[x - \frac{x^3}{2} + \frac{x^5}{40} + \frac{3x^7}{560} + \cdots \right] \quad (13.56)$$

which contains two arbitrary constants, so that it must be the complete general solution with the series in brackets constituting two linearly independent solutions of (13.51). One can easily show, using the ratio test, that both series in (13.56) converge for $|x| < 1$.

When $\sigma = 1$, it follows from (13.54) that $a_1 = 0$ and from (13.55) that

$$a_{m+2} = -\frac{1 + 3(m+1) - (m+1)^2}{(m+2)(m+3)} a_m$$

Then,

$$a_3 = a_5 = a_7 = \cdots = a_{2n+1} = 0$$

$$a_2 = -\frac{a_0}{2} \quad a_4 = -\frac{a_2}{20} = \frac{a_0}{40} \quad a_6 = \frac{3a_4}{14} = \frac{3a_0}{560}$$

Hence,

$$y = a_0 x \left[1 - \frac{x^2}{2} + \frac{x^4}{40} + \frac{3x^6}{560} + \cdots \right] \quad (13.57)$$

which is a constant multiple of the second solution in (13.56). As mentioned before, (13.56) is the general solution of (13.51).

Next, we consider a case that needs a modification of the above procedure. Thus, we consider Bessel's equation of order unity

$$x^2 y'' + xy' + (x^2 - 1)y = 0 \quad (13.58)$$

As before, substituting (13.45) into (13.58) gives

338 LINEAR EQUATIONS WITH VARIABLE COEFFICIENTS

$$\sum_{n=0}^{\infty} (\sigma+n)(\sigma+n-1)a_n x^{\sigma+n} + \sum_{n=0}^{\infty} (\sigma+n)a_n x^{\sigma+n}$$

$$+ (x^2-1) \sum_{n=0}^{\infty} a_n x^{\sigma+n} = 0$$

which can be rewritten as

$$\sum_{n=0}^{\infty} [(\sigma+n)^2 - 1]a_n x^{\sigma+n} + \sum_{n=0}^{\infty} a_n x^{\sigma+n+2} = 0$$

or

$$(\sigma^2 - 1)a_0 x^{\sigma} + \sigma(\sigma+2)a_1 x^{\sigma+1} + \sum_{n=2}^{\infty} [(\sigma+n)^2 - 1]a_n x^{\sigma+n}$$

$$+ \sum_{n=0}^{\infty} a_n x^{\sigma+n+2} = 0 \quad (13.59)$$

Setting the coefficients of the dominant terms x^{σ} and $x^{\sigma+1}$ in (13.59) equal to zero, we have

$$(\sigma^2 - 1)a_0 = (\sigma-1)(\sigma+1)a_0 = 0 \quad (13.60a)$$

$$[(\sigma+1)^2 - 1]a_1 = \sigma(\sigma+2)a_1 = 0 \quad (13.60b)$$

Then, (13.59) becomes

$$\sum_{n=2}^{\infty} [(\sigma+n)^2 - 1]a_n x^{\sigma+n} + \sum_{n=0}^{\infty} a_n x^{\sigma+n+2} = 0 \quad (13.61)$$

The powers of x in these summations can be made identical by letting $n-2 = m$ in the first summation and $n = m$ in the second summation. The result is

$$\sum_{m=0}^{\infty} \{[(\sigma+m+2)^2 - 1]a_{m+2} + a_m\} x^{\sigma+m+2} = 0$$

Equating the coefficient of each of the powers of x to zero leads to

$$a_{m+2} = -\frac{a_m}{(\sigma+m+2)^2 - 1} = -\frac{a_m}{(\sigma+m+1)(\sigma+m+3)} \quad (13.62)$$

It follows from (13.60a) that either $\sigma = 1$ or -1 if $a_0 \neq 0$. Then, it follows from (13.60b) that $a_1 = 0$, and hence, $a_3 = a_5 = a_7 = \cdots = a_{2n+1} = 0$. Putting $\sigma = -1$ in (13.62) yields

$$a_{m+2} = -\frac{a_m}{m(m+2)} \quad (13.63)$$

SOLUTIONS NEAR REGULAR SINGULAR POINTS

Letting $m = 0$ in (13.63), we find that a_2 is infinite. Thus, the above procedure does not yield a solution when $\sigma = -1$ without modification.

When $\sigma = 1$, it follows from (13.62) that

$$a_{m+2} = -\frac{a_m}{(m+2)(m+4)}$$

Hence,

$$a_2 = -\frac{a_0}{2 \cdot 4} \qquad a_4 = -\frac{a_2}{4 \cdot 6} = \frac{a_0}{2 \cdot 4^2 \cdot 6} \qquad a_6 = -\frac{a_4}{6 \cdot 8} = -\frac{a_0}{2 \cdot 4^2 \cdot 6^2 \cdot 8}$$

Thus,

$$y_1(x) = a_0 x \left[1 - \frac{x^2}{2 \cdot 4} + \frac{x^4}{2 \cdot 4^2 \cdot 6} - \frac{x^6}{2 \cdot 4^2 \cdot 6^2 \cdot 8} + \cdots \right] \qquad (13.64)$$

which is one of the solutions of (13.58), and it is usually denoted by $J_1(x)$ when $a_0 = \frac{1}{2}$.

To determine a second linearly independent solution for (13.58), we may follow one of the procedures used to determine the second solution of the Euler equation in the case of equal indices. Here, we use the second procedure and solve for the a_m in terms of a_0 without substituting for σ. From (13.62),

$$a_2 = -\frac{a_0}{(\sigma+1)(\sigma+3)} \qquad a_4 = -\frac{a_2}{(\sigma+3)(\sigma+5)} = \frac{a_0}{(\sigma+1)(\sigma+3)^2(\sigma+5)}$$

$$a_6 = -\frac{a_4}{(\sigma+5)(\sigma+7)} = -\frac{a_0}{(\sigma+1)(\sigma+3)^2(\sigma+5)^2(\sigma+7)}$$

Hence,

$$y(x) = x^\sigma \left[a_0 - \frac{a_0 x^2}{(\sigma+1)(\sigma+3)} + \frac{a_0 x^4}{(\sigma+1)(\sigma+3)^2(\sigma+5)} \right.$$
$$\left. - \frac{a_0 x^6}{(\sigma+1)(\sigma+3)^2(\sigma+5)^2(\sigma+7)} + \cdots \right] \qquad (13.65)$$

As before, if we set $\sigma = -1$ in (13.65), the coefficients become infinite, owing to the factor $\sigma + 1$ in the denominators. To circumvent this difficulty, we replace a_0 with $b(\sigma + 1)$ and obtain from (13.65) that

$$y = bx^\sigma \left[(\sigma+1) - \frac{x^2}{\sigma+3} + \frac{x^4}{(\sigma+3)^2(\sigma+5)} - \frac{x^6}{(\sigma+3)^2(\sigma+5)^2(\sigma+7)} + \cdots \right] \qquad (13.66)$$

which, when substituted into the differential equation (13.58), yields

340 LINEAR EQUATIONS WITH VARIABLE COEFFICIENTS

$$x^2 y'' + xy' + (x^2 - 1)y = b(\sigma + 1)^2 (\sigma - 1)x^\sigma \tag{13.67}$$

Alternatively, one can arrive at (13.67) from (13.59) by putting $a_0 = (\sigma + 1)b$, noting that $a_1 = 0$, and using (13.61). As in the case of the Euler equation with equal indices, the occurrence of the squared factor $(\sigma + 1)^2$ on the right-hand side of (13.67) makes y and $\partial y / \partial \sigma$ satisfy the differential equation when $\sigma = -1$. Setting $\sigma = -1$ in (13.66) and putting $b = 1$, we obtain

$$y_1(x) = x^{-1} \left[-\frac{x^2}{2} + \frac{x^4}{2^2 \cdot 4} - \frac{x^6}{2^2 \cdot 4^2 \cdot 6} + \cdots \right] \tag{13.68a}$$

which is a multiple of (13.64). Differentiating (13.66) with respect to σ and setting $b = 1$ yields

$$y_2 = \frac{\partial v}{\partial \sigma} = \ln x \, y_1 + x^\sigma \left\{ 1 + \frac{x^2}{(\sigma + 3)^2} - \left[\frac{2}{(\sigma + 3)^3 (\sigma + 5)} \right. \right.$$

$$\left. \left. + \frac{1}{(\sigma + 3)^2 (\sigma + 5)^2} \right] x^4 + \cdots \right\}$$

which, upon putting $\sigma = -1$, becomes

$$y_2(x) = y_1(x) \ln x + x^{-1} \left[1 + \frac{x^2}{2^2} - \frac{5x^4}{2^2 \cdot 4^2} + \cdots \right] \tag{13.68b}$$

Therefore, (13.64) and (13.68b) are two linearly independent solutions of (13.58). Using the ratio test, one can easily show that the series in y_1 and y_2 converge for all values of x.

THE CASE $\sigma_2 = \sigma_1$

As an example, we consider Bessel's equation of order zero

$$xy'' + y' + xy = 0 \tag{13.69}$$

Again, substituting (13.45) into (13.69) leads to

$$\sum_{n=0}^{\infty} (\sigma + n)(\sigma + n - 1)a_n x^{\sigma + n - 1} + \sum_{n=0}^{\infty} (\sigma + n)a_n x^{\sigma + n - 1}$$

$$+ \sum_{n=0}^{\infty} a_n x^{\sigma + n + 1} = 0$$

or

$$\sum_{n=0}^{\infty} (\sigma + n)^2 a_n x^{\sigma + n - 1} + \sum_{n=0}^{\infty} a_n x^{\sigma + n + 1} = 0$$

or

$$\sigma^2 a_0 x^{\sigma-1} + (\sigma+1)^2 a_1 x^\sigma + \sum_{n=2}^{\infty}(\sigma+n)^2 a_n x^{\sigma+n-1} + \sum_{n=0}^{\infty} a_n x^{\sigma+n+1} = 0 \tag{13.70}$$

Setting the coefficients of the first two dominant terms $x^{\sigma-1}$ and x^σ equal to zero, we have

$$\sigma^2 a_0 = 0 \tag{13.71a}$$

$$(\sigma+1)^2 a_1 = 0 \tag{13.71b}$$

Then, (13.70) becomes

$$\sum_{n=2}^{\infty}(\sigma+n)^2 a_n x^{\sigma+n-1} + \sum_{n=0}^{\infty} a_n x^{\sigma+n+1} = 0 \tag{13.72}$$

which, upon putting $n-2=m$ in the first summation and $n=m$ in the second summation, can be rewritten as

$$\sum_{m=0}^{\infty} [(\sigma+m+2)^2 a_{m+2} + a_m] x^{\sigma+m+1} = 0$$

Equating the coefficient of each power of x to zero yields the recurrence relation

$$a_{m+2} = -\frac{a_m}{(\sigma+m+2)^2} \tag{13.73}$$

It follows from (13.71a) that the two indices are equal to zero if $a_0 \neq 0$ and it follows from (13.71b) that $a_1 = 0$. Hence, only one solution of (13.69) can be obtained by the straightforward Frobenius procedure. Thus, we first determine y as a function of σ. To this end, we find from (13.73) that

$$a_3 = a_5 = a_7 = \cdots = a_{2n+1} = 0$$

and

$$a_2 = -\frac{a_0}{(\sigma+2)^2} \quad a_4 = -\frac{a_2}{(\sigma+4)^2} = \frac{a_0}{(\sigma+2)^2(\sigma+4)^2}$$

Hence,

$$y = a_0 x^\sigma \left[1 - \frac{x^2}{(\sigma+2)^2} + \frac{x^4}{(\sigma+2)^2(\sigma+4)^2} - \frac{x^6}{(\sigma+2)^2(\sigma+4)^2(\sigma+6)^2} + \cdots \right] \tag{13.74}$$

Substituting (13.74) into (13.69) yields

$$xy'' + y' + xy = a_0 \sigma^2 x^{\sigma-1} \tag{13.75}$$

342 LINEAR EQUATIONS WITH VARIABLE COEFFICIENTS

Alternatively, one can arrive at (13.75) from (13.70) by noting that $a_1 = 0$ and using (13.72). As before, the occurrence of the squared factor σ^2 in (13.75) makes y and $\partial y/\partial \sigma$ satisfy the differential equation when $\sigma = 0$.

Putting $\sigma = 0$ and setting $a_0 = 1$ in (13.74) leads to the following solution:

$$y_1 = J_0(x) = 1 - \frac{x^2}{2^2} + \frac{x^4}{2^2 \cdot 4^2} - \frac{x^6}{2^2 \cdot 4^2 \cdot 6^2} + \cdots \quad (13.76)$$

Differentiating (13.74) with respect to σ and setting $a_0 = 1$, we obtain

$$\frac{\partial y}{\partial \sigma} = y \ln x + x^\sigma \left\{ \frac{2x^2}{(\sigma+2)^3} - \left[\frac{2}{(\sigma+2)^3 (\sigma+4)^2} + \frac{2}{(\sigma+2)^2 (\sigma+4)^3} \right] x^4 + \cdots \right\}$$

which, upon setting $\sigma = 0$, yields the second solution

$$y_2 = y_1 \ln x + \frac{x^2}{4} - \frac{3x^4}{128} + \cdots \quad (13.77)$$

Using the ratio test, one can easily show that the series in y_1 and y_2 converge for all values of x.

13.4. Singularity at Infinity

In Sections 13.2 and 13.3, we obtained series solutions for differential equations in the neighborhood of a finite point. In this section, we consider solutions in the neighborhood of infinity. To determine whether infinity is an ordinary or a singular point and the type of the singularity, one usually transforms infinity into the origin by letting $z = x^{-1}$. Then,

$$\frac{d}{dx} = \frac{d}{dz}\frac{dz}{dx} = -\frac{1}{x^2}\frac{d}{dz} = -z^2 \frac{d}{dz}$$

$$\frac{d^2}{dx^2} = \frac{2}{x^3}\frac{d}{dz} + \frac{1}{x^4}\frac{d^2}{dz^2} = 2z^3 \frac{d}{dz} + z^4 \frac{d^2}{dz^2}$$

Then, (13.23) becomes

$$2z^3 \frac{dy}{dz} + z^4 \frac{d^2 y}{dz^2} - z^2 p\left(\frac{1}{z}\right) \frac{dy}{dz} + q\left(\frac{1}{z}\right) y = 0$$

or

$$\frac{d^2 y}{dz^2} + \left[\frac{2}{z} - \frac{1}{z^2}p\left(\frac{1}{z}\right)\right] \frac{dy}{dz} + \frac{1}{z^4} q\left(\frac{1}{z}\right) y = 0 \quad (13.78)$$

Therefore, infinity is an ordinary point of (13.23) if $z = 0$ is an ordinary point of (13.78). Thus, infinity is an ordinary point of (13.23) if

$$\frac{2}{z} - \frac{1}{z^2} p\left(\frac{1}{z}\right) \quad \text{and} \quad \frac{1}{z^4} q\left(\frac{1}{z}\right)$$

are analytic functions, regular in the neighborhood of $z = 0$; otherwise, infinity is a singular point of the original equation. These conditions correspond to

$$2x - x^2 p(x) \quad \text{and} \quad x^4 q(x)$$

being expandable in convergent Taylor series in inverse powers of x^{-1} (i.e., Laurent series).

Consider for example

$$y'' + \left(\frac{2}{x} + \frac{3}{x^2}\right) y' + \left(\frac{1}{x^4} + \frac{2}{x^5}\right) y = 0 \qquad (13.79)$$

Under the transformation $z = x^{-1}$, it becomes

$$\frac{d^2 y}{dz^2} - 3 \frac{dy}{dz} + (2z + 1) y = 0 \qquad (13.80)$$

Since $z = 0$ is an ordinary point of (13.80), infinity is an ordinary point of (13.79). Therefore, the solutions of (13.79) can be expressed in the form

$$y = \sum_{n=0}^{\infty} \frac{a_n}{x^n} \qquad (13.81)$$

where the a_n are constants and the series converges.

As a second example, consider

$$y'' + \left(\frac{1}{x} + \frac{3}{x^2}\right) y' + \left(\frac{1}{x^3} + \frac{2}{x^4}\right) y = 0 \qquad (13.82)$$

Under the transformation $z = x^{-1}$, it becomes

$$\frac{d^2 y}{dz^2} + \left(\frac{1}{z} - 3\right) \frac{dy}{dz} + \left(\frac{1}{z} + 2\right) y = 0 \qquad (13.83)$$

Since $z = 0$ is a regular singular point of (13.83), infinity is a regular singular point of (13.82). Therefore, at least one of the solutions of (13.82) can be expressed in the form

$$y = x^\sigma \sum_{n=0}^{\infty} \frac{a_n}{x^n} \qquad (13.84)$$

where the a_n are constants and the series converges. The second solution has the form (13.84) or it can be determined from (13.84) as in the preceding section.

As a third example, we consider Bessel's equation of order zero

$$xy'' + y' + xy = 0 \qquad (13.85)$$

344 LINEAR EQUATIONS WITH VARIABLE COEFFICIENTS

Under the transformation $z = x^{-1}$, it becomes

$$\frac{d^2y}{dz^2} + \frac{1}{z}\frac{dy}{dz} + \frac{1}{z^4}y = 0 \tag{13.86}$$

Since $z = 0$ is an irregular singular point of (13.86), infinity is an irregular singular point of (13.85). Approximations to the solutions of second-order differential equations in the neighborhood of an irregular singular point are discussed in the next section.

13.5. Solutions Near an Irregular Singular Point

As discussed in Section 13.1, if $x = 0$ is an irregular singular point of (13.23), its solutions have the form

$$y(x) = e^{\Lambda(x)} x^{\sigma} u(x) \tag{13.87}$$

where $u(x)$ can be expressed as a power series in $x^{m/n}$, which need not be convergent, and $\Lambda(x)$ is a polynomial in $x^{-m/n}$, where m and n are prime integers. When $n = 1$, the solution is called a *normal solution,* whereas when $n \neq 1$ the solution is called a *subnormal solution.* In this section, we consider three examples—one having a normal solution, one having a subnormal solution, and one having an irregular singular point at infinity.

EXAMPLE 1
As a first example, we consider the following equation, which possesses a normal solution near the origin:

$$x^2 y'' - (1 - 3x) y' + y = 0 \tag{13.88}$$

The origin is an irregular singular point. To determine the form of $\Lambda(x)$, we determine first its leading term. Assuming the leading term in $\Lambda(x)$ to be $\lambda x^{-\nu}$, we let

$$y \sim e^{\lambda x^{-\nu}} \tag{13.89}$$

Then,

$$y' \sim -\frac{\lambda \nu}{x^{\nu+1}} e^{\lambda x^{-\nu}} \qquad y'' \sim \left[\frac{\lambda^2 \nu^2}{x^{2\nu+2}} + \frac{\lambda \nu(\nu+1)}{x^{\nu+2}}\right] e^{\lambda x^{-\nu}} \tag{13.90}$$

Substituting for y, y', and y'' in (13.88) and dividing by x^2, we obtain

$$\frac{\lambda^2 \nu^2}{x^{2\nu+2}} + \frac{\lambda \nu(\nu+1)}{x^{\nu+2}} + \frac{\lambda \nu}{x^{\nu+3}} - \frac{3\lambda \nu}{x^{\nu+2}} + \frac{1}{x^2} + \cdots = 0 \tag{13.91}$$

Since $x^{-2\nu-2} \gg x^{-\nu-2}$ and $x^{-\nu-3} \gg x^{-\nu-2}$ as $x \to 0$, extracting the dominant terms in (13.91) leads to

SOLUTIONS NEAR AN IRREGULAR SINGULAR POINT

$$\frac{\lambda^2 \nu^2}{x^{2\nu+2}} + \frac{\lambda \nu}{x^{\nu+3}} + \frac{1}{x^2} + \cdots = 0 \tag{13.92}$$

Next, we need to equate the coefficients of the powers of x. The result depends on the value of ν. In this case, it appears that there are three choices:

$$2\nu + 2 = \nu + 3 \quad \text{or} \quad 2\nu + 2 = 2 \quad \text{or} \quad \nu + 3 = 2$$

Thus, it appears that

$$\nu = 1 \quad \text{or} \quad 0 \quad \text{or} \quad -1$$

However, only the largest value should be chosen; otherwise, the dominant term in (13.92) will not be balanced. For example, if we choose $\nu = 0$, then the first term is $O(x^{-2})$, the second term is $O(x^{-3})$, and the third term is $O(x^{-2})$. Consequently, one cannot choose λ to eliminate the dominant term, which in this case is the second term. The choice $\nu = -1$ would make $\Lambda(x) \to 0$ as $x \to 0$, and hence, make $\exp[\Lambda(x)]$ expressible in powers of x, which can be absorbed into $u(x)$.

With $\nu = 1$, the first two terms in (13.92) are the dominant terms. They can be balanced by setting

$$\lambda^2 \nu^2 + \lambda \nu = 0 \quad \text{or} \quad \lambda^2 + \lambda = 0$$

so that either $\lambda = 0$ or -1.

When $\lambda = 0$, we substitute

$$y(x) = \sum_{n=0}^{\infty} a_n x^{\sigma+n} \tag{13.93}$$

in (13.88) and obtain

$$\sum_{n=0}^{\infty} (\sigma+n)(\sigma+n-1) a_n x^{\sigma+n} - (1-3x) \sum_{n=0}^{\infty} (\sigma+n) a_n x^{\sigma+n-1}$$
$$+ \sum_{n=0}^{\infty} a_n x^{\sigma+n} = 0$$

or

$$\sum_{n=0}^{\infty} (\sigma+n+1)^2 a_n x^{\sigma+n} - \sum_{n=0}^{\infty} (\sigma+n) a_n x^{\sigma+n-1} = 0 \tag{13.94}$$

The dominant term is proportional to $x^{\sigma-1}$, which corresponds to $n = 0$ in the second summation. Setting the coefficient of $x^{\sigma-1}$ equal to zero, we have

$$\sigma a_0 = 0 \tag{13.95}$$

Then, (13.94) becomes

$$\sum_{n=0}^{\infty} (\sigma + n + 1)^2 a_n x^{\sigma+n} - \sum_{n=1}^{\infty} (\sigma + n) a_n x^{\sigma+n-1} = 0$$

which, upon putting $n = m$ in the first summation and $m = n - 1$ in the second summation, becomes

$$\sum_{m=0}^{\infty} [(\sigma + m + 1)^2 a_m - (\sigma + m + 1) a_{m+1}] x^{\sigma+m} = 0$$

Equating the coefficient of each power of x to zero, we obtain

$$a_{m+1} = (\sigma + m + 1) a_m \tag{13.96}$$

It follows from (13.95) that $\sigma = 0$ if $a_0 \neq 0$. Then, it follows from (13.96) that

$$a_{m+1} = (m + 1) a_m$$

Hence,

$$a_1 = a_0 \quad a_2 = 2a_1 = 2a_0 \quad a_3 = 3a_2 = 3!a_0,$$
$$a_4 = 4a_3 = 4!a_0 \quad a_5 = 5a_4 = 5!a_0$$

or

$$y_1(x) = a_0 [1 + x + 2!x^2 + 3!x^3 + 4!x^4 + 5!x^5 + \cdots]$$

Thus,

$$y_1(x) \sim a_0 \sum_{m=0}^{\infty} m! x^m \tag{13.97}$$

Using the ratio test, one can easily show that the series in (13.97) diverges for all values of x, and hence, an asymptotic sign instead of an equality sign was used. The divergence of the series is not surprising because the origin is an irregular singular point.

When $\lambda = -1$, we put

$$y(x) = v(x) \exp\left(-\frac{1}{x}\right) \tag{13.98}$$

in (13.88) so that

$$y' = \left(v' + \frac{1}{x^2} v\right) \exp\left(-\frac{1}{x}\right)$$

$$y'' = \left(v'' + \frac{2}{x^2} v' - \frac{2}{x^3} v + \frac{1}{x^4} v\right) \exp\left(-\frac{1}{x}\right)$$

SOLUTIONS NEAR AN IRREGULAR SINGULAR POINT

Substituting for y, y', and y'' in (13.88), we have

$$x^2 v'' + 2v' - \frac{2}{x}v + \frac{1}{x^2}v - v' - \frac{1}{x^2}v + 3xv' + \frac{3}{x}v + v = 0$$

or

$$x^2 v'' + (1 + 3x)v' + \left(\frac{1}{x} + 1\right)v = 0 \qquad (13.99)$$

Since the exponential part $\exp[\Lambda(x)]$ was taken out of $y(x)$, the solution for $v(x)$ is sought in the Frobenius form

$$v(x) = \sum_{n=0}^{\infty} a_n x^{\sigma+n} \qquad (13.100)$$

Substituting (13.100) into (13.99) yields

$$\sum_{n=0}^{\infty} (\sigma+n)(\sigma+n-1)a_n x^{\sigma+n} + \sum_{n=0}^{\infty} (\sigma+n)a_n x^{\sigma+n-1}$$

$$+ 3 \sum_{n=0}^{\infty} (\sigma+n)a_n x^{\sigma+n} + \sum_{n=0}^{\infty} a_n x^{\sigma+n-1} + \sum_{n=0}^{\infty} a_n x^{\sigma+n} = 0$$

or

$$\sum_{n=0}^{\infty} (\sigma+n+1)a_n x^{\sigma+n-1} + \sum_{n=0}^{\infty} (\sigma+n+1)^2 a_n x^{\sigma+n} = 0 \qquad (13.101)$$

Setting the leading term in (13.101) equal to zero yields

$$(\sigma+1)a_0 = 0 \qquad (13.102)$$

Then, (13.101) becomes

$$\sum_{n=1}^{\infty} (\sigma+n+1)a_n x^{\sigma+n-1} + \sum_{n=0}^{\infty} (\sigma+n+1)^2 a_n x^{\sigma+n} = 0$$

which, upon putting $n - 1 = m$ in the first summation and $n = m$ in the second summation, becomes

$$\sum_{m=0}^{\infty} [(\sigma+m+2)a_{m+1} + (\sigma+m+1)^2 a_m] x^{\sigma+m} = 0$$

Equating the coefficient of each power of x to zero, we have

$$a_{m+1} = -\frac{(\sigma+m+1)^2}{\sigma+m+2} a_m \qquad (13.103)$$

It follows from (13.102) that $\sigma = -1$ when $a_0 \neq 0$, and then it follows from (13.103) that

$$a_{m+1} = -\frac{m^2}{m+1} a_m$$

Hence, $a_1 = a_2 = a_3 = \cdots = a_m = 0$ for $m \neq 0$, and the series (13.100) terminates. Therefore, when $a_0 = 1$,

$$y_2(x) = \frac{1}{x} \exp\left(-\frac{1}{x}\right) \tag{13.104}$$

Since the exponent in (13.104) involves inverse powers of x, $y_2(x)$ is a normal solution.

EXAMPLE 2

As a second example, we consider the following equation that possesses subnormal solutions near the origin:

$$x^3 y'' - x^2 y' - y = 0 \tag{13.105}$$

Since the origin is an irregular singular point, the solutions of (13.105) have the form (13.87). To determine the form of $\Lambda(x)$, we assume that the leading term has the form (13.89). Substituting (13.89) and (13.90) into (13.105) leads to

$$\frac{\lambda^2 \nu^2}{x^{2\nu-1}} + \frac{\lambda\nu(\nu+1)}{x^{\nu-1}} + \frac{\lambda\nu}{x^{\nu-1}} - 1 + \cdots = 0 \tag{13.106}$$

whose dominant terms are the first term and perhaps the last. Balancing these terms demands that $\nu = \frac{1}{2}$ and $\lambda^2 \nu^2 = 1$. Hence, $\lambda = \pm 2$. Setting $\nu = \frac{1}{2}$ in (13.106) confirms our stipulation that the first and last terms are the dominant terms.

With the above choice, y has the form

$$y = e^{\pm 2x^{-1/2}} v(x) \tag{13.107}$$

Since the exponent in (13.107) involves fractional inverse powers of x, the solution is called a subnormal solution. It follows from (13.107) that

$$y' = (v' \mp x^{-3/2} v) \exp(\pm 2x^{-1/2})$$
$$y'' = (v'' \mp 2x^{-3/2} v' + x^{-3} v \pm \tfrac{3}{2} x^{-5/2} v) \exp(\pm 2x^{-1/2})$$

Substituting for y, y', and y'' in (13.105), we obtain

$$x^3 v'' \mp 2x^{3/2} v' + v \pm \tfrac{3}{2} x^{1/2} v - x^2 v' \pm x^{1/2} v - v = 0$$

or

$$x^3 v'' - (x^2 \pm 2x^{3/2}) v' \pm \tfrac{5}{2} x^{1/2} v = 0 \tag{13.108}$$

Next, we seek the solutions of (13.108) in powers of $x^{1/2}$ as

SOLUTIONS NEAR AN IRREGULAR SINGULAR POINT 349

$$v \sim \sum_{n=0}^{\infty} a_n x^{(\sigma+n)/2} \tag{13.109}$$

which, when substituted into (13.108), gives

$$\sum_{n=0}^{\infty} \tfrac{1}{4}(\sigma+n)(\sigma+n-2)a_n x^{(\sigma+n+2)/2} - \sum_{n=0}^{\infty} \tfrac{1}{2}(\sigma+n)a_n x^{(\sigma+n+2)/2}$$

$$\mp 2 \sum_{n=0}^{\infty} \tfrac{1}{2}(\sigma+n)a_n x^{(\sigma+n+1)/2} \pm \tfrac{5}{2} \sum_{n=0}^{\infty} a_n x^{(\sigma+n+1)/2} = 0$$

or

$$\sum_{n=0}^{\infty} \tfrac{1}{4}(\sigma+n)(\sigma+n-4)a_n x^{(\sigma+n+2)/2} + \sum_{n=0}^{\infty} (\pm \tfrac{5}{2} \mp \sigma \mp n)a_n x^{(\sigma+n+1)/2} = 0$$

$$\tag{13.110}$$

The dominant term in (13.110) is the one proportional to $x^{(\sigma+1)/2}$, corresponding to $n = 0$ in the second summation. Setting this term equal to zero yields

$$(\pm \tfrac{5}{2} \mp \sigma) a_0 = 0 \tag{13.111}$$

Then, letting $m = n$ in the first summation and $n - 1 = m$ in the second summation, we rewrite (13.110) as

$$\sum_{m=0}^{\infty} \{\tfrac{1}{4}(\sigma+m)(\sigma+m-4)a_m + (\pm \tfrac{5}{2} \mp \sigma \mp m \mp 1)a_{m+1}\} x^{(\sigma+m+2)/2} = 0$$

Equating the coefficient of each power of x to zero yields

$$a_{m+1} = -\frac{(\sigma+m)(\sigma+m-4)}{4(\pm \tfrac{5}{2} \mp \sigma \mp m \mp 1)} a_m \tag{13.112}$$

It follows from (13.111) that $\sigma = \tfrac{5}{2}$ when $a_0 \neq 0$. Then, (13.112) becomes

$$a_{m+1} = \pm \frac{(m + \tfrac{5}{2})(m - \tfrac{3}{2})}{4(m+1)} a_m \tag{13.113}$$

Hence,

$$a_1 = \mp \frac{3 \cdot 5}{16} a_0 \qquad a_2 = \mp \frac{7 \cdot 1}{16 \cdot 2} a_1 = \frac{3 \cdot 5 \cdot 7}{16^2 \cdot 2} a_0$$

$$a_3 = \pm \frac{9 \cdot 1}{16 \cdot 3} a_2 = \pm \frac{3 \cdot 5 \cdot 7 \cdot 9}{16^3 \cdot 3!} a_0$$

$$a_4 = \pm \frac{11 \cdot 3}{16 \cdot 4} a_3 = \frac{3^2 \cdot 5 \cdot 7 \cdot 9 \cdot 11}{16^4 \cdot 4!} a_0$$

Hence,

$$y_1(x) \sim e^{2x^{-1/2}} x^{5/4} \left[1 - \frac{3 \cdot 5 x^{1/2}}{16} + \frac{3 \cdot 5 \cdot 7 x}{16^2 \cdot 2} + \frac{3 \cdot 5 \cdot 7 \cdot 9 x^{3/2}}{16^3 \cdot 3!} \right.$$
$$\left. + \frac{3^2 \cdot 5 \cdot 7 \cdot 9 \cdot 11 x^2}{16^4 \cdot 4!} + \cdots \right] \quad (13.114)$$

$$y_2(x) \sim e^{-2x^{-1/2}} x^{5/4} \left[1 + \frac{3 \cdot 5 x^{1/2}}{16} + \frac{3 \cdot 5 \cdot 7 x}{16^2 \cdot 2} - \frac{3 \cdot 5 \cdot 7 \cdot 9 x^{3/2}}{16^3 \cdot 3!} \right.$$
$$\left. + \frac{3^2 \cdot 5 \cdot 7 \cdot 9 \cdot 11 x^2}{16^4 \cdot 4!} + \cdots \right] \quad (13.115)$$

Using the ratio test in (13.109), we have

$$\lim_{m \to \infty} \frac{(m+1)\text{th term}}{m\text{th term}} = \lim_{m \to \infty} \frac{a_{m+1} x^{(\sigma+m+1)/2}}{a_m x^{(\sigma+m)/2}} = x^{1/2} \lim_{m \to \infty} \frac{a_{m+1}}{a_m} \to \pm\infty$$

according to (13.113). Hence, the series (13.114) and (13.115) are asymptotic and for this reason an asymptotic rather than an equality sign was used.

EXAMPLE 3

As a third example, we consider an approximation for the solutions of Bessel's equation of order zero

$$xy'' + y' + xy = 0 \quad (13.116)$$

for large x. As shown in Section 13.4, infinity is an irregular singular point. Hence, the solutions of (13.116) for large x have the form

$$y(x) = e^{\Lambda(x)} x^\sigma u(x) \quad (13.117)$$

where Λ is a polynomial in $x^{m/n}$ with m and n being prime integers and $u(x)$ can be expressed in a series in inverse powers of $x^{m/n}$.

To determine the form of Λ, we assume that the leading term has the form

$$y \sim e^{\lambda x^\nu} \quad (13.118)$$

Then,

$$y' \sim \lambda \nu x^{\nu-1} e^{\lambda x^\nu} \qquad y'' \sim [\lambda^2 \nu^2 x^{2\nu-2} + \lambda \nu (\nu-1) x^{\nu-2}] e^{\lambda x^\nu}$$

Substituting for y, y', and y'' in (13.116) leads to

$$\lambda^2 \nu^2 x^{2\nu-1} + \lambda \nu (\nu-1) x^{\nu-1} + \lambda \nu x^{\nu-1} + x + \cdots = 0 \quad (13.119)$$

As $x \to \infty$, the dominant terms are the first and last terms in (13.119), which, when balanced, yield

SOLUTIONS NEAR AN IRREGULAR SINGULAR POINT

$$\nu = 1 \quad \text{and} \quad \lambda = \pm i \tag{13.120}$$

With $\lambda = i$, we put

$$y = e^{ix}v(x)$$

in (13.116) and obtain

$$x(v'' + 2iv' - v) + v' + iv + xv = 0$$

or

$$xv'' + (2ix + 1)v' + iv = 0 \tag{13.121}$$

Next, we seek a solution for (13.121) in the form

$$v \sim \sum_{n=0}^{\infty} a_n x^{-\sigma - n} \tag{13.122}$$

and obtain

$$\sum_{n=0}^{\infty} (\sigma + n)(\sigma + n + 1) a_n x^{-\sigma - n - 1} - 2i \sum_{n=0}^{\infty} (\sigma + n) a_n x^{-\sigma - n}$$

$$- \sum_{n=0}^{\infty} (\sigma + n) a_n x^{-\sigma - n - 1} + i \sum_{n=0}^{\infty} a_n x^{-\sigma - n} = 0$$

or

$$\sum_{n=0}^{\infty} (\sigma + n)^2 a_n x^{-\sigma - n - 1} + i \sum_{n=0}^{\infty} (1 - 2\sigma - 2n) a_n x^{-\sigma - n} = 0 \tag{13.123}$$

The leading term is proportional to $x^{-\sigma}$, corresponding to $n = 0$ in the second summation, which, when set equal to zero, yields

$$(1 - 2\sigma) a_0 = 0 \tag{13.124}$$

Then, letting $m = n$ in the first summation and $n - 1 = m$ in the second summation, we rewrite (13.123) as

$$\sum_{m=0}^{\infty} [(\sigma + m)^2 a_m - i(2\sigma + 2m + 1) a_{m+1}] x^{-\sigma - m - 1} = 0$$

Equating the coefficient of each power of x to zero yields

$$a_{m+1} = -\frac{i(\sigma + m)^2}{2\sigma + 2m + 1} a_m \tag{13.125}$$

It follows from (13.124) that $\sigma = \frac{1}{2}$ when $a_0 \neq 0$, and then it follows from (13.125) that

352 LINEAR EQUATIONS WITH VARIABLE COEFFICIENTS

$$a_{m+1} = -\frac{i(m+\tfrac{1}{2})^2}{2(m+1)}a_m$$

Hence,

$$a_1 = -\frac{ia_0}{4\cdot 2} \qquad a_2 = -\frac{3^2 ia_1}{4\cdot 2^2} = -\frac{1\cdot 3^2 a_0}{4^2 \cdot 2^3}$$

$$a_3 = -\frac{5^2 ia_2}{4\cdot 3!} = \frac{1\cdot 3^2 \cdot 5^2 ia_0}{4^3 \cdot 2^3 \cdot 3!}$$

$$a_4 = -\frac{7^2 ia_3}{4\cdot 2 \cdot 4} = \frac{1\cdot 3^2 \cdot 5^2 \cdot 7^2 a_0}{4^4 \cdot 2^4 \cdot 4!}$$

Therefore, for $a_0 = 1$

$$y = e^{ix}x^{-1/2}\left[1 - \frac{1}{4\cdot 2x}i - \frac{1\cdot 3^2}{4^2 \cdot 2^2 \cdot 2!x^2} + \frac{1\cdot 3^2 \cdot 5^2}{4^3 \cdot 2^3 \cdot 3!x^3}i\right.$$
$$\left.+ \frac{1\cdot 3^2 \cdot 5^2 \cdot 7^2}{4^4 \cdot 2^4 \cdot 4!x^4} + \cdots\right] \quad \text{as} \quad x \to \infty \quad (13.126)$$

Since the ratio of two successive terms

$$-\frac{i(2m+1)^2}{8(m+1)x} \to -i\infty \quad \text{as} \quad m \to \infty$$

the series in (13.126) diverges for all values of x. However, for large values of x, it is an asymptotic expansion because the leading terms diminish very rapidly as m increases. Using $\lambda = -i$ yields the complex conjugate.

Separating (13.126) into real and imaginary parts yields the following two linearly independent solutions of (13.116):

$$y_1 \sim x^{-1/2}(u\cos x + v\sin x) \qquad (13.127)$$

$$y_2 \sim x^{-1/2}(u\sin x - v\cos x) \qquad (13.128)$$

where

$$u(x) = 1 - \frac{1^2 \cdot 3^2}{4^2 \cdot 2^2 \cdot 2!x^2} + \frac{1^2 \cdot 3^2 \cdot 5^2 \cdot 7^2}{4^4 \cdot 2^4 \cdot 4!x^4} + \cdots \qquad (13.129)$$

$$v(x) = \frac{1}{4\cdot 2x} - \frac{1^2 \cdot 3^2 \cdot 5^2}{4^3 \cdot 2^3 \cdot 3!x^3} + \cdots \qquad (13.130)$$

Therefore,

$$J_0(x) \sim Ay_1 + By_2 \quad \text{as} \quad x \to \infty \qquad (13.131)$$

where A and B are constants to be determined from the usual initial conditions

SOLUTIONS NEAR AN IRREGULAR SINGULAR POINT 353

$J_0(0) = 1$ and $J_0'(0) = 0$. However, (13.131) is not valid for small x and, in fact, it blows up as $x \to 0$. Hence, the initial conditions cannot be applied directly to the above asymptotic development. To circumvent difficulties of this type, one usually tries to determine an integral representation of the function under consideration that satisfies the initial conditions. Then, by determining the leading term in the asymptotic development of the integral, one can relate the constants in representations, such as in (13.131), to the initial conditions.

An integral representation of $J_0(x)$ can be found as follows. We expand the function $\exp(ix \sin \theta)$ in ascending powers of x and obtain

$$e^{ix \sin \theta} = 1 + \frac{ix \sin \theta}{1!} + \frac{(ix \sin \theta)^2}{2!} + \frac{(ix \sin \theta)^3}{3!} + \cdots$$

$$= \sum_{n=0}^{\infty} \frac{(ix \sin \theta)^n}{n!} \qquad (13.132)$$

Since

$$\lim_{n \to \infty} \frac{n\text{th term}}{(n-1)\text{th term}} = \lim_{n \to \infty} \frac{(ix \sin \theta)^n (n-1)!}{n!(ix \sin \theta)^{n-1}} \to 0$$

the series (13.132) converges for all values of x. Integrating both sides with respect to θ from 0 to 2π gives

$$\frac{1}{2\pi}\int_0^{2\pi} e^{ix \sin \theta}\, d\theta = \sum_{n=0}^{\infty} \frac{(ix)^n}{n!} \frac{1}{2\pi}\int_0^{2\pi} \sin^n \theta\, d\theta \qquad (13.133)$$

But, it follows from (A38) that

$$\frac{1}{2\pi}\int_0^{2\pi} \sin^n \theta\, d\theta = 0 \quad \text{if } n \text{ is odd}$$

$$= 1 \text{ if } n = 0$$

$$= \frac{(n-1)(n-3)\cdots 3 \cdot 1}{n(n-2)\cdots 4 \cdot 2} \quad \text{if } n \text{ is even}$$

Then, it follows from (13.133) that

$$\frac{1}{2\pi}\int_0^{2\pi} e^{ix \sin \theta}\, d\theta = 1 - \frac{x^2}{2^2} + \frac{x^4}{2^2 \cdot 4^2} - \frac{x^6}{2^2 \cdot 4^2 \cdot 6^2} + \cdots \qquad (13.134)$$

Since the series on the right-hand side of (13.134) is $J_0(x)$ according to (13.76),

$$J_0(x) = \frac{1}{2\pi}\int_0^{2\pi} e^{ix \sin \theta}\, d\theta \qquad (13.135)$$

354 LINEAR EQUATIONS WITH VARIABLE COEFFICIENTS

which is the desired integral representation.

Next, we determine the leading term in the asymptotic development of (13.135) for large x. To accomplish this, we use the method of stationary phase and note that the stationary points are given by $\cos\theta = 0$ or $\theta = \pm\frac{1}{2}\pi$. Since in the neighborhood of $\theta = \pm\frac{1}{2}\pi$

$$\sin\theta = \pm[1 - \tfrac{1}{2}(\theta \mp \tfrac{1}{2}\pi)^2 + \cdots] \tag{13.136}$$

the leading term in the asymptotic development of the integral is given by

$$J_0(x) \sim \frac{1}{2\pi} e^{ix} \int_{-\infty}^{\infty} e^{-(1/2)ixt^2}\, dt + \frac{1}{2\pi} e^{-ix} \int_{-\infty}^{\infty} e^{(1/2)ixt^2}\, dt \tag{13.137}$$

where $t = \theta \pm \frac{1}{2}\pi$, according to Section 3.4. The integrals in (13.137) can be transformed into Laplace integrals by rotating the contours of integration by the angles $-\frac{1}{4}\pi$ and $\frac{1}{4}\pi$, respectively. The result is

$$J_0(x) \sim \frac{1}{2\pi} e^{ix} \int_{-\infty e^{-(1/4)i\pi}}^{\infty e^{-(1/4)i\pi}} e^{-(1/2)ixt^2}\, dt + \frac{1}{2\pi} e^{-ix} \int_{-\infty e^{(1/4)i\pi}}^{\infty e^{(1/4)i\pi}} e^{(1/2)ixt^2}\, dt \tag{13.138}$$

Letting $t = \sqrt{2}\, r x^{-1/2} \exp(-\tfrac{1}{4}i\pi)$ in the first integral and $t = \sqrt{2}\, \tau x^{-1/2} \exp(\tfrac{1}{4}i\pi)$ in the second integral, we rewrite (13.138) as

$$J_0(x) \sim \frac{1}{\pi\sqrt{2x}} e^{i[x - (1/4)\pi]} \int_{-\infty}^{\infty} e^{-r^2}\, dr + \frac{1}{\pi\sqrt{2x}} e^{-i[x - (1/4)\pi]} \int_{-\infty}^{\infty} e^{-\tau^2}\, d\tau$$

or

$$J_0(x) \sim \frac{1}{\sqrt{2\pi x}} \{e^{i[x - (1/4)\pi]} + e^{-i[x - (1/4)\pi]}\}$$

on account of (3.25). Hence,

$$J_0(x) \sim \sqrt{\frac{2}{\pi x}} \cos(x - \tfrac{1}{4}\pi) \quad \text{as } x \to \infty \tag{13.139}$$

As $x \to \infty$, it follows from (13.127) through (13.130) that

$$y_1 \sim x^{-1/2} \cos x \quad \text{and} \quad y_2 \sim x^{-1/2} \sin x$$

Hence, it follows from (13.131) that

$$J_0 \sim x^{-1/2}(A \cos x + B \sin x) \tag{13.140}$$

Equating (13.139) and (13.140), we have

$$\sqrt{\frac{2}{\pi}}(\cos x \cos \tfrac{1}{4}\pi + \sin x \sin \tfrac{1}{4}\pi) = A \cos x + B \sin x$$

so that

$$A = \sqrt{\frac{2}{\pi}} \cos \tfrac{1}{4}\pi \qquad B = \sqrt{\frac{2}{\pi}} \sin \tfrac{1}{4}\pi$$

Substituting for A and B in (13.131) and using (13.127) and (13.128), we obtain

$$J_0(x) \sim \sqrt{\frac{2}{\pi x}} [u(\cos x \cos \tfrac{1}{4}\pi + \sin x \sin \tfrac{1}{4}\pi) + v(\sin x \cos \tfrac{1}{4}\pi - \cos x \sin \tfrac{1}{4}\pi)]$$

or

$$J_0(x) \sim \sqrt{\frac{2}{\pi x}} [u \cos(x - \tfrac{1}{4}\pi) + v \sin(x - \tfrac{1}{4}\pi)] \qquad \text{as} \qquad x \to \infty \qquad (13.141)$$

Equations (13.76) and (13.141) are two different representations of the same function $J_0(x)$. The first representation involves a series that converges for all values of x, whereas the second series diverges for all values of x. As mentioned in Chapter 1, although the convergent series (13.76) provides an excellent representation for small values of x, it is useless from the computational point of view for very large values of x, owing to the finite word length of modern computers. In fact, any attempt to evaluate $J_0(x)$ for large x from the convergent series using a computer fails beyond a given value of x; this value depends on the skill of the programmer. However, although the divergent series (13.141) is useless for small values of x, it is an excellent representation of $J_0(x)$ for large x. In fact, its accuracy increases with increasing x.

Exercises

13.1. Determine two linearly independent solutions of each of the following equations:

(a) $xy'' + y' = 0$
(b) $x^2 y'' - y = 0$
(c) $x^2 y'' + xy' - y = 0$
(d) $x^2 y'' + 2xy' - 4y = 0$
(e) $x^2 y'' - xy' + y = 0$

13.2. Determine three linearly independent solutions of each of the following equations:

(a) $x^2 y''' + 2xy'' - 2y' = 0$
(b) $x^3 y''' - 3xy' + 3y = 0$
(c) $x^3 y''' + 2x^2 y'' - 3xy' + 3y = 0$
(d) $x^3 y''' - 6x^2 y'' + 7xy' - 7y = 0$

356 LINEAR EQUATIONS WITH VARIABLE COEFFICIENTS

13.3. Determine two linearly independent solutions of each of the following equations near the origin and determine the radius of convergence for each series:

(a) $y'' - xy = 0$
(b) $y'' + xy' - 2y = 0$
(c) $y'' + y' - 2xy = 0$
(d) $y'' - xy' - y = 0$

13.4. Consider the Hermite equation

$$y'' - 2xy' + \gamma y = 0$$

Determine two linearly independent solutions of this equation in power series near the origin and show that one of them terminates if $\gamma = 2n$, where $n = 0, 1, 2, \cdots$.

13.5. Consider the Legendre equation

$$(1 - x^2)y'' - 2xy' + \gamma y = 0$$

Determine two linearly independent solutions of this equation in power series near the origin and show that one of them terminates if $\gamma = n(n + 1)$, $n = 0, 1, 2, \cdots$.

13.6. Determine two linearly independent solutions in power series near the origin of the Tschebycheff equation

$$(1 - x^2)y'' - xy' + \gamma y = 0$$

and show that one of them terminates if $\gamma = n^2$, $n = 0, 1, 2, \cdots$.

13.7. Determine two linearly independent solutions in power series near the origin for each of the following equations:

(a) $4xy'' + 2y' - y = 0$
(b) $(2x + x^2)y'' + y' - 6xy = 0$
(c) $9x(1 - x)y'' - 12y' + 4y = 0$
(d) $2x(1 - x)y'' + (1 - x)y' + 3y = 0$

13.8. Determine two linearly independent solutions in power series near the origin for each of the following equations:

(a) $x^2 y'' + x(x - 1)y' - xy = 0$
(b) $x^2 y'' + xy' + (x^2 - 4)y = 0$
(c) $(1 - x^2)y'' - 2xy' + 2y = 0$
(d) $x(1 - x)y'' - 3xy' - y = 0$
(e) $y'' + x^2 y = 0$
(f) $(2 + x^2)y'' + xy' + (1 + x)y = 0$

13.9. Determine two linearly independent solutions in power series near the origin for each of the following equations:

(a) $xy'' + (1 + x)y' + 2y = 0$
(b) $(x - x^2)y'' + (1 - x)y' - y = 0$

(c) $(x - x^2)y'' + (1 - 5x)y' - 4y = 0$
(d) $4(x^4 - x^2)y'' + 8x^3y' - y = 0$

13.10. Determine two linearly independent solutions in power series near the origin for each of the following equations:

(a) $x^2y'' + x^2y' - 2y = 0$
(b) $xy'' - (1 + x)y' + 2(1 - x)y = 0$

13.11. Show that the origin is an apparent singularity for the equation

$$x^2y'' - (4x + \lambda_1 x^2)y' + (4 - \lambda_2 x)y = 0$$

when (a) $\lambda_2 = -\lambda_1$, (b) $\lambda_2 = -2\lambda_1$, and (c) $\lambda_3 = -3\lambda_1$. (Hint: Show that none of the solutions are singular at $x = 0$.)

13.12. Show that

$$y_1 = x^{-1/3}\left[1 + \frac{3}{3!}x + \frac{9}{6!}x^2 + \frac{27}{9!}x^3 + \cdots\right]$$

$$y_2 = \frac{1}{1!} + \frac{3}{4!}x + \frac{9}{7!}x^2 + \frac{27}{10!}x^3 + \cdots$$

$$y_3 = x^{1/3}\left[\frac{1}{2!} + \frac{3}{5!}x + \frac{9}{8!}x^2 + \frac{27}{11!}x^3 + \cdots\right]$$

are linearly independent solutions of

$$9x^2y''' + 27xy'' + 8y' - y = 0$$

13.13. Show that

$$y_1 = 1 + x + \frac{x^2}{2^2} + \frac{x^3}{2^2 \cdot 3^2} + \cdots$$

$$y_2 = y_1 \ln x + 2\left[-x - \frac{1}{2^2}\left(1 + \frac{1}{2}\right)x^2 - \frac{1}{2^2 \cdot 3^2}\left(1 + \frac{1}{2} + \frac{1}{3}\right)x^3 + \cdots\right]$$

$$y_3 = 2y_2 \ln x - y_1 (\ln x)^2 + \left[6x + \left(\frac{6}{2^2} + \frac{8}{2^3} + \frac{6}{2^4}\right)x^2 + \cdots\right]$$

are three linearly independent solutions of

$$x^2y''' + 3xy'' + (1 - x)y' - y = 0$$

13.14. Show that

$$y = c_1 \exp\left(x + \frac{1}{x}\right) + c_2 \exp\left(-x - \frac{1}{x}\right)$$

is the general solution of

$$x^4(1 - x^2)y'' + 2x^3y' - (1 - x^2)^3 y = 0$$

358 LINEAR EQUATIONS WITH VARIABLE COEFFICIENTS

Show that both the origin and infinity are irregular singular points of this equation.

13.15. Determine two linearly independent solutions near the origin for each of the following equations:

(a) $x^3 y'' + x(1 - 2x)y' - 2y = 0$
(b) $x^4 y'' + 2x^3 y' - y = 0$
(c) $x^2 y'' + 2(1 - x)y' - y = 0$

13.16. Determine two linearly independent solutions for large x for each of the following equations:

(a) $16x^2 y'' + 32xy' - (4x + 5)y = 0$
(b) $xy'' + 2(1 - x)y' - y = 0$
(c) $4x^2 y'' + 8xy' - (4x^2 + 3)y = 0$

13.17. Consider the modified Bessel equation of zeroth order

$$xy'' + y' - xy = 0$$

(a) Show that it has the following bounded solution at the origin:

$$I_0(x) = 1 + \frac{x^2}{2^2} + \frac{x^4}{2^2 \cdot 4^2} + \frac{x^6}{2^2 \cdot 4^2 \cdot 6^2} + \cdots$$

Determine the second solution.

(b) Determine an asymptotic solution for large x; it involves an arbitrary constant.

(c) Show that

$$I_0(x) = \frac{1}{2\pi} \int_0^{2\pi} e^{x \sin \theta} \, d\theta$$

(d) Determine the asymptotic expansion for large x of the integral in (c) and use it to determine the constant in (b).

13.18. Consider Bessel's equation of order one

$$x^2 y'' + xy' + (x^2 - 1)y = 0$$

(a) Show that one of the solutions of this equation can be expressed as

$$J_1(x) = \frac{x}{2}\left[1 - \frac{x^2}{2 \cdot 4} + \frac{x^4}{2 \cdot 4^2 \cdot 6} - \cdots\right]$$

which converges for all x. Determine the second solution.

(b) Expand $\sin \theta \exp(ix \sin \theta)$ in powers of x, integrate the result with respect to θ from 0 to 2π, and obtain

$$J_1(x) = \frac{-i}{2\pi} \int_0^{2\pi} \sin \theta \, e^{ix \sin \theta} \, d\theta$$

(c) Determine the asymptotic expansion of the solution of the equation for large x; it involves two arbitrary constants.

(d) Determine the leading term in the asymptotic expansion of the integral in (b) for large x and use it to determine the constants in (c).

13.19. Determine approximations to three linearly independent solutions for large x for each of the following equations:

(a) $x^6 y''' + 6x^5 y'' - y = 0$
(b) $xy''' - (2x + 1)y'' - (1 + x)y' + (2x + 3)y = 0$

13.20. Consider Bessel's equation of order ν

$$x^2 y'' + xy' + (x^2 - \nu^2)y = 0$$

(a) Show that

$$J_\nu(x) = \sum_{n=0}^{\infty} \frac{(-1)^n (x/2)^{2n+\nu}}{n!\Gamma(\nu + n + 1)}$$

When ν is different from an integer show that

$$J_{-\nu} = \sum_{n=0}^{\infty} \frac{(-1)^n (x/2)^{2n-\nu}}{n!\Gamma(-\nu + n + 1)}$$

(b) Show that for large x

$$y \sim Ay_1 + By_2$$

where

$$y_1 \sim \frac{1}{\sqrt{x}} e^{ix} \left[1 + \frac{c_1}{x} + \frac{c_2}{x^2} + \cdots \right] \qquad y_2 = \bar{y}_1$$

(c) Use the integral representation

$$J_\nu(x) = \frac{2(\tfrac{1}{2}x)^\nu}{\sqrt{\pi}\,\Gamma(\nu + \tfrac{1}{2})} \int_0^1 (1 - t^2)^{\nu - (1/2)} \cos xt \, dt \qquad \nu > -\tfrac{1}{2}$$

and show that

$$J_\nu(x) \sim \sqrt{\frac{2}{\pi x}} \cos(x - \tfrac{1}{2}\nu\pi - \tfrac{1}{4}\pi) \qquad \text{as} \qquad x \to \infty$$

(d) Use the result in (c) to determine A and B in (b).

CHAPTER 14

Differential Equations with a Large Parameter

In this chapter, we discuss approximations to the solutions of homogeneous second-order differential equations containing a large parameter having the form

$$y'' + p(x; \lambda)y' + q(x; \lambda)y = 0 \qquad (14.1)$$

where λ is a dimensionless parameter that is much much bigger than one. Equation (14.1) can be transformed into one without the first derivative by making the substitution

$$y(x) = P(x)u(x)$$

so that

$$y' = P'u + Pu' \qquad y'' = P''u + 2P'u' + Pu''$$

Substituting for y, y', and y'' in (14.1) leads to

$$P''u + 2P'u' + Pu'' + pP'u + pPu' + qPu = 0 \qquad (14.2)$$

Setting the coefficient of u' equal to zero, we have

$$2P' + pP = 0 \qquad (14.3)$$

Then, (14.2) becomes

$$u'' + \left[\frac{P''}{P} + \frac{pP'}{P} + q\right]u = 0 \qquad (14.4)$$

Separating variables in (14.3), we have

$$\frac{dP}{P} = -\tfrac{1}{2} p\, dx$$

which, upon integration, gives

$$\ln P = -\tfrac{1}{2} \int p\, dx$$

or
$$P = \exp[-\tfrac{1}{2}\int p\,dx] \qquad (14.5)$$

Then,
$$P' = -\tfrac{1}{2}p\exp[-\tfrac{1}{2}\int p\,dx],$$
$$P'' = (-\tfrac{1}{2}p' + \tfrac{1}{4}p^2)\exp[-\tfrac{1}{2}\int p\,dx]$$

and (14.4) becomes
$$u'' + [q - \tfrac{1}{4}p^2 - \tfrac{1}{2}p']u = 0 \qquad (14.6)$$

Thus, we need only to investigate the solutions of the standard form
$$y'' + q(x;\lambda)y = 0 \qquad (14.7)$$
because equations of the form (14.1) can be transformed into this standard form. As an example, consider Bessel's equation of zeroth order
$$xy'' + y' + xy = 0 \qquad (14.8)$$
In this case, $p = x^{-1}$ and $q = 1$, and (14.6) becomes
$$u'' + \left(1 + \frac{1}{4x^2}\right)u = 0 \qquad (14.9)$$

In fact, the standard form is convenient for obtaining the asymptotic solution for large x.

In this chapter, we investigate the special class
$$y'' + [\lambda^2 q_1(x) + q_2(x)]y = 0 \qquad (14.10)$$
of (14.7), which is usually referred to as *Liouville's problem*. We start with the WKB pproximation, then the Liouville-Green transformation, and finally the Langer transformation in case of turning or transition points.

14.1. The WKB Approximation

We assume that $q_1(x)$ is differentiable and $q_2(x)$ is continuous in the interval of interest. Dividing (14.10) by λ^2, we have
$$\frac{1}{\lambda^2}y'' + q_1 y + \frac{1}{\lambda^2}q_2 y = 0 \qquad (14.11)$$

Letting $\lambda \to \infty$ in (14.11) leads to
$$q_1 y = 0 \qquad (14.12)$$
which yields the trivial solution. Therefore, we cannot determine an approximation to the solution of (14.10) by seeking solutions in the form

$$y(x) = y_0(x) + \frac{1}{\lambda} y_1(x) + \cdots \tag{14.13}$$

as we did in the preceding chapters.

To motivate the method that can be used, we consider the simple problem in which q_1 is a constant and $q_2 = 0$. Then, the solution of (14.10) can be written as

$$y = c_1 e^{i\lambda \sqrt{q_1} x} + c_2 e^{-i\lambda \sqrt{q_1} x} \tag{14.14}$$

where c_1 and c_2 are constants. When q_1 is negative, the solution of (14.10) can be written as

$$y_1 = c_1 e^{\lambda \sqrt{-q_1} x} + c_2 e^{-\lambda \sqrt{-q_1} x} \tag{14.15}$$

In either case, two linearly independent solutions of (14.10) can be expressed in exponential forms with the parameter λ appearing in the exponents. Hence, instead of seeking a straightforward expansion in inverse powers of λ, we seek an expansion for y in (14.10) in the form

$$y = e^{\lambda G(x, \lambda)} \tag{14.16}$$

where G has a straightforward expansion in inverse powers of λ. Differentiating (14.16) twice yields

$$y' = \lambda G' e^{\lambda G} \qquad y'' = (\lambda^2 G'^2 + \lambda G'') e^{\lambda G}$$

Substituting for y, y', and y'' in the linear equation (14.10), we transform it into the following nonlinear equation:

$$\lambda^2 G'^2 + \lambda G'' + \lambda^2 q_1 + q_2 = 0$$

or

$$G'^2 + q_1 + \frac{1}{\lambda} G'' + \frac{1}{\lambda^2} q_2 = 0 \tag{14.17}$$

As discussed above, we seek a straightforward expansion for G in the form

$$G(x; \lambda) = G_0(x) + \frac{1}{\lambda} G_1(x) + \cdots \tag{14.18}$$

Substituting (14.18) into (14.17), we have

$$\left(G_0' + \frac{1}{\lambda} G_1' + \cdots \right)^2 + q_1 + \frac{1}{\lambda} \left(G_0'' + \frac{1}{\lambda} G_1'' + \cdots \right) + \frac{1}{\lambda^2} q_2 = 0$$

or

$$G_0'^2 + \frac{2}{\lambda} G_0' G_1' + q_1 + \frac{1}{\lambda} G_0'' + \cdots = 0 \tag{14.19}$$

Equating the coefficients of λ^0 and λ^{-1} to zero, we obtain

$$G_0'^2 + q_1 = 0 \tag{14.20}$$

$$G_0'' + 2G_0'G_1' = 0 \tag{14.21}$$

It follows from (14.20) that $G_0'^2 = -q_1$, so that

$$G_0' = \begin{cases} \pm i\sqrt{q_1} & \text{if } q_1 > 0 \\ \pm\sqrt{-q_1} & \text{if } q_1 < 0 \end{cases} \tag{14.22}$$

Then,

$$G_0 = \begin{cases} \pm i\int\sqrt{q_1}\, dx & \text{if } q_1 > 0 \\ \pm\int\sqrt{-q_1}\, dx & \text{if } q_1 < 0 \end{cases} \tag{14.23}$$

To solve (14.21), we divide it first by $2G_0'$ and obtain

$$\frac{1}{2}\frac{G_0''}{G_0'} + G_1' = 0$$

which, upon integration, gives

$$G_1 + \tfrac{1}{2} \ln G_0' = 0$$

The constant of integration is not needed as it becomes clear below. Hence,

$$G_1 = -\ln\sqrt{G_0'} \tag{14.24}$$

Substituting for G_0 and G_1 in (14.18) gives

$$G = \pm i \int \sqrt{q_1}\, dx - \frac{1}{\lambda}[\ln\sqrt{\pm i} + \ln\sqrt[4]{q_1}] + \cdots \text{ if } q_1 > 0 \tag{14.25}$$

and

$$G = \pm \int \sqrt{-q_1}\, dx - \frac{1}{\lambda}[\ln\sqrt{\pm 1} + \ln\sqrt[4]{-q_1}] + \cdots \text{ if } q_1 < 0 \tag{14.26}$$

Substituting (14.25) into (14.16), we have

$$y = \exp\{\pm i\lambda \int\sqrt{q_1}\, dx - [\ln\sqrt{\pm i} + \ln\sqrt[4]{q_1}] + \cdots\}$$

which, upon using the fact that $\exp(-\ln z) = z^{-1}$, can be rewritten as

$$y \approx \frac{\exp[\pm i\lambda \int\sqrt{q_1}\, dx]}{\sqrt{\pm i}\sqrt[4]{q_1}} \tag{14.27}$$

Equation (14.27) provides two linearly independent approximate solutions of (14.10). Expressing the exponentials in terms of trigonometric functions, we can write an approximation to the general solution of (14.10) in the form

$$y \approx \frac{c_1 \cos \left[\lambda \int \sqrt{q_1} \, dx\right] + c_2 \sin \left[\lambda \int \sqrt{q_1} \, dx\right]}{\sqrt[4]{q_1}} \qquad (14.28)$$

where c_1 and c_2 are arbitrary constants. Substituting (14.26) into (14.16) and following steps similar to the above, we can write an approximation to the general solution of (14.10) for the case $q_1 < 0$ as

$$y \approx \frac{c_1 \exp \left[\lambda \int \sqrt{-q_1} \, dx\right] + c_2 \exp \left[-\lambda \int \sqrt{-q_1} \, dx\right]}{\sqrt[4]{-q_1}} \qquad (14.29)$$

Equations (14.28) and (14.29) are usually referred to as the *WKB approximation* after Wentzel, Kramers, and Brillouin. We note that (14.28) and (14.29) break down at or near the zeros of $q_1(x)$. These zeros are called *turning* or *transition points*. Turning point problems are discussed in Sections 14.5 and 14.6.

As an example, we consider

$$y'' + \lambda^2 (1+x)^2 y = 0 \qquad (14.30)$$

so that $q_1 = (1+x)^2 > 0$ and (14.28) yields

$$y \approx \frac{c_1 \cos \left[\lambda (x + \tfrac{1}{2} x^2)\right] + c_2 \sin \left[\lambda (x + \tfrac{1}{2} x^2)\right]}{\sqrt{1+x}} \qquad (14.31)$$

As a second example, we consider

$$y'' - \lambda^2 (1+x)^2 y = 0 \qquad (14.32)$$

so that $q_1 = -(1+x)^2$ and (14.29) yields

$$y \approx \frac{c_1 \exp \left[\lambda (x + \tfrac{1}{2} x^2)\right] + c_2 \exp \left[-\lambda (x + \tfrac{1}{2} x^2)\right]}{\sqrt{1+x}} \qquad (14.33)$$

14.2. The Liouville-Green Transformation

In this section, we consider an alternative to the derivation in the preceding section. This involves the use of the so-called *Liouville-Green transformation* to transform both the dependent and independent variables as

$$z = \phi(x) \qquad v(z) = \psi(x) y(x) \qquad (14.34)$$

where ϕ and ψ are to be chosen so that (14.10) is transformed into an equation whose dominant part has constant coefficients. It follows from (14.34) that

$$y(x) = \frac{v(z)}{\psi(x)} \qquad (14.35)$$

$$\frac{dy}{dx} = -\frac{\psi'}{\psi^2} v + \frac{1}{\psi} \frac{dv}{dx} = -\frac{\psi'}{\psi^2} v + \frac{1}{\psi} \frac{dv}{dz} \frac{dz}{dx}$$

$$= -\frac{\psi'}{\psi^2}v + \frac{\phi'}{\psi}\frac{dv}{dz} \tag{14.36}$$

$$\frac{d^2y}{dx^2} = -\frac{d}{dx}\left(\frac{\psi'}{\psi^2}\right)v - \frac{\psi'}{\psi^2}\frac{dv}{dz}\frac{dz}{dx} + \frac{d}{dx}\left(\frac{\phi'}{\psi}\right)\frac{dv}{dz} + \frac{\phi'}{\psi}\frac{d^2v}{dz^2}\frac{dz}{dx}$$

$$= \frac{\phi'^2}{\psi}\frac{d^2v}{dz^2} + \left(\frac{\phi''}{\psi} - \frac{2\phi'\psi'}{\psi^2}\right)\frac{dv}{dz} - \left(\frac{\psi''}{\psi^2} - \frac{2\psi'^2}{\psi^3}\right)v \tag{14.37}$$

Substituting for y and y'' in (14.10), we obtain

$$\frac{\phi'^2}{\psi}\frac{d^2v}{dz^2} + \left(\frac{\phi''}{\psi} - \frac{2\phi'\psi'}{\psi^2}\right)\frac{dv}{dz} - \left(\frac{\psi''}{\psi^2} - \frac{2\psi'^2}{\psi^3}\right)v + (\lambda^2 q_1 + q_2)\frac{v}{\psi} = 0$$

or

$$\frac{d^2v}{dz^2} + \frac{1}{\phi'^2}\left(\phi'' - \frac{2\phi'\psi'}{\psi}\right)\frac{dv}{dz} + \left[\frac{\lambda^2 q_1}{\phi'^2} + \frac{q_2}{\phi'^2} - \frac{\psi''}{\psi\phi'^2} + \frac{2\psi'^2}{\psi^2\phi'^2}\right]v = 0 \tag{14.38}$$

As mentioned above, the functions ϕ and ψ are to be chosen so that the dominant part of the transformed equation (14.38) has constant coefficients. To this end, we force the coefficient of dv/dz to be zero by putting

$$\phi'' - \frac{2\phi'\psi'}{\psi} = 0 \tag{14.39}$$

Equation (14.39) can be solved by separating the variables; that is,

$$\frac{\phi''}{2\phi'} = \frac{\psi'}{\psi}$$

Hence,

$$\tfrac{1}{2} \ln \phi' = \ln \psi$$

or

$$\psi = \sqrt{\phi'} \tag{14.40}$$

Then, (14.38) becomes

$$\frac{d^2v}{dz^2} + \left(\frac{\lambda^2 q_1}{\phi'^2} + \delta\right)v = 0 \tag{14.41}$$

where

$$\delta = \frac{q_2}{\phi'^2} - \frac{\psi''}{\psi\phi'^2} + \frac{2\psi'^2}{\psi^2\phi'^2} \tag{14.42}$$

Equation (14.41) has two variable-coefficient terms, namely q_1/ϕ'^2 and δ.

Since we have imposed only the condition (14.40) on ϕ and ψ, we still have the freedom of imposing another condition. We use the second condition to make the dominant term in (14.41), namely $\lambda^2 q_1/\phi'^2$, constant, thereby transforming (14.10) into an equation whose dominant part has constant coefficients. Thus, we put $\lambda^2 q_1/\phi'^2$ = constant. Without loss of generality, the constant can be taken to be 1 if $q_1 > 0$ and -1 if $q_1 < 0$. Hence,

$$\phi'^2 = \begin{cases} \lambda^2 q_1 & \text{if } q_1 > 0 \\ -\lambda^2 q_1 & \text{if } q_1 < 0 \end{cases}$$

Then,

$$z = \phi = \begin{cases} \pm \lambda \int \sqrt{q_1}\, dx & \text{if } q_1 > 0 \\ \pm \lambda \int \sqrt{-q_1}\, dx & \text{if } q_1 < 0 \end{cases} \quad (14.43)$$

and it follows from (14.40) that

$$\psi = \begin{cases} \lambda^{1/2} \sqrt[4]{q_1} & \text{if } q_1 > 0 \\ \lambda^{1/2} \sqrt[4]{-q_1} & \text{if } q_1 < 0 \end{cases} \quad (14.44)$$

With the above choice, (14.41) becomes

$$\frac{d^2 v}{dz^2} \pm v = -\delta v \quad (14.45)$$

where the positive and negative signs correspond to the cases of positive and negative q_1, respectively.

To the first approximation, we can neglect the small term $-\delta v$ on the right-hand side of (14.45) and obtain

$$\frac{d^2 v}{dz^2} \pm v \approx 0$$

whose general solution can be expressed as

$$v = c_1 \cos z + c_2 \sin z \quad \text{if } q_1 > 0 \quad (14.46)$$

$$v = c_1 e^z + c_2 e^{-z} \quad \text{if } q_1 < 0 \quad (14.47)$$

Substituting for z, ψ, and v in (14.35), we obtain to the first approximation (14.28) when $q_1 > 0$ and (14.29) when $q_1 < 0$.

14.3. Eigenvalue Problems

In this section, we use the WKB approximation to determine the eigenvalues and eigenfunctions of some second-order differential equations with variable coefficients. As an example, we consider the eigenvalue problem

$$y'' + \lambda^2 q_1(x) y = 0 \quad (14.48)$$

$$y(0) = 0 \quad y(1) = 0 \tag{14.49}$$

for large λ and $q_1 > 0$ in $[0, 1]$. Since $q_1 > 0$, the general solution of (14.48) is given to the first approximation by (14.28). Next, we need to satisfy the boundary conditions in (14.49). To facilitate the satisfaction of these conditions, we take the lower limits of the integrals in (14.28), without loss of generality, to be the left end of the interval of interest; in this case, it is $x = 0$. Thus, we rewrite (14.28) as

$$y \approx \frac{c_1 \cos\left[\lambda \int_0^x \sqrt{q_1(\tau)}\, d\tau\right] + c_2 \sin\left[\lambda \int_0^x \sqrt{q_1(\tau)}\, d\tau\right]}{\sqrt[4]{q_1(x)}} \tag{14.50}$$

where τ is a dummy variable of integration. Putting $y(0) = 0$ in (14.50) gives

$$0 \approx \frac{c_1}{\sqrt[4]{q_1(0)}} \quad \text{or} \quad c_1 = 0$$

so that (14.50) becomes

$$y \approx c_2 q_1^{-1/4} \sin\left[\lambda \int_0^x \sqrt{q_1(\tau)}\, d\tau\right]$$

Imposing the boundary condition $y(1) = 0$ yields

$$0 = c_2 [q_1(1)]^{-1/4} \sin\left[\lambda \int_0^1 \sqrt{q_1(\tau)}\, d\tau\right]$$

For a nontrivial solution, $c_2 \neq 0$, and hence,

$$\sin\left[\lambda \int_0^1 \sqrt{q_1(\tau)}\, d\tau\right] = 0$$

or

$$\lambda \int_0^1 \sqrt{q_1(\tau)}\, d\tau = n\pi \quad n = 1, 2, 3, \cdots$$

where $n = 0$ is excluded because it corresponds to the trivial solution. Hence,

$$\lambda_n = n\pi \left[\int_0^1 \sqrt{q_1(\tau)}\, d\tau\right]^{-1} \tag{14.51}$$

corresponding to the eigenfunction

$$y_n \approx q_1^{-1/4} \sin\left[\lambda_n \int_0^x \sqrt{q_1(\tau)}\,d\tau\right] \tag{14.52}$$

When $q_1(x) = (1+x)^2$

$$\lambda_n = n\pi\left[\int_0^1 (1+\tau)\,d\tau\right]^{-1} = \tfrac{2}{3} n\pi$$

Table 14-1 compares the approximate eigenvalues with those obtained by a combination of a numerical integration of the original problem and a Newton-Raphson iteration technique. The agreement is very good even for the lowest eigenvalue, which is about 2.0604, a not very large number. As expected, the accuracy of the perturbation solution improves as the eigenvalue increases.

As a second example, we consider (14.48) subject to the boundary conditions

$$y'(0) = 0 \quad y(1) = 0 \tag{14.53}$$

To impose the first boundary condition, we differentiate (14.50) with respect to x and obtain

$$y' \approx -\lambda \sqrt[4]{q_1}\left\{c_1 \sin\left[\lambda \int_0^x \sqrt{q_1(\tau)}\,d\tau\right]\right.$$

$$\left. - c_2 \cos\left[\lambda \int_0^x \sqrt{q_1(\tau)}\,d\tau\right]\right\} + O(1) \tag{14.54}$$

We note that the terms that arise from differentiating $q_1^{-1/4}$ are $O(1)$, and hence, small compared with the terms that arise from differentiating the circular functions because they are $O(\lambda)$. Imposing the condition $y'(0) = 0$ demands that $c_2 = 0$, so that (14.50) becomes

$$y \approx c_1 q_1^{-1/4} \cos\left[\lambda \int_0^x \sqrt{q_1(\tau)}\,d\tau\right]$$

Then, imposing the condition $y(1) = 0$ demands that

TABLE 14-1. Comparison of Perturbationally and Numerically Calculated Eigenvalues

Eigenvalue Number	1	2	3	4	5	6	7
Perturbation Result	2.0944	4.1888	6.2832	8.3776	10.4720	12.5664	14.6608
Numerical Result	2.0604	4.1686	6.2691	8.3668	10.4632	12.5590	14.6545
Error (%)	1.65	0.49	0.23	0.13	0.08	0.06	0.04

EQUATIONS WITH SLOWLY VARYING COEFFICIENTS 369

TABLE 14-2. Comparison of Perturbationally and Numerically Calculated Eigenvalues

Eigenvalue Number	1	2	3	4	5	6	7
Perturbation Result	1.0472	3.1416	5.2360	7.3304	9.4248	11.5192	13.6136
Numerical Result	1.1879	3.2089	5.2793	7.3621	9.4497	11.5397	13.6310
Error (%)	11.84	2.10	0.82	0.43	0.26	0.18	0.13

$$\cos\left[\lambda \int_0^1 \sqrt{q_1(\tau)}\, d\tau\right] = 0$$

or

$$\lambda \int_0^1 \sqrt{q_1(\tau)}\, d\tau = (n + \tfrac{1}{2})\pi \quad n = 0, 1, 2, \cdots$$

Hence,

$$\lambda_n = (n + \tfrac{1}{2})\pi \left[\int_0^1 \sqrt{q_1(\tau)}\, d\tau\right]^{-1} \tag{14.55}$$

corresponding to the eigenfunction

$$y_n = q_1^{-1/4} \cos\left[\lambda_n \int_0^x \sqrt{q_1(\tau)}\, d\tau\right] \tag{14.56}$$

For the case $q_1 = (1 + x)^2$,

$$\lambda_n = (n + \tfrac{1}{2})\pi \left[\int_0^1 (1 + \tau)\, d\tau\right]^{-1} = \tfrac{2}{3}(n + \tfrac{1}{2})\pi$$

Table 14-2 compares the approximate eigenvalues with those obtained by a combination of a numerical integration technique and a Newton-Raphson iteration technique. As expected, the accuracy of the approximate solution improves as the eigenvalue increases. In fact, the error in the approximate third eigenvalue is less than 1%.

14.4. Equations with Slowly Varying Coefficients

In this section, we consider equations of the form

$$\frac{d^2 y}{dt^2} + q_1(\tau)y = 0 \tag{14.57}$$

where $\tau = \epsilon t$ and ϵ is a small dimensionless parameter. Thus, τ is a slow variable compared with t. Changing the independent variable from t to τ, we find that

370 DIFFERENTIAL EQUATIONS WITH A LARGE PARAMETER

$$\frac{dy}{dt} = \frac{dy}{d\tau}\frac{d\tau}{dt} = \epsilon \frac{dy}{d\tau}$$

$$\frac{d^2y}{dt^2} = \epsilon^2 \frac{d^2y}{d\tau^2}$$

so that (14.57) becomes

$$\epsilon^2 \frac{d^2y}{d\tau^2} + q_1(\tau)y = 0 \tag{14.58}$$

Letting $\epsilon = \lambda^{-1}$, we rewrite (14.58) as

$$\frac{d^2y}{d\tau^2} + \lambda^2 q_1(\tau)y = 0 \tag{14.59}$$

which has the same form as (14.10) so that the WKB approximation in Section 14.1 holds for this case. Since

$$\lambda \int \sqrt{q_1}\, d\tau = \epsilon\lambda \int \sqrt{q_1}\, dt = \int \sqrt{q_1}\, dt$$

we rewrite (14.28) and (14.29) as

$$y \approx q_1^{-1/4}\{c_1 \cos[\int \sqrt{q_1}\, dt] + c_2 \sin[\int \sqrt{q_1}\, dt]\} \quad q_1 > 0 \tag{14.60}$$

$$y \approx (-q_1)^{-1/4}[c_1 e^{\int \sqrt{-q_1}\, dt} + c_2 e^{-\int \sqrt{-q_1}\, dt}] \quad q_1 < 0 \tag{14.61}$$

14.5 Turning-Point Problems

As discussed in Section 14.1, the WKB approximations (14.28) and (14.29) break down at or near the zeros of $q_1(x)$. These points are called *turning* or *transition points*. For example, when $q_1 = 1 - x^3$, (14.28) and (14.29) become

$$y^{(1)} \approx \frac{c_1 \cos\left[\lambda \int_x^1 \sqrt{1-\tau^3}\, d\tau\right] + c_2 \sin\left[\lambda \int_x^1 \sqrt{1-\tau^3}\, d\tau\right]}{\sqrt[4]{1-x^3}} \quad \text{if } x < 1 \tag{14.62}$$

where the upper limit is taken to be 1 so that the integrals will be positive and

$$y^{(2)} \approx \frac{a_1 \exp\left[\lambda \int_1^x \sqrt{\tau^3-1}\, d\tau\right] + a_2 \exp\left[-\lambda \int_1^x \sqrt{\tau^3-1}\, d\tau\right]}{\sqrt[4]{x^3-1}} \quad \text{if } x > 1 \tag{14.63}$$

where a_1, a_2, c_1, and c_2 are constants and the lower limit is taken to be 1, so that the integrals will be positive. In turning-point problems, it is convenient to set one of the limits in the integrals to be the location of the turning point. Thus, (14.62) and (14.63) provide two different representations of the same function $y(x)$, one valid for $x > 1$ and one valid for $x < 1$. Since our differential equation is of second order, it can only support two constants, and hence, the a_n and c_n must be related. One method of determining the relations between these constants involves determining an expansion valid near $x = 1$ (an inner expansion) and matching it to (14.62) and (14.63) separately. An alternate method involves the Langer transformation and it is discussed in the next section.

To determine an expansion valid near $x = 1$, we introduce a stretching transformation, so that the region close to $x = 1$ is magnified to $O(1)$. To this end, we let

$$\xi = (x - 1)\lambda^\nu \quad \text{so that} \quad x = 1 + \lambda^{-\nu}\xi \tag{14.64}$$

where ν must be greater than zero for (14.64) to be a stretching transformation. The value of ν will be chosen later. The original derivatives become

$$\frac{dy}{dx} = \frac{dy}{d\xi}\frac{d\xi}{dx} = \lambda^\nu \frac{dy}{d\xi}$$

$$\frac{d^2y}{dx^2} = \lambda^{2\nu} \frac{d^2y}{d\xi^2}$$

and (14.10) with $q_1 = 1 - x^3$ and $q_2 = 0$ becomes

$$\lambda^{2\nu} \frac{d^2y}{d\xi^2} + \lambda^2 [1 - (1 + \lambda^{-\nu}\xi)^3] y = 0$$

or

$$\frac{d^2y}{d\xi^2} + \lambda^{2-2\nu}(-3\lambda^{-\nu}\xi - 3\lambda^{-2\nu}\xi^2 - \lambda^{-3\nu}\xi^3) y = 0$$

or

$$\frac{d^2y}{d\xi^2} - 3\lambda^{2-3\nu}\xi(1 + \lambda^{-\nu}\xi + \tfrac{1}{3}\lambda^{-2\nu}\xi^2) y = 0 \tag{14.65}$$

As $\lambda \to \infty$, the second and third terms in the parentheses tend to zero because $\nu > 0$ and (14.65) becomes

$$\frac{d^2y}{d\xi^2} - 3\lambda^{2-3\nu}\xi y = 0 \tag{14.66}$$

As $\lambda \to \infty$, there are three possibilities. If $\nu > \tfrac{2}{3}$, $\lambda^{2-3\nu} \to 0$ and (14.66) tends to

$$\frac{d^2y}{d\xi^2} = 0 \tag{14.67}$$

If $\nu < \frac{2}{3}$, $\lambda^{2-3\nu} \to \infty$ and (14.66) tends to

$$\xi y = 0 \tag{14.68}$$

If $\nu = \frac{2}{3}$, $\lambda^{2-3\nu} = 1$ and (14.66) tends to

$$\frac{d^2y}{d\xi^2} - 3\xi y = 0 \tag{14.69}$$

The last limit is the least degenerate limit and it includes the first two cases as special cases. Moreover, the first two cases correspond to the indefinite limits $\nu > \frac{2}{3}$ and $\nu < \frac{2}{3}$, whereas the third case corresponds to the definite (distinguished) limit $\nu = \frac{2}{3}$. Therefore, the appropriate limit is the third (14.69), corresponding to $\nu = \frac{2}{3}$.

Apart from the factor 3, (14.69) is the so-called *Airy equation* whose solutions are well-known. To use these solutions, we find it convenient to modify the transformation (14.64), so that the factor 3 disappears from the limiting equation. Thus, instead of (14.64), we use

$$z = 3^{1/3}(x-1)\lambda^{2/3} \quad \text{or} \quad x = 1 + 3^{-1/3}\lambda^{-2/3}z \tag{14.70}$$

so that (14.69) is replaced with the standard form of the Airy equation

$$\frac{d^2y}{dz^2} - zy = 0 \tag{14.71}$$

The general solution of (14.71) is usually expressed as

$$y^{(3)} = b_1 Ai(z) + b_2 Bi(z) \tag{14.72}$$

where b_1 and b_2 are arbitrary constants and Ai and Bi are the Airy functions of the first and second kind, respectively. This solution needs to be matched with the WKB solutions (14.62) and (14.63). This requires the asymptotic expansions of $Ai(z)$ and $Bi(z)$ for large positive z (i.e., $x > 1$) for matching with (14.63) and for large negative z (i.e., $x < 1$) for matching with (14.62).

If we determine the asymptotic expansions from the differential equation (14.71), we find that the leading terms in y are exponentially growing and decaying for large positive z and sinusoidally varying for large negative z. In fact, using the WKB approximation (14.28) and (14.29) with $q_1 = -z$, we have

$$y \sim z^{-1/4}[\tilde{c}_1 \cos(\tfrac{2}{3}z^{3/2}) + \tilde{c}_2 \sin(\tfrac{2}{3}z^{3/2})] \quad \text{as} \quad z \to -\infty$$

and

$$y \sim z^{-1/4}[\tilde{a}_1 e^{(2/3)z^{3/2}} + \tilde{a}_2 e^{-(2/3)z^{3/2}}] \quad \text{as} \quad z \to \infty$$

But again, we do not know the relations among the $\tilde{a}_n$ and $\tilde{c}_n$ and the a_n and c_n on the one hand and the b_n on the other hand. To circumvent this difficulty, one usually represents the solutions of (14.71) in integral form, and then determines the leading asymptotic terms of these integrals as we did in Example 3 of Section 13.5.

Using a modified Laplace transform, one can obtain the integral representations of two linearly independent solutions of the Airy equations. The Airy functions Ai and Bi are usually defined as

$$Ai(z) = \frac{1}{\pi} \int_0^\infty \cos(\tfrac{1}{3} t^3 + zt) \, dt \qquad (14.73)$$

$$Bi(z) = \frac{1}{\pi} \int_0^\infty [\exp(-\tfrac{1}{3} t^3 + zt) + \sin(\tfrac{1}{3} t^3 + zt)] \, dt \qquad (14.74)$$

The leading term in the asymptotic expansion of $Ai(z)$ as $z \to \infty$ is obtained in Section 3.5 by using the saddle-point method. Thus, it follows from (3.239) that

$$Ai(z) \sim \frac{e^{-(2/3)z^{3/2}}}{2\sqrt{\pi} \, z^{1/4}} \qquad \text{as} \quad z \to \infty \qquad (14.75)$$

Using the method of stationary phase, one can obtain the leading term in the asymptotic expansion of $Ai(z)$ as $z \to -\infty$. It follows from Exercise 3.21 that

$$Ai(z) \sim \frac{1}{\sqrt{\pi} (-z)^{1/4}} \sin[\tfrac{2}{3}(-z)^{3/2} + \tfrac{1}{4}\pi] \qquad \text{as} \quad z \to -\infty \qquad (14.76)$$

Using Laplace's method and integration by parts to determine the leading terms in the integrals in (14.74), we find that (Exercise 3.31)

$$Bi(z) \sim \frac{e^{(2/3)z^{3/2}}}{\sqrt{\pi} \, z^{1/4}} \qquad \text{as} \quad z \to \infty \qquad (14.77)$$

Using Laplace's method and the method of stationary phase to determine the leading terms in the integrals in (14.74), we find that (Exercise 3.31)

$$Bi(z) \sim \frac{1}{\sqrt{\pi} (-z)^{1/4}} \cos[\tfrac{2}{3}(-z)^{3/2} + \tfrac{1}{4}\pi] \qquad \text{as} \quad z \to -\infty \qquad (14.78)$$

The asymptotic expansions (14.75) through (14.78) provide the relations connecting the $\tilde{c}_n$, $\tilde{a}_n$, and b_n. Letting $z \to \infty$ in (14.72) and using (14.75) and (14.77), we have

$$y \sim \frac{b_1 e^{-(2/3)z^{3/2}}}{2\sqrt{\pi} \, z^{1/4}} + \frac{b_2 e^{(2/3)z^{3/2}}}{\sqrt{\pi} \, z^{1/4}} \qquad (14.79)$$

whereas, letting $z \to -\infty$ in (14.72) and using (14.76) and (14.78), we have

$$y \sim \frac{b_1}{\sqrt{\pi}(-z)^{1/4}} \sin\left[\tfrac{2}{3}(-z)^{3/2} + \tfrac{1}{4}\pi\right] + \frac{b_2}{\sqrt{\pi}(-z)^{1/4}} \cos\left[\tfrac{2}{3}(-z)^{3/2} + \tfrac{1}{4}\pi\right]$$
(14.80)

To determine the relations connecting the a_n and c_n in the WKB approximations (14.62) and (14.63), we need to match these expansions with (14.72), which is valid near the turning point. To match (14.62) with (14.72), we express the former in terms of z and let $\lambda \to \infty$. In this case $x < 1$; hence, we express x in terms of z and let $\lambda \to \infty$ with z being fixed. Thus, we write

$$\lambda \int_x^1 \sqrt{1 - \tau^3}\, d\tau = \lambda \int_{1 + 3^{-1/3}\lambda^{-2/3}z}^1 \sqrt{1 - \tau^3}\, d\tau$$

which, upon putting $\tau = 1 + 3^{-1/3}\lambda^{-2/3}t$, becomes

$$\lambda \int_x^1 \sqrt{1 - \tau^3}\, d\tau = \lambda \int_z^0 [-3^{2/3}\lambda^{-2/3}t - 3^{1/3}\lambda^{-4/3}t^2 + \cdots]^{1/2} 3^{-1/3}\lambda^{-2/3}\, dt$$

$$\approx \int_z^0 (-t)^{1/2}\, dt = \tfrac{2}{3}(-z)^{3/2}$$

Then,

$$\lim_{\substack{\lambda \to \infty \\ z\ \text{fixed}}} y^{(1)} = \frac{c_1 \cos\left[\tfrac{2}{3}(-z)^{3/2}\right] + c_2 \sin\left[\tfrac{2}{3}(-z)^{3/2}\right]}{3^{1/6}\lambda^{-1/6}(-z)^{1/4}}$$
(14.81)

To apply the matching condition, we need to express $y^{(3)}$ in (14.72) in terms of x for $x < 1$ and take the limit as $\lambda \to \infty$ with x being fixed. This process is the same as taking the limit of $y^{(3)}$ as $z \to -\infty$; the result is (14.80). Equating (14.81) and (14.80) according to the matching principle, we obtain

$$\pi^{1/2}\lambda^{1/6}3^{-1/6}\{c_1 \cos\left[\tfrac{2}{3}(-z)^{3/2}\right] + c_2 \sin\left[\tfrac{2}{3}(-z)^{3/2}\right]\}$$
$$= b_1 \sin\left[\tfrac{2}{3}(-z)^{3/2} + \tfrac{1}{4}\pi\right] + b_2 \cos\left[\tfrac{2}{3}(-z)^{3/2} + \tfrac{1}{4}\pi\right]$$

Using trigonometric identities for the sum of two angles and equating the coefficients of $\cos\left[\tfrac{2}{3}(-z)^{2/3}\right]$ and $\sin\left[\tfrac{2}{3}(-z)^{2/3}\right]$ on both sides, we obtain

$$c_1 = 3^{1/6}\pi^{-1/2}\lambda^{-1/6}[b_1 \sin \tfrac{1}{4}\pi + b_2 \cos \tfrac{1}{4}\pi]$$
$$c_2 = 3^{1/6}\pi^{-1/2}\lambda^{-1/6}[b_1 \cos \tfrac{1}{4}\pi - b_2 \sin \tfrac{1}{4}\pi]$$
(14.82)

To match (14.63) with (14.72), we express x in (14.63) in terms of z and let $\lambda \to \infty$ with z being fixed but positive to correspond to $x > 1$. To this end, we write

$$\lambda \int_1^x \sqrt{\tau^3 - 1}\, d\tau = \lambda \int_1^{1+3^{-1/3}\lambda^{-2/3}z} \sqrt{\tau^3 - 1}\, d\tau$$

which, upon putting $\tau = 1 + 3^{-1/3}\lambda^{-2/3}t$, becomes

$$\lambda \int_1^x \sqrt{\tau^3 - 1}\, d\tau = \lambda \int_0^z [3^{2/3}\lambda^{-2/3}t + 3^{1/3}\lambda^{-4/3}t^2 + \cdots]^{1/2} 3^{-1/3}\lambda^{-2/3}\, dt$$

$$\approx \int_0^z t^{1/2}\, dt = \tfrac{2}{3} z^{3/2}$$

Then, it follows from (14.63) that

$$\lim_{\substack{\lambda \to \infty \\ z \text{ fixed}}} y^{(2)} = 3^{-1/6} \lambda^{1/6} z^{-1/4} [a_1 e^{(2/3)z^{3/2}} + a_2 e^{-(2/3)z^{3/2}}] \qquad (14.83)$$

Next, we need to express z in (14.72) in terms of x and take the limit as $\lambda \to \infty$ with x being fixed but larger than 1. This is equivalent to taking the limit of $y^{(3)}$ as $z \to \infty$. The result is given by (14.79). Equating (14.79) and (14.83) according to the matching principle, we obtain

$$3^{-1/6}\lambda^{1/6}\pi^{1/2}[a_1 e^{(2/3)z^{3/2}} + a_2 e^{-(2/3)z^{3/2}}] = \tfrac{1}{2} b_1 e^{-(2/3)z^{3/2}} + b_2 e^{(2/3)z^{3/2}}$$

which, upon equating the coefficients of each of the exponentials on both sides, leads to

$$a_1 = 3^{1/6}\lambda^{-1/6}\pi^{-1/2} b_2 \qquad a_2 = \tfrac{1}{2} 3^{1/6}\lambda^{-1/6}\pi^{-1/2} b_1 \qquad (14.84)$$

In summary, the solution to our problem is given by the three separate expansions (14.62) valid for $x < 1$, (14.63) valid for $x > 1$, and (14.72) valid in the neighborhood of $x = 1$. These expansions were matched using the asymptotic developments of the integral representations of the Airy functions. This matching produced the relations (14.82) and (14.84) connecting the coefficients a_n, b_n, and c_n in the three expansions. One can form a composite expansion as in Chapter 12 to provide a single uniformly valid expansion. However, the present procedure is clumsy and one can use alternatively the Langer transformation to produce a single expansion in terms of the Airy functions that is valid everywhere including the neighborhood of the turning point. This method is discussed next.

14.6. The Langer Transformation

The gist of Langer's transformation is to transform the dependent and independent variables as in (14.34) and to choose ϕ and ψ so that the dominant part of the transformed equation has the simplest possible form and, at the same time, its solutions have qualitatively the same behavior as the solutions of the original equation. For example, when $q_1 > 0$ everywhere in the interval of interest, the solutions of the original equation (14.10) are oscillatory, and hence, ϕ and ψ must be chosen so that the dominant part of the transformed equation is

$$\frac{d^2v}{dz^2} + v = 0 \qquad (14.85a)$$

which is the simplest possible equation with oscillatory solutions. When $q_1 < 0$ everywhere in the interval of interest, one of the solutions of the original equation (14.10) grows exponentially with x, whereas the other decays exponentially with x. Hence, ϕ and ψ must be chosen so that the dominant part of the transformed equation is

$$\frac{d^2v}{dz^2} - v = 0 \qquad (14.85b)$$

which is the simplest possible equation with exponentially growing and decaying solutions. However, when q_1 changes sign once in the interval of interest, such as the case $1 - x^3$ discussed in the preceding section, the solutions of the original equation (14.10) are oscillatory for $x < 1$ and exponentially growing and decaying for $x > 1$. Hence, ϕ and ψ must be chosen so that the dominant part of the transformed equation has solutions whose behavior changes from oscillatory to exponentially growing and decaying at a given point. The simplest possible equation with these properties is the Airy equation

$$\frac{d^2v}{dz^2} - zv = 0 \qquad (14.86)$$

discussed in the preceding section. When $z > 0$, the solutions of (14.86) are growing and decaying with z, whereas when $z < 0$ its solutions are oscillatory.

The above discussion provides an explanation of the breakdown of the WKB approximation or Liouville-Green transformation. In Sections 14.1 and 14.2, we insisted on representing the solutions of (14.10) in terms of either the elementary circular functions or the elementary exponential functions. Since neither of these elementary functions represents the solutions of turning-point problems, the WKB approximations must fail in regions containing the turning points. A uniformly valid expansion for all x must be expressed in terms of *nonelementary functions* that have the same qualitative behavior as the solutions of the original equation.

Alternatively, the breakdown of the WKB approximation in the neighborhood of a turning point may be explained by the fact that the transformation (14.34) is singular at a turning point. According to the Liouville-Green transformation, $\psi \propto \sqrt[4]{q_1}$. Since q_1 vanishes at a turning point, ψ also vanishes there. Consequently, (14.35) ceases to be valid near a turning point because we are dividing by a zero. Thus, to obtain a uniformly valid expansion, we have to insist on the transformation being regular everywhere in the interval of interest. Hence ψ must be regular and have no zeros in the interval of interest. Then, it follows from (14.40) that ϕ' must be regular and have no zeros in the interval of interest. Consequently, we set

$$\lambda^2 q_1 = \phi'^2 \zeta(z) \tag{14.87}$$

so that (14.41) becomes

$$\frac{d^2 v}{dz^2} + \zeta(z) v = -\delta v \tag{14.88}$$

and choose the simplest possible function $\zeta(z)$ that yields a nonsingular transformation. In order that ϕ' be regular and have no zeros in the interval of interest, $\zeta(z)$ must have the same number, type, and order of singularities and zeros as q_1.

For example, if $q_1(x)$ is regular and has only a simple zero (simple turning point) such as $1 - x^3$, then $\zeta(z)$ must be chosen to be regular and have only a simple zero. The simplest possible function that satisfies these requirements is $\zeta(z) = z$. If $q_1(x)$ is regular and has only a double zero at a point in the interval of interest (i.e., turning point of order 2), $\zeta(z)$ must be chosen to be regular and have only a double zero. The simplest possible function satisfying these requirements is $\zeta(z) = z^2$. If $q_1(x)$ is regular and has only a zero of order n (i.e., turning point of order n), $\zeta(z)$ must be chosen to be z^n. If $q_1(x)$ has two zeros at $x = a$ and b, where $b > a$, of order m and n, then one puts

$$\zeta(z) = z^m (1 - z)^n$$

A third explanation of the breakdown of the WKB approximation and the Liouville-Green transformation is as follows. The term δv in (14.45) is small compared with the other terms only away from a turning point. At a turning point, δ is singular and hence

$$\frac{d^2 v}{dz^2} \pm v = 0$$

is not the dominant part of the transformed equation, as we have assumed. In the example discussed in the preceding section, $q_1 = 1 - x^3$ so that

$$\psi = O[\lambda^{1/2}(x - 1)^{1/4}] \quad \text{and} \quad \phi'^2 = O[\lambda^2 (x - 1)] \quad \text{as} \quad x \to 1 \quad \text{and} \quad \lambda \to \infty$$

according to (14.43) and (14.44). Then, it follows from (14.42) that

$$\delta = O\left[\frac{1}{\lambda^2 (x - 1)^{3/2}}\right] \quad \text{as} \quad x \to 1 \quad \text{and} \quad \lambda \to \infty$$

Hence, δ is small compared with 1 as $\lambda \to \infty$ only when x is away from 1, that is the turning point. Thus, to ensure the uniformity of the resulting expansion, one inspects the neglected terms in the transformed equation to make sure that they are small compared with the kept terms.

The above discussion shows that, in the case of a simple turning point (i.e., q_1 has only a simple zero in the interval of interest), $\zeta(z)$ must be chosen to be regular and have only a simple zero. The simplest possible choice is $\zeta = \pm z$. Here, we use $\zeta = -z$ so that the dominant equation will be the standard Airy equation (14.86). Thus, it follows from (14.87) that

$$\phi\phi'^2 = -\lambda^2 q_1 \qquad (14.89)$$

because $z = \phi$ according to (14.34). To solve (14.89), we first take the square root of both sides and obtain

$$\phi^{1/2}\phi' = \pm\lambda\sqrt{-q_1(x)}$$

which, upon separation of variables, becomes

$$\phi^{1/2}\,d\phi = \pm\lambda\sqrt{-q_1(x)}\,dx$$

Integrating once, we have

$$\tfrac{2}{3}\phi^{3/2} = \pm\lambda\int^x \sqrt{-q_1(\tau)}\,d\tau$$

where τ is a dummy variable of integration. It is convenient to use the location of the turning point, say $x = \mu$, as one of the limits of integration. Then, one writes

$$\tfrac{2}{3}z^{3/2} = \tfrac{2}{3}\phi^{3/2} = \pm\lambda\int_\mu^x \sqrt{-q_1(\tau)}\,d\tau \qquad (14.90)$$

It follows from (14.89) that

$$\phi'^2 = -\lambda^2\phi^{-1}q_1$$

and hence,

$$\psi = \lambda^{1/2}\phi^{-1/4}(-q_1)^{1/4} \qquad (14.91)$$

according to (14.40).

As before, the general solution of (14.86) can be expressed in terms of the Airy functions as

$$v(z) = \tilde{c}_1 Ai(z) + \tilde{c}_2 Bi(z) \qquad (14.92)$$

Substituting (14.91) and (14.92) into (14.35) yields

$$y \sim z^{1/4}(-q_1)^{-1/4}[c_1 Ai(z) + c_2 Bi(z)] \qquad (14.93)$$

where the factor $\lambda^{-1/2}$ has been absorbed into the constants of integration c_1 and c_2.

Next, we check the regularity of the transformation and the order of magnitude of the neglected term δ in the transformed equation (14.41). As $x \to \mu$, $q_1 = O(x - \mu)$ because μ is assumed to be a simple turning point. Then, it follows from (14.90) that

$$\tfrac{2}{3}\phi^{3/2} \to \pm\lambda \int_\mu^x [-q_1'(\mu)(\tau - \mu) + \cdots]^{1/2}\, d\tau \approx \pm\lambda\sqrt{-q_1'(\mu)} \int_\mu^x (\tau - \mu)^{1/2}\, d\tau$$

$$= \pm\tfrac{2}{3}\lambda\sqrt{-q_1'(\mu)}\,(x - \mu)^{3/2}$$

Hence,

$$\phi = O[\lambda^{2/3}(x - \mu)]$$

and hence

$$\phi' = O(\lambda^{2/3}) \quad \text{as} \quad x \to \mu \quad \text{and} \quad \lambda \to \infty$$

Then,

$$\psi = \sqrt{\phi'} = O(\lambda^{1/3}) \quad \text{as} \quad x \to \mu \quad \text{and} \quad \lambda \to \infty$$

proving that the transformation is regular everywhere including the turning point. Substituting the above estimates for ϕ and ψ in (14.42), we find that $\delta = O(\lambda^{-4/3})$ for all values of x, making it small compared with the kept terms in (14.86). Therefore, the single expansion (14.93) with z given by (14.90) is valid everywhere including the turning point.

14.7. Eigenvalue Problems with Turning Points

In this section, we apply the results of the preceding section to two eigenvalue problems. First, we consider the problem

$$y'' + \lambda^2(1 - x^3)y = 0 \tag{14.94}$$

$$y(0) = 0 \quad \text{and} \quad y \to 0 \quad \text{as} \quad x \to \infty \tag{14.95}$$

which is a prototype of problems that arise in quantum mechanics. Here, $q_1 = 1 - x^3$ and (14.90) becomes

$$\tfrac{2}{3} z^{3/2} = \tfrac{2}{3}\phi^{3/2} = \lambda \int_1^x \sqrt{\tau^3 - 1}\, d\tau \tag{14.96}$$

where the positive sign was used so that z is positive when $x > 1$. Hence, as $x \to \infty$, $z \to \infty$, and (14.93) tends to

$$y \sim \frac{(x^3 - 1)^{-1/4}}{\sqrt{\pi}} [\tfrac{1}{2} c_1 e^{-(2/3)z^{3/2}} + c_2 e^{(2/3)z^{3/2}}]$$

according to (14.75) and (14.77). Since $\exp[\tfrac{2}{3} z^{3/2}]$ tends to infinity much faster than $(x^3 - 1)^{-1/4}$ tends to zero as $x \to \infty$, the boundedness condition in (14.95) demands that $c_2 = 0$. Hence, (14.93) becomes

$$y \sim c_1 z^{1/4}(x^3 - 1)^{-1/4} Ai(z) \tag{14.97}$$

Imposing the boundary condition $y(0) = 0$, we have

$$c_1 [z(0)]^{1/4} Ai[z(0)] = 0$$

Hence,

$$Ai[z(0)] = 0 \tag{14.98}$$

Since z is a function of λ, the roots of (14.98) provide the desired eigenvalues. It follows from (14.96) that

$$\tfrac{2}{3} [z(0)]^{3/2} = \lambda \int_1^0 \sqrt{\tau^3 - 1}\, d\tau = -i\lambda \int_0^1 \sqrt{1 - \tau^3}\, d\tau$$

so that

$$z(0) = -\lambda^{2/3} \left[\tfrac{3}{2} \int_0^1 \sqrt{1 - \tau^3}\, d\tau \right]^{2/3} \tag{14.99}$$

Hence, $z(0) \to -\infty$ as $\lambda \to \infty$ and

$$Ai[z(0)] \sim \frac{1}{\sqrt{\pi}\,[-z(0)]^{1/4}} \sin\{\tfrac{2}{3} [-z(0)]^{3/2} + \tfrac{1}{4}\pi\} \tag{14.100}$$

according to (14.76). Using (14.99) and (14.100) in (14.98), we have

$$\sin\left[\lambda \int_0^1 \sqrt{1 - \tau^3}\, d\tau + \tfrac{1}{4}\pi\right] = 0$$

or

$$\lambda \int_0^1 \sqrt{1 - \tau^3}\, d\tau + \tfrac{1}{4}\pi = n\pi \qquad n = 0, 1, 2, \cdots$$

which yields

$$\lambda_n = (n - \tfrac{1}{4})\pi \left[\int_0^1 \sqrt{1 - \tau^3}\, d\tau \right]^{-1} \tag{14.101}$$

Therefore, the eigenfunctions are given by (14.97) with z defined by (14.96) and the eigenvalues are defined by (14.101). Table 14-3 compares the approximate expression (14.101) for the eigenvalues with those obtained by a combination

TABLE 14-3. Comparison of Perturbationally and Numerically Calculated Eigenvalues

Eigenvalue Number	1	2	3	4
Perturbation Result	2.892	6.535	10.27	14.00
Numerical Result	2.807	6.540	10.27	14.00

of a numerical integration technique and a Newton-Raphson iteration technique. The agreement is very good, even for the lowest eigenvalue, which is about 2.807, a not very large number. The error in the first eigenvalue is 3%, whereas that in the second eigenvalue is 0.08%. The third eigenvalue is correct to four significant figures.

As a second example, we consider the problem

$$y'' + \lambda^2(1 - x^2)y = 0 \qquad (14.102)$$

$$y(0) = 0 \quad y(1) = 0 \qquad (14.103)$$

which arises from the problem of heat transfer in a two-dimensional duct carrying a fully developed laminar flow. In this case, $q_1 = 1 - x^2$ and (14.102) has two turning points, one at $x = 1$ and the other at $x = -1$. However, there is only one turning point, namely $x = 1$, within the interval $[0, 1]$ of interest. Since the interval of interest is finite, it is more convenient to express the solution in terms of the Bessel functions rather than the Airy functions.

In this case, $q_1 = 1 - x^2$ and we put

$$\lambda^2(1 - x^2) = \phi'^2 \phi \qquad (14.104)$$

so that the dominant part of the transformed equation becomes

$$\frac{d^2 v}{dz^2} + zv = 0 \qquad (14.105)$$

The solution of (14.104) is taken as

$$\tfrac{2}{3} z^{3/2} = \tfrac{2}{3} \phi^{3/2} = -\lambda \int_1^x \sqrt{1 - \tau^2} \, d\tau = \lambda \int_x^1 \sqrt{1 - \tau^2} \, d\tau = \lambda H \qquad (14.106)$$

where the negative sign is taken so that z will be positive in $[0, 1]$. The general solution of (14.105) can be expressed in terms of Bessel's functions of order $\tfrac{1}{3}$ as in Exercise 14.3. The result is

$$v = \sqrt{z} \, [\tilde{c}_1 J_{-1/3}(\tfrac{2}{3} z^{3/2}) + \tilde{c}_2 J_{1/3}(\tfrac{2}{3} z^{3/2})] \qquad (14.107)$$

It follows from (14.104) that

$$\phi' = \lambda \phi^{-1/2} \sqrt{1 - x^2}$$

and hence,

$$\psi = \lambda^{1/2} z^{-1/4} (1 - x^2)^{1/4} \tag{14.108}$$

according to (14.40). Substituting (14.107) and (14.108) into (14.35) and using (14.106), we obtain

$$y \sim H^{1/2}(1 - x^2)^{-1/4} [c_1 J_{-1/3}(\lambda H) + c_2 J_{1/3}(\lambda H)] \tag{14.109}$$

To impose the condition $y(1) = 0$, we need the limit of (14.109) as $x \to 1$. To this end, we note that as $x \to 1$

$$H = \int_x^1 \sqrt{1 - \tau^2} \, d\tau \to \sqrt{2} \int_x^1 (1 - \tau)^{1/2} \, d\tau = \tfrac{2}{3} \sqrt{2} (1 - x)^{3/2}$$

Moreover, it follows from Exercise 13.20 that

$$J_{-1/3} = O(H^{-1/3}) \quad \text{and} \quad J_{1/3} = O(H^{1/3}) \quad \text{as} \quad H \to 0$$

Hence,

$$J_{-1/3} = O(1 - x)^{-1/2} \quad \text{and} \quad J_{1/3} = O(1 - x)^{1/2} \quad \text{as} \quad x \to 1$$

Consequently,

$$y = O\{(1 - x)^{3/4}(1 - x)^{-1/4}[c_1(1 - x)^{-1/2} + c_2(1 - x)^{1/2}]\}$$

or

$$y = c_1 O(1) + c_2 O(1 - x) \quad \text{as} \quad x \to 1$$

Hence, the condition $y(1) = 0$ demands that $c_1 = 0$, and (14.109) becomes

$$y \sim c_2 H^{1/2}(1 - x^2)^{-1/4} J_{1/3}(\lambda H) \tag{14.110}$$

Imposing the condition $y(0) = 0$ yields the eigenvalues λ_n where λ_n is a root of

$$J_{1/3}\left[\lambda \int_0^1 \sqrt{1 - \tau^2} \, d\tau \right] = 0$$

or

$$J_{1/3}(\tfrac{1}{4} \pi \lambda) = 0 \tag{14.111}$$

Since λ is large, the argument of the Bessel function is large and

$$J_{1/3}(z) \sim \sqrt{\frac{2}{\pi z}} \cos(z - \tfrac{5}{12}\pi) \quad \text{as} \quad z \to \infty$$

according to Exercise 13.20. Then, in place of (14.111), we have

$$\tfrac{1}{4} \pi \lambda - \tfrac{5}{12} \pi = (n + \tfrac{1}{2})\pi \quad n = 0, 1, 2, \cdots$$

TABLE 14-4. Comparison of Perturbationally and Numerically Calculated Eigenvalues

Eigenvalue Number	1	2	3	4	5	6
Perturbation Result	3.6667	7.6667	11.6667	15.6667	19.6667	23.6667
Numerical Result	3.6723	7.6688	11.6679	15.6675	19.6673	23.6672
Error (%)	0.152	0.027	0.010	0.005	0.003	0.002

Therefore, the eigenfunctions are given by (14.110), whereas the eigenvalues are defined by

$$\lambda_n = 4(n + \tfrac{11}{12}) \tag{14.112}$$

Table 14-4 compares the approximate eigenvalues with those obtained by a combination of a numerical integration technique and a Newton-Raphson iteration technique. The agreement is excellent even for the lowest eigenvalue, which is about 3.6723, a not very large number. As expected, the accuracy of the approximate solution increases rapidly as the eigenvalue increases. In fact, the fifth approximate eigenvalue is correct to five significant figures.

Exercises

14.1. Consider Bessel's equation of order $\tfrac{1}{2}$

$$x^2 y'' + xy' + (x^2 - \tfrac{1}{4})y = 0$$

Introduce a transformation to eliminate the first derivative and obtain

$$u'' + u = 0$$

Then, show that

$$J_{1/2}(x) = x^{-1/2}(c_1 \sin x + c_2 \cos x)$$

14.2. Consider the differential equation

$$x^2 \zeta'' + x\zeta' + (x^2 - \nu^2)\zeta = 0$$

governing the cylindrical function $\zeta_\nu(x)$. Let $x = \gamma z^\beta$ and $\zeta_\nu = z^{\alpha - \beta\nu} u(z)$ so that

$$u = z^{\beta\nu - \alpha} \zeta_\nu(\gamma z^\beta)$$

Show that

$$\frac{d\zeta}{dx} = \left[z^{\alpha - \beta\nu} \frac{du}{dz} + (\alpha - \beta\nu)z^{\alpha - \beta\nu - 1} u \right] \frac{dz}{dx}$$

$$\frac{dz}{dx} = \frac{z^{1-\beta}}{\gamma\beta}$$

384 DIFFERENTIAL EQUATIONS WITH A LARGE PARAMETER

$$\frac{d^2\zeta}{dx^2} = \left[z^{\alpha-\beta\nu} \frac{d^2u}{dz^2} + 2(\alpha-\beta\nu)z^{\alpha-\beta\nu-1} \frac{du}{dz} \right.$$

$$+ (\alpha-\beta\nu)(\alpha-\beta\nu-1)z^{\alpha-\beta\nu-2}u \left] \frac{z^{2-2\beta}}{\gamma^2\beta^2} \right.$$

$$+ \left[z^{\alpha-\beta\nu} \frac{du}{dz} + (\alpha-\beta\nu)z^{\alpha-\beta\nu-1}u \right] \frac{(1-\beta)}{\gamma^2\beta^2} z^{1-2\beta}$$

Then, show that u satisfies the differential equation

$$z^2 \frac{d^2u}{dz^2} + (2\alpha - 2\beta\nu + 1)z \frac{du}{dz} + [\beta^2\gamma^2 z^{2\beta} + \alpha(\alpha - 2\nu\beta)] u = 0$$

and hence, u can be expressed in terms of cylindrical functions as above.

14.3. Consider the Airy equation

$$\frac{d^2u}{dz^2} + zu = 0$$

Use the results of the preceding exercise to express the general solution of this equation as

$$u = \sqrt{z}\, [c_1 J_{-1/3}(\tfrac{2}{3} z^{3/2}) + c_2 J_{1/3}(\tfrac{2}{3} z^{3/2})]$$

Hint: Rewrite equation as

$$z^2 \frac{d^2u}{dz^2} + z^3 u = 0$$

Compare this equation with that in the preceding exercise and put

$$2\alpha - 2\nu\beta + 1 = 0 \qquad \beta^2\gamma^2 = 1 \qquad 2\beta = 3 \qquad \alpha(\alpha - 2\nu\beta) = 0$$

14.4. Use the results of Exercise 14.2 to show that the general solution of

$$\frac{d^2u}{dz^2} + z^n u = 0$$

is

$$u = \sqrt{z}\left[c_1 J_\nu\left(\frac{2}{n+2} z^{(n+2)/2}\right) + c_2 J_{-\nu}\left(\frac{2}{n+2} z^{(n+2)/2}\right) \right]$$

where $\nu = (n+2)^{-1}$.

14.5. Consider the problem

$$\epsilon^2 y'' + (x^2 + 2x + 2)y = 0 \qquad 0 < \epsilon \ll 1$$

$$y(0) = 0 \qquad y(1) = 0$$

Show that

$$\epsilon_n = \frac{1}{n\pi} \int_0^1 \sqrt{x^2 + 2x + 2}\, dx$$

14.6. Show that the large eigenvalues of

$$u'' + \lambda^2 f(x) u = 0 \qquad f(x) > 0$$
$$u(0) = 0 \qquad u(1) = 0$$

are given by

$$\lambda_n = n\pi \left[\int_0^1 \sqrt{f(x)}\, dx \right]^{-1}$$

14.7. Show that the large eigenvalues of

$$u'' + \lambda^2 f(x) u = 0 \qquad f(x) > 0$$
$$u(0) = 0 \qquad u'(1) = 0$$

are given by

$$\lambda_n = (n + \tfrac{1}{2})\pi \left[\int_0^1 \sqrt{f(x)}\, dx \right]^{-1}$$

14.8. Consider the problem

$$y'' + \lambda^2 x^2 y = 0$$
$$y(1) = 0 \qquad y(2) = 0$$

Show that

$$\lambda_n = \tfrac{2}{3} n\pi$$

14.9. Show that the large eigenvalues of

$$u'' + \lambda^2 f(x) u = 0 \qquad f(x) > 0$$
$$u'(a) = 0 \qquad u(b) = 0 \qquad b > a$$

are given by

$$\lambda_n = (n + \tfrac{1}{2})\pi \left[\int_a^b \sqrt{f(x)}\, dx \right]^{-1}$$

14.10. Consider the problem

$$y'' + \lambda^2 (1 - x^2) y = 0$$
$$y'(0) = 0 \qquad y(1) = 0$$

Show that the eigenvalues are given by

$$\lambda_n = 4(n + \tfrac{5}{12})$$

14.11. Apply the WKB method directly to

$$xy'' + y' + \lambda^2 x(1 - x^2)y = 0 \qquad \lambda \gg 1$$

Indicate the validity of the resulting approximation.

14.12. Consider the problem

$$y'' + \lambda^2(1 - x^2)f(x)y = 0$$

$$y(0) = 0 \qquad y(1) = 0$$

where $f(x) = f(-x) > 0$ in $[0, 1]$. This problem describes heat transfer in a two-dimensional duct carrying fully developed turbulent flow. Show that

$$\lambda_n = (n + \tfrac{11}{12})\pi \left[\int_0^1 \sqrt{(1 - \tau^2)f(\tau)} \, d\tau \right]^{-1}$$

14.13. Consider the problem in the preceding exercise but with the boundary conditions

$$y'(0) = y(1) = 0$$

Show that

$$\lambda_n = (n + \tfrac{5}{12})\pi \left[\int_0^1 \sqrt{(1 - \tau^2)f(\tau)} \, d\tau \right]^{-1}$$

14.14. Consider the equation

$$y'' + \lambda^2(1 - x^2)^2 y = 0$$

Show that, as $\lambda \to \infty$,

$$y \sim H^{1/2}(1 - x^2)^{-1/2}[c_1 J_{1/4}(\lambda H) + c_2 J_{-1/4}(\lambda H)]$$

for $x > -1$ where

$$H = \int_1^x (1 - \tau^2) \, d\tau = x - 1 - \tfrac{1}{3}(x - 1)^3$$

14.15. Consider the equation

$$y'' + \lambda^2 q(x) y = 0$$

where $q(\mu) = q'(\mu) = 0$ but $q''(\mu) \neq 0$. Show that, as $\lambda \to \infty$,

$$y \sim H^{1/2} q^{-1/4}[c_1 J_{1/4}(\lambda H) + c_2 J_{-1/4}(\lambda H)]$$

where

$$H = \int_\mu^x \sqrt{q} \, dx$$

14.16. Consider
$$\epsilon y'' + (2x + 1)y' + 2y = 0 \quad 0 < \epsilon \ll 1$$
Introduce a transformation to eliminate the middle term and then apply the WKB approximation to the resulting equation.

14.17. Consider
$$\epsilon y'' + (2x + 1)y' + 2y = 0 \quad \text{for} \quad 0 < \epsilon \ll 1$$
Seek a solution in the form $y = \exp[\epsilon^{-1} G(x; \epsilon)]$ and determine two terms in the expansion of G.

CHAPTER 15

Solvability Conditions

In applying perturbation methods such as the method of multiple scales, one obtains problems that need to be solved in succession. Usually the first-order problem is homogeneous, whereas the higher-order problems are linear and inhomogeneous. To determine the dependence on the slow scales, one investigates the higher-order problems and imposes conditions that make the expansion uniform. For simple nonlinear vibration problems, the above process leads us to the elimination of secular and small-divisor terms. When we were dealing with only one-degree-of-freedom systems, it was easy for us to determine the conditions for the elimination of the secular and small-divisor terms. All we had to do was to set each of the coefficients of the terms that produce secular terms equal to zero. However, for mutidegree-of-freedom systems where the governing equations are coupled, eliminating the terms that lead to secular terms is a little bit more involved. This problem and its application to two-degree-of-freedom systems constitutes the first part of this chapter.

In other problems, the nonuniformity in the expansion may manifest itself in our inability to satisfy all the boundary conditions, leading to inconsistencies. The inconsistency is eliminated by imposing certain conditions, which are referred to as *solvability* or *consistency* or *integrability* or *compatibility* conditions. These conditions are derived for the case of second-order inhomogeneous differential equations with various boundary conditions. The results are then applied to two simple eigenvalue problems, waves in ducts with sinusoidal walls and vibrations of near circular membranes. In Sections 15-10 and 15-11, the solvability conditions are derived for the case of fourth-order inhomogeneous equations with various boundary conditions. The results are applied to two problems; one arises from the problem of stability of boundary layers and the other arises from the problem of vibrations of near annular plates. Then, in Section 15-12, the theory is applied to a fourth-order degenerate eigenvalue problem. A differential system of equations is treated in Section 15-13, general systems of first-order differential equations are treated in Section 15-14, and a differential system with interfacial boundary conditions is treated in Section

15-15. Integral equations are treated in Section 15-16, and partial-differential equations are treated in Section 15-17.

15.1. Algebraic Equations

Let us consider the system of two algebraic equations

$$x_1 - x_2 = b_1 \tag{15.1}$$

$$2x_1 - 2x_2 = b_2 \tag{15.2}$$

It is clear that this system of equations will not have a solution unless $b_2 = 2b_1$. For multiplying the first equation by 2, one obtains

$$2x_1 - 2x_2 = 2b_1 \tag{15.3}$$

Comparing (15.2) with (15.3), we conclude that $b_2 = 2b_1$ if (15.1) and (15.2) are to be *consistent*. If $b_2 \neq 2b_1$, (15.1) and (15.2) are contradictory, and hence, they do not possess a solution. However, when $b_2 = 2b_1$, (15.1) and (15.2) are redundant and only one of them is needed. Then, one can solve say for x_2 and obtain

$$x_2 = x_1 - b_1 \tag{15.4}$$

showing that there are an infinite number of possible solutions.

If we set the $b_n = 0$ in (15.1) and (15.2), we find that the homogeneous equations

$$x_1 - x_2 = 0$$

$$2x_1 - 2x_2 = 0$$

have *nontrivial solutions* consisting of $x_2 = x_1$.

Let us modify the system of equations so that the homogeneous equations do not have nontrivial solutions and take as an example

$$x_1 - x_2 = b_1 \tag{15.5}$$

$$2x_1 + 2x_2 = b_2 \tag{15.6}$$

Multiplying (15.5) by 2 gives

$$2x_1 - 2x_2 = 2b_1 \tag{15.7}$$

Adding (15.6) and (15.7) yields

$$4x_1 = b_2 + 2b_1$$

whereas subtracting (15.7) from (15.6) yields

$$4x_2 = b_2 - 2b_1$$

390 SOLVABILITY CONDITIONS

Hence, the solution of (15.5) and (15.6) is

$$x_1 = \tfrac{1}{4}(b_2 + 2b_1) \qquad x_2 = \tfrac{1}{4}(b_2 - 2b_1)$$

which exists for all values of b_1 and b_2. In this case, the homogeneous equations (15.5) and (15.6) are

$$x_1 - x_2 = 0$$
$$2x_1 + 2x_2 = 0$$

which have only the *trivial* solution. The system (15.1) and (15.2) requires a solvability condition on the constants b_1 and b_2 for solutions to exist and the homogeneous system admits a nontrivial solution. However, the system (15.5) and (15.6) does not require a solvability condition on the constants b_1 and b_2 and the homogeneous system admits only the trivial solution.

Next, let us consider the third-order system

$$x_1 + 2x_2 - 3x_3 = b_1 \qquad (15.8)$$
$$-2x_1 + x_2 + x_3 = b_2 \qquad (15.9)$$
$$3x_1 + x_2 - 4x_3 = b_3 \qquad (15.10)$$

Adding (15.9) and (15.10) yields

$$x_1 + 2x_2 - 3x_3 = b_2 + b_3 \qquad (15.11)$$

Comparing (15.8) and (15.11), we conclude that they are consistent only when

$$b_2 + b_3 = b_1$$

which is the solvability condition. Then, (15.8) through (15.10) are redundant and only two of them are needed. Taking (15.8) and (15.9), we can solve for x_1 and x_2 in terms of x_3. The result is

$$x_1 = x_3 + \tfrac{1}{5}(b_1 - 2b_2) \qquad x_2 = x_3 + \tfrac{1}{5}(2b_1 + b_2) \qquad (15.12)$$

which yields an infinite number of possible solutions. Substituting (15.12) into (15.10) shows that $b_1 - b_2 = b_3$, which is the above derived solvability condition. In this case, the homogeneous equations (15.8) through (15.10) have the nontrivial solutions $x_1 = x_2 = x_3$.

The preceding discussion indicates that, when the homogeneous equations have a nontrivial solution, the inhomogeneous equations have a solution if and only if the inhomogeneous parts satisfy a solvability condition. Next, we illustrate this for the general system of algebraic equations

$$A\mathbf{x} = \mathbf{b} \qquad (15.13)$$

where A is an $N \times N$ matrix and $\mathbf{x}$ and $\mathbf{b}$ are $N \times 1$ column vectors. This system has a unique solution for all **b**'s if and only if the homogeneous system

$$Ax = 0 \tag{15.14}$$

has only the trivial solution. If the homogeneous system has a nontrivial solution, then the inhomogeneous system will not have a solution unless the components of **b** satisfy certain solvability conditions. These solvability conditions can be determined by the process of elimination. Alternatively, they can be determined by manipulating the so-called *augmented matrix B*. It is defined as the $N \times (N + 1)$ matrix consisting of inserting $b_1, b_2, b_3, \ldots, b_N$ as a new column into the so-called coefficient matrix A. Thus, if

$$A = \begin{bmatrix} a_{11} & a_{12} & \cdots & a_{1N} \\ a_{21} & a_{22} & \cdots & a_{2N} \\ \cdot & \cdot & \cdots & \cdot \\ a_{N1} & a_{N2} & \cdots & a_{NN} \end{bmatrix} \tag{15.15}$$

then,

$$B = \begin{bmatrix} a_{11} & a_{12} & \cdots & a_{1N} & b_1 \\ a_{21} & a_{22} & \cdots & a_{2N} & b_2 \\ \cdot & \cdot & \cdots & \cdot & \cdot \\ a_{N1} & a_{N2} & \cdots & a_{NN} & b_N \end{bmatrix} \tag{15.16}$$

The solvability condition can be stated as follows: *A set of linear algebraic equations possesses a solution if and only if the rank of the augmented matrix B is equal to the rank of the coefficient matrix A.*

We note that the homogeneous system (15.14) has a nontrivial solution if and only if the determinant of the coefficient matrix vanishes, that is,

$$|A| = 0$$

Then, the rank of A is less than N, and hence, the rank of B is less than N. Consequently, the determinant of any $N \times N$ matrix formed from B by eliminating one of the columns must be zero. This condition, for example,

$$\begin{vmatrix} a_{12} & a_{13} & \cdots & a_{1N} & b_1 \\ a_{22} & a_{23} & \cdots & a_{2N} & b_2 \\ \cdot & \cdot & \cdots & \cdot & \cdot \\ a_{N2} & a_{N3} & \cdots & a_{NN} & b_N \end{vmatrix} = 0 \tag{15.17}$$

yields a solvability condition. This solvability condition can be interpreted in the following alternate way. If $|A| \neq 0$, one can use Cramer's rule to solve for $x_1, x_2, \ldots, x_N$. For example,

$$x_1 = \frac{\begin{vmatrix} b_1 & a_{12} & a_{13} & \cdots & a_{1N} \\ b_2 & a_{22} & a_{23} & \cdots & a_{2N} \\ \cdot & \cdot & \cdot & \cdots & \cdot \\ b_N & a_{N2} & a_{N3} & \cdots & a_{NN} \end{vmatrix}}{|A|} \tag{15.18}$$

If $|A| = 0$, (15.18) shows that x_1 is infinite, and hence, the system of equations (15.13) is not solvable unless the determinant in the numerator also vanishes. In this case, we have a zero over zero, which is indeterminate.

The solvability condition of (15.13) when the homogeneous system has a nontrivial solution can be expressed in the following alternate way. We form the transpose of (15.13) and multiply the result from the right with the column vector $\bar{\mathbf{u}}$, where the overbar indicates the complex conjugate and $\mathbf{u}$ is an $N \times 1$ column vector called the *adjoint* and defined below. The result is

$$(A\mathbf{x})^T \bar{\mathbf{u}} = \mathbf{b}^T \bar{\mathbf{u}} \tag{15.19}$$

where the superscript T indicates the transpose, that is,

$$\begin{bmatrix} a_{11} & a_{12} & \cdots & a_{1N} \\ a_{21} & a_{22} & \cdots & a_{2N} \\ \vdots & & \cdots & \vdots \\ a_{N1} & a_{N2} & \cdots & a_{NN} \end{bmatrix}^T = \begin{bmatrix} a_{11} & a_{21} & \cdots & a_{N1} \\ a_{12} & a_{22} & \cdots & a_{N2} \\ \vdots & & \cdots & \vdots \\ a_{1N} & a_{2N} & \cdots & a_{NN} \end{bmatrix}$$

$$\begin{bmatrix} b_1 \\ b_2 \\ \vdots \\ b_N \end{bmatrix}^T = [b_1 \quad b_2 \quad \cdots \quad b_N]$$

Since $(A\mathbf{x})^T = \mathbf{x}^T A^T$, (15.19) can be rewritten as

$$\mathbf{x}^T A^T \bar{\mathbf{u}} = \mathbf{b}^T \bar{\mathbf{u}}$$

Hence,

$$\mathbf{x}^T \overline{A^T \mathbf{u}} = \mathbf{b}^T \bar{\mathbf{u}}$$

or

$$\mathbf{x}^T \overline{A^* \mathbf{u}} = \mathbf{b}^T \bar{\mathbf{u}} \tag{15.20}$$

where $A^* = \bar{A}^T$ is referred to as the *adjoint matrix* of A. The matrix A is called *self-adjoint* if $A^* = A$, that is, it is either a Hermitian matrix or a symmetric matrix, depending on whether A is complex or real. If the homogeneous system (15.13) has a nontrivial solution $\mathbf{x}$, then $|A| = |A^*| = 0$ demands that

$$A^* \mathbf{u} = 0 \tag{15.21}$$

have a nontrivial solution.

Having defined $\mathbf{u}$ by (15.21), we return to the inhomogeneous system (15.13), that is, $\mathbf{b} \neq 0$. Using (15.21) in (15.20), we obtain the following form for the solvability conditions:

$$\mathbf{b}^T \bar{\mathbf{u}} = 0 \tag{15.22}$$

where $\mathbf{u}$ is any solution of the adjoint system. In other words, the solvability

ALGEBRAIC EQUATIONS 393

conditions demand that the right-hand side of (15.13) be orthogonal to every solution of the adjoint homogeneous problem.

If we define the inner product of the column vectors **u** and **v** by

$$(\mathbf{u}, \mathbf{v}) = \mathbf{u}^T \overline{\mathbf{v}} \tag{15.23}$$

then, (15.19) and (15.20) can be rewritten as

$$(A\mathbf{x}, \mathbf{u}) = (\mathbf{x}, A^*\mathbf{u}) = (\mathbf{b}, \mathbf{u}) \tag{15.24a}$$

Moreover, (15.22) can be rewritten as

$$(\mathbf{b}, \mathbf{u}) = 0 \tag{15.24b}$$

Although we have only shown necessity, (15.24b) is a sufficient condition for (15.13) to have a solution. As an example, we consider (15.8) through (15.10). In this case,

$$A = \begin{bmatrix} 1 & 2 & -3 \\ -2 & 1 & 1 \\ 3 & 1 & -4 \end{bmatrix} \tag{15.25}$$

$$A^T = \begin{bmatrix} 1 & -2 & 3 \\ 2 & 1 & 1 \\ -3 & 1 & -4 \end{bmatrix} \tag{15.26}$$

and $A^* = A^T$ because A is real. Then, the adjoint system is

$$\begin{bmatrix} 1 & -2 & 3 \\ 2 & 1 & 1 \\ -3 & 1 & -4 \end{bmatrix} \begin{bmatrix} u_1 \\ u_2 \\ u_3 \end{bmatrix} = 0 \tag{15.27}$$

or

$$u_1 - 2u_2 + 3u_3 = 0 \tag{15.28a}$$

$$2u_1 + u_2 + u_3 = 0 \tag{15.28b}$$

$$-3u_1 + u_2 - 4u_3 = 0 \tag{15.28c}$$

Equations (15.28) have a nontrivial solution because the homogeneous system (15.8) through (15.10) has a nontrivial solution. Adding (15.28a) to 2 times (15.28b), we have

$$5u_1 + 5u_3 = 0$$

whereas subtracting (15.28c) from (15.28b), we have

$$5u_1 + 5u_3 = 0$$

Hence, $u_3 = -u_1$. Then, it follows from (15.28b) that $u_2 = -u_1$, and hence, the solution of the adjoint problem can be expressed as

$$\mathbf{u} = c \begin{bmatrix} 1 \\ -1 \\ -1 \end{bmatrix}$$

Imposing the condition that **b** be orthogonal to the solution of the adjoint homogeneous problem, we find that the solvability condition is

$$b_1 - b_2 - b_3 = 0$$

in agreement with that obtained above by using elimination of variables.

We note that the form (15.17) is convenient for our use, and consequently it is frequently used in this book.

Next we apply the above results to two problems involving the vibration of two-degree-of-freedom gyroscopic systems.

15.2. Nonlinear Vibrations of Two-Degree-of-Freedom Gyroscopic Systems

We consider the free oscillations of a two-degree-of-freedom gyroscopic system with quadratic nonlinearities. Specifically, we consider

$$\ddot{u}_1 + \dot{u}_2 + 2u_1 = 2u_1 u_2$$
$$\ddot{u}_2 - \dot{u}_1 + 2u_2 = u_1^2$$
(15.29)

for small but finite amplitudes.

We seek a uniform expansion by using the method of multiple scales in the form

$$u_1 = \epsilon u_{11}(T_0, T_1) + \epsilon^2 u_{12}(T_0, T_1) + \cdots$$
$$u_2 = \epsilon u_{21}(T_0, T_1) + \epsilon^2 u_{22}(T_0, T_1) + \cdots$$
(15.30)

where $T_0 = t$, $T_1 = \epsilon t$, and ϵ is a small dimensionless parameter that characterizes the amplitude of oscillation. Substituting (15.30) into (15.29) and using (5.45), we obtain

$$(D_0^2 + 2\epsilon D_0 D_1)(\epsilon u_{11} + \epsilon^2 u_{12}) + (D_0 + \epsilon D_1)(\epsilon u_{21} + \epsilon^2 u_{22})$$
$$+ 2(\epsilon u_{11} + \epsilon^2 u_{12}) + \cdots = 2(\epsilon u_{11} + \epsilon^2 u_{12})(\epsilon u_{21} + \epsilon^2 u_{22}) + \cdots$$
$$(D_0^2 + 2\epsilon D_0 D_1)(\epsilon u_{21} + \epsilon^2 u_{22}) - (D_0 + \epsilon D_1)(\epsilon u_{11} + \epsilon^2 u_{12})$$
$$+ 2(\epsilon u_{21} + \epsilon^2 u_{22}) + \cdots = (\epsilon u_{11} + \epsilon^2 u_{12})^2 + \cdots$$

Equating the coefficients of like powers of ϵ in these equations, we have

Order ϵ

$$D_0^2 u_{11} + D_0 u_{21} + 2u_{11} = 0 \qquad (15.31a)$$

TWO-DEGREE-OF-FREEDOM GYROSCOPIC SYSTEMS

$$D_0^2 u_{21} - D_0 u_{11} + 2u_{21} = 0 \tag{15.31b}$$

Order ϵ^2

$$D_0^2 u_{12} + D_0 u_{22} + 2u_{12} = -2D_0 D_1 u_{11} - D_1 u_{21} + 2u_{11} u_{21}$$
$$D_0^2 u_{22} - D_0 u_{12} + 2u_{22} = -2D_0 D_1 u_{21} + D_1 u_{11} + u_{11}^2 \tag{15.32}$$

Equations (15.31) constitute a system of two coupled differential equations with constant coefficients. Hence, their solutions can be obtained by letting

$$u_{11} = c_1 e^{i\omega T_0} \qquad u_{21} = c_2 e^{i\omega T_0} \tag{15.33}$$

Substituting (15.33) into (15.31) yields

$$(2 - \omega^2) c_1 + i\omega c_2 = 0$$
$$-i\omega c_1 + (2 - \omega^2) c_2 = 0 \tag{15.34}$$

For a nontrivial solution, the determinant of the coefficient matrix in (15.34) must be zero; that is,

$$\begin{vmatrix} 2 - \omega^2 & i\omega \\ -i\omega & 2 - \omega^2 \end{vmatrix} = 0$$

Hence,

$$(2 - \omega^2)^2 - \omega^2 = 0$$

or

$$\omega^4 - 5\omega^2 + 4 = 0 = (\omega^2 - 4)(\omega^2 - 1)$$

so that $\omega = 1$ and 2, where the frequencies are defined with the positive sign. When $\omega = 1$, it follows from the first equation in (15.34) that

$$c_1 + ic_2 = 0 \quad \text{or} \quad c_2 = ic_1$$

When $\omega = 2$, it follows from the first equation in (15.34) that

$$-2c_1 + 2ic_2 = 0 \quad \text{or} \quad c_2 = -ic_1$$

Therefore, the general solution of (15.31) can be written as

$$u_{11} = A_1(T_1) e^{iT_0} + \bar{A}_1(T_1) e^{-iT_0} + A_2(T_1) e^{2iT_0} + \bar{A}_2(T_1) e^{-2iT_0}$$
$$u_{21} = iA_1(T_1) e^{iT_0} - i\bar{A}_1(T_1) e^{-iT_0} - iA_2(T_1) e^{2iT_0} + i\bar{A}_2(T_1) e^{-2iT_0} \tag{15.35}$$

As in the one-degree-of-freedom case, A_1 and A_2 are undetermined at this level of approximation; they are determined at the next level of approximation by imposing the solvability conditions.

Substituting (15.35) into (15.32), we have

396 SOLVABILITY CONDITIONS

$$D_0^2 u_{12} + D_0 u_{22} + 2u_{12} = -2iA_1' e^{iT_0} - 4iA_2' e^{2iT_0} - iA_1' e^{iT_0}$$
$$+ iA_2' e^{2iT_0} + cc + 2(A_1 e^{iT_0} + \bar{A}_1 e^{-iT_0}$$
$$+ A_2 e^{2iT_0} + \bar{A}_2 e^{-2iT_0})(iA_1 e^{iT_0}$$
$$- i\bar{A}_1 e^{-iT_0} - iA_2 e^{2iT_0} + i\bar{A}_2 e^{-2iT_0})$$

$$D_0^2 u_{22} - D_0 u_{12} + 2u_{22} = 2A_1' e^{iT_0} - 4A_2' e^{2iT_0} + A_1' e^{iT_0}$$
$$+ A_2' e^{2iT_0} + cc + (A_1 e^{iT_0} + \bar{A}_1 e^{-iT_0}$$
$$+ A_2 e^{2iT_0} + \bar{A}_2 e^{-2iT_0})^2$$

After some algebraic manipulations and simplifications, we have

$$D_0^2 u_{12} + D_0 u_{22} + 2u_{12} = -i(3A_1' + 4A_2 \bar{A}_1) e^{iT_0}$$
$$+ i(-3A_2' + 2A_1^2) e^{2iT_0} + cc + \text{NST} \quad (15.36)$$

$$D_0^2 u_{22} - D_0 u_{12} + 2u_{22} = (3A_1' + 2A_2 \bar{A}_1) e^{iT_0}$$
$$+ (-3A_2' + A_1^2) e^{2iT_0} + cc + \text{NST} \quad (15.37)$$

Since the homogeneous parts of (15.36) and (15.37) have solutions proportional to $\exp(\pm iT_0)$ and $\exp(\pm 2iT_0)$, the inhomogeneous terms proportional to $\exp(\pm iT_0)$ and $\exp(\pm 2iT_0)$ will produce secular terms in u_{12} and u_{22}. We note that each of (15.36) and (15.37) contains terms proportional to $\exp(\pm iT_0)$ and $\exp(\pm 2iT_0)$, and hence it is not necessary to set the coefficient of each of these terms equal to zero. In fact, if we do that, we will end up with four complex incompatible equations governing A_1 and A_2. Therefore, to eliminate these secular terms (i.e., to determine the solvability conditions), we seek a particular solution free of secular terms corresponding to $\exp(iT_0)$ and $\exp(2iT_0)$ in the form

$$u_{12} = P_1(T_1) e^{iT_0} + P_2 e^{2iT_0}$$
$$u_{22} = Q_1(T_1) e^{iT_0} + Q_2 e^{2iT_0} \quad (15.38)$$

Substituting (15.38) into (15.36) and (15.37), we have

$$(P_1 + iQ_1) e^{iT_0} + (-2P_2 + 2iQ_2) e^{2iT_0} = -i(3A_1' + 4A_2 \bar{A}_1) e^{iT_0}$$
$$+ i(-3A_2' + 2A_1^2) e^{2iT_0}$$

$$(-iP_1 + Q_1) e^{iT_0} + (-2iP_2 - 2Q_2) e^{2iT_0} = (3A_1' + 2A_2 \bar{A}_1) e^{iT_0}$$
$$+ (-3A_2' + A_1^2) e^{2iT_0}$$

Equating the coefficients of each of $\exp(iT_0)$ and $\exp(2iT_0)$ on both sides, we obtain

$$P_1 + iQ_1 = -i(3A_1' + 4A_2 \bar{A}_1) \quad (15.39a)$$

$$-iP_1 + Q_1 = 3A'_1 + 2A_2\bar{A}_1 \qquad (15.39b)$$

$$-2P_2 + 2iQ_2 = i(-3A'_2 + 2A_1^2)$$
$$-2iP_2 - 2Q_2 = -3A'_2 + A_1^2 \qquad (15.40)$$

Equations (15.39) constitute a system of two inhomogeneous algebraic equations for P_1 and Q_1. Their homogeneous parts have a nontrivial solution because the determinant of their coefficient matrix

$$\begin{vmatrix} 1 & i \\ -i & 1 \end{vmatrix} = 0$$

Then, their solvability condition can be written either as

$$\begin{vmatrix} 1 & -i(3A'_1 + 4A_2\bar{A}_1) \\ -i & 3A'_1 + 2A_2\bar{A}_1 \end{vmatrix} = 0 \quad \text{or} \quad \begin{vmatrix} -i(3A'_1 + 4A_2\bar{A}_1) & i \\ 3A'_1 + 2A_2\bar{A}_1 & 1 \end{vmatrix} = 0$$

In the two-dimensional case, the two conditions yield the same result. But for a higher-dimensional space, they may not. In this case, either condition yields

$$3A'_1 + 2A_2\bar{A}_1 + 3A'_1 + 4A_2\bar{A}_1 = 0$$

or

$$A'_1 = -A_2\bar{A}_1 \qquad (15.41)$$

Similarly, the determinant of the coefficient matrix of the system of equations (15.40) is zero, and hence they have a solution if and only if the following solvability condition is satisfied:

$$\begin{vmatrix} -2 & i(-3A'_2 + 2A_1^2) \\ -2i & -3A'_2 + A_1^2 \end{vmatrix} = 0$$

or

$$A'_2 = \tfrac{1}{2} A_1^2 \qquad (15.42)$$

Equations (15.41) and (15.42) are the equations describing the modulation of A_1 and A_2 with T_1. As in the case of one-degree-of-freedom systems, A_1 and A_2 are usually expressed in polar form and (15.41) and (15.42) are separated into real and imaginary parts. Once these equations are solved, the first-order solution is completed. We will not present these details in here and refer the reader to the book of Nayfeh and Mook (1979).

15.3. Parametrically Excited Gyroscopic Systems

We consider the parametrically excited simple linear two-degree-of-freedom system. Thus, we consider the solution of

398 SOLVABILITY CONDITIONS

$$\ddot{u}_1 + \dot{u}_2 + 2u_1 + 2\epsilon \cos \Omega t (f_{11} u_1 + f_{12} u_2) = 0$$
$$\ddot{u}_2 - \dot{u}_1 + 2u_2 + 2\epsilon \cos \Omega t (f_{21} u_1 + f_{22} u_2) = 0$$
(15.43)

for small ϵ. Here, Ω is assumed to be positive for definiteness. Using the method of multiple scales, we seek a uniform expansion in the form

$$u_1 = u_{10}(T_0, T_1) + \epsilon u_{11}(T_0, T_1) + \cdots$$
$$u_2 = u_{20}(T_0, T_1) + \epsilon u_{21}(T_0, T_1) + \cdots$$
(15.44)

Substituting (15.44) into (15.43), using (5.45), and expressing $\cos \Omega t$ as $\cos \Omega T_0$, we obtain

$$(D_0^2 + 2\epsilon D_0 D_1)(u_{10} + \epsilon u_{11}) + (D_0 + \epsilon D_1)(u_{20} + \epsilon u_{21})$$
$$+ 2(u_{10} + \epsilon u_{11}) + 2\epsilon \cos \Omega T_0 (f_{11} u_{10} + f_{12} u_{20}) + \cdots = 0$$

$$(D_0^2 + 2\epsilon D_0 D_1)(u_{20} + \epsilon u_{21}) - (D_0 + \epsilon D_1)(u_{10} + \epsilon u_{11})$$
$$+ 2(u_{20} + \epsilon u_{21}) + 2\epsilon \cos \Omega T_0 (f_{21} u_{10} + f_{22} u_{20}) + \cdots = 0$$

Equating coefficients of like powers of ϵ we have

Order ϵ^0

$$D_0^2 u_{10} + D_0 u_{20} + 2 u_{10} = 0$$
$$D_0^2 u_{20} - D_0 u_{10} + 2 u_{20} = 0$$
(15.45)

Order ϵ

$$D_0^2 u_{11} + D_0 u_{21} + 2 u_{11} = -2 D_0 D_1 u_{10} - D_1 u_{20} - 2 \cos \Omega T_0 (f_{11} u_{10} + f_{12} u_{20})$$
(15.46)

$$D_0^2 u_{21} - D_0 u_{11} + 2 u_{21} = -2 D_0 D_1 u_{20} + D_1 u_{10} - 2 \cos \Omega T_0 (f_{21} u_{10} + f_{22} u_{20})$$
(15.47)

As in the preceding section, the general solution of (15.45) can be expressed as

$$u_{10} = A_1(T_1) e^{iT_0} + \bar{A}_1(T_1) e^{-iT_0} + A_2(T_1) e^{2iT_0} + \bar{A}_2(T_1) e^{-2iT_0}$$
$$u_{20} = i A_1 e^{iT_0} - i \bar{A}_1 e^{-iT_0} - i A_2 e^{2iT_0} + i \bar{A}_2 e^{-2iT_0}$$
(15.48)

where A_1 and A_2 are to be determined from the solvability conditions at the next level of approximation. Substituting (15.48) into (15.46) and (15.47) yields

$$D_0^2 u_{11} + D_0 u_{21} + 2 u_{11} = -3i A_1' e^{iT_0} - 3i A_2' e^{2iT_0} + cc$$
$$- (e^{i\Omega T_0} + e^{-i\Omega T_0})(f_{11} A_1 e^{iT_0}$$
$$+ f_{11} A_2 e^{2iT_0} + i f_{12} A_1 e^{iT_0}$$

PARAMETRICALLY EXCITED GYROSCOPIC SYSTEMS 399

$$- if_{12}A_2 e^{2iT_0} + cc)$$

$$D_0^2 u_{21} - D_0 u_{11} + 2u_{21} = 3A_1' e^{iT_0} - 3A_2' e^{2iT_0} + cc$$
$$- (e^{i\Omega T_0} + e^{-i\Omega T_0})(f_{21}A_1 e^{iT_0}$$
$$+ f_{21}A_2 e^{2iT_0} + if_{22}A_1 e^{iT_0}$$
$$- if_{22}A_2 e^{2iT_0} + cc)$$

or

$$D_0^2 u_{11} + D_0 u_{21} + 2u_{11} = -3iA_1' e^{iT_0} - 3iA_2' e^{2iT_0} - (f_{11} + if_{12})A_1$$
$$\cdot [e^{i(1+\Omega)T_0} + e^{i(1-\Omega)T_0}] - (f_{11} - if_{12})A_2$$
$$\cdot [e^{i(2+\Omega)T_0} + e^{i(2-\Omega)T_0}] + cc \quad (15.49)$$

$$D_0^2 u_{21} - D_0 u_{11} + 2u_{21} = 3A_1' e^{iT_0} - 3A_2' e^{2iT_0} - (f_{21} + if_{22})A_1$$
$$\cdot [e^{i(1+\Omega)T_0} + e^{i(1-\Omega)T_0}] - (f_{21} - if_{22})A_2$$
$$\cdot [e^{i(2+\Omega)T_0} + e^{i(2-\Omega)T_0}] + cc \quad (15.50)$$

When A_1 and A_2 are constants, as in the straightforward-expansion case, the right-hand sides of (15.49) and (15.50) contain terms that may lead to secular or small-divisor terms, depending on the value of Ω. Secular terms will appear in u_{11} and u_{21} if any of the exponents in (15.49) and (15.50) is equal to either ± 1 or ± 2, because $\exp(\pm iT_0)$ and $\exp(\pm 2iT_0)$ are solutions of the homogeneous problem. Thus, secular terms will appear in u_{11} and u_{21} if

$$1 + \Omega = \pm 2 \quad 1 - \Omega = \pm 2 \quad 1 + \Omega = \pm 1 \quad 1 - \Omega = \pm 1$$
$$2 + \Omega = \pm 2 \quad 2 - \Omega = \pm 2 \quad 2 + \Omega = \pm 1 \quad 2 - \Omega = \pm 1$$

or if $\Omega = 0$ or 1 or 2 or 3 or 4. When the equality sign is replaced with an approximate sign, small-divisor rather than secular terms will appear in the solution. Next, we consider the case $\Omega \approx 3$.

To eliminate the small-divisor terms when $\Omega \approx 3$, we first transform them into secular terms by introducing the detuning parameter σ defined by

$$\Omega = 3 + \epsilon\sigma \quad (15.51)$$

Thus, we write

$$\Omega T_0 = 3T_0 + \epsilon\sigma T_0 = 3T_0 + \sigma T_1 \quad (15.52)$$

Substituting (15.52) into (15.49) and (15.50), we have

$$D_0^2 u_{11} + D_0 u_{21} + 2u_{11} = -[3iA_1' + (f_{11} + if_{12})\bar{A}_2 e^{i\sigma T_1}] e^{iT_0}$$
$$- [3iA_2' + (f_{11} - if_{12})\bar{A}_1 e^{i\sigma T_1}] e^{2iT_0}$$
$$+ \text{terms proportional to } (e^{4iT_0}, e^{5iT_0}) + cc \quad (15.53)$$

$$D_0^2 u_{21} - D_0 u_{11} + 2u_{21} = [3A_1' - (f_{21} + if_{22})\bar{A}_2 e^{i\sigma T_1}]e^{iT_0}$$
$$- [3A_2' + (f_{21} - if_{22})\bar{A}_1 e^{i\sigma T_1}]e^{2iT_0}$$
$$+ \text{terms proportional to } (e^{4iT_0}, e^{5iT_0}) + cc$$

$$(15.54)$$

To determine the solvability conditions, we seek a particular solution free of secular terms corresponding to the terms proportional to $\exp(iT_0)$ and $\exp \cdot (2iT_0)$ in (15.53) and (15.54) in the form

$$u_{11} = P_1(T_1)e^{iT_0} + P_2(T_1)e^{2iT_0}$$
$$u_{21} = Q_1(T_1)e^{iT_0} + Q_2(T_1)e^{2iT_0}$$

$$(15.55)$$

As in the preceding section, substituting (15.55) into (15.53) and (15.54) and equating each of the coefficients of $\exp(iT_0)$ and $\exp(2iT_0)$ on both sides, we obtain

$$P_1 + iQ_1 = -3iA_1' - (f_{11} + if_{12})\bar{A}_2 e^{i\sigma T_1}$$
$$-iP_1 + Q_1 = 3A_1' - (f_{21} + if_{22})\bar{A}_2 e^{i\sigma T_1}$$

$$(15.56)$$

$$-2P_2 + 2iQ_2 = -3iA_2' - (f_{11} - if_{12})\bar{A}_1 e^{i\sigma T_1}$$
$$-2iP_2 - 2Q_2 = -3A_2' - (f_{21} - if_{22})\bar{A}_1 e^{i\sigma T_1}$$

$$(15.57)$$

Since the homogeneous equations (15.56) have a nontrivial solution, the inhomogeneous equations have a solution if and only if the following solvability condition is satisfied:

$$\begin{vmatrix} 1 & -3iA_1' - (f_{11} + if_{12})]\bar{A}_2 e^{i\sigma T_1} \\ -i & 3A_1' - (f_{21} + if_{22})\bar{A}_2 e^{i\sigma T_1} \end{vmatrix} = 0$$

This condition can be rewritten as

$$A_1' = \tfrac{1}{6}[f_{21} - f_{12} + i(f_{11} + f_{22})]\bar{A}_2 e^{i\sigma T_1} \tag{15.58}$$

The solvability condition for (15.57) is

$$\begin{vmatrix} -2 & -3iA_2' - (f_{11} - if_{12})\bar{A}_1 e^{i\sigma T_1} \\ -2i & -3A_2' - (f_{21} - if_{22})\bar{A}_1 e^{i\sigma T_1} \end{vmatrix} = 0$$

which can be rewritten as

$$A_2' = -\tfrac{1}{6}[f_{21} - f_{12} - i(f_{11} + f_{22})]\bar{A}_1 e^{i\sigma T_1} \tag{15.59}$$

Equations (15.58) and (15.59) describe the modulation of A_1 and A_2 with T_1. We refer the reader to the book of Nayfeh and Mook (1979) for the details of their analysis.

15.4. Second-Order Differential Systems

In this section, we consider the solvability conditions for inhomogeneous linear second-order differential equations subject to general inhomogeneous boundary conditions. Thus, we consider

$$p_2(x)y'' + p_1(x)y' + p_0(x)y = f(x) \quad a \leq x \leq b \tag{15.60}$$

$$\alpha_{11}y'(a) + \alpha_{12}y(a) + \alpha_{13}y'(b) + \alpha_{14}y(b) = \beta_1$$
$$\alpha_{21}y'(a) + \alpha_{22}y(a) + \alpha_{23}y'(b) + \alpha_{24}y(b) = \beta_2 \tag{15.61}$$

where the boundary conditions are linearly independent; that is, the matrix

$$\begin{bmatrix} \alpha_{11} & \alpha_{12} & \alpha_{13} & \alpha_{14} \\ \alpha_{21} & \alpha_{22} & \alpha_{23} & \alpha_{24} \end{bmatrix}$$

has a rank of two, and hence, there exists at least one 2×2 nonsingular submatrix. In other words, at least one of the determinants

$$\Delta_{12} = \begin{vmatrix} \alpha_{11} & \alpha_{12} \\ \alpha_{21} & \alpha_{22} \end{vmatrix} \quad \Delta_{13} = \begin{vmatrix} \alpha_{11} & \alpha_{13} \\ \alpha_{21} & \alpha_{23} \end{vmatrix} \quad \Delta_{14} = \begin{vmatrix} \alpha_{11} & \alpha_{14} \\ \alpha_{21} & \alpha_{24} \end{vmatrix}$$

$$\Delta_{23} = \begin{vmatrix} \alpha_{12} & \alpha_{13} \\ \alpha_{22} & \alpha_{23} \end{vmatrix} \quad \Delta_{24} = \begin{vmatrix} \alpha_{12} & \alpha_{14} \\ \alpha_{22} & \alpha_{24} \end{vmatrix} \quad \Delta_{34} = \begin{vmatrix} \alpha_{13} & \alpha_{14} \\ \alpha_{23} & \alpha_{24} \end{vmatrix}$$

is different from zero. The boundary conditions (15.61) are mixed or nonseparable. A boundary condition is called mixed or nonseparable if it involves the value of the function, its derivative, or both at both ends.

In this section, we consider the case in which $\Delta_{13} \neq 0$. Solving (15.61) for $y'(a)$ and $y'(b)$, we have

$$y'(a) = \gamma_{11}y(a) + \gamma_{12}y(b) + \delta_1$$
$$y'(b) = \gamma_{21}y(a) + \gamma_{22}y(b) + \delta_2 \tag{15.62}$$

where

$$\gamma_{11} = -\frac{\Delta_{23}}{\Delta_{13}} \quad \gamma_{12} = \frac{\Delta_{34}}{\Delta_{13}} \quad \gamma_{21} = -\frac{\Delta_{12}}{\Delta_{13}} \quad \gamma_{22} = -\frac{\Delta_{14}}{\Delta_{13}}$$

$$\delta_1 = \frac{\beta_1 \alpha_{23} - \beta_2 \alpha_{13}}{\Delta_{13}} \quad \delta_2 = \frac{\beta_2 \alpha_{11} - \beta_1 \alpha_{21}}{\Delta_{13}} \tag{15.63}$$

To motivate the discussion, we consider the simple problem

$$y'' + \pi^2 y = \pi \sin \pi x$$
$$y(0) = \beta_1 \quad y(1) = \beta_2 \tag{15.64}$$

In this case, the boundary conditions are separable. The homogeneous problem

$$y'' + \pi^2 y = 0$$
$$y(0) = y(1) = 0 \tag{15.65}$$

has the nontrivial solution $y = \sin \pi x$. Hence, the inhomogeneous problem (15.64) will not have a solution unless a solvability condition is satisfied. To determine this solvability condition, we proceed to find the solution of (15.64) if it exists. As in Appendix B, the general solution of the equation in (15.64) consists of the superposition of a particular solution and the homogeneous solution. The result is

$$y = c_1 \sin \pi x + c_2 \cos \pi x - \tfrac{1}{2} x \cos \pi x \tag{15.66}$$

where c_1 and c_2 are arbitrary constants. Imposing the boundary conditions in (15.64), we have

$$c_2 = \beta_1$$
$$-c_2 + \tfrac{1}{2} = \beta_2$$

These equations are inconsistent and hence the original problem (15.64) does not have a solution unless

$$\beta_1 + \beta_2 = \tfrac{1}{2} \tag{15.67}$$

which is the desired solvability condition. Then, the solution of (15.64) is

$$y = c_1 \sin \pi x + \beta_1 \cos \pi x - \tfrac{1}{2} x \cos \pi x \tag{15.68}$$

In general, one need not follow the above procedure to determine the solvability conditions, especially if one is not interested in actually determining the solutions, as is the case in perturbation-type problems. In such cases, we use the concept of adjoint as follows. We multiply (15.60) by the function $u(x)$, which is called an *adjoint solution* and is specified later, and obtain

$$p_2 u y'' + p_1 u y' + p_0 u y = f u \tag{15.69}$$

Integrating (15.69) term by term from $x = a$ to $x = b$ (i.e., the interval of interest where the boundary conditions are enforced) yields

$$\int_a^b p_2 u y'' \, dx + \int_a^b p_1 u y' \, dx + \int_a^b p_0 u y \, dx = \int_a^b f u \, dx \tag{15.70}$$

Next, we integrate by parts the integrals involving derivatives of y to transfer the derivatives to u. Thus,

$$\int_a^b p_2 u y'' \, dx = p_2 u y' \Big|_a^b - \int_a^b (p_2 u)' y' \, dx$$

and

$$\int_a^b (p_2 u)' y' \, dx = (p_2 u)' y \Big|_a^b - \int_a^b (p_2 u)'' y \, dx$$

so that

$$\int_a^b p_2 u y'' \, dx = [p_2 u y' - (p_2 u)' y]_a^b + \int_a^b (p_2 u)'' y \, dx \qquad (15.71)$$

Moreover,

$$\int_a^b p_1 u y' \, dx = p_1 u y \Big|_a^b - \int_a^b (p_1 u)' y \, dx \qquad (15.72)$$

Substituting (15.71) and (15.72) into (15.70), we have

$$[p_2 u y' - (p_2 u)' y + p_1 u y]_a^b + \int_a^b (p_2 u)'' y \, dx - \int_a^b (p_1 u)' y \, dx$$

$$+ \int_a^b p_0 u y \, dx = \int_a^a f u \, dx$$

or

$$\int_a^b [p_2 u'' + (2p_2' - p_1) u' + (p_0 + p_2'' - p_1') u] y \, dx$$

$$+ \{p_2 u y' + [(p_1 - p_2') u - p_2 u'] y\}_a^b = \int_a^b f u \, dx \qquad (15.73)$$

The equation governing the adjoint u of the homogeneous part of (15.60) is defined by setting the coefficient of y in the integrand of the left-hand side of (15.73) equal to zero. That is,

$$p_2 u'' + (2p_2' - p_1) u' + (p_0 + p_2'' - p_1') u = 0 \qquad (15.74)$$

which is usually called the *adjoint equation* of the homogeneous equation (15.60). To determine the boundary conditions needed to define u, we consider the homogeneous problem in which $f = 0$ and $\delta_1 = \delta_2 = 0$. Then, (15.73) and (15.62) become

$$\{p_2 u y' + [(p_1 - p_2') u - p_2 u'] y\}_{x=b} - \{p_2 u y' + [(p_1 - p_2') u - p_2 u'] y\}_{x=a} = 0$$

$$(15.75)$$

$$y'(a) = \gamma_{11} y(a) + \gamma_{12} y(b)$$
$$y'(b) = \gamma_{21} y(a) + \gamma_{22} y(b) \qquad (15.76)$$

Substituting for $y'(a)$ and $y'(b)$ in (15.75) and collecting the coefficients of $y(a)$ and $y(b)$, we obtain

$$[\gamma_{21} p_2 u|_{x=b} - (\gamma_{11} p_2 + p_1 - p_2') u|_{x=a} + p_2 u'|_{x=a}] y(a)$$
$$- [\gamma_{12} p_2 u|_{x=a} - (\gamma_{22} p_2 + p_1 - p_2') u|_{x=b} + p_2 u'|_{x=b}] y(b) = 0 \qquad (15.77)$$

We choose the adjoint boundary conditions such that the coefficients of $y(a)$ and $y(b)$ in (15.77) vanish independently, that is,

$$\gamma_{21} p_2 u|_{x=b} - (\gamma_{11} p_2 + p_1 - p_2') u|_{x=a} + p_2 u'|_{x=a} = 0$$
$$\gamma_{12} p_2 u|_{x=a} - (\gamma_{22} p_2 + p_1 - p_2') u|_{x=b} + p_2 u'|_{x=b} = 0 \qquad (15.78)$$

Therefore, the adjoint u is defined by the adjoint system consisting of (15.74) and the boundary conditions (15.78).

The homogeneous differential equation (15.60) is said to be self-adjoint if it is the same as its adjoint (15.74). They are the same if

$$2p_2' - p_1 = p_1 \qquad p_2'' - p_1' = 0 \qquad (15.79)$$

or $p_1 = p_2'$. In this case the homogeneous part of (15.60) is

$$p_2 y'' + p_2' y' + p_0 y = 0$$

or

$$(p_2 y')' + p_0 y = 0 \qquad (15.80)$$

Its adjoint (15.74) can be written as

$$(p_2 u')' + p_0 u = 0 \qquad (15.81)$$

and the boundary conditions (15.78) become

$$u'(a) = \gamma_{11} u(a) - \gamma_{21} p_2(b) p_2^{-1}(a) u(b)$$
$$u'(b) = -\gamma_{12} p_2(a) p_2^{-1}(b) u(a) + \gamma_{22} u(b) \qquad (15.82)$$

We note that the boundary conditions on u are in general different from the homogeneous boundary conditions (15.76) on y unless

$$-\gamma_{21} p_2(b) p_2^{-1}(a) = \gamma_{12} \quad \text{or} \quad \gamma_{21} p_2(b) = -\gamma_{12} p_2(a) \qquad (15.83)$$

With the conditions (15.79) and (15.83) being satisfied, the adjoint differential equation (15.81) and its boundary conditions (15.82) are identical to the original homogeneous differential equation (15.60) and its homogeneous boundary conditions (15.62). Such systems are called *self-adjoint systems*.

SECOND-ORDER DIFFERENTIAL SYSTEMS 405

If a second-order homogeneous differential equation is not self-adjoint (i.e., $p_1 \neq p_2'$), we can always transform it into a self-adjoint equation by multiplying it with an appropriate function v. To determine v, we multiply the homogeneous part of (15.60) with v and obtain

$$p_2 v y'' + p_1 v y' + p_0 v y = 0 \qquad (15.84)$$

In order that (15.84) be self-adjoint,

$$p_1 v = (p_2 v)' = p_2 v' + p_2' v$$

Hence,

$$\frac{v'}{v} = \frac{p_1}{p_2} - \frac{p_2'}{p_2}$$

or

$$\ln v = \int \frac{p_1}{p_2} dx - \ln p_2$$

Therefore,

$$v = \frac{1}{p_2} e^{\int (p_1/p_2) dx} \qquad (15.85)$$

We should mention that it is not always possible to transform a differential equation of order higher than two into a self-adjoint equation.

Having defined the adjoint system (15.74) and (15.78), we return to the inhomogeneous system (15.60) and (15.62), to determine the solvability condition. With u satisfying (15.74), (15.73) becomes

$$[p_2 u y' + (p_1 - p_2')uy - p_2 u' y]_{x=b} - [p_2 u y' + (p_1 - p_2')uy$$

$$- p_2 u' y]_{x=a} = \int_a^b fu\, dx \quad (15.86)$$

Substituting for $y'(a)$ and $y'(b)$ from (15.62) into (15.86), we have

$$\delta_2 p_2 u|_{x=b} - \delta_1 p_2 u|_{x=a} + [\gamma_{21} p_2 u|_{x=b} - (\gamma_{11} p_2 + p_1 - p_2')u|_{x=a}$$

$$+ p_2 u'|_{x=a}] y(a) - [\gamma_{12} p_2 u|_{x=a} - (\gamma_{22} p_2 + p_1 - p_2')u|_{x=b}$$

$$+ p_2 u'|_{x=b}] y(b) = \int_a^b fu\, dx \qquad (15.87)$$

Since the terms in the square brackets vanish according to (15.78), (15.87) reduces to the desired solvability condition

$$\delta_2 p_2(b)u(b) - \delta_1 p_2(a)u(a) = \int_a^b f(x)u(x)\,dx \tag{15.88}$$

where u is a solution of the adjoint system consisting of (15.74) and (15.78).

As a special case, let us consider the inhomogeneous Sturm-Liouville problem

$$[p(x)y']' + q(x)y - \lambda r(x)y = f(x) \tag{15.89a}$$

$$y'(a) = \gamma_{11} y(a) + \gamma_{12} y(b) \tag{15.89b}$$

$$y'(b) = \gamma_{21} y(a) + \gamma_{22} y(b) \tag{15.89c}$$

where the condition (15.83) is satisfied and $r(x) > 0$ on $[a, b]$. If λ is not an eigenvalue of the homogeneous problem (i.e., the homogeneous problem has only the trivial solution), the inhomogeneous problem has a unique solution for every continuous $f(x)$. On the other hand, if λ is an eigenvalue of the homogeneous problem (i.e., the homogeneous problem has a nontrivial solution), the inhomogeneous problem does not have a solution unless

$$\int_a^b f(x)u(x)\,dx = 0 \tag{15.90}$$

That is, $f(x)$ is orthogonal to the eigenfunctions $u(x)$ corresponding to the eigenvalue λ. These results constitute the so-called Fredholm's alternative theorem:

For a given value of λ, either the inhomogeneous problem (15.89) has a unique solution for each continuous f, or else the homogeneous problem has a nontrivial solution.

In the following section, we consider the general boundary conditions (15.61). In Sections 15.6 and 15.7, we apply the theory to two eigenvalue problems, and in Section 15.8 we apply the theory to sound waves in a duct with sinusoidal walls. In Section 15.9, we modify the present theory to treat a case with a regular singular point and apply it to the vibrations of nearly circular membranes.

15.5. General Boundary Conditions

ADJOINT OPERATOR

We let L be the second-order differential operator that is defined by the second-order differential equation (15.60), that is,

$$L(y) \equiv \left[p_2(x) \frac{d^2}{dx^2} + p_1(x) \frac{d}{dx} + p_0 \right] y \tag{15.91}$$

where p_2'', p_1', and p_0 are continuous over the interval $[a, b]$. If $y(x)$ and $u(x)$ are any two functions possessing two continuous derivatives over $[a, b]$, we have

$$\int_a^x uL(y)\,dx = \int_a^x [(p_2 u)y'' + (p_1 u)y' + (p_0 u)y]\,dx \qquad a \leq x \leq b \qquad (15.92)$$

As in the preceding section, integrating (15.92) by parts to transfer the derivatives from y to u, we have

$$\int_a^x uL(y)\,dx = [p_2 uy' - (p_2 u)'y + p_1 uy]_a^x + \int_a^x [(p_2 u)'' - (p_1 u)' + p_0 u]y\,dx$$

$$(15.93)$$

We denote the operator in the integrand on the right-hand side of (15.93) by L^*, that is,

$$L^*(u) \equiv (p_2 u)'' - (p_1 u)' + p_0 u = \left[p_2 \frac{d^2}{dx^2} + (2p_2' - p_1)\frac{d}{dx} + p_0 + p_2'' - p_1' \right] u$$

$$(15.94)$$

Then, (15.93) can be rewritten as

$$\int_a^x [uL(y) - yL^*(u)]\,dx = [p_2(uy' - u'y) + (p_1 - p_2')uy]_a^x \qquad (15.95)$$

The operator L^* is called the *adjoint operator* corresponding to the operator L. Multiplying (15.94) by y and integrating the result by parts to transfer the derivatives from u to y, one can easily show that L is the adjoint operator corresponding to the operator L^*. Thus, L and L^* are adjoint to each other.

As in the preceding section, we call the differential equation

$$L^*(u) = 0 \qquad (15.96)$$

the adjoint of the differential equation

$$L(y) = 0 \qquad (15.97)$$

and vice versa.

If $L = L^*$, we say that the operator L is self-adjoint and the differential equation $L(y) = 0$ is self-adjoint. Comparing (15.91) and (15.94), we conclude that $L = L^*$ if and only if

$$2p_2' - p_1 = p_1 \quad \text{and} \quad p_2'' - p_1' = 0$$

Thus, L is self-adjoint if and only if $p_1 = p_2'$. Then,

$$L = L^* = p_2 \frac{d^2}{dx^2} + p_2' \frac{d}{dx} + p_0$$

408 SOLVABILITY CONDITIONS

As in the preceding section, any second-order differential equation of the form (15.97) can be put in a self-adjoint form by multiplying it by v defined in (15.85).

Differentiating (15.95) with respect to x, we obtain Lagrange's identity

$$uL(y) - yL^*(u) = \frac{d}{dx}[p_2(uy' - u'y) + (p_1 - p_2')uy] \qquad (15.98)$$

The expression in the square brackets is called the *bilinear concomitant* of u and y because, for a given y, it is linear in u whereas, for a given u, it is linear in y. Putting $x = b$ in (15.95), we obtain *Green's identity*

$$\int_a^b [uL(y) - yL^*(u)] \, dx = [p_2(uy' - u'y) + (p_1 - p_2')uy]_a^b \qquad (15.99)$$

The right-hand side of (15.99) can be written as

$$R = [p_2(uy' - u'y) + (p_1 - p_2')uy]_a^b = p_2(b)u(b)y'(b) - p_2(b)u'(b)y(b)$$
$$+ [p_1(b) - p_2'(b)]u(b)y(b) - p_2(a)u(a)y'(a) + p_2(a)u'(a)y(a)$$
$$- [p_1(a) - p_2'(a)]u(a)y(a) = \mathbf{u}_b^T P \mathbf{y}_b \qquad (15.100)$$

where

$$\mathbf{u}_b = \begin{bmatrix} u'(a) \\ u(a) \\ u'(b) \\ u(b) \end{bmatrix} \qquad \mathbf{y}_b = \begin{bmatrix} y'(a) \\ y(a) \\ y'(b) \\ y(b) \end{bmatrix}$$

$$P = \begin{bmatrix} 0 & p_2(a) & 0 & 0 \\ -p_2(a) & p_2'(a) - p_1(a) & 0 & 0 \\ 0 & 0 & 0 & -p_2(b) \\ 0 & 0 & p_2(b) & p_1(b) - p_2'(b) \end{bmatrix} \qquad (15.101)$$

We note that

$$|P| = [p_2(a)p_2(b)]^2$$

and hence P is a nonsingular matrix. Substituting (15.100) into (15.99) yields

$$\int_a^b [uL(y) - yL^*(u)] \, dx = \mathbf{u}_b^T P \mathbf{y}_b \qquad (15.102)$$

ADJOINT HOMOGENEOUS SYSTEM

Since the general boundary conditions (15.61) involve linear combinations of the components $y'(a)$, $y(a)$, $y'(b)$, and $y(b)$ of $\mathbf{y}_b$, we introduce a linear nonsingular transformation from y_b to $\mathbf{Y}$ according to

$$\mathbf{Y} = \mathcal{C} \mathbf{y}_b \tag{15.103}$$

where

$$\mathbf{Y} = \begin{bmatrix} Y_1 \\ Y_2 \\ Y_3 \\ Y_4 \end{bmatrix} \qquad \mathcal{C} = \begin{bmatrix} \alpha_{11} & \alpha_{12} & \alpha_{13} & \alpha_{14} \\ \alpha_{21} & \alpha_{22} & \alpha_{23} & \alpha_{24} \\ \alpha_{31} & \alpha_{32} & \alpha_{33} & \alpha_{34} \\ \alpha_{41} & \alpha_{42} & \alpha_{43} & \alpha_{44} \end{bmatrix} \tag{15.104}$$

We note that the above transformation can be accomplished in an infinite number of ways, depending on the choice of $\mathcal{C}$. We already narrowed down this choice by requiring the first two rows in $\mathcal{C}$ to be the same as the α_{ij} in (15.61). This choice left the last two rows in $\mathcal{C}$ arbitrary, except that they are linearly independent from each other and from the first two rows so that $|\mathcal{C}| \neq 0$. For a given nonzero y_b, the last two rows in $\mathcal{C}$ can be chosen to yield any desired nonzero values for Y_3 and Y_4. This observation is used later to determine the adjoint boundary conditions.

Since $|\mathcal{C}| \neq 0$, we can invert the transformation (15.103) and obtain

$$\mathbf{y}_b = \mathcal{C}^{-1} \mathbf{Y}$$

Then, (15.102) can be rewritten as

$$\int_a^b [uL(y) - yL^*(u)] \, dx = \mathbf{u}_b^T P \mathcal{C}^{-1} \mathbf{Y}$$

or

$$\int_a^b [uL(y) - yL^*(u)] \, dx = \mathbf{U}^T \mathbf{Y} = U_1 Y_1 + U_2 Y_2 + U_3 Y_3 + U_4 Y_4 \tag{15.105}$$

where

$$\mathbf{U}^T = \mathbf{u}_b^T P \mathcal{C}^{-1} \qquad \text{or} \qquad \mathbf{U} = (\mathcal{C}^{-1})^T P^T \mathbf{u}_b \tag{15.106}$$

The bilinear form $\mathbf{U}^T \mathbf{Y}$ in (15.105) is referred to as the *canonical representation* of the bilinear form on the right-hand side of (15.102).

To determine the boundary conditions defining the adjoint u of y, we put $L(y) = 0$ and $L^*(u) = 0$ in (15.105) and obtain

$$U_1 Y_1 + U_2 Y_2 + U_3 Y_3 + U_4 Y_4 = 0 \tag{15.107}$$

It follows from (15.103) and (15.61) that the homogeneous boundary conditions on y are

$$Y_1 = \alpha_{11} y'(a) + \alpha_{12} y(a) + \alpha_{13} y'(b) + \alpha_{14} y(b) = 0 \tag{15.108a}$$

$$Y_2 = \alpha_{21} y'(a) + \alpha_{22} y(a) + \alpha_{23} y'(b) + \alpha_{24} y(b) = 0 \qquad (15.108\text{b})$$

Hence, (15.107) becomes

$$U_3 Y_3 + U_4 Y_4 = 0 \qquad (15.109)$$

As mentioned above, if $y_b \neq 0$, the last two rows in $\mathcal{Q}$ can be chosen so that Y_3 and Y_4 can assume any desired values other than both zero. In particular, the last rows in $\mathcal{Q}$ can be chosen so that $Y_3 = 1$ and $Y_4 = 0$, and hence, it follows from (15.109) that $U_3 = 0$. Similarly, the last two rows in $\mathcal{Q}$ can be chosen so that $Y_3 = 0$ and $Y_4 = 1$, and hence, it follows from (15.109) that $U_4 = 0$. Therefore, the system adjoint to

$$L(y) = 0 \quad \text{and} \quad Y_1 = Y_2 = 0 \qquad (15.110)$$

is

$$L^*(u) = 0 \quad \text{and} \quad U_3 = U_4 = 0 \qquad (15.111)$$

where U_3 and U_4 are related to the components $u'(a)$, $u(a)$, $u'(b)$, and $u(b)$ of u_b by (15.106). The system (15.110) is said to be self-adjoint if and only if $L = L^*$ and each of U_3 and U_4 is a linear combination of $Y_1(u_b)$ and $Y_2(u_b)$; in other words $u(x) \propto y(x)$.

Since the two boundary conditions (15.61) are linearly independent, at least one of the determinants

$$\Delta_{ij} = \alpha_{1i} \alpha_{2j} - \alpha_{1j} \alpha_{2i} \qquad i \neq j$$

of the 2 × 2 submatrices

$$\begin{bmatrix} \alpha_{1i} & \alpha_{1j} \\ \alpha_{2i} & \alpha_{2j} \end{bmatrix}$$

must be different from zero. Let us assume that $\Delta_{13} \neq 0$ so that we can compare the results with those in the preceding section. Then, we choose Y_3 and Y_4 such that the rows of $\mathcal{Q}$ are linearly independent. For instance, we let $Y_3 = y(a)$ and $Y_4 = -y(b)$ so that the matrix $\mathcal{Q}$ becomes

$$\mathcal{Q} = \begin{bmatrix} \alpha_{11} & \alpha_{12} & \alpha_{13} & \alpha_{14} \\ \alpha_{21} & \alpha_{22} & \alpha_{23} & \alpha_{24} \\ 0 & 1 & 0 & 0 \\ 0 & 0 & 0 & -1 \end{bmatrix} \qquad (15.112)$$

It follows from (15.112) that $|\mathcal{Q}| = -\Delta_{13} \neq 0$. Then,

$$(\mathcal{Q}^{-1})^T = -\frac{1}{\Delta_{13}} \begin{bmatrix} -\alpha_{23} & 0 & \alpha_{21} & 0 \\ \alpha_{13} & 0 & -\alpha_{11} & 0 \\ \Delta_{23} & -\Delta_{13} & \Delta_{12} & 0 \\ \Delta_{34} & 0 & -\Delta_{14} & \Delta_{13} \end{bmatrix} \qquad (15.113)$$

Substituting (15.101) and (15.113) into (15.106) we have

$$U = -\frac{1}{\Delta_{13}} \begin{bmatrix} 0 & \alpha_{23}p_2(a) & 0 & \alpha_{21}p_2(b) \\ 0 & -\alpha_{13}p_2(a) & 0 & -\alpha_{11}p_2(b) \\ -\Delta_{13}p_2(a) & -\Delta_{23}p_2(a) - \Delta_{13}[p_2'(a) - p_1(a)] & 0 & \Delta_{12}p_2(b) \\ 0 & -\Delta_{34}p_2(a) & -\Delta_{13}p_2(b) & -\Delta_{14}p_2(b) + \Delta_{13}[p_1(b) - p_2'(b)] \end{bmatrix} u_b$$

(15.114)

Therefore, the boundary conditions on u are

$$U_3 = p_2(a)u'(a) + \left[p_2(a) - p_1(a) + \frac{\Delta_{23}}{\Delta_{13}} p_2(a) \right] u(a) - \frac{\Delta_{12}}{\Delta_{13}} p_2(b)u(b) = 0$$

(15.115a)

$$U_4 = p_2(b)u'(b) + \frac{\Delta_{34}}{\Delta_{13}} p_2(a)u(a) - \left[p_1(b) - p_2'(b) - \frac{\Delta_{14}}{\Delta_{13}} p_2(b) \right] u(b) = 0$$

(15.115b)

which are identical with (15.78) obtained in the preceding section on account of (15.63).

In order that the system (15.60) and (15.61) be self-adjoint, $L = L^*$ and each of U_3 and U_4 is a linear combination of $Y_1(u_b)$ and $Y_2(u_b)$. As discussed above, $L = L^*$ if and only if $p_1 = p_2'$. Then, (15.115) become

$$u'(a) + \frac{\Delta_{23}}{\Delta_{13}} u(a) - \frac{\Delta_{12}p_2(b)}{\Delta_{13}p_2(a)} u(b) = 0$$

(15.116)

$$u'(b) + \frac{\Delta_{34}p_2(a)}{\Delta_{13}p_2(b)} u(a) + \frac{\Delta_{14}}{\Delta_{13}} u(b) = 0$$

Solving (15.61) for $y'(a)$ and $y'(b)$ when $\beta_1 = \beta_2 = 0$ and replacing y with u, we obtain

$$u'(a) = -\frac{\Delta_{23}}{\Delta_{13}} u(a) + \frac{\Delta_{34}}{\Delta_{13}} u(b)$$

(15.117)

$$u'(b) = -\frac{\Delta_{12}}{\Delta_{13}} u(a) - \frac{\Delta_{14}}{\Delta_{13}} u(b)$$

Comparing (15.116) and (15.117), we conclude that they are identical if and only if

$$p_2(a)\Delta_{34} = p_2(b)\Delta_{12}$$

(15.118)

which is the condition (15.83) obtained in the preceding section. Therefore, the system is self-adjoint if and only if $p_1 = p_2'$ and condition (15.118) is satisfied.

SOLVABILITY CONDITION

Having defined the adjoint system, we return to the inhomogeneous system (15.60) and (15.61). Using the definition (15.111) of the adjoint, we rewrite Green's identity (15.105) as

$$\int_a^b uL(y)\,dx = Y_1 U_1 + Y_2 U_2 \qquad (15.119)$$

But $L(y) = f(x)$, $Y_1 = \beta_1$, and $Y_2 = \beta_2$; therefore, it follows from (15.119) that the solvability condition is

$$\beta_1 U_1 + \beta_2 U_2 = \int_a^b f(x)u(x)\,dx \qquad (15.120)$$

To compare (15.120) with the solvability condition (15.88) obtained in the preceding section, we use the relations of U_1 and U_2 to u_b from (15.114), that is,

$$U_1 = -\frac{\alpha_{23} p_2(a)}{\Delta_{13}} u(a) - \frac{\alpha_{21} p_2(b)}{\Delta_{13}} u(b)$$

$$U_2 = \frac{\alpha_{13} p_2(a)}{\Delta_{13}} u(a) + \frac{\alpha_{11} p_2(b)}{\Delta_{13}} u(b) \qquad (15.121)$$

Then, (15.120) becomes

$$\frac{\alpha_{11}\beta_2 - \alpha_{21}\beta_1}{\Delta_{13}} p_2(b)u(b) - \frac{\alpha_{23}\beta_1 - \alpha_{13}\beta_2}{\Delta_{13}} p_2(a)u(a) = \int_a^b f(x)u(x)\,dx$$

which is identical with (15.88) on account of (15.63).

Comparing the development in this section with that in the preceding section, we conclude that the algebra is less involved in the preceding section. Consequently, when one faces general boundary conditions, such as those given by (15.61), we recommend that one solve for two of the end values in terms of the other two as we did in (15.62) and then proceed as in the preceding section.

15.6. A Simple Eigenvalue Problem

We consider the eigenvalue problem

$$y'' + [\lambda + \epsilon g(x)]y = 0 \qquad \epsilon \ll 1$$
$$y(0) = 0 \qquad y(\pi) = 0 \qquad (15.122)$$

where λ is an eigenvalue. Since λ is a function of the parameter ϵ, we seek a first-order uniform expansion by using the method of strained parameters and expand both λ and y in powers of ϵ as

$$y = y_0(x) + \epsilon y_1(x) + \cdots$$
$$\lambda = \lambda_0 + \epsilon \lambda_1 + \cdots \tag{15.123}$$

Substituting (15.123) into (15.122), we have

$$y_0'' + \epsilon y_1'' + [\lambda_0 + \epsilon \lambda_1 + \epsilon g](y_0 + \epsilon y_1) + \cdots = 0$$
$$y_0(0) + \epsilon y_1(0) + \cdots = 0 \qquad y_0(\pi) + \epsilon y_1(\pi) + \cdots = 0$$

Equating coefficients of like powers of ϵ yields

Order ϵ^0

$$y_0'' + \lambda_0 y_0 = 0$$
$$y_0(0) = y_0(\pi) = 0 \tag{15.124}$$

Order ϵ

$$y_1'' + \lambda_0 y_1 = -g y_0 - \lambda_1 y_0 \tag{15.125}$$
$$y_1(0) = y_1(\pi) = 0 \tag{15.126}$$

The general solution of the differential equation in (15.124) can be written as

$$y_0 = c_1 \cos \sqrt{\lambda_0}\, x + c_2 \sin \sqrt{\lambda_0}\, x$$

Imposing the boundary condition $y_0(0) = 0$ leads to $c_1 = 0$. Then, imposing the boundary condition $y_0(\pi) = 0$ leads to

$$c_2 \sin \sqrt{\lambda_0}\, \pi = 0$$

Hence, for a nontrivial solution (i.e., $c_2 \neq 0$)

$$\sin \sqrt{\lambda_0}\, \pi = 0 \quad \text{or} \quad \sqrt{\lambda_0}\, \pi = n\pi \quad n = 1, 2, \cdots$$

Hence, to zeroth order, the eigenfunctions are

$$y_0 = \sin nx \tag{15.127}$$

and the eigenvalues are

$$\lambda_0 = n^2 \tag{15.128}$$

Substituting (15.127) and (15.128) into (15.125), we have

$$y_1'' + n^2 y_1 = -g(x) \sin nx - \lambda_1 \sin nx \tag{15.129}$$

Since the homogeneous first-order problem (15.129) and (15.126) has a nontrivial solution, the inhomogeneous problem has a solution only if a solvability condition is satisfied. Instead of applying the general results of the preceding section, we find it more instructive to derive the condition again. In (15.129),

$p_2 = 1$ and $p_1 = 0$ so that $p'_2 = p_1$ and (15.118) is satisfied so that the problem is self-adjoint. Hence, the solution u of the adjoint problem can be taken as $u = y_0 = \sin nx$. Multiplying (15.129) with u and integrating the result by parts from $x = 0$ to $x = \pi$, we obtain

$$\int_0^\pi y_1(u'' + n^2 u)\, dx + [y'_1 u - y_1 u']_0^\pi = - \int_0^\pi [g(x) + \lambda_1] u \sin nx\, dx$$

(15.130)

Since $u = \sin nx$ and $y_1(0) = 0$ and $y_1(\pi) = 0$, the left-hand side vanishes and the solvability condition becomes

$$\int_0^\pi g(x) \sin^2 nx\, dx + \lambda_1 \int_0^\pi \sin^2 nx\, dx = 0$$

Since the second integral is $\frac{1}{2}\pi$,

$$\lambda_1 = -\frac{2}{\pi} \int_0^\pi g(x) \sin^2 nx\, dx \qquad (15.131)$$

Substituting (15.127), (15.128), and (15.131) into (15.123), we find that to the first approximation

$$y = \sin nx + O(\epsilon)$$

$$\lambda = n^2 - \frac{2\epsilon}{\pi} \int_0^\pi g(x) \sin^2 nx\, dx + O(\epsilon^2) \qquad (15.132)$$

15.7. A Degenerate Eigenvalue Problem

We consider the eigenvalue problem

$$y'' + [\lambda + \epsilon f(x)] y = 0, \quad \epsilon \ll 1$$
$$y(0) = y(1), \quad y'(0) = y'(1) \qquad (15.133)$$

As in the preceding section, we seek a first-order uniform expansion by expanding both y and λ as

$$y(x; \epsilon) = y_0(x) + \epsilon y_1(x) + \cdots$$
$$\lambda = \lambda_0 + \epsilon \lambda_1 + \cdots \qquad (15.134)$$

Substituting (15.134) into (15.133) and equating coefficients of like powers of ϵ, we obtain

Order ϵ^0

$$y_0'' + \lambda_0 y_0 = 0 \tag{15.135}$$

$$y_0(0) = y_0(1) \quad y_0'(0) = y_0'(1)$$

Order ϵ

$$y_1'' + \lambda_0 y_1 = -(\lambda_1 + f) y_0 \tag{15.136}$$

$$y_1(0) = y_1(1) \quad y_1'(0) = y_1'(1)$$

The general solution of the equation in (15.135) can be written as

$$y_0 = a_1 \cos \sqrt{\lambda_0}\, x + a_2 \sin \sqrt{\lambda_0}\, x$$

Imposing the boundary conditions in (15.135), we have

$$a_1 = a_1 \cos \sqrt{\lambda_0} + a_2 \sin \sqrt{\lambda_0}$$

$$a_2 = -a_1 \sin \sqrt{\lambda_0} + a_2 \cos \sqrt{\lambda_0}$$

or

$$(\cos \sqrt{\lambda_0} - 1) a_1 + \sin \sqrt{\lambda_0}\, a_2 = 0$$
$$-\sin \sqrt{\lambda_0}\, a_1 + (\cos \sqrt{\lambda_0} - 1) a_2 = 0 \tag{15.137}$$

For a nontrivial solution, the determinant of the coefficient matrix in (15.137) must be zero, that is,

$$(\cos \sqrt{\lambda_0} - 1)^2 + \sin^2 \sqrt{\lambda_0} = 0$$

which for real λ_0 demands that

$$\sin \sqrt{\lambda_0} = 0 \quad \text{and} \quad \cos \sqrt{\lambda_0} = 1$$

Hence,

$$\lambda_0 = 4n^2 \pi^2 \quad n = 0, 1, 2, \cdots \tag{15.138}$$

and it follows from (15.137) that a_1 and a_2 are arbitrary. Then,

$$y_0 = a_1 \cos 2n\pi x + a_2 \sin 2n\pi x \tag{15.139}$$

for arbitrary a_1 and a_2.

Thus for every $\lambda_0 = 4n^2 \pi^2$, where $n \geq 1$, there are two different eigenfunctions, namely $\cos 2n\pi x$ and $\sin 2n\pi x$. Eigenvalue problems having two or more eigenfunctions corresponding to the same eigenvalue are called *degenerate eigenvalue problems*. The degeneracy is a result of the symmetry of the problem and it may be removed by the presence of an asymmetry in the problem. As discussed below, the term $f(x) y(x)$ in the present example may produce such an asymmetry to remove the degeneracy.

416 SOLVABILITY CONDITIONS

Substituting (15.139) into the equation in (15.136) and recalling that $\lambda_0 = 4n^2\pi^2$, we have

$$y_1'' + 4n^2\pi^2 y_1 = -(\lambda_1 + f)(a_1 \cos 2n\pi x + a_2 \sin 2n\pi x) \qquad (15.140)$$

Since the homogeneous problem governing y_1 is the same as (15.135) and since the latter has a nontrivial solution, the inhomogeneous problem governing y_1 has a solution only if solvability conditions are satisfied. To determine these solvability conditions, we multiply (15.140) by $u(x)$, integrate the result by parts from $x = 0$ to $x = 1$ to transfer the derivatives from y_1 to u, and obtain

$$[y_1' u - y_1 u']_0^1 + \int_0^1 y_1(u'' + 4n^2\pi^2 u)\, dx = -\int_0^1 u(\lambda_1 + f)(a_1 \cos 2n\pi x$$

$$+ a_2 \sin 2n\pi x)\, dx \qquad (15.141)$$

To define the adjoint function u, we first consider the homogeneous problem, that is, we put $\lambda_1 = 0$ and $f = 0$. Then, the adjoint equation is

$$u'' + 4n^2\pi^2 u = 0 \qquad (15.142)$$

and (15.141) becomes

$$y_1'(1)u(1) - y_1(1)u'(1) - y_1'(0)u(0) + y_1(0)u'(0) = 0 \qquad (15.143)$$

But $y_1(0) = y_1(1)$ and $y_1'(0) = y'(1)$; hence, (15.143) can be rewritten as

$$[u(1) - u(0)]y_1'(1) - [u'(1) - u'(0)]y_1(1) = 0 \qquad (15.144)$$

We choose the adjoint boundary conditions such that each of the coefficients of $y_1(1)$ and $y_1'(1)$ vanishes independently, that is,

$$u(1) = u(0) \qquad u'(1) = u'(0) \qquad (15.145)$$

Thus, the homogeneous system (15.136) is self-adjoint. Hence,

$$u(x) = \sin 2n\pi x \quad \text{or} \quad \cos 2n\pi x \qquad (15.146)$$

Having defined the adjoint problem, we return to the inhomogeneous system. Using (15.142) and (15.145), we reduce Green's identity (15.141) to

$$\int_0^1 u(x)[\lambda_1 + f(x)](a_1 \cos 2n\pi x + a_2 \sin 2n\pi x)\, dx = 0 \qquad (15.147)$$

which yields the desired solvability conditions. We note that (15.147) should hold for all possible values of $u(x)$ that satisfy the adjoint problem. In the present example, when $n \geq 1$, $u(x) = \sin 2n\pi x$ or $\cos 2n\pi x$. Hence, the inhomogeneous problem governing y_1 has a solution only if (15.147) is satisfied when $u(x) = \sin 2n\pi x$ or $\cos 2n\pi x$.

Putting $u(x) = \sin 2n\pi x$ in (15.147), we have

A DEGENERATE EIGENVALUE PROBLEM

$$f_{12}a_1 + (\lambda_1 + f_{11})a_2 = 0 \qquad (15.148)$$

whereas putting $u(x) = \cos 2n\pi x$ in (15.147), we have

$$(\lambda_1 + f_{22})a_1 + f_{12}a_2 = 0 \qquad (15.149)$$

where

$$f_{11} = 2\int_0^1 f(x)\sin^2 2n\pi x\, dx \qquad f_{22} = 2\int_0^1 f(x)\cos^2 2n\pi x\, dx,$$

$$f_{12} = \int_0^1 f(x)\sin 4n\pi x\, dx$$

For a nontrivial solution, the determinant of the coefficient matrix in (15.148) and (15.149) must be equal to zero, that is,

$$(\lambda_1 + f_{11})(\lambda_1 + f_{22}) - f_{12}^2 = 0$$

or

$$\lambda_1^2 + (f_{11} + f_{22})\lambda_1 + f_{11}f_{22} - f_{12}^2 = 0$$

Hence,

$$\lambda_1 = \lambda_1^{(1)} \text{ or } \lambda_1^{(2)} = -\tfrac{1}{2}(f_{11} + f_{22}) \mp \tfrac{1}{2}[(f_{11} - f_{22})^2 + 4f_{12}^2]^{1/2} \qquad (15.150)$$

Then, it follows from (15.149) that

$$a_2 = -\frac{\lambda_1^{(m)} + f_{22}}{f_{12}}a_1 \qquad (15.151)$$

Therefore, to the first approximation either

$$y^{(1)} = \cos 2n\pi x - \frac{\lambda_1^{(1)} + f_{22}}{f_{12}}\sin 2n\pi x + O(\epsilon)$$

$$\lambda^{(1)} = 4n^2\pi^2 - \tfrac{1}{2}\epsilon\{f_{11} + f_{22} + [(f_{11} - f_{22})^2 + 4f_{12}^2]^{1/2}\} + O(\epsilon^2)$$

$$(15.152)$$

or

$$y^{(2)} = \cos 2n\pi x - \frac{\lambda_1^{(2)} + f_{22}}{f_{12}}\sin 2n\pi x + O(\epsilon)$$

$$\lambda^{(2)} = 4n^2\pi^2 - \tfrac{1}{2}\epsilon\{f_{11} + f_{22} - [(f_{11} - f_{22})^2 + 4f_{12}^2]^{1/2}\} + O(\epsilon^2)$$

$$(15.153)$$

Consequently, if $\lambda_1^{(1)}$ is different from $\lambda_1^{(2)}$ (i.e., $f_{11} \neq f_{22}$ and $f_{12} \neq 0$), the degeneracy will be removed because corresponding to each eigenvalue $\lambda^{(m)}$ there

418 SOLVABILITY CONDITIONS

is only one eigenfunction $y^{(m)}$. If $\lambda_1^{(1)} = \lambda_1^{(2)}$, the degeneracy may be removed at higher order.

15.8. Acoustic Waves in a Duct with Sinusoidal Walls

We consider the porblem of linear harmonic acoustic wave propagation in a two-dimensional duct (waveguide) with sinusoidally varying walls. The problem can be stated mathematically as

$$\phi_{xx} + \phi_{yy} + \omega^2 \phi = 0 \tag{15.154}$$

$$\phi_y = 0 \quad \text{at} \quad y = 0 \tag{15.155}$$

$$\phi_y = \epsilon k_w \phi_x \cos k_w x \quad \text{at} \quad y = 1 + \epsilon \sin k_w x \tag{15.156}$$

where ω and k_w are constants and ϵ is a small dimensionless parameter. Except for the form of the boundary conditions, this problem is a prototype for electromagnetic and elastic waves in waveguides.

As a first step to obtaining a uniform first-order expansion, we carry out a straightforward expansion in the form

$$\phi = \phi_0(x, y) + \epsilon \phi_1(x, y) + \cdots \tag{15.157}$$

We note that the boundary condition (15.156) is imposed at $y = 1 + \epsilon \sin k_w x$, and hence, ϵ appears in the argument of ϕ as well as in the coefficients. Since the usual procedure in perturbation methods is to equate coefficients of equal powers of ϵ, we will not be able to do that unless ϵ is removed from the argument. To accomplish this, we perform what is usually referred to as the *transfer of the boundary condition*. In this case, we transfer the boundary condition from $y = 1 + \epsilon \sin k_w x$ to $y = 1$ by a Taylor series expansion. We write ϕ_y at $y = 1 + \epsilon \sin k_w x$ in (15.156) as

$$\phi_y(x, 1 + \epsilon \sin k_w x)$$

Expanding it in a Taylor series about $y = 1$, we have

$$\phi_y(x, 1 + \epsilon \sin k_w x) = \phi_y(x, 1) + \phi_{yy}(x, 1)\epsilon \sin k_w x$$

$$+ \frac{1}{2!} \phi_{yyy}(x, 1)\epsilon^2 \sin^2 k_w x$$

$$+ \frac{1}{3!} \phi_{yyyy}(x, 1)\epsilon^3 \sin^3 k_w x + \cdots$$

Similarly, we expand ϕ_x at $y = 1 + \epsilon \sin k_w x$ as

$$\phi_x(x, 1 + \epsilon \sin k_w x) = \phi_x(x, 1) + \phi_{xy}(x, 1)\epsilon \sin k_w x$$

$$+ \frac{1}{2!} \phi_{xyy}(x, 1)\epsilon^2 \sin^2 k_w x$$

$$+ \frac{1}{3!} \phi_{xyyy}(x, 1)\epsilon^3 \sin^3 k_w x + \cdots$$

Substituting these Taylor-series expansions into (15.156), we obtain

$$\phi_y(x, 1) + \epsilon \phi_{yy}(x, 1) \sin k_w x = \epsilon k_w \phi_x(x, 1) \cos k_w x + \cdots \quad (15.158)$$

thereby transferring the boundary condition from $y = 1 + \epsilon \sin k_w x$ to $y = 1$ and removing ϵ from the arguments of ϕ_y and ϕ_x. Now, we are ready to carry out the straightforward expansion.

Substituting (15.157) into (15.154), (15.155), and (15.158), we have

$$\phi_{0xx} + \epsilon \phi_{1xx} + \phi_{0yy} + \epsilon \phi_{1yy} + \omega^2 \phi_0 + \epsilon \omega^2 \phi_1 + \cdots = 0$$

$$\phi_{0y}(x, 0) + \epsilon \phi_{1y}(x, 0) + \cdots = 0$$

$$\phi_{0y}(x, 1) + \epsilon \phi_{1y}(x, 1) + \epsilon \phi_{0yy}(x, 1) \sin k_w x = \epsilon k_w \phi_{0x}(x, 1) \cos k_w x + \cdots$$

Equating coefficients of like powers of ϵ leads to

Order ϵ^0

$$\phi_{0xx} + \phi_{0yy} + \omega^2 \phi_0 = 0 \quad (15.159)$$

$$\phi_{0y}(x, 0) = 0 \quad (15.160)$$

$$\phi_{0y}(x, 1) = 0 \quad (15.161)$$

Order ϵ

$$\phi_{1xx} + \phi_{1yy} + \omega^2 \phi_1 = 0 \quad (15.162)$$

$$\phi_{1y}(x, 0) = 0 \quad (15.163)$$

$$\phi_{1y}(x, 1) = -\phi_{0yy}(x, 1) \sin k_w x + k_w \phi_{0x}(x, 1) \cos k_w x \quad (15.164)$$

Since (15.159) through (15.161) have constant coefficients, they can be solved by separation of variables. To this end, we let

$$\phi_0 = X(x)Y(y) \quad (15.165)$$

and obtain

$$X''Y + XY'' + \omega^2 XY = 0 \quad (15.166)$$

$$X(x)Y'(0) = 0 \quad (15.167)$$

$$X(x)Y'(1) = 0 \quad (15.168)$$

Dividing (15.166) by XY and rearranging, we have

$$-\frac{X''}{X} = \frac{Y''}{Y} + \omega^2 \qquad (15.169)$$

Since the left-hand side of (15.169) is a function of x only and its right-hand side is a function of y only, we conclude that each side must be a constant, that is,

$$-\frac{X''}{X} = b \qquad (15.170)$$

$$\frac{Y''}{Y} + \omega^2 = b \qquad (15.171)$$

Equation (15.170) can be rewritten as

$$X'' + bX = 0$$

For propagating waves, X must be sinusoidal, and hence, b must be positive. Usually one puts $b = k^2$ so that

$$X = e^{\pm ikx} \qquad (15.172)$$

and k is called the *wavenumber*. Putting $b = k^2$, we rewrite (15.171) as

$$Y'' + (\omega^2 - k^2)Y = 0$$

whose general solution is

$$Y = c_1 \sin \sqrt{\omega^2 - k^2}\, y + c_2 \cos \sqrt{\omega^2 - k^2}\, y \qquad (15.173)$$

Since $X(x) \neq 0$, it follows from (15.167) and (15.168) that

$$Y'(0) = Y'(1) = 0 \qquad (15.174)$$

Imposing these conditions in (15.173), we find that $c_1 = 0$ and

$$\sqrt{\omega^2 - k^2} = n\pi \quad \text{so that} \quad k_n^2 = \omega^2 - n^2\pi^2 \quad n = 0, 1, 2, \cdots \qquad (15.175)$$

Then,

$$Y = \cos n\pi y \qquad (15.176)$$

Substituting (15.172) and (15.176) into (15.165), we have

$$\phi_0 = e^{ik_n x} \cos n\pi y \quad \text{or} \quad e^{-ik_n x} \cos n\pi y \qquad (15.177)$$

where one of these solutions corresponds to a wave propagating in one direction and the other solution corresponds to a wave propagating in the opposite direction. The solution corresponding to a given n is called the nth mode. Let us take the case with the positive sign.

ACOUSTIC WAVES IN A DUCT WITH SINUSOIDAL WALLS   421

Substituting (15.177) with the positive sign into (15.164), we have

$$\phi_{1y}(x,1) = (-1)^n n^2 \pi^2 e^{ik_n x} \sin k_w x + i(-1)^n k_n k_w e^{ik_n x} \cos k_w x$$
$$= (-1)^n n^2 \pi^2 e^{ik_n x} \cdot -\tfrac{1}{2}i(e^{ik_w x} - e^{-ik_w x})$$
$$+ i(-1)^n k_n k_w e^{ik_n x} \cdot \tfrac{1}{2}(e^{ik_w x} + e^{-ik_w x})$$
$$= \zeta_1 e^{i(k_n + k_w)x} + \zeta_2 e^{i(k_n - k_w)x} \quad (15.178)$$

where

$$\zeta_1 = \tfrac{1}{2}(-1)^n i(k_n k_w - n^2 \pi^2) \qquad \zeta_2 = \tfrac{1}{2}(-1)^n i(k_n k_w + n^2 \pi^2) \quad (15.179)$$

To find the solution of (15.162), (15.163), and (15.178) for ϕ_1, we note that the boundary condition (15.178) is inhomogeneous and its form suggests that the variables be separated as follows:

$$\phi_1 = \Phi_1(y) e^{i(k_n + k_w)x} + \Phi_2(y) e^{i(k_n - k_w)x} \quad (15.180)$$

Substituting (15.180) into (15.162), (15.163), and (15.178), we have

$$[\Phi_1'' + \alpha_1^2 \Phi_1] e^{i(k_n + k_w)x} + [\Phi_2'' + \alpha_2^2 \Phi_2] e^{i(k_n - k_w)x} = 0$$
$$\Phi_1'(0) e^{i(k_n + k_w)x} + \Phi_2'(0) e^{i(k_n - k_w)x} = 0$$
$$\Phi_1'(1) e^{i(k_n + k_w)x} + \Phi_2'(1) e^{i(k_n - k_w)x} = \zeta_1 e^{i(k_n + k_w)x}$$
$$+ \zeta_2 e^{i(k_n - k_w)x}$$

where

$$\alpha_1^2 = \omega^2 - (k_n + k_w)^2 \qquad \alpha_2^2 = \omega^2 - (k_n - k_w)^2 \quad (15.181)$$

Equating the coefficients of each of the exponentials on both sides, we obtain

$$\Phi_1'' + \alpha_1^2 \Phi_1 = 0 \quad (15.182)$$
$$\Phi_1'(0) = 0 \qquad \Phi_1'(1) = \zeta_1$$

$$\Phi_2'' + \alpha_2^2 \Phi_2 = 0 \quad (15.183)$$
$$\Phi_2'(0) = 0 \qquad \Phi_2'(1) = \zeta_2$$

It follows from (15.182) that the general solution for Φ_1 is

$$\Phi_1 = c_1 \cos \alpha_1 y + c_2 \sin \alpha_1 y$$

Imposing the boundary conditions in (15.182), we find that

$$c_2 = 0 \quad \text{and} \quad c_1 = -\frac{\zeta_1}{\alpha_1 \sin \alpha_1}$$

Hence,

$$\Phi_1 = -\frac{\zeta_1 \cos \alpha_1 y}{\alpha_1 \sin \alpha_1}$$

Similarly, the solution of (15.183) can be found to be

$$\Phi_2 = -\frac{\zeta_2 \cos \alpha_2 y}{\alpha_2 \sin \alpha_2}$$

Therefore,

$$\phi_1 = -\frac{\zeta_1 \cos \alpha_1 y}{\alpha_1 \sin \alpha_1} e^{i(k_n + k_w)x} - \frac{\zeta_2 \cos \alpha_2 y}{\alpha_2 \sin \alpha_2} e^{i(k_n - k_w)x} \quad (15.184)$$

Substituting (15.177) and (15.184) into (15.157), we obtain

$$\phi = \cos n\pi y\, e^{ik_n x} - \epsilon \left[\frac{\zeta_1 \cos \alpha_1 y}{\alpha_1 \sin \alpha_1} e^{i(k_n + k_w)x} \right.$$
$$\left. + \frac{\zeta_2 \cos \alpha_2 y}{\alpha_2 \sin \alpha_2} e^{i(k_n - k_w)x} \right] + \cdots \quad (15.185)$$

Inspection of the expansion (15.185) shows that it breaks down if $\sin \alpha_1 = O(\epsilon)$ or $\sin \alpha_2 = O(\epsilon)$. If either $\sin \alpha_1$ or $\sin \alpha_2 = 0$, the second term tends to infinity. When $\sin \alpha_l = O(\epsilon)$, a small-divisor term appears, making the expansion nonuniform. Since $\sin \alpha_l = 0$ implies that $\alpha_l = m\pi$, $m = 0, 1, 2, \ldots$, the straightforward expansion breaks down when

$$\omega^2 - (k_n + k_w)^2 \approx m^2 \pi^2 \quad \text{or} \quad \omega^2 - (k_n - k_w)^2 \approx m^2 \pi^2 \quad (15.186)$$

according to (15.181). But $\omega^2 - m^2 \pi^2 = k_m^2$ according to (15.175); hence, (15.186) can be expressed as

$$(k_n + k_w)^2 \approx k_m^2 \quad \text{or} \quad (k_n - k_w)^2 \approx k_m^2$$

which can be rewritten as

$$k_w \approx \pm k_n \pm k_m \quad (15.187)$$

In other words, the straightforward expansion breaks down whenever the wavenumber of the wall undulations is approximately equal to the sum or difference of the wavenumbers k_m and k_n of two propagating modes, that is, whenever a combination resonance exists.

To determine an expansion valid when $k_w \approx k_n - k_m$, we introduce a detuning parameter σ according to

$$k_w = k_n - k_m + \epsilon \sigma \quad (15.188)$$

Moreover, we use the method of multiple scales and seek the expansion in the form

$$\phi(x, y; \epsilon) = \phi(x_0, x_1, y; \epsilon) = \phi_0(x_0, x_1, y) + \epsilon\phi_1(x_0, x_1, y) + \cdots \quad (15.189)$$

where $x_0 = x$ and $x_1 = \epsilon x$. Thus,

$$\frac{\partial}{\partial x} = \frac{\partial}{\partial x_0} + \epsilon \frac{\partial}{\partial x_1} + \cdots \qquad \frac{\partial^2}{\partial x^2} = \frac{\partial^2}{\partial x_0^2} + 2\epsilon \frac{\partial^2}{\partial x_0 \partial x_1} + \cdots \quad (15.190)$$

Substituting (15.190) and (15.189) into (15.154), (15.155), and (15.158) and equating coefficients of like powers of ϵ, we obtain

Order ϵ^0

$$\frac{\partial^2 \phi_0}{\partial x_0^2} + \frac{\partial^2 \phi_0}{\partial y^2} + \omega^2 \phi_0 = 0$$

$$\frac{\partial \phi_0}{\partial y} = 0 \quad \text{at} \quad y = 0 \qquad \frac{\partial \phi_0}{\partial y} = 0 \quad \text{at} \quad y = 1 \quad (15.191)$$

Order ϵ

$$\frac{\partial^2 \phi_1}{\partial x_0^2} + \frac{\partial^2 \phi_1}{\partial y^2} + \omega^2 \phi_1 = -2 \frac{\partial^2 \phi_0}{\partial x_0 \partial x_1} \quad (15.192)$$

$$\frac{\partial \phi_1}{\partial y} = 0 \quad \text{at} \quad y = 0 \quad (15.193)$$

$$\frac{\partial \phi_1}{\partial y} = -\frac{\partial^2 \phi_0}{\partial y^2} \sin k_w x_0 + k_w \frac{\partial \phi_0}{\partial x_0} \cos k_w x_0 \quad \text{at} \quad y = 1 \quad (15.194)$$

where $\sin k_w x$ and $\cos k_w x$ are expressed in terms of x_0, implying that k_w is assumed to be away from zero.

The solution of (15.191) can be obtained by separating variables as done above. However, instead of making ϕ_0 contain only one mode, we make ϕ_0 contain the two interacting modes, namely the mth and nth modes, and hence, write

$$\phi_0 = A_n(x_1) \cos n\pi y \, e^{ik_n x_0} + A_m(x_1) \cos m\pi y \, e^{ik_m x_0} \quad (15.195)$$

where k_m and k_n are defined by (15.175) and A_n and A_m are determined by imposing the solvability conditions at the next level of approximation. Substituting (15.195) into (15.192) and (15.194), we have

$$\frac{\partial^2 \phi_1}{\partial x_0^2} + \frac{\partial^2 \phi_1}{\partial y^2} + \omega^2 \phi_1 = -2ik_n A_n' \cos n\pi y \, e^{ik_n x_0} - 2ik_m A_m' \cos m\pi y \, e^{ik_m x_0}$$

$$(15.196)$$

$$\frac{\partial \phi_1}{\partial y} = \zeta_{1n} A_n e^{i(k_n + k_w) x_0} + \zeta_{2n} A_n e^{i(k_n - k_w) x_0}$$

$$+ \zeta_{1m}A_m e^{i(k_m + k_w)x_0} + \zeta_{2m}A_m e^{i(k_m - k_w)x_0} \quad \text{at} \quad y = 1 \quad (15.197)$$

where ζ_{1n} and ζ_{2n} are defined by (15.179) and ζ_{1m} and ζ_{2m} are also defined by (15.179) if n is replaced by m.

To determine the solvability conditions for the first-order problem, we first substitute (15.188) into (15.197) to convert any small-divisor terms into secular terms and obtain

$$\frac{\partial \phi_1}{\partial y} = \zeta_{1n}A_n e^{i(2k_n - k_m + \epsilon\sigma)x_0} + \zeta_{2n}A_n e^{i(k_m - \epsilon\sigma)x_0} + \zeta_{1m}A_m e^{i(k_n + \epsilon\sigma)x_0}$$

$$+ \zeta_{2m}A_m e^{i(2k_m - k_n - \epsilon\sigma)x_0} \quad \text{at} \quad y = 1$$

or

$$\frac{\partial \phi_1}{\partial y} = \zeta_{1n}A_n e^{i\sigma x_1} e^{i(2k_n - k_m)x_0} + \zeta_{2n}A_n e^{-i\sigma x_1} e^{ik_m x_0} + \zeta_{1m}A_m e^{i\sigma x_1} e^{ik_n x_0}$$

$$+ \zeta_{2m}A_m e^{-i\sigma x_1} e^{i(2k_m - k_n)x_0} \quad \text{at} \quad y = 1 \quad (15.198)$$

We note that only the terms proportional to $\exp(ik_m x_0)$ and $\exp(ik_n x_0)$ in (15.196) and (15.198) may lead to inconsistencies or incompatabilities and solvability conditions must be imposed on them. These solvability conditions can be obtained by seeking a particular solution corresponding to these terms in the form

$$\phi_1 = \Phi_n(x_1, y) e^{ik_n x_0} + \Phi_m(x_1, y) e^{ik_m x_0} \quad (15.199)$$

Substituting (15.199) into (15.196), (15.193), and (15.198) and equating the coefficients of $\exp(ik_n x_0)$ and $\exp(ik_m x_0)$ on both sides, we obtain

$$\frac{\partial^2 \Phi_n}{\partial y^2} + n^2 \pi^2 \Phi_n = -2ik_n A'_n \cos n\pi y$$

$$\frac{\partial \Phi_n}{\partial y} = 0 \quad \text{at} \quad y = 0 \qquad \frac{\partial \Phi_n}{\partial y} = \zeta_{1m}A_m e^{i\sigma x_1} \quad \text{at} \quad y = 1$$

(15.200)

$$\frac{\partial^2 \Phi_m}{\partial y^2} + m^2 \pi^2 \Phi_m = -2ik_m A'_m \cos m\pi y$$

$$\frac{\partial \Phi_m}{\partial y} = 0 \quad \text{at} \quad y = 0 \qquad \frac{\partial \Phi_m}{\partial y} = \zeta_{2n}A_n e^{-i\sigma x_1} \quad \text{at} \quad y = 1$$

(15.201)

where use has been made of (15.175). Thus, determining the solvability conditions for ϕ_1 has been transformed into determining the solvability conditions for Φ_n and Φ_m.

The equation in (15.200) is self-adjoint because $p_2 = 1$ and $p_1 = 0$. The solution u of the adjoint problem can be taken as $\cos n\pi y$. Multiplying the equation in (15.200) by $u(y)$ and integrating the result by parts from $y = 0$ to $y = 1$ to transfer the derivatives from Φ_n to u, we obtain

$$\int_0^1 (u'' + n^2\pi^2 u)\Phi_n \, dy + \left[\frac{\partial \Phi_n}{\partial y} u - \Phi_n u'\right]_0^1 = -2ik_n A_n' \int_0^1 u \cos n\pi y \, dy$$

(15.202)

Since $u = \cos n\pi y$, (15.202) simplifies to

$$\left.\frac{\partial \Phi_n}{\partial y} \cos n\pi y\right|_0^1 = -i\delta k_n A_n'$$

where

$$\delta = 1 \quad \text{if} \quad n \geq 1 \quad \text{and} \quad \delta = 2 \quad \text{if} \quad n = 0$$

Using the boundary conditions in (15.200), we have

$$(-1)^n \zeta_{1m} A_m e^{i\sigma x_1} = -i\delta k_n A_n'$$

or

$$A_n' = (-1)^n i\zeta_{1m} k_n^{-1} \delta^{-1} A_m e^{i\sigma x_1} \qquad (15.203)$$

Similarly, if $m \neq 0$, the solvability condition for (15.201) can be found to be

$$A_m' = (-1)^m i\zeta_{2n} k_m^{-1} A_n e^{-i\sigma x_1} \qquad (15.204)$$

If we let

$$A_n = a_n e^{i\gamma_1 x_1} \qquad A_m = a_m e^{i\gamma_2 x_1} \qquad (15.205)$$

where a_n, a_m, γ_1, and γ_2 are constants, then it follows from (15.204) and (15.205) that

$$i\gamma_1 a_n = (-1)^n i\zeta_{1m} k_n^{-1} \delta^{-1} a_m \qquad (15.206)$$

$$i\gamma_2 a_m = (-1)^m i\zeta_{2n} k_m^{-1} a_n \qquad (15.207)$$

$$\gamma_2 = \gamma_1 - \sigma \qquad (15.208)$$

Eliminating γ_2 and a_m from (15.206) through (15.208) yields

$$\gamma_1(\gamma_1 - \sigma) = (-1)^{n+m} (k_n k_m \delta)^{-1} \zeta_{1m} \zeta_{2n}$$

or

$$\gamma_1^2 - \sigma\gamma_1 - (-1)^{n+m} (k_n k_m \delta)^{-1} \zeta_{1m} \zeta_{2n} = 0$$

Hence,

$$\gamma_1 = \tfrac{1}{2}\sigma \mp [\tfrac{1}{4}\sigma^2 + (-1)^{n+m} (k_n k_m \delta)^{-1} \zeta_{1m} \zeta_{2n}]^{1/2} \qquad (15.209)$$

15.9. Vibrations of Nearly Circular Membranes

We consider the linear vibrations of nearly circular membranes. In dimensionless quantities, the mathematical statement of the problem is

$$\frac{\partial^2 w}{\partial r^2} + \frac{1}{r}\frac{\partial w}{\partial r} + \frac{1}{r^2}\frac{\partial^2 w}{\partial \theta^2} - \frac{\partial^2 w}{\partial t^2} = 0 \tag{15.210}$$

$$w < \infty \quad \text{at} \quad r = 0 \tag{15.211}$$

$$w = 0 \quad \text{at} \quad r = 1 + \epsilon f(\theta) \tag{15.212}$$

where 1 is the mean radius so that

$$\int_0^{2\pi} f(\theta)\,d\theta = 0 \tag{15.213}$$

For time harmonic variations, we let

$$w(r, \theta, t) = \phi(r, \theta) \cos(\omega t + \tau) \tag{15.214}$$

where ω is the dimensionless frequency. Substituting (15.214) into (15.210) through (15.212), we separate the time variations and obtain

$$\frac{\partial^2 \phi}{\partial r^2} + \frac{1}{r}\frac{\partial \phi}{\partial r} + \frac{1}{r^2}\frac{\partial^2 \phi}{\partial \theta^2} + \omega^2 \phi = 0 \tag{15.215}$$

$$\phi < \infty \quad \text{at} \quad r = 0 \tag{15.216}$$

$$\phi = 0 \quad \text{at} \quad r = 1 + \epsilon f(\theta) \tag{15.217}$$

As in the preceding section, we need to transfer the boundary condition (15.217) from $r = 1 + \epsilon f(\theta)$ to $r = 1$ by using a Taylor-series expansion. Thus, we rewrite (15.217) as

$$\phi(1, \theta) + \epsilon \frac{\partial \phi}{\partial r}(1, \theta) f(\theta) + \cdots = 0 \tag{15.218}$$

To determine a uniform first-order expansion for ϕ, we use the method of strained parameters and expand both ϕ and ω as follows:

$$\begin{aligned} \phi &= \phi_0(r, \theta) + \epsilon \phi_1(r, \theta) + \cdots \\ \omega &= \omega_0 + \epsilon \omega_1 + \cdots \end{aligned} \tag{15.219}$$

Substituting (15.219) into (15.215), (15.216), and (15.218) and equating coefficients of like powers of ϵ, we obtain

Order ϵ^0

$$\frac{\partial^2 \phi_0}{\partial r^2} + \frac{1}{r}\frac{\partial \phi_0}{\partial r} + \frac{1}{r^2}\frac{\partial^2 \phi_0}{\partial \theta^2} + \omega_0^2 \phi_0 = 0 \tag{15.220}$$

$$\phi_0(0, \theta) < \infty \tag{15.221}$$

$$\phi_0(1, \theta) = 0 \tag{15.222}$$

Order ϵ

$$\frac{\partial^2 \phi_1}{\partial r^2} + \frac{1}{r}\frac{\partial \phi_1}{\partial r} + \frac{1}{r^2}\frac{\partial^2 \phi_1}{\partial \theta^2} + \omega_0^2 \phi_1 = -2\omega_0 \omega_1 \phi_0 \tag{15.223}$$

$$\phi_1(0, \theta) < \infty \tag{15.224}$$

$$\phi_1(1, \theta) = -\frac{\partial \phi_0}{\partial r}(1, \theta) f(\theta) \tag{15.225}$$

The solution of the zeroth-order problem can be obtained by separating the variables. Thus, we let

$$\phi_0 = R(r)\Theta(\theta) \tag{15.226}$$

in (15.220) and obtain

$$R''\Theta + \frac{1}{r}R'\Theta + \frac{1}{r^2}R\Theta'' + \omega_0^2 R\Theta = 0$$

which can be rewritten as

$$\frac{1}{R}(r^2 R'' + rR' + r^2 \omega_0^2 R) + \frac{\Theta''}{\Theta} = 0$$

Hence,

$$\frac{\Theta''}{\Theta} = -\beta \qquad \frac{1}{R}(r^2 R'' + rR' + \omega_0^2 r^2 R) = \beta$$

The solution for Θ is

$$\Theta = c_1 \cos \sqrt{\beta}\,\theta + c_2 \sin \sqrt{\beta}\,\theta$$

In order that Θ, and hence, ϕ_0 be single-valued functions, $\sqrt{\beta} = n$ where n is an integer. Hence,

$$\Theta = \tilde{c}_1 \cos n\theta + \tilde{c}_2 \sin n\theta \tag{15.227}$$

and the equation governing R becomes

$$r^2 R'' + rR' + (\omega_0^2 r^2 - n^2)R = 0$$

which is a Bessel's equation of order n whose general solution is

$$R = c_3 J_n(\omega_0 r) + c_4 Y_n(\omega_0 r) \tag{15.228}$$

Substituting (15.226) into (15.221) and (15.222), we find that $R(0) < \infty$ and $R(1) = 0$. Since $Y_n \to \infty$ as $r \to 0$ (see Section 13.3), the boundary condition $R(0) < \infty$ demands that $c_4 = 0$. Then, the boundary condition $R(1) = 0$ demands that

$$J_n(\omega_0) = 0 \tag{15.229}$$

Hence,

$$\omega_0 = \Omega_{nm} \tag{15.230}$$

where the Ω_{nm} are the roots of $J_n(\Omega) = 0$. Therefore,

$$\phi_0 = J_n(\Omega_{nm} r)[c_1 \cos n\theta + c_2 \sin n\theta]$$

or

$$\phi_0 = J_n(\Omega_{nm} r)(A_{nm} e^{in\theta} + \overline{A}_{nm} e^{-in\theta}) \tag{15.231}$$

where the A_{nm} are complex constants. The solution for a given n and m is called the nmth mode and the total solution involves summation of the contributions for all modes. Here, we determine the effect of the deviation from a circular geometry on the frequency Ω_{nm} of the nmth mode.

We note that, for a given frequency Ω_{nm}, there are two possible modes of oscillation, namely

$$J_n(\Omega_{nm} r) \cos n\theta \quad \text{and} \quad J_n(\Omega_{nm} r) \sin n\theta$$

Consequently, we speak of the circular membrane as a degenerate system because there are more than one eigenfunction (mode shape) corresponding to a given eigenvalue (frequency). The degeneracy is a result of the symmetry and it can be removed by introducing asymmetries as shown below.

Substituting (15.231) into (15.223) and (15.225) and setting $\omega_0 = \Omega_{nm}$, we have

$$\frac{\partial^2 \phi_1}{\partial r^2} + \frac{1}{r} \frac{\partial \phi_1}{\partial r} + \frac{1}{r^2} \frac{\partial^2 \phi_1}{\partial \theta^2} + \Omega_{nm}^2 \phi_1 = -2\Omega_{nm} \omega_1 J_n(\Omega_{nm} r)(A_{nm} e^{in\theta} + \overline{A}_{nm} e^{-in\theta}) \tag{15.232}$$

$$\phi_1(1, \theta) = -\Omega_{nm} J_n'(\Omega_{nm}) f(\theta) [A_{nm} e^{in\theta} + \overline{A}_{nm} e^{-in\theta}] \tag{15.233}$$

To proceed further, we expand $f(\theta)$ in a Fourier series as

$$f(\theta) = \sum_{-\infty}^{\infty} f_q e^{iq\theta} \quad f_q = \frac{1}{2\pi} \int_0^{2\pi} f(\theta) e^{-iq\theta} \, d\theta \tag{15.234}$$

where $f_0 = 0$ on account of (15.213). We separate the θ variations from ϕ_1 by expanding it in a Fourier series as

$$\phi_1 = \sum_{-\infty}^{\infty} \Phi_k(r) e^{ik\theta} \qquad (15.235)$$

Substituting (15.235) into (15.232), we have

$$\sum_k \left[\Phi_k'' + \frac{1}{r}\Phi_k' + \left(\Omega_{nm}^2 - \frac{k^2}{r^2}\right)\Phi_k \right] e^{ik\theta}$$
$$= -2\Omega_{nm}\omega_1 J_n(\Omega_{nm}r)[A_{nm}e^{in\theta} + \overline{A}_{nm}e^{-in\theta}]$$

which, upon multiplying by $\exp(-is\theta)$ and integrating from $\theta = 0$ to $\theta = 2\pi$, yields

$$\Phi_n'' + \frac{1}{r}\Phi_n' + \left(\Omega_{nm}^2 - \frac{n^2}{r^2}\right)\Phi_n = -2\Omega_{nm}\omega_1 A_{nm} J_n(\Omega_{nm}r) \qquad (15.236)$$

$$\Phi_s'' + \frac{1}{r}\Phi_s' + \left(\Omega_{nm}^2 - \frac{s^2}{r^2}\right)\Phi_s = 0 \quad \text{for} \quad s \neq n \qquad (15.237)$$

Substituting (15.234) and (15.235) into (15.233), we have

$$\sum_k \Phi_k(1) e^{ik\theta} = -\Omega_{nm} J_n'(\Omega_{nm}) \sum_q [A_{nm} f_q e^{i(q+n)\theta} + \overline{A}_{nm} f_q e^{i(q-n)\theta}]$$

which, upon multiplying by $\exp(-is\theta)$ and integrating from $\theta = 0$ to $\theta = 2\pi$, yields

$$\Phi_s(1) = -\Omega_{nm} J_n'(\Omega_{nm})[A_{nm} f_{s-n} + \overline{A}_{nm} f_{s+n}] \qquad (15.238)$$

Substituting (15.235) into (15.224), we conclude that

$$\Phi_s(0) < \infty \qquad (15.239)$$

When $s \neq \pm n$, one can uniquely solve (15.237) through (15.239) for Φ_s because the homogeneous problem has only the trivial solution. Since we are stopping at this order, we need not solve for Φ_s. When $s = n$, the homogeneous problem (15.236), (15.238), and (15.239) has a nontrivial solution, and hence the inhomogeneous problem has a solution only if a solvability condition is satisfied. To determine this solvability condition, we first multiply (15.236) by r to make it self-adjoint. The result can be rewritten in the self-adjoint form

$$(r\Phi_n')' + \left(\Omega_{nm}^2 r - \frac{n^2}{r}\right)\Phi_n = -2\Omega_{nm}\omega_1 A_{nm} r J_n(\Omega_{nm}r) \qquad (15.240)$$

Moreover, putting $s = n$ in (15.238), we have

$$\Phi_n(1) = -\Omega_{nm} f_{2n} J_n'(\Omega_{nm}) \overline{A}_{nm} \qquad (15.241)$$

430 SOLVABILITY CONDITIONS

because $f_0 = 0$ according to (15.213). We note that (15.240) has a regular singular point at the origin and that the boundary condition (15.239) is a boundedness rather than a definite condition. Hence, we present the details of determining the adjoint and the solvability condition.

Multiplying (15.240) by $u(r)$ and integrating the result by parts from $r = 0$ to $r = 1$, we obtain

$$\int_0^1 \Phi_n \left[(ru')' + \left(\Omega_{nm}^2 r - \frac{n^2}{r} \right) u \right] dr + [ru\Phi_n' - ru'\Phi_n]_0^1$$

$$= -2\Omega_{nm} \omega_1 A_{nm} \int_0^1 r u J_n(\Omega_{nm} r) \, dr \quad (15.242)$$

To determine the adjoint, we first consider the homogeneous problem. As before, we set the coefficient of Φ_n in the integrand on the left-hand side of (15.242) equal to zero, that is,

$$(ru')' + \left(\Omega_{nm}^2 r - \frac{n^2}{r} \right) u = 0 \quad (15.243)$$

Then for the homogeneous case, (15.242) becomes

$$[ru\Phi_n' - ru'\Phi_n]_0^1 = 0$$

But for the homogeneous problem, $\Phi_n(1) = 0$ and $\Phi_n(0) < \infty$; hence,

$$u(1)\Phi_n'(1) - \lim_{r \to 0} [ru\Phi_n' - ru'\Phi_n] = 0 \quad (15.244)$$

We choose the adjoint boundary conditions such that the two terms in (15.244) vanish independently, that is,

$$u(1) = 0 \quad \text{and} \quad \lim_{r \to 0} [ru\Phi_n' - ru'\Phi_n] = 0$$

The second condition is satisfied if $u(0) < \infty$. Hence, the boundary conditions on u are

$$u(1) = 0 \quad \text{and} \quad u(0) < \infty \quad (15.245)$$

Thus, the problem is self-adjoint and the solution of the adjoint problem can be taken to be $u = J_n(\Omega_{nm} r)$.

Returning to the inhomogeneous problem, we obtain from (15.242) that

$$\Omega_{nm}^2 J_n'^2(\Omega_{nm}) f_{2n} \overline{A}_{nm} = -2\Omega_{nm} \omega_1 A_{nm} \int_0^1 r J_n^2(\Omega_{nm} r) \, dr \quad (15.246)$$

But

$$\int_0^r rJ_n^2(\alpha r)\,dr = \tfrac{1}{2} r^2 J_n'^2(\alpha r) + \tfrac{1}{2} r^2 \left(1 - \frac{n^2}{\alpha^2 r^2}\right) J_n^2(\alpha r) \quad (15.247)$$

Putting $\alpha = \Omega_{nm}$ and using the fact that $J_n(\Omega_{nm}) = 0$, we obtain from (15.247) that

$$\int_0^1 rJ_n^2(\Omega_{nm}r)\,dr = \tfrac{1}{2} J_n'^2(\Omega_{nm}) \quad (15.248)$$

Using (15.248) in (15.246) yields

$$\Omega_{nm} f_{2n} \overline{A}_{nm} = -\omega_1 A_{nm} \quad (15.249)$$

To analyze (15.249), we express A_{nm} and f_{2n} in polar form as

$$A_{nm} = \tfrac{1}{2} a_{nm} e^{i\beta_{nm}} \quad f_{2n} = F_{2n} e^{i\nu_{2n}} \quad (15.250)$$

and obtain

$$\omega_1 = -\Omega_{nm} F_{2n} e^{i(\nu_{2n} - 2\beta_{nm})} \quad (15.251)$$

Since ω_1 is real, it follows from (15.251) that

$$\nu_{2n} - 2\beta_{nm} = 0 \quad \text{or} \quad \pi$$

Hence,

$$\beta_{nm} = \tfrac{1}{2} \nu_{2n} \quad \text{or} \quad \tfrac{1}{2}(\nu_{2n} - \pi) \quad (15.252)$$

and it follows from (15.251) that

$$\omega_1 = -\Omega_{nm} F_{2n} \quad \text{or} \quad \Omega_{nm} F_{2n} \quad (15.253)$$

Substituting (15.250), (15.252), and (15.253) into (15.219), (15.231), and (15.214) and noting that $\omega_0 = \Omega_{nm}$, we obtain to the first approximation

$$w^{(1)} = a_{nm} J_n(\Omega_{nm} r) \cos(n\theta + \tfrac{1}{2}\nu_{2n}) \cos(\omega^{(1)} t + \tau) + \cdots \quad (15.254)$$

$$w^{(2)} = a_{nm} J_n(\Omega_{nm} r) \sin(n\theta + \tfrac{1}{2}\nu_{2n}) \cos(\omega^{(2)} t + \tau) + \cdots \quad (15.255)$$

where

$$\begin{aligned} \omega^{(1)} &= \Omega_{nm}(1 - \epsilon F_{2n}) + \cdots \\ \omega^{(2)} &= \Omega_{nm}(1 + \epsilon F_{2n}) + \cdots \end{aligned} \quad (15.256)$$

We note that, in the circular case, β_{nm} is arbitrary, and hence, there are two eigenfunctions $\sin n\theta$ and $\cos n\theta$ corresponding to the same eigenvalue Ω_{nm}. However, in the near circular case, there are two different eigenfunctions corresponding to the two different eigenvalue $\omega^{(1)}$ and $\omega^{(2)}$ if $F_{2n} \neq 0$. If $F_{2n} = 0$, one needs to continue the expansion to higher order in order that the degeneracy may be removed.

432 SOLVABILITY CONDITIONS

15.10. A Fourth-Order Differential System

We consider in this section the adjoint and solvability conditions for problems consisting of linear inhomogeneous fourth-order ordinary-differential equations and inhomogeneous boundary conditions. Thus, we consider

$$p_4(x)\phi^{iv} + p_3(x)\phi''' + p_2(x)\phi'' + p_1(x)\phi' + p_0(x)\phi = f(x) \quad (15.257)$$

$$\phi(0) = \beta_1 \quad \phi'(0) = \beta_2 \quad \phi(1) = \beta_3 \quad \phi'(1) = \beta_4 \quad (15.258)$$

A system consisting of a fourth-order differential equation and general mixed boundary conditions is discussed in the next section, whereas a fourth-order eigenvalue problem is discussed in Section 15.12.

To determine the solvability conditions for (15.257) and (15.258), we multiply (15.257) by the adjoint $u(x)$, which is specified below, integrate the result term by term from $x = 0$ to $x = 1$ (i.e., the interval of interest where the boundary conditions are enforced), and obtain

$$\int_0^1 p_4 u \phi^{iv} \, dx + \int_0^1 p_3 u \phi''' \, dx + \int_0^1 p_2 u \phi'' \, dx + \int_0^1 p_1 u \phi' \, dx + \int_0^1 p_0 u \phi \, dx$$

$$= \int_0^1 u f \, dx \quad (15.259)$$

Next, we integrate by parts the integrals in (15.259) to transfer the derivatives from ϕ to u. To this end, we note that

$$\int_0^1 p_4 u \phi^{iv} \, dx = p_4 u \phi''' \Big|_0^1 - \int_0^1 (p_4 u)' \phi''' \, dx = [p_4 u \phi''' - (p_4 u)' \phi'']_0^1$$

$$+ \int_0^1 (p_4 u)'' \phi'' \, dx = [p_4 u \phi''' - (p_4 u)' \phi'' + (p_4 u)'' \phi']_0^1$$

$$- \int_0^1 (p_4 u)''' \phi' \, dx = [p_4 u \phi''' - (p_4 u)' \phi'' + (p_4 u)'' \phi'$$

$$- (p_4 u)''' \phi]_0^1 + \int_0^1 (p_4 u)^{iv} \phi \, dx$$

$$\int_0^1 p_3 u \phi''' \, dx = [p_3 u \phi'' - (p_3 u)' \phi' + (p_3 u)'' \phi]_0^1 - \int_0^1 (p_3 u)''' \phi \, dx$$

$$\int_0^1 p_2 u \phi'' \, dx = [p_2 u \phi' - (p_2 u)' \phi]_0^1 + \int_0^1 (p_2 u)'' \phi \, dx$$

$$\int_0^1 p_1 u\phi' \, dx = p_1 u\phi \Big|_0^1 - \int_0^1 (p_1 u)' \phi \, dx$$

With these expressions, we can rewrite (15.259) as

$$\int_0^1 \phi[(p_4 u)^{iv} - (p_3 u)''' + (p_2 u)'' - (p_1 u)' + p_0 u] \, dx$$

$$+ \{p_4 u\phi''' - [(p_4 u)' - p_3 u]\phi'' + [(p_4 u)'' - (p_3 u)' + p_2 u]\phi'$$

$$- [(p_4 u)''' - (p_3 u)'' + (p_2 u)' - p_1 u]\phi\}_0^1 = \int_0^1 fu \, dx \quad (15.260)$$

As before, the differential equation describing the adjoint u is obtained by setting the coefficient of ϕ in the integrand on the left-hand side of (15.260) equal to zero. The result is

$$(p_4 u)^{iv} - (p_3 u)''' + (p_2 u)'' - (p_1 u)' + p_0 u = 0 \quad (15.261)$$

which is called the adjoint homogeneous differential equation to (15.257).

In order that (15.257) be self-adjoint, (15.261) must be the same as the homogeneous equation (15.257). Expanding the derivatives in (15.261) and rearranging, we have

$$p_4 u^{iv} + (4p_4' - p_3)u''' + (6p_4'' - 3p_3' + p_2)u'' + (4p_4''' - 3p_3'' + 2p_2' - p_1)u'$$

$$+ (p_4^{iv} - p_3''' + p_2'' - p_1' + p_0)u = 0 \quad (15.262)$$

In order that (15.262) be the same as the homogeneous equation (15.257),

$$4p_4' - p_3 = p_3 \qquad 6p_4'' - 3p_3' + p_2 = p_2$$

$$4p_4''' - 3p_3'' + 2p_2' - p_1 = p_1 \qquad p_4^{iv} - p_3''' + p_2'' - p_1' + p_0 = p_0$$

or

$$p_3 = 2p_4' \qquad p_3' = 2p_4'' \qquad p_1 = 2p_4''' - \tfrac{3}{2} p_3'' + p_2' = -p_4''' + p_2'$$

Then, (15.257) becomes

$$p_4 \phi^{iv} + 2p_4' \phi''' + p_2 \phi'' + (p_2' - p_4''')\phi' + p_0 \phi = 0$$

which can be rewritten as

$$\frac{d^2}{dx^2}\left(p_4 \frac{d^2\phi}{dx^2}\right) + \frac{d}{dx}\left[(p_2 - p_4'') \frac{d\phi}{dx}\right] + p_0 \phi = 0 \quad (15.263)$$

Hence, any fourth-order self-adjoint homogeneous differential equation can be written in the form

$$\frac{d^2}{dx^2}\left(A_2 \frac{d^2\phi}{dx^2}\right) + \frac{d}{dx}\left(A_1 \frac{d\phi}{dx}\right) + A_0 \phi = 0 \qquad (15.264)$$

We note that, whereas a second-order differential equation can be multiplied by a factor to make it self-adjoint as in (15.84), a differential equation of order higher than two cannot always be made self-adjoint.

To determine the boundary conditions needed to specify u, we consider the homogeneous problem (i.e., $f = 0$ and $\beta_n = 0$). Thus, using (15.261), we obtain from (15.260) that

$$\{p_4 u \phi''' - [(p_4 u)' - p_3 u]\phi'' + [(p_4 u)'' - (p_3 u)' + p_2 u]\phi'$$
$$- [(p_4 u)''' - (p_3 u)'' + (p_2 u)' - p_1 u]\phi\}_0^1 = 0$$

But for the homogeneous problem, $\phi(0) = \phi'(0) = \phi(1) = \phi'(1) = 0$; hence

$$p_4 u|_1 \phi'''(1) - [(p_4 u)' - p_3 u]|_1 \phi''(1) - p_4 u|_0 \phi'''(0)$$
$$+ [(p_4 u)' - p_3 u]|_0 \phi''(0) = 0 \qquad (15.265)$$

We choose the adjoint boundary conditions such that each of the coefficients of $\phi''(0), \phi''(1), \phi'''(0)$, and $\phi'''(1)$ in (15.265) vanish independently. The result is

$$u(0) = 0 \quad u(1) = 0 \quad u'(0) = 0 \quad u'(1) = 0 \qquad (15.266)$$

Thus, the adjoint problem is defined by (15.261) subject to the boundary conditions (15.266).

To determine the solvability conditions, we return to (15.260), use the definition of u, and obtain

$$[p_4 u'' \phi' - (p_4 u''' + 3 p_4' u'' - p_3 u'')\phi]_0^1 = \int_0^1 fu\, dx \qquad (15.267)$$

Using (15.258), we rewrite (15.267) as

$$(p_4 u'' \beta_4 - p_4 u''' \beta_3 - 3 p_4' u'' \beta_3 + p_3 u'' \beta_3)|_1 - (p_4 u'' \beta_2 - p_4 u''' \beta_1$$
$$- 3 p_4' u'' \beta_1 + p_3 u'' \beta_1)|_0 = \int_0^1 fu\, dx \qquad (15.268)$$

For every nontrivial solution u of the adjoint homogeneous problem, (15.268) provides a solvability condition. Clearly, if the adjoint homogeneous problem has only the trivial solution, (15.268) is satisfied identically for all values of β_n and $f(x)$. Next, we apply the above results to two examples.

EXAMPLE 1

Working with the streamfunction and using the method of multiple scales to determine the stability of nonparallel flows over a flat surface, one obtains the

A FOURTH-ORDER DIFFERENTIAL SYSTEM

inhomogeneous Orr-Sommerfeld problem

$$\left(\frac{d^2}{dy^2} - k^2\right)^2 \phi - iR(Uk - \omega)\left(\frac{d^2\phi}{dy^2} - k^2\phi\right) + ikR\frac{d^2U}{dy^2}\phi = f(y) \quad (15.269)$$

$$\phi(0) = 0 \quad \frac{d\phi}{dy}(0) = 0$$
$$\phi, \frac{d\phi}{dy} \to 0 \quad \text{as} \quad y \to \infty \quad (15.270)$$

where R, k, and ω are independent of y, and U and f are known functions of y. Moreover, the homogeneous problem is an eigenvalue problem having a nontrivial solution. Hence, the inhomogeneous problem has a solution only if a solvability condition is satisfied.

Expanding the derivatives, we rewrite (15.269) as

$$\phi^{iv} - (2k^2 + ikRU - i\omega R)\phi'' + (k^4 + ik^3RU - i\omega k^2 R + ikRU'')\phi = f \quad (15.271)$$

Hence,

$$p_4 = 1 \quad p_3 = 0 \quad p_2 = -(2k^2 + ikRU - i\omega R) \quad p_1 = 0$$
$$p_0 = k^4 + ik^3RU - i\omega k^2 R + ikRU''$$

Consequently, it follows from (15.261) that the adjoint homogeneous equation is

$$u^{iv} - [(2k^2 + ikRU - i\omega R)u]'' + (k^4 + ik^3RU - i\omega k^2 R + ikRU'')u = 0$$

or

$$u^{iv} - (2k^2 + ikRU - i\omega R)u'' - 2ikRU'u' + (k^4 + ik^3RU - i\omega k^2 R)u = 0$$

which can be rearranged to

$$\left(\frac{d^2}{dy^2} - k^2\right)^2 u - iR(kU - \omega)\left(\frac{d^2 u}{dy^2} - k^2 u\right) - 2ikR\frac{dU}{dy}\frac{du}{dy} = 0 \quad (15.272)$$

The boundary conditions on u are

$$u(0) = 0 \quad u'(0) = 0$$
$$u, u' \to 0 \quad \text{as} \quad y \to \infty \quad (15.273)$$

Having defined the adjoint u, we substitute the boundary conditions (15.270) into (15.268) and take the upper limit to be ∞ instead of 1. Then, the solvability condition reduces to

436 SOLVABILITY CONDITIONS

$$\int_0^\infty fu\, dy = 0 \qquad (15.274)$$

EXAMPLE 2

The analysis of the vibration of nearly annular plates with clamped edges by using the method of strained parameters leads to the following inhomogeneous problem:

$$\left[\left(\frac{d^2}{dr^2}+\frac{1}{r}\frac{d}{dr}-\frac{n^2}{r^2}\right)^2 - \omega_{nm}^2\right]\phi = f(r) \qquad (15.275)$$

$$\phi(a) = \beta_1 \quad \phi'(a) = \beta_2 \quad \phi(b) = \beta_3 \quad \phi'(b) = \beta_4 \qquad (15.276)$$

where $b > a$ and the β_n, n, and ω_{nm} are constants. The homogeneous problem has a nontrivial solution so that the inhomogeneous problem has a solution only if a solvability condition is satisfied.

Instead of expanding the operator in (15.275) and applying the solvability condition developed in this section, we show that it is more convenient to work directly with (15.275) because it can be made self-adjoint by its multiplication with the factor r. Thus, we multiply (15.275) with $ru(r)$, where $u(r)$ is a solution of the adjoint homogeneous problem to be specified below, integrate the result from $r = a$ to $r = b$, and obtain

$$\int_a^b ru\left[\left(\frac{d^2}{dr^2}+\frac{1}{r}\frac{d}{dr}-\frac{n^2}{r^2}\right)^2 - \omega_{nm}^2\right]\phi\, dr = \int_a^b ruf\, dr \qquad (15.277)$$

If we let

$$\frac{d^2\phi}{dr^2}+\frac{1}{r}\frac{d\phi}{dr}-\frac{n^2}{r^2}\phi = \psi \qquad (15.278)$$

then,

$$\int_a^b ru\left(\frac{d^2}{dr^2}+\frac{1}{r}\frac{d}{dr}-\frac{n^2}{r^2}\right)\psi\, dr = \int_a^b \left[u\frac{d}{dr}\left(r\frac{d\psi}{dr}\right)-\frac{n^2 u\psi}{r}\right]dr$$

$$= \left[ru\frac{d\psi}{dr}-r\frac{du}{dr}\psi\right]_a^b$$

$$+ \int_a^b \psi\left[\frac{d}{dr}\left(r\frac{du}{dr}\right)-\frac{n^2 u}{r}\right]dr \qquad (15.279)$$

$$\int_a^b \psi\left[\frac{d}{dr}\left(r\frac{du}{dr}\right)-\frac{n^2 u}{r}\right]dr = \int_a^b \left[\frac{d}{dr}\left(r\frac{d\phi}{dr}\right)-\frac{n^2}{r}\phi\right]\left[\frac{1}{r}\frac{d}{dr}\left(r\frac{du}{dr}\right)-\frac{n^2}{r^2}u\right]dr$$

A FOURTH-ORDER DIFFERENTIAL SYSTEM 437

$$= \int_a^b \left[\frac{d}{dr}\left(r\frac{d\phi}{dr}\right) - \frac{n^2}{r}\phi\right]\chi\, dr$$

$$= \left[r\frac{d\phi}{dr}\chi - r\phi\frac{d\chi}{dr}\right]_a^b$$

$$+ \int_a^b \phi\left[\frac{d}{dr}\left(r\frac{d\chi}{dr}\right) - \frac{n^2}{r}\chi\right]dr \qquad (15.280)$$

where

$$\chi = \frac{1}{r}\frac{d}{dr}\left(r\frac{du}{dr}\right) - \frac{n^2 u}{r^2} = \left(\frac{d^2}{dr^2} + \frac{1}{r}\frac{d}{dr} - \frac{n^2}{r^2}\right)u \qquad (15.281)$$

Substituting (15.281) into (15.280), we have

$$\int_a^b \psi\left[\frac{d}{dr}\left(r\frac{du}{dr}\right) - \frac{n^2 u}{r}\right]dr = \left[r\frac{d\phi}{dr}\left(\frac{d^2 u}{dr^2} + \frac{1}{r}\frac{du}{dr} - \frac{n^2 u}{r^2}\right)\right.$$

$$\left. - r\phi\frac{d}{dr}\left(\frac{d^2 u}{dr^2} + \frac{1}{r}\frac{du}{dr} - \frac{n^2 u}{r^2}\right)\right]_a^b$$

$$+ \int_a^b r\phi\left(\frac{d^2}{dr^2} + \frac{1}{r}\frac{d}{dr} - \frac{n^2}{r^2}\right)^2 u\, dr \qquad (15.282)$$

Substituting (15.282) into (15.279) and then substituting the result into (15.277), we obtain

$$\int_a^b r\phi\left[\left(\frac{d^2}{dr^2} + \frac{1}{r}\frac{d}{dr} - \frac{n^2}{r^2}\right)^2 - \omega_{nm}^2\right]u\, dr + \left[ru\frac{d}{dr}\left(\frac{d^2\phi}{dr^2} + \frac{1}{r}\frac{d\phi}{dr} - \frac{n^2}{r^2}\phi\right)\right.$$

$$\left. - r\frac{du}{dr}\left(\frac{d^2\phi}{dr^2} + \frac{1}{r}\frac{d\phi}{dr} - \frac{n^2\phi}{r^2}\right) + r\frac{d\phi}{dr}\left(\frac{d^2 u}{dr^2} + \frac{1}{r}\frac{du}{dr} - \frac{n^2 u}{r^2}\right)\right.$$

$$\left. - r\phi\frac{d}{dr}\left(\frac{d^2 u}{dr^2} + \frac{1}{r}\frac{du}{dr} - \frac{n^2 u}{r^2}\right)\right]_a^b = \int_a^b ruf\, dr \qquad (15.283)$$

To specify the adjoint u, we consider first the homogeneous problem for which $f = 0$ and $\beta_n = 0$. Then, setting the coefficient of ϕ in the integrand in (15.283) equal to zero yields

$$\left[\left(\frac{d^2}{dr^2} + \frac{1}{r}\frac{d}{dr} - \frac{n^2}{r^2}\right)^2 - \omega_{nm}^2\right]u = 0 \qquad (15.284)$$

which is identical to the homogeneous equation (15.275), and hence, it is self-adjoint. Putting the $\beta_n = 0$ in (15.276) and substituting the result into (15.283) with $f = 0$, we have

$$[ru\phi''' + u\phi'' - ru'\phi'']_a^b = 0$$

or

$$bu(b)\phi'''(b) + [u(b) - bu'(b)]\phi''(b) - au(a)\phi'''(a) - [u(a) - au'(a)]\phi''(a) = 0$$

(15.285)

As before, we choose the adjoint boundary conditions such that each of the coefficients of $\phi''(a)$, $\phi''(b)$, $\phi'''(a)$, and $\phi'''(b)$ in (15.285) vanish independently, and hence,

$$u(a) = u(b) = 0 \quad u'(a) = u'(b) = 0 \qquad (15.286)$$

Comparing (15.284) and (15.286) with (15.275) and (15.276), we conclude that the homogeneous problem is self-adjoint.

Having defined the adjoint homogeneous problem, we return to the inhomogeneous problem to determine the solvability condition. Substituting (15.284), (15.286), and (15.276) into (15.283), we obtain the solvability condition.

$$b\beta_4 u''(b) - b\beta_3 u'''(b) - \beta_3 u''(b) - a\beta_2 u''(a) + a\beta_1 u'''(a) + \beta_1 u''(a) = \int_a^b ruf\, dr$$

(15.287)

15.11. General Fourth-Order Differential Systems

In this section, we consider the solvability condition of

$$p_4(x)\phi^{iv} + p_3(x)\phi''' + p_2(x)\phi'' + p_1(x)\phi' + p_0\phi = f(x) \qquad a \leqslant x \leqslant b \quad (15.288)$$

$$\sum_{j=1}^{8} \alpha_{ij}\phi_j = \beta_i \quad \text{for} \quad i = 1, 2, 3, \text{ and } 4 \qquad (15.289)$$

where the ϕ_j are the components $\phi'''(a)$, $\phi''(a)$, $\phi'(a)$, $\phi(a)$, $\phi'''(b)$, $\phi''(b)$, $\phi'(b)$, and $\phi(b)$ of the column vector ϕ_b. We assume that the boundary conditions in (15.289) are linearly independent, that is, there exists at least one 4×4 nonsingular submatrix of $[\alpha_{ij}]$.

To determine the adjoint problem, we denote the operator on the left-hand side of (15.288) by L so that

$$L(\phi) = p_4 \phi^{iv} + p_3 \phi''' + p_2 \phi'' + p_1 \phi' + p_0 \phi \qquad a \leqslant x \leqslant b \quad (15.290)$$

where p_4^{iv}, p_3''', p_2'', p_1', and p_0 are continuous over the interval $[a, b]$. If $\phi(x)$

and $u(x)$ are any two functions possessing four continuous derivatives over $[a, b]$, we have

$$\int_a^b uL(\phi)\, dx = \int_a^b [p_4 u\phi^{iv} + p_3 u\phi''' + p_2 u\phi'' + p_1 u\phi' + p_0 u\phi]\, dx \quad (15.291)$$

Integrating (15.291) by parts to transfer the derivatives from ϕ to u, we obtain

$$\int_a^b uL(\phi)\, dx = \int_a^b \phi[(p_4 u)^{iv} - (p_3 u)''' + (p_2 u)'' - (p_1 u)' + p_0 u]\, dx$$

$$+ \{p_4 u\phi''' - [(p_4 u)' - p_3 u]\phi'' + [(p_4 u)'' - (p_3 u)' + p_2 u]\phi'$$

$$- [(p_4 u)''' - (p_3 u)'' + (p_2 u)' - p_1 u]\phi\}_a^b \quad (15.292)$$

We denote the operator in the integrand on the right-hand side of (15.292) by L^*, that is,

$$L^*(u) = (p_4 u)^{iv} - (p_3 u)''' + (p_2 u)'' - (p_1 u)' + p_0 u \quad (15.293)$$

Then, (15.292) can be rewritten as the following Green's identity:

$$\int_a^b [uL(\phi) - \phi L^*(u)]\, dx = \{p_4 u\phi''' - [p_4 u' + (p_4' - p_3)u]\phi''$$

$$+ [p_4 u'' + (2p_4' - p_3)u' + (p_4'' - p_3' + p_2)u]\phi'$$

$$- [p_4 u''' + (3p_4' - p_3)u'' + (3p_4'' - 2p_3' + p_2)u'$$

$$+ (p_4''' - p_3'' + p_2' - p_1)u]\phi\}_a^b \quad (15.294)$$

The operator L^* is called the adjoint operator corresponding to the operator L. One can easily verify that L is adjoint to L^* so that L and L^* are adjoint to each other. As in the preceding section, we call the differential equation

$$L^*(u) = 0 \quad (15.295)$$

the adjoint of the differential equation

$$L(\phi) = 0 \quad (15.296)$$

and vice versa. If $L = L^*$, we say that the operator L is self-adjoint, and the differential equation $L(\phi) = 0$ is self-adjoint. Comparing (15.290) and (15.293), we conclude, as in the preceding section, that L is self-adjoint if and only if

$$p_3 = 2p_4' \quad \text{and} \quad p_1 = p_2' - p_4'''$$

Then, (15.290) can be rewritten as

$$L(\phi) = \frac{d^2}{dx^2}\left(p_4 \frac{d^2\phi}{dx^2}\right) + \frac{d}{dx}\left[(p_2 - p_4'')\frac{d\phi}{dx}\right] + p_0\phi \tag{15.297}$$

The right-hand side of (15.294) is called a bilinear concomitant of ϕ and u because for a given ϕ it is linear in u and for a given u it is linear in ϕ. Using vector and matrix notation, we rewrite (15.294) as

$$\int_a^b [uL(\phi) - \phi L^*(u)] \, dx = \mathbf{u}_b^T P \phi_b \tag{15.298}$$

where ϕ_b has been defined earlier, $\mathbf{u}_b$ is a column vector whose components are $u'''(a), u''(a), u'(a), u(a), u'''(b), u''(b), u'(b)$, and $u(b)$, and P is defined by

$$P = \begin{bmatrix} \Lambda_a & 0 \\ 0 & -\Lambda_b \end{bmatrix} \tag{15.299}$$

where

$$\Lambda_{a,b} = \begin{bmatrix} 0 & 0 & 0 & p_4 \\ 0 & 0 & -p_4 & 3p_4' - p_3 \\ 0 & p_4 & p_3 - 2p_4' & 3p_4'' - 2p_3' + p_2 \\ -p_4 & p_4' - p_3 & p_3' - p_4'' - p_2 & p_4''' - p_3'' + p_2' - p_1 \end{bmatrix}_{x=a,b} \tag{15.300}$$

We note that $|P| = [p_4(a)p_4(b)]^4 \neq 0$, and hence, P is a nonsingular matrix.

To determine the adjoint boundary conditions, we transform the right-hand side into a canonical bilinear form. To this end, we introduce a linear nonsingular transformation from ϕ_b to Φ according to

$$\Phi = \mathcal{Q}\phi_b \tag{15.301}$$

where $\mathcal{Q}$ is an 8×8 constant-coefficient matrix whose elements are α_{ij}. We choose the first four rows to be the same as those in (15.289). The last four rows are arbitrary, except that they are linearly independent of each other and of the first four rows. As a consequence, for a given nonzero ϕ_b, the last four rows can be chosen to produce any desired nonzero values for Φ_5, Φ_6, Φ_7, and Φ_8. This fact will be used later to define the adjoint of the above homogeneous system. The transformation (15.301) can be inverted to yield

$$\phi_b = \mathcal{Q}^{-1}\Phi$$

Then,

$$\int_a^b [uL(\phi) - \phi L^*(u)] \, dx = \mathbf{u}_b^T P \mathcal{Q}^{-1} \Phi = \mathbf{U}^T \Phi \tag{15.302}$$

where

$$\mathbf{U}^T = \mathbf{u}_b^T P \mathcal{Q}^{-1} \quad \text{or} \quad \mathbf{U} = (\mathcal{Q}^{-1})^T P^T \mathbf{u}_b \tag{15.303}$$

It follows from (15.289) and (15.301) that

$$\Phi_i = \sum_{j=1}^{8} \alpha_{ij}\phi_j = \beta_i \quad i = 1, 2, 3, 4 \tag{15.304}$$

To determine the adjoint problem, we consider the homogeneous case (i.e., $\beta_i = 0$) and set the left-hand side of (15.302) equal to zero. The result is

$$\sum_{i=1}^{8} U_i \Phi_i = 0 \quad \Phi_i = 0 \quad \text{for} \quad i = 1, 2, 3, 4$$

Hence,

$$U_5 \Phi_5 + U_6 \Phi_6 + U_7 \Phi_7 + U_8 \Phi_8 = 0 \tag{15.305}$$

As mentioned above, for a given nonzero ϕ_b, the last four rows in $\mathcal{C}$ can be chosen to produce any desired values for Φ_5, Φ_6, Φ_7, and Φ_8 in (15.301). Thus, we can choose the last four rows in $\mathcal{C}$ so that $\Phi_5 = 1$ and $\Phi_6 = \Phi_7 = \Phi_8 = 0$. Then, it follows from (15.305) that $U_5 = 0$. Similarly, we can choose the last four rows in $\mathcal{C}$ so that $\Phi_6 = 1$ and $\Phi_5 = \Phi_7 = \Phi_8 = 0$. Then, it follows from (15.305) that $U_6 = 0$. Using similar arguments, we can show that $U_7 = U_8 = 0$. Therefore, the problem adjoint to

$$L(\phi) = 0 \quad \Phi_1 = \Phi_2 = \Phi_3 = \Phi_4 = 0 \tag{15.306}$$

is

$$L^*(u) = 0 \quad U_5 = U_6 = U_7 = U_8 = 0 \tag{15.307}$$

Having defined the adjoint problem, we return to the inhomogeneous problem to determine the solvability condition. With (15.307), Green's identity (15.302) becomes

$$\int_a^b uL(\phi)\,dx = \sum_{i=1}^{4} U_i \Phi_i \tag{15.308}$$

But it follows from (15.288) and (15.304) that $L(\phi) = f(x)$ and $\Phi_i = \beta_i$ for $i = 1, 2, 3$, and 4; therefore, it follows from (15.308) that the solvability condition is

$$\int_a^b u(x)f(x)\,dx = \sum_{i=1}^{4} \beta_i U_i \tag{15.309}$$

15.12. A Fourth-Order Eigenvalue Problem

As an application of the theory in the preceding two sections, we consider the eigenvalue problem

442 SOLVABILITY CONDITIONS

$$\phi^{iv} + 10\phi'' + [\lambda + \epsilon f(x)]\phi = 0 \quad \epsilon \ll 1$$
$$\phi(0) = \phi''(0) = \phi(\pi) = \phi''(\pi) = 0 \tag{15.310}$$

We seek a first-order uniform expansion to (15.310) by using the method of strained parameters in the form

$$\phi(x;\epsilon) = \phi_0(x) + \epsilon\phi_1(x) + \cdots$$
$$\lambda = \lambda_0 + \epsilon\lambda_1 + \cdots \tag{15.311}$$

Substituting (15.311) into (15.310) and equating coefficients of like powers of ϵ, we obtain

Order ϵ^0

$$\phi_0^{iv} + 10\phi_0'' + \lambda_0\phi_0 = 0$$
$$\phi_0(0) = \phi_0''(0) = \phi_0(\pi) = \phi_0''(\pi) = 0 \tag{15.312}$$

Order ϵ

$$\phi_1^{iv} + 10\phi_1'' + \lambda_0\phi_1 = -(\lambda_1 + f)\phi_0 \tag{15.313}$$
$$\phi_1(0) = \phi_1''(0) = \phi_1(\pi) = \phi_1''(\pi) = 0 \tag{15.314}$$

One can easily verify that the solution of (15.312) is

$$\phi_0 = \sin nx \quad \lambda_0 = n^2(10 - n^2) \tag{15.315}$$

We note that $n = 1$ and 3 produce the same eigenvalue $\lambda_0 = 9$. Hence, $\sin x$ and $\sin 3x$ correspond to the same eigenvalue $\lambda_0 = 9$ and the problem is degenerate when $n = 1$ or 3. For all other n, there is only one eigenfunction corresponding to each eigenvalue and the problem is nondegenerate. Both cases are considered below.

Since the homogeneous first-order problem (15.313) and (15.314) is the same as the zeroth-order problem (15.312) and since the latter has a nontrivial solution, the inhomogeneous first-order problem has a solution only if solvability conditions are satisfied. To determine the solvability conditions, we need first to determine the adjoint problem. To this end, we multiply (15.313) by $u(x)$, integrate the result by parts from $x = 0$ to $x = \pi$ to transfer the derivatives from ϕ_1 to u, and obtain

$$\int_0^\pi \phi_1(u^{iv} + 10u'' + \lambda_0 u)\, dx + [\phi_1''' u - \phi_1'' u' + \phi_1' u'' - \phi_1 u''']_0^\pi$$

$$+ 10[\phi_1' u - \phi_1 u']_0^\pi = -\int_0^\pi (\lambda_1 + f)\phi_0 u\, dx \tag{15.316}$$

Hence, the adjoint equation is

$$u^{iv} + 10u'' + \lambda_0 u = 0 \qquad (15.317)$$

which is identical to the homogeneous equation (15.313). This is not surprising because the homogeneous equation (15.313) has the form (15.263). To determine the adjoint boundary conditions, we consider the homogeneous problem (i.e., $\lambda_1 = 0$ and $f = 0$) so that (15.316) becomes

$$[\phi_1''' u - \phi_1'' u' + \phi_1'(u'' + 10u) - \phi_1(u''' + 10u')]_0^\pi = 0 \qquad (15.318)$$

Using the boundary conditions (15.314) in (15.318), we have

$$\phi_1'''(\pi)u(\pi) + \phi_1'(\pi)[u''(\pi) + 10u(\pi)] - \phi_1'''(0)u(0)$$
$$- \phi_1'(0)[u''(0) + 10u(0)] = 0 \qquad (15.319)$$

As before, we choose the adjoint boundary conditions such that each of the coefficients of $\phi_1'''(\pi)$, $\phi_1'(\pi)$, $\phi_1'''(0)$, and $\phi_1'(0)$ in (15.319) vanish independently, that is,

$$u(0) = u''(0) = u(\pi) = u''(\pi) = 0 \qquad (15.320)$$

Thus, the first-order problem is self-adjoint.

We return to the inhomogeneous problem, use the definition of the adjoint problem, and obtain from (15.316) that

$$\int_0^\pi (\lambda_1 + f)\phi_0 u \, dx = 0 \qquad (15.321)$$

for every solution u of the adjoint problem. Next, we consider the degenerate and nondegenerate cases starting with the second case.

NONDEGENERATE CASE

In this case, $n \neq 1$ or 3 and the solution of the zeroth-order problem is given by (15.315). Then, the solution of the adjoint problem is $u = \sin nx$, which when put in (15.321) yields

$$\int_0^\pi [\lambda_1 + f(x)] \sin^2 nx \, dx = 0$$

Hence,

$$\lambda_1 = -\frac{2}{\pi} \int_0^\pi f(x) \sin^2 nx \, dx$$

and it follows from (15.311) that, to the first approximation,

$$\phi = \sin nx + O(\epsilon) \qquad (15.322a)$$

$$\lambda = n^2(10 - n^2) - \frac{2\epsilon}{\pi} \int_0^\pi f(x) \sin^2 nx \, dx + O(\epsilon^2) \qquad (15.322b)$$

DEGENERATE CASE

In this case, $n = 1$ or 3 and the solution of the zeroth-order problem is

$$\phi_0 = a_1 \sin x + a_3 \sin 3x \qquad \lambda_0 = 9 \qquad (15.323)$$

where the constants a_1 and a_3 are independent at this level of approximation, a manifestation of the degeneracy. With $\lambda_0 = 9$, the solution of the adjoint problem is $u = \sin x$ or $\sin 3x$. Putting (15.323) in (15.321), we have

$$\int_0^\pi [\lambda_1 + f(x)](a_1 \sin x + a_3 \sin 3x)u(x) \, dx = 0 \qquad (15.324)$$

for all possible solutions of the adjoint problem. In this case, there are two possible solutions. Putting $u = \sin x$ and $\sin 3x$, respectively, in (15.324), we obtain

$$\int_0^\pi [\lambda_1 + f(x)](a_1 \sin x + a_3 \sin 3x) \sin x \, dx = 0$$

$$\int_0^\pi [\lambda_1 + f(x)](a_1 \sin x + a_3 \sin 3x) \sin 3x \, dx = 0$$

Hence,

$$(\lambda_1 + f_{11})a_1 + f_{13}a_3 = 0$$
$$f_{13}a_1 + (\lambda_1 + f_{33})a_3 = 0 \qquad (15.325)$$

where

$$f_{11} = \frac{2}{\pi} \int_0^\pi f \sin^2 x \, dx \qquad f_{33} = \frac{2}{\pi} \int_0^\pi f \sin^2 3x \, dx$$

$$f_{13} = \frac{2}{\pi} \int_0^\pi f \sin x \sin 3x \, dx$$

For a nontrivial solution, the determinant of the coefficient matrix in (15.325) must vanish, that is,

$$\begin{vmatrix} \lambda_1 + f_{11} & f_{13} \\ f_{13} & \lambda_1 + f_{33} \end{vmatrix} = 0$$

Hence,

$$\lambda_1^2 + (f_{11} + f_{33})\lambda_1 + f_{11}f_{33} - f_{13}^2 = 0$$

or

$$\lambda_1^{(1)}, \lambda_1^{(2)} = -\tfrac{1}{2}(f_{11} + f_{33}) \pm \tfrac{1}{2}[(f_{11} - f_{33})^2 + 4f_{13}^2]^{1/2} \quad (15.326)$$

Then,

$$a_3 = -\frac{\lambda_1 + f_{11}}{f_{13}} a_1$$

Therefore, it follows from (15.311) that, to the first approximation,

$$\phi^{(1)} = \sin x - \frac{\lambda_1^{(1)} + f_{11}}{f_{13}} \sin 3x + \cdots$$

$$\lambda^{(1)} = 9 - \tfrac{1}{2}\epsilon(f_{11} + f_{33}) + \tfrac{1}{2}\epsilon[(f_{11} - f_{33})^2 + 4f_{13}^2]^{1/2} + \cdots \quad (15.327)$$

and

$$\phi^{(2)} = \sin x - \frac{\lambda_1^{(2)} + f_{11}}{f_{13}} \sin 3x + \cdots$$

$$\lambda^{(2)} = 9 - \tfrac{1}{2}\epsilon(f_{11} + f_{33}) - \tfrac{1}{2}\epsilon[(f_{11} - f_{33})^2 + 4f_{13}^2]^{1/2} + \cdots \quad (15.328)$$

Thus, the degeneracy is removed from the problem at first order if $\lambda_1^{(1)} \neq \lambda_1^{(2)}$, that is, if $f_{11} \neq f_{33}$ or $f_{13} \neq 0$.

15.13. A Differential System of Equations

In this section, we consider the solvability conditions for a special system of first-order ordinary-differential equations. In the next section, we consider a general system of first-order equations, whereas in Section 15.15 we consider two systems of differential equations with interfacial boundary conditions.

In analyzing the propagation and attenuation of sound waves in an annular duct carrying compressible mean flows, one may encounter the following inhomogeneous problem:

$$-i(\omega - ku_0)\phi_4 + ik\rho_0\phi_1 + \frac{im\rho_0}{r}\phi_3 + \frac{1}{r}\frac{d}{dr}(r\rho_0\phi_2) = f_1(r) \quad (15.329)$$

$$-i\rho_0(\omega - ku_0)\phi_1 + \rho_0 u_0' \phi_2 + ik\phi_5 = f_2(r) \quad (15.330)$$

$$-i\rho_0(\omega - ku_0)\phi_2 + \phi_5' = f_3(r) \quad (15.331)$$

$$-i\rho_0(\omega - ku_0)\phi_3 + \frac{im}{r}\phi_5 = f_4(r) \quad (15.332)$$

$$-i\rho_0(\omega - ku_0)\phi_6 + \rho_0 T_0' \phi_2 + i(\gamma - 1)(\omega - ku_0)\phi_5 = f_5(r) \quad (15.333)$$

446 SOLVABILITY CONDITIONS

$$\frac{\phi_5}{\rho_0} = \frac{\phi_4}{\rho_0} + \frac{\phi_6}{T_0} \quad (15.334)$$

$$\phi_2 - \beta_1 \phi_5 = \alpha_1 \quad \text{at} \quad r = R_1$$
$$\phi_2 - \beta_2 \phi_5 = \alpha_2 \quad \text{at} \quad r = R_2 \quad (15.335)$$

where u_0, ρ_0, p_0, T_0, and $f_n(r)$ are known functions of r and $\omega, m, k, \alpha_n, \beta_n$, and γ are independent of r. We note that four of the equations are algebraic equations. In the analysis, the homogeneous problem has a nontrivial solution so that the inhomogeneous problem has a solution only if a solvability condition is satisfied.

To determine the solvability condition, we multiply (15.329) with ψ_1, (15.330) with ψ_2, (15.331) with ψ_3, (15.332) with ψ_4, (15.333) with ψ_5, and (15.334) with ψ_6. Then, we add the resulting equations, integrate the result by parts from $r = R_1$ to $r = R_2$ to transfer the derivatives from the ϕ_n to the ψ_n, and obtain

$$\int_{R_1}^{R_2} i\rho_0 \phi_1 [-\hat{\omega}\psi_2 + k\psi_1] \, dr + \int_{R_1}^{R_2} \rho_0 \phi_2 \left[-i\hat{\omega}\psi_3 + u_0'\psi_2 - r\frac{d}{dr}\left(\frac{\psi_1}{r}\right) \right.$$

$$\left. + T_0'\psi_5 \right] dr + \int_{R_1}^{R_2} i\rho_0 \phi_3 \left[-\hat{\omega}\psi_4 + \frac{m}{r}\psi_1 \right] dr + \int_{R_1}^{R_2} \phi_4 [-i\hat{\omega}\psi_1$$

$$- T_0 \rho_0 \psi_6] \, dr + \int_{R_1}^{R_2} \phi_5 \left[ik\psi_2 - \psi_3' + \frac{im}{r}\psi_4 + i(\gamma - 1)\hat{\omega}\psi_5 \right.$$

$$\left. + \rho_0 T_0 \psi_6 \right] dr + \int_{R_1}^{R_2} \rho_0 \phi_6 [-i\hat{\omega}\psi_5 - p_0 \psi_6] \, dr + [\rho_0 \phi_2 \psi_1 + \phi_5 \psi_3]_{R_1}^{R_2}$$

$$= \sum_{n=1}^{5} \int_{R_1}^{R_2} \psi_n f_n \, dr \quad (15.336)$$

where $\hat{\omega} = \omega - ku_0$. As before, to define the adjoint, we first consider the homogeneous problem (i.e., $\alpha_n = 0$ and $f_n = 0$). We note that the integrands have been arranged so that the coefficients of the ϕ_n are separated. This is because the adjoint equations are defined by setting the coefficient of each ϕ_n equal to zero. The result is

$$-\hat{\omega}\psi_2 + k\psi_1 = 0 \quad (15.337)$$

$$-i\hat{\omega}\psi_3 + u_0'\psi_2 - r\frac{d}{dr}\left(\frac{\psi_1}{r}\right) + T_0'\psi_5 = 0 \quad (15.338)$$

$$-\hat{\omega}\psi_4 + \frac{m}{r}\psi_1 = 0 \quad (15.339)$$

GENERAL DIFFERENTIAL SYSTEMS OF FIRST-ORDER EQUATIONS

$$i\hat{\omega}\psi_1 + T_0 p_0 \psi_6 = 0 \qquad (15.340)$$

$$ik\psi_2 - \psi_3' + \frac{im}{r}\psi_4 + i(\gamma - 1)\hat{\omega}\psi_5 + p_0 T_0 \psi_6 = 0 \qquad (15.341)$$

$$i\hat{\omega}\psi_5 + p_0 \psi_6 = 0 \qquad (15.342)$$

Equations (15.337) through (15.342) are the equations adjoint to the homogeneous equations (15.329) through (15.334). To determine the boundary conditions on the ψ_n, we set the $f_n = 0$ in (15.336) and use (15.337) through (15.342). The result is

$$[\rho_0 \phi_2 \psi_1 + \phi_5 \psi_3]_{R_1}^{R_2} = 0 \qquad (15.343)$$

Putting $\alpha_n = 0$ in (15.335) and then substituting for ϕ_2 in terms of ϕ_5, we rewrite (15.343) as

$$(\rho_0 \beta_2 \psi_1 + \psi_3)_{r=R_2} \phi_5(R_2) - (\rho_0 \beta_1 \psi_1 + \psi_3)_{r=R_1} \phi_5(R_1) = 0 \qquad (15.344)$$

As before, we choose the adjoint boundary conditions such that each of the coefficients of $\phi_5(R_2)$ and $\phi_5(R_1)$ in (15.344) vanish independently. The result is

$$\begin{aligned}\psi_3 + \rho_0 \beta_2 \psi_1 &= 0 \quad \text{at} \quad r = R_2 \\ \psi_3 + \rho_0 \beta_1 \psi_1 &= 0 \quad \text{at} \quad r = R_1 \end{aligned} \qquad (15.345)$$

Thus, the adjoint homogeneous problem is specified by equations (15.337) through (15.342) subject to the boundary conditions (15.345).

Returning to the inhomogeneous problem, we use (15.337) through (15.342) in (15.336) and obtain

$$[\rho_0 \phi_2 \psi_1 + \phi_5 \psi_3]_{R_1}^{R_2} = \sum_{n=1}^{5} \int_{R_1}^{R_2} \psi_n f_n \, dr \qquad (15.346)$$

Substituting for ψ_3 from (15.345) in terms of ψ_1 and substituting for ϕ_2 from (15.335) in terms of ϕ_5, we rewrite (15.346) as

$$\alpha_2 \rho_0(R_2)\psi_1(R_2) - \alpha_1 \rho_0(R_1)\psi_1(R_1) = \sum_{n=1}^{5} \int_{R_1}^{R_2} f_n \psi_n \, dr \qquad (15.347)$$

which is the required solvability condition.

15.14. General Differential Systems of First-Order Equations

We consider the solvability conditions for the problem

$$\frac{d\phi}{dx} - A(x)\phi = f(x) \qquad (15.348)$$

448 SOLVABILITY CONDITIONS

$$\phi_i(0) = \beta_i \quad \text{for} \quad i = 1, 2, \ldots, m$$
$$\phi_i(1) = \beta_i \quad \text{for} \quad i = m+1, m+2, \ldots, n \tag{15.349}$$

where ϕ and f are column vectors with n components, A is an $n \times n$ matrix, and the β_i are known constants. The boundary conditions need not be disjoint as in (15.349). The theory is applicable to any n linearly independent combinations of $\phi_i(0)$ and $\phi_i(1)$. Two general cases are considered at the end of this section.

We assume that the homogeneous problem has a nontrivial solution so that the inhomogeneous problem will have a solution only if solvability conditions are satisfied. To determine the solvability conditions of (15.348) and (15.349), we multiply the ith equation in (15.348) with ψ_i and add the results. This step is equivalent to multiplying (15.348) from the left with ψ^T, where ψ^T is the transpose of the adjoint column vector ψ with n components. Thus, we have

$$\psi^T \frac{d\phi}{dx} - \psi^T A \phi = \psi^T f$$

which, upon integration from $x = 0$ to $x = 1$, gives

$$\int_0^1 \psi^T \frac{d\phi}{dx} dx - \int_0^1 \psi^T A \phi \, dx = \int_0^1 \psi^T f \, dx \tag{15.350}$$

Integrating by parts the first integral on the left-hand side of (15.350) to transfer the derivative from ϕ to ψ^T, we obtain

$$\psi^T \phi \Big|_0^1 - \int_0^1 \frac{d\psi^T}{dx} \phi \, dx - \int_0^1 \psi^T A \phi \, dx = \int_0^1 \psi^T f \, dx$$

or

$$\psi^T \phi \Big|_0^1 - \int_0^1 \left(\frac{d\psi^T}{dx} + \psi^T A \right) \phi \, dx = \int_0^1 \psi^T f \, dx \tag{15.351}$$

The adjoint equations are defined by setting the coefficient of ϕ in the integrand in (15.351) equal to zero and obtaining

$$\frac{d\psi^T}{dx} + \psi^T A = 0$$

Taking the transpose, we have

$$\left(\frac{d\psi^T}{dx} \right)^T + (\psi^T A)^T = 0$$

or

GENERAL DIFFERENTIAL SYSTEMS OF FIRST-ORDER EQUATIONS 449

$$\frac{d\psi}{dx} + A^T \psi = 0 \tag{15.352}$$

Comparing (15.348) and (15.352), we conclude that the differential equations are self-adjoint if $A = -A^T$. To determine the boundary conditions on ψ, we consider the homogeneous problem. Setting $\mathbf{f} = 0$ in (15.351) we obtain

$$\psi^T \phi \big|_0^1 = 0 \tag{15.353}$$

or

$$[\psi_1 \phi_1 + \psi_2 \phi_2 + \cdots + \psi_n \phi_n]_0^1 = 0 \tag{15.354}$$

Putting $\beta_i = 0$ in (15.349) and substituting the result into (15.354), we have

$$\psi_1(1)\phi_1(1) + \psi_2(1)\phi_2(1) + \cdots + \psi_m(1)\phi_m(1) - \psi_{m+1}(0)\phi_{m+1}(0)$$
$$- \psi_{m+2}(0)\phi_{m+2}(0) - \cdots - \psi_n(0)\phi_n(0) = 0 \tag{15.355}$$

As before, we define the adjoint boundary conditions such that each of the coefficients of the $\phi_i(1)$ for $i = 1, 2, \ldots, m$ and the $\phi_i(0)$ for $i = m+1, m+2, \ldots, n$ vanish independently. The result is

$$\begin{aligned} \psi_i(0) &= 0 \quad \text{for} \quad i = m+1, m+2, \ldots, n \\ \psi_i(1) &= 0 \quad \text{for} \quad i = 1, 2, \ldots, m \end{aligned} \tag{15.356}$$

Returning to the inhomogeneous problem, we substitute (15.349), (15.352), and (15.356) into (15.351) and obtain

$$\beta_{m+1} \psi_{m+1}(1) + \beta_{m+2} \psi_{m+2}(1) + \cdots + \beta_n \psi_n(1)$$
$$- \beta_1 \psi_1(0) - \beta_2 \psi_2(0) - \cdots - \beta_m \psi_m(0) = \int_0^1 \psi^T \mathbf{f}\, dx \tag{15.357}$$

as the desired solvability condition.

Instead of the boundary conditions (15.349), let us consider the general boundary conditions

$$\phi(0) = \mathcal{C}\phi(1) + \beta \tag{15.358}$$

where $\mathcal{C}$ is an $n \times n$ constant matrix and β is a column vector with n components. In this case, all preceding equations up to (15.354) still hold. Thus, considering the homogeneous problem (i.e., $\mathbf{f} = 0$ and $\beta = 0$), we find from (15.353) and (15.358) that

$$\psi^T(1)\phi(1) - \psi^T(0)\phi(0) = \psi^T(1)\phi(1) - \psi^T(0)\mathcal{C}\phi(1) = 0$$

or

$$[\psi^T(1) - \psi^T(0)\mathcal{C}]\phi(1) = 0 \tag{15.359}$$

As before, we choose the adjoint boundary conditions such that each of the coefficients of the $\phi_i(1)$ vanish independently. The result is

$$\psi^T(1) - \psi^T(0)\mathcal{C} = 0$$

which, upon taking the transpose, becomes

$$\psi(1) = \mathcal{C}^T \psi(0) \qquad (15.360)$$

Returning to the inhomogeneous problem, we use (15.352) in (15.351) and obtain

$$\psi^T(1)\phi(1) - \psi^T(0)\phi(0) = \int_0^1 \psi^T \mathbf{f}\, dx \qquad (15.361)$$

Substituting for $\phi(0)$ from (15.358) into (15.361), we have

$$\psi^T(1)\phi(1) - \psi^T(0)\mathcal{C}\phi(1) - \psi^T(0)\beta = \int_0^1 \psi^T \mathbf{f}\, dx \qquad (15.362)$$

Using (15.360), we find that the first two terms on the left-hand side of (15.362) vanish, and hence, it reduces to

$$-\psi^T(0)\beta = \int_0^1 \psi^T \mathbf{f}\, dx \qquad (15.363)$$

which is the desired solvability condition. Problems with similar boundary conditions can be treated the same way.

Finally, we consider the most general linear boundary conditions

$$\sum_{j=1}^{2n} \alpha_{ij}\phi_j = \beta_i \quad i = 1, 2, \ldots, n \qquad (15.364)$$

where the ϕ_j are the $2n$ components of the boundary column vector ϕ_b whose first n components are given by $\phi(1)$ and whose last n components are given by $\phi(0)$. Again in this case, all equations up to (15.354) still hold. To determine the adjoint boundary conditions, we introduce a linear nonsingular transformation from ϕ_b to Φ according to

$$\Phi = \mathcal{C}\phi_b \qquad (15.365)$$

where $\mathcal{C}$ is a $2n \times 2n$ constant coefficient matrix whose first n rows are the same as those in (15.364) so that

$$\Phi_i = \sum_{j=1}^{2n} \alpha_{ij}\phi_j = \beta_i \quad i = 1, 2, \ldots, n \qquad (15.366)$$

GENERAL DIFFERENTIAL SYSTEMS OF FIRST-ORDER EQUATIONS 451

As before, the last n rows are arbitrary, except that they must be linearly independent of each other and of the first n rows. Thus, if $\phi_b \neq 0$, they can be chosen to produce any desired values for Φ_i, $i \geq n+1$.

It follows from (15.353) that

$$\psi^T \phi \big|_0^1 = \psi^T(1) I \phi(1) - \psi^T(0) I \phi(0) = 0$$

or

$$[\psi^T(1)\ \psi^T(0)] \begin{bmatrix} I & 0 \\ 0 & -I \end{bmatrix} \begin{bmatrix} \phi(1) \\ \phi(0) \end{bmatrix} = 0$$

or

$$\psi^T \phi \big|_0^1 = \psi_b^T P \phi_b = 0 \tag{15.367}$$

where I is the $n \times n$ identity matrix and

$$P = \begin{bmatrix} I & 0 \\ 0 & -I \end{bmatrix}$$

Solving (15.365) for ϕ_b, we have

$$\phi_b = \mathcal{G}^{-1} \Phi$$

Hence, (15.367) becomes

$$\psi^T \phi \big|_0^1 = \psi_b^T P \mathcal{G}^{-1} \Phi = 0$$

or

$$\psi^T \phi \big|_0^1 = \Psi^T \Phi = 0 \tag{15.368}$$

where

$$\Psi^T = \psi_b^T P \mathcal{G}^{-1} \quad \text{or} \quad \Psi = (\mathcal{G}^{-1})^T P^T \psi_b \tag{15.369}$$

It follows from (15.366) that the homogeneous boundary conditions are $\Phi_i = 0$ for $i = 1, 2, \ldots, n$. Hence, (15.368) becomes

$$\sum_{i=n+1}^{2n} \Psi_i \Phi_i = 0 \tag{15.370}$$

Since the last n rows in $\mathcal{G}$ can be chosen so that the last Φ_i, $i = n+1, \ldots, 2n$ can assume any nonzero values, it follows from (15.370) that

$$\Psi_i = 0 \quad \text{for} \quad i = n+1, n+2, \ldots, 2n \tag{15.371}$$

Having defined the adjoint equations (15.352) and boundary conditions (15.371), we return to the inhomogeneous problem to determine the solvability conditions. Using (15.352) and the fact that $\psi^T \phi \big|_0^1 = \Psi^T \Phi$, we obtain from (15.351) that

452 SOLVABILITY CONDITIONS

$$\Psi^T \Phi = \int_0^1 \psi^T \mathbf{f}\, dx \qquad (15.372)$$

Substituting (15.366) and (15.370) into (15.372), we obtain the desired solvability condition

$$\sum_{i=1}^n \beta_i \Psi_i = \int_0^1 \psi^T \mathbf{f}\, dx$$

for every possible solution of the adjoint problem.

15.15. Differential Systems with Interfacial Boundary Conditions

In analyzing the propagation of waves on the interfaces of composite materials, one may encounter inhomogeneous differential equations and inhomogeneous conditions imposed at the interfaces. In this section, we consider, as an example, the solvability condition for the inhomogeneous problem that arises from the coupling of torsional modes in a clad-rod having a sinusoidally perturbed core-cladding interface. Thus, we consider the problem

$$\frac{d^2\phi_1}{dr^2} + \frac{1}{r}\frac{d\phi_1}{dr} + \left(\alpha_n^2 - \frac{1}{r^2}\right)\phi_1 = f_1(r) \qquad 0 \leqslant r \leqslant a \qquad (15.373)$$

$$\frac{d^2\phi_2}{dr^2} + \frac{1}{r}\frac{d\phi_2}{dr} - \left(\gamma_n^2 + \frac{1}{r^2}\right)\phi_2 = f_2(r) \qquad a \leqslant r \leqslant b \qquad (15.374)$$

$$\phi_1(0) < \infty \qquad \phi_2(b) - b\phi_2'(b) = 0$$

$$\phi_1 - \phi_2 = \beta_1 \qquad \phi_2' - \mu_1\phi_1' - \mu_2\phi_1 = \beta_2 \qquad \text{at} \qquad r = a \qquad (15.375)$$

where $\alpha_n^2, \gamma_n^2, \mu_n, \beta_n, a$, and b are constants.

As discussed in Section 15.9, the homogeneous equations in (15.373) and (15.374) can be made self-adjoint by multiplying each with r. Thus, to determine the solvability condition for (15.373) through (15.375), we multiply (15.373) by $ru_1(r)$, integrate the result from $r = 0$ to $r = a$, and obtain

$$\int_0^a \left[u_1(r\phi_1')' + \left(\alpha_n^2 r - \frac{1}{r}\right)\phi_1 u_1 \right] dr = \int_0^a ru_1 f_1\, dr \qquad (15.376)$$

which, upon integration by parts, gives

$$[ru_1\phi_1' - ru_1'\phi_1]_0^a + \int_0^a \phi_1 \left[(ru_1')' + \left(\alpha_n^2 r - \frac{1}{r}\right)u_1 \right] dr = \int_0^a ru_1 f_1\, dr$$

$$(15.377)$$

DIFFERENTIAL SYSTEMS WITH INTERFACIAL BOUNDARY CONDITIONS 453

Similarly, we multiply (15.374) by $ru_2(r)$, integrate the result from $r = a$ to $r = b$, and obtain

$$\int_a^b \left[u_2(r\phi_2')' - \left(\gamma_n^2 r + \frac{1}{r} \right) \phi_2 u_2 \right] dr = \int_a^b r u_2 f_2 \, dr$$

which, upon integration by parts, gives

$$[r u_2 \phi_2' - r u_2' \phi_2]_a^b + \int_a^b \phi_2 \left[(r u_2')' - \left(\gamma_n^2 r + \frac{1}{r} \right) u_2 \right] dr = \int_a^b r u_2 f_2 \, dr \tag{15.378}$$

Next, we add (15.377) and (15.378) to obtain

$$[r u_2 \phi_2' - r u_2' \phi_2]_a^b + [r u_1 \phi_1' - r u_1' \phi_1]_0^a + \int_0^a \phi_1 \left[(r u_1')' + \left(\alpha_n^2 r - \frac{1}{r} \right) u_1 \right] dr$$

$$\left[+ \int_a^b \phi_2 \left[(r u_2')' - \left(\gamma_n^2 r + \frac{1}{r} \right) u_2 \right] dr = \int_0^a r u_1 f_1 \, dr + \int_a^b r u_2 f_2 \, dr \right. \tag{15.379}$$

To define the adjoint homogeneous problem, we consider the homogeneous problem (i.e., $f_n = 0$ and $\beta_n = 0$). Then, the adjoint equations are defined by setting the integrands in (15.379) equal to zero, that is,

$$(r u_1')' + \left(\alpha_n^2 r - \frac{1}{r} \right) u_1 = 0 \tag{15.380}$$

$$(r u_2')' - \left(\gamma_n^2 r + \frac{1}{r} \right) u_2 = 0 \tag{15.381}$$

Then, (15.379) becomes

$$b u_2(b) \phi_2'(b) - b u_2'(b) \phi_2(b) - \lim_{r \to 0} [r u_1 \phi_1' - r u_1' \phi_1]$$
$$+ a u_1(a) \phi_1'(a) - a u_1'(a) \phi_1(a) - a u_2(a) \phi_2'(a) + a u_2'(a) \phi_2(a) = 0 \tag{15.382}$$

Since (15.380) possesses a bounded solution and an unbounded solution and since $\phi_1(0) < \infty$, the term involving the limit as $r \to 0$ in (15.382) vanishes if

$$u_1(0) < \infty \tag{15.383}$$

For the homogeneous problem, it follows from (15.375) that

$$\phi_2(b) = b \phi_2'(b)$$

$$\phi_2(a) = \phi_1(a), \quad \phi_2'(a) = \mu_1 \phi_1'(a) + \mu_2 \phi_1(a)$$

Hence, (15.382) becomes

$$b[u_2(b) - bu_2'(b)]\phi_2'(b) + a[u_1(a) - \mu_1 u_2(a)]\phi_1'(a)$$
$$+ a[u_2'(a) - u_1'(a) - \mu_2 u_2(a)]\phi_1(a) = 0 \quad (15.384)$$

Again, we choose the adjoint boundary conditions such that each of the coefficients of $\phi_2'(b)$, $\phi_1'(a)$, and $\phi_1(a)$ vanish independently. Hence,

$$u_2(b) = bu_2'(b) \qquad u_1(a) = \mu_1 u_2(a)$$
$$u_2'(a) - u_1'(a) - \mu_2 u_2(a) = 0 \quad (15.385)$$

Therefore, the adjoint problem is defined by (15.380) and (15.381) and boundary conditions (15.383) and (15.385). Comparing the adjoint problem with the original homogeneous problem, we conclude that, although the differential equations are self-adjoint, the homogeneous problem is not self-adjoint unless $\mu_1 = 1$.

Having defined the adjoint problem, we return to the inhomogeneous problem to determine the solvability condition. Using (15.380), (15.381), and (15.383) in (15.379), we obtain

$$bu_2(b)\phi_2'(b) - bu_2'(b)\phi_2(b) + au_1(a)\phi_1'(a) - au_1'(a)\phi_1(a) - au_2(a)\phi_2'(a)$$
$$+ au_2'(a)\phi_2(a) = \int_0^a ru_1 f_1 \, dr + \int_a^b ru_2 f_2 \, dr \quad (15.386)$$

Substituting for $\phi_2(b)$, $\phi_2(a)$, and $\phi_2'(a)$ from (15.375) into (15.386), we have

$$b[u_2(b) - bu_2'(b)]\phi_2'(b) + a[u_1(a) - \mu_1 u_2(a)]\phi_1'(a)$$
$$+ a[u_2'(a) - u_1'(a) - \mu_2 u_2(a)]\phi_1(a) - a\beta_2 u_2(a) - a\beta_1 u_2'(a)$$
$$= \int_0^a ru_1 f_1 \, dr + \int_a^b ru_2 f_2 \, dr$$

which, upon using (15.385), simplifies to

$$-a\beta_2 u_2(a) - a\beta_1 u_2'(a) = \int_0^a ru_1 f_1 \, dr + \int_a^b ru_2 f_2 \, dr \quad (15.387)$$

Equation (15.387) is the desired solvability condition.

15.16. Integral Equations

In this section, we consider the solvability condition of Fredholm's integral equation

$$\phi(s) = f(s) + \lambda \int_a^b K(s, t)\phi(t)\, dt \qquad (15.388)$$

when the homogeneous problem (i.e., $f = 0$) has a nontrivial solution. To determine the solvability condition for (15.388), we multiply it by $\psi(s)$, integrate the result from $s = a$ to $s = b$, and obtain

$$\int_a^b \phi(s)\psi(s)\, ds = \int_a^b f(s)\psi(s)\, ds + \lambda \int_a^b \left[\int_a^b K(s, t)\phi(t)\, dt \right] \psi(s)\, ds$$

or

$$\int_a^b \phi(s)\psi(s)\, ds = \int_a^b f(s)\psi(s)\, ds + \lambda \int_a^b \int_a^b K(s, t)\phi(t)\psi(s)\, dt\, ds$$

which, upon interchanging the integrations with respect to s and t in the double integral, gives

$$\int_a^b \phi(s)\psi(s)\, ds - \lambda \int_a^b \int_a^b K(s, t)\psi(s)\phi(t)\, ds\, dt = \int_a^b f(s)\psi(s)\, ds$$

$$(15.389)$$

Since s and t are dummy variables of integration, they can be interchanged in the double integral so that (15.389) can be rewritten as

$$\int_a^b \psi(s)\phi(s)\, ds - \lambda \int_a^b \int_a^b K(t, s)\psi(t)\phi(s)\, dt\, ds = \int_a^b f(s)\psi(s)\, ds$$

or

$$\int_a^b \left[\psi(s) - \lambda \int_a^b K(t, s)\psi(t)\, dt \right] \phi(s)\, ds = \int_a^b f(s)\psi(s)\, ds \qquad (15.390)$$

To define the adjoint equation, we consider the homogeneous problem (i.e., $f = 0$). Then, (15.390) becomes

$$\int_a^b \left[\psi(s) - \lambda \int_a^b K(t, s)\psi(t)\, dt \right] \phi(s)\, ds = 0 \qquad (15.391)$$

We choose the adjoint equation such that

$$\psi(s) = \lambda \int_a^b K(t, s)\psi(t)\, dt \qquad (15.392)$$

Comparing (15.392) with the homogeneous equation (15.388), we conclude that the adjoint equation differs from the homogeneous equation in that the integration is performed with respect to the first variable in the kernel $K(s, t)$. Therefore, the homogeneous equation is self-adjoint if and only if the kernel $K(s, t)$ is symmetric, that is, $K(s, t) = K(t, s)$.

Having defined the adjoint equation, we return to the inhomogeneous problem. Using (15.392) in (15.390), we obtain the desired solvability condition

$$\int_a^b f(s)\psi(s)\, ds = 0 \qquad (13.393)$$

which states that the inhomogeneous problem is solvable only if the inhomogeneous term $f(s)$ is orthogonal to every solution of the adjoint problem. It turns out, in this case, that (13.393) is also a sufficient condition for the inhomogeneous problem to have a solution. If the homogeneous problem does not have a nontrivial solution, (15.393) is satisfied automatically because $\psi(s) \equiv 0$. These statements constitute Fredholm's alternative theorem: either the integral equation (15.388) is soluble for any $f(s)$ and the homogeneous equation has only a trivial solution or the homogeneous equation has nontrivial solutions and the inhomogeneous equation is soluble if and only if the inhomogeneity is orthogonal to every solution of the adjoint homogeneous problem.

As an application of the theory, we consider the case of a degenerate kernel. The kernel $K(s, t)$ is said to be degenerate if it can be expressed as a finite sum of products of functions of s and t, that is,

$$K(s, t) = \sum_{i=1}^{N} \alpha_i(s)\beta_i(t) \qquad (15.394)$$

In this case, (15.388) can be rewritten as

$$\phi(s) = f(s) + \lambda \sum_{i=1}^{N} \alpha_i(s) \int_a^b \beta_i(t)\phi(t)\, dt \qquad (15.395)$$

To solve (15.395), we put

$$\int_a^b \beta_i(t)\phi(t)\, dt = x_i \qquad i = 1, 2, \ldots, N \qquad (15.396)$$

Then, (15.395) becomes

$$\phi(s) = f(s) + \lambda \sum_{i=1}^{N} x_i \alpha_i(s) \qquad (15.397)$$

We multiply (15.397) by $\beta_j(s)$, integrate the result from $s = a$ to $s = b$, and obtain

$$\int_a^b \beta_j(s)\phi(s)\,ds = \int_a^b \beta_j(s)f(s)\,ds + \lambda \sum_{i=1}^{N} x_i \int_a^b \alpha_i(s)\beta_j(s)\,ds \quad (15.398)$$

It follows from (15.396) that the term on the left-hand side of (15.398) is x_j so that

$$x_j = f_j + \lambda \sum_{i=1}^{N} a_{ji} x_i \quad j = 1, 2, \ldots, N \quad (15.399)$$

where

$$f_j = \int_a^b \beta_j(s) f(s)\,ds \qquad a_{ji} = \int_a^b \beta_j(s) \alpha_i(s)\,ds \quad (15.400)$$

Equation (15.399) can be rewritten in matrix notation as

$$[I - \lambda A]\mathbf{x} = \mathbf{f} \quad (15.401)$$

where I is an $N \times N$ identity matrix, A in an $N \times N$ matrix whose elements are the a_{ji}, and $\mathbf{x}$ and $\mathbf{f}$ are column vectors with the components x_i and f_i.

We note that the problem of solving the integral equation (15.388) has been transformed into the problem of solving the linear algebraic system of equations (15.401). As discussed in Section 15.1, if the homogeneous system of equations (15.401) has only the trivial solution, the inhomogeneous system has a unique solution for every $\mathbf{f}$. On the other hand, if the homogeneous system (15.401) has a nontrivial solution (i.e., $|I - \lambda A| = 0$), the inhomogeneous problem has a solution if and only if

$$(\mathbf{u}, \mathbf{f}) = 0 \quad (15.402)$$

where $\mathbf{u}$ is any solution of the adjoint system

$$[I - \lambda A]^* \mathbf{u} = 0 \quad (15.403)$$

Since A is real, (15.403) can be rewritten as

$$[I - \lambda A^T]\mathbf{u} = 0$$

or in tensor notation as

$$u_j - \lambda \sum_{i=1}^{N} a_{ij} u_i = 0 \quad (15.404)$$

Thus, the problem is self-adjoint if and only if A is symmetric, that is, $a_{ij} = a_{ji}$, which happens if, for example,

$$\alpha_i(s) = \beta_i(s)$$

458 SOLVABILITY CONDITIONS

15.17. Partial-Differential Equations

In this section, we consider the solvability conditions for systems governed by partial- rather than ordinary-differential equations. The procedure in this case is similar to that followed in the case of ordinary-differential equations except that the integration is carried out over more than one dimension. We describe the procedure by applying it to two examples.

EXAMPLE 1

In analyzing the propagation of acoustic waves in acoustically lined rectangular ducts with varying cross sections, one encounters the inhomogeneous problem

$$\frac{\partial^2 \phi}{\partial y^2} + \frac{\partial^2 \phi}{\partial z^2} + \lambda \phi = f(y, z) \quad 0 \leq y \leq a \quad 0 \leq z \leq b \quad (15.405)$$

$$\phi(0, z) = 0, \, \phi(y, 0) = 0 \quad (15.406)$$

$$\frac{\partial \phi}{\partial y} - \alpha_1 \phi = \beta_1(z) \quad \text{at} \quad y = a$$

$$\frac{\partial \phi}{\partial z} - \alpha_2 \phi = \beta_2(y) \quad \text{at} \quad z = b \quad (15.407)$$

To determine the solvability condition for (15.405) through (15.407), we multiply (15.405) by $u(y, z)$ and integrate the result over the domain of interest; that is, $0 \leq y \leq a$ and $0 \leq z \leq b$. The result is

$$\int_0^b \int_0^a \left[u \frac{\partial^2 \phi}{\partial y^2} + u \frac{\partial^2 \phi}{\partial z^2} + \lambda u \phi \right] dy \, dz = \int_0^b \int_0^a uf \, dy \, dz \quad (15.408)$$

Next, we integrate (15.408) by parts to transfer the derivatives from ϕ to u. Thus,

$$\int_0^b \int_0^a u \frac{\partial^2 \phi}{\partial y^2} dy \, dz = \int_0^b \left[\int_0^a u \frac{\partial^2 \phi}{\partial y^2} dy \right] dz$$

$$= \int_0^b \left[\left(u \frac{\partial \phi}{\partial y} - \frac{\partial u}{\partial y} \phi \right) \bigg|_0^a + \int_0^a \phi \frac{\partial^2 u}{\partial y^2} dy \right] dz \quad (15.409)$$

$$\int_0^b \int_0^a u \frac{\partial^2 \phi}{\partial z^2} dy \, dz = \int_0^a \left[\int_0^b u \frac{\partial^2 \phi}{\partial z^2} dz \right] dy$$

$$= \int_0^a \left[\left(u \frac{\partial \phi}{\partial z} - \frac{\partial u}{\partial z} \phi \right) \bigg|_0^b + \int_0^b \phi \frac{\partial^2 u}{\partial z^2} dz \right] dy \quad (15.410)$$

Using (15.409) and (15.410), we have

$$\int_0^b \int_0^a \left\{ u \left[\frac{\partial^2 \phi}{\partial y^2} + \frac{\partial^2 \phi}{\partial z^2} + \lambda \phi \right] - \phi \left[\frac{\partial^2 u}{\partial y^2} + \frac{\partial^2 u}{\partial z^2} + \lambda u \right] \right\} dy\, dz$$

$$= \int_0^b \left(u \frac{\partial \phi}{\partial y} - \frac{\partial u}{\partial y} \phi \right) \bigg|_0^a dz + \int_0^a \left(u \frac{\partial \phi}{\partial z} - \frac{\partial u}{\partial z} \phi \right) \bigg|_0^b dy \quad (15.411)$$

which can be obtained alternatively from the general Green's identity

$$\iint_S (u \nabla^2 \phi - \phi \nabla^2 u)\, dS = \oint_\Gamma \left(u \frac{\partial \phi}{\partial n} - \phi \frac{\partial u}{\partial n} \right) ds \quad (15.412)$$

where Γ is the boundary of S.

As before, we define the adjoint equation by setting the coefficient of ϕ in the integrand on the left-hand side of (15.411) equal to zero, that is,

$$\frac{\partial^2 u}{\partial y^2} + \frac{\partial^2 u}{\partial z^2} + \lambda u = 0 \quad (15.413)$$

To define the adjoint boundary conditions, we consider the homogeneous problem (i.e., $f = 0$ and $\beta_n = 0$). With $f = 0$, (15.411) becomes

$$\int_0^b \left(u \frac{\partial \phi}{\partial y} - \frac{\partial u}{\partial y} \phi \right) \bigg|_0^a dz + \int_0^a \left(u \frac{\partial \phi}{\partial z} - \frac{\partial u}{\partial z} \phi \right) \bigg|_0^b dy = 0 \quad (15.414)$$

The homogeneous boundary conditions (15.407) can be rewritten as

$$\frac{\partial \phi}{\partial y} = \alpha_1 \phi \quad \text{at} \quad y = a \quad \text{and} \quad \frac{\partial \phi}{\partial z} = \alpha_2 \phi \quad \text{at} \quad z = b \quad (15.415)$$

Using (15.406) and (15.415) in (15.414), we have

$$\int_0^b \left[\left(\alpha_1 u - \frac{\partial u}{\partial y} \right) \phi \bigg|_a - u \frac{\partial \phi}{\partial y} \bigg|_0 \right] dz + \int_0^a \left[\left(\alpha_2 u - \frac{\partial u}{\partial z} \right) \phi \bigg|_b \right.$$

$$\left. - u \frac{\partial \phi}{\partial z} \bigg|_0 \right] dy = 0 \quad (15.416)$$

Since (15.416) should hold for all $\phi(y, z)$ satisfying the homogeneous equation (15.405) and homogeneous boundary conditions (15.406) and (15.407), each of the coefficients of

$$\phi(a, z) \quad \frac{\partial \phi}{\partial y}(0, z) \quad \phi(y, b) \quad \frac{\partial \phi}{\partial z}(y, 0)$$

must vanish independently; that is,

$$u(0, z) = 0 \quad u(y, 0) = 0 \tag{15.417}$$

$$\frac{\partial u}{\partial y} - \alpha_1 u = 0 \quad \text{at} \quad y = a$$

$$\frac{\partial u}{\partial z} - \alpha_2 u = 0 \quad \text{at} \quad z = b \tag{15.418}$$

Having defined the adjoint problem, we return to the inhomogeneous problem. Using (15.405) and (15.413) in (15.411), we have

$$\int_0^b \int_0^a uf \, dy \, dz = \int_0^b \left(u \frac{\partial \phi}{\partial y} - \frac{\partial u}{\partial y} \phi \right) \bigg|_0^a dz + \int_0^a \left(u \frac{\partial \phi}{\partial z} - \frac{\partial u}{\partial z} \phi \right) \bigg|_0^b dy$$

which, upon using (15.406), (15.407), (15.417), and (15.418), becomes

$$\int_0^b \int_0^a uf \, dy \, dz = \int_0^b \beta_1(z) u(a, z) \, dz + \int_0^a \beta_2(y) u(y, b) \, dy \tag{15.419}$$

Therefore, the inhomogeneous problem has a solution only if (15.419) is satisfied for all possible solutions $u(y, z)$ of the adjoint problem defined by (15.413), (15.417), and (15.418). Comparing the adjoint problem with the original homogeneous problem, we find that they are identical, and hence, the problem is self-adjoint.

EXAMPLE 2

As a second example, we consider the solvability condition of the inhomogeneous problem consisting of equation (15.223) and the boundary conditions (15.224) and (15.225). Rather than expanding ϕ_1 and f in Fourier series and reducing the governing equation into an ordinary-differential equation, we work directly with the partial-differential equation. To this end, we rewrite (15.223) as

$$\nabla^2 \phi_1 + \omega_0^2 \phi_1 = -2\omega_0 \omega_1 \phi_0 \tag{15.420}$$

Either we multiply (15.420) by $u(r, \theta)$ and integrate the result by parts or we use Green's identity (15.412). Since we performed the integration by parts in the preceding example, we apply Green's identity (15.412). Since there are solutions of (15.420) that are unbounded at the origin, we take S to consist of the area bounded by $r = r_0$ and $r = 1$, where r_0 is very small. Then, $dS = r \, dr \, d\theta$ and $ds = r \, d\theta$. Moreover, $\partial \phi / \partial n = \partial \phi / \partial r$ on $r = 1$ and $\partial \phi / \partial n = -\partial \phi / \partial r$ on $r = r_0$ because $\mathbf{n}$ is the outward normal. Then, (15.412) becomes

$$\int_0^{2\pi} \int_{r_0}^1 (u \nabla^2 \phi_1 - \phi_1 \nabla^2 u) r \, dr \, d\theta = \int_0^{2\pi} \left(u \frac{\partial \phi_1}{\partial r} - \phi_1 \frac{\partial u}{\partial r} \right) \bigg|_{r=1} d\theta$$

$$-\int_0^{2\pi}\left(u\frac{\partial\phi_1}{\partial r}-\phi_1\frac{\partial u}{\partial r}\right)\bigg|_{r=r_0}r_0\,d\theta \qquad (15.421)$$

Substituting for $\nabla^2\phi_1$ from (15.420) into (15.421), we have

$$\int_0^{2\pi}\int_{r_0}^1 -2\omega_0\omega_1 r\phi_0 u\,dr\,d\theta - \int_0^{2\pi}\int_{r_0}^1 \phi_1[\nabla^2 u+\omega_0^2 u]\,r\,dr\,d\theta$$

$$=\int_0^{2\pi}\left(u\frac{\partial\phi_1}{\partial r}-\phi_1\frac{\partial u}{\partial r}\right)\bigg|_{r=1}d\theta - \int_0^{2\pi}\left(u\frac{\partial\phi_1}{\partial r}-\phi_1\frac{\partial u}{\partial r}\right)\bigg|_{r=r_0}r_0\,d\theta \qquad (15.422)$$

As before, the adjoint equation is defined by setting the coefficient of ϕ_1 in the integrand on the left-hand side of (15.422) equal to zero, that is,

$$\nabla^2 u + \omega_0^2 u = 0 \qquad (15.423)$$

which is identical to the homogeneous equation (15.420), and hence, the latter is self-adjoint. Using (15.423) and considering the homogeneous problem (i.e., $\omega_1 = 0$ and $f = 0$), we obtain from (15.422) that

$$\int_0^{2\pi} u\frac{\partial\phi_1}{\partial r}\bigg|_{r=1}d\theta - \lim_{r_0\to 0}\int_0^{2\pi}\left(u\frac{\partial\phi_1}{\partial r}-\phi_1\frac{\partial u}{\partial r}\right)\bigg|_{r=r_0}r_0\,d\theta = 0 \qquad (15.424)$$

Since $\phi_1(0,\theta)<\infty$, the last term in (15.424) vanishes if $u(0,\theta)<\infty$. Since (15.424) should hold for all possible $\phi_1(r,\theta)$ satisfying the homogeneous equation (15.223) and homogeneous boundary conditions (15.224) and (15.225), the coefficient of $\partial\phi_1/\partial r$ must vanish, that is, $u(1,\theta)=0$. Thus the adjoint boundary conditions are

$$u(0,\theta)<\infty \quad\text{and}\quad u(1,\theta)=0 \qquad (15.425)$$

Comparing (15.423) and (15.425) with the homogeneous equations (15.223) through (15.225), we conclude that the problem is self-adjoint.

To determine the solvability condition, we return to the inhomogeneous problem. Using (15.423) and (15.425) and taking the limit as $r_0 \to 0$, we obtain from (15.422) that

$$2\omega_0\omega_1\int_0^{2\pi}\int_0^1 r\phi_0 u\,dr\,d\theta = \int_0^{2\pi}\phi_1\frac{\partial u}{\partial r}\bigg|_{r=1}d\theta$$

which, upon using (15.225), becomes

$$2\omega_0\omega_1\int_0^{2\pi}\int_0^1 r\phi_0 u\,dr\,d\theta = -\int_0^{2\pi}\frac{\partial\phi_0}{\partial r}\frac{\partial u}{\partial r}\bigg|_{r=1}f(\theta)\,d\theta \qquad (15.426)$$

Substituting for ϕ_0 from (15.231) into (15.426) and recalling that $\omega_0 = \Omega_{nm}$, we have

$$2\omega_1 \int_0^{2\pi} \int_0^1 rJ_n(\Omega_{nm}r)(A_{nm}e^{in\theta} + \bar{A}_{nm}e^{-in\theta})u \, dr \, d\theta$$

$$= -J_n'(\Omega_{nm}) \int_0^{2\pi} (A_{nm}e^{in\theta} + \bar{A}_{nm}e^{-in\theta}) \frac{\partial u}{\partial r}\bigg|_{r=1} f(\theta) \, d\theta \quad (15.427)$$

Since the problem is self-adjoint,

$$u = J_n(\Omega_{nm}r)e^{in\theta} \quad \text{or} \quad J_n(\Omega_{nm}r)e^{-in\theta}$$

and (15.427) should hold when u is replaced by any one of them. Using the second solution for u, we obtain from (15.427) that

$$2\omega_1 \int_0^{2\pi} \int_0^1 rJ_n^2(\Omega_{nm}r)(A_{nm} + \bar{A}_{nm}e^{-2in\theta}) \, dr \, d\theta$$

$$= -\Omega_{nm}J_n'^2(\Omega_{nm}) \int_0^{2\pi} [A_{nm}f(\theta) + \bar{A}_{nm}e^{-2in\theta}f(\theta)] \, d\theta \quad (15.428)$$

Using (15.248) and the fact that

$$\int_0^{2\pi} e^{im\theta} \, d\theta = 0$$

when m is an integer, we obtain from (15.428) that

$$\omega_1 A_{nm} = -\Omega_{nm} f_{2n} \bar{A}_{nm}$$

which is identical with (15.249).

Exercises

15.1. Determine the solvability conditions for

$$\ddot{u}_1 + \dot{u}_2 + 2u_1 = \sum_{n=1}^{4} P_n e^{int}$$

$$\ddot{u}_2 - \dot{u}_1 + 2u_2 = \sum_{n=1}^{4} Q_n e^{int}$$

where the P_n and Q_n are constants.

15.2. Determine the solvability conditions for

$$\ddot{x} + \dot{y} + x = P_1 e^{2it} + P_2 e^{i2^{-1/2}t}$$
$$\dot{y} - \tfrac{3}{2}\dot{x} + 2y = Q_1 e^{2it} + Q_2 e^{i2^{-1/2}t}$$

where the P_n and Q_n are constants.

15.3. Determine the solvability conditions for

(a) $u'' + \tfrac{1}{4}u = f(x)$
$u'(0) = a \quad u(\pi) = b$

(b) $u'' + \dfrac{1}{x} u' + k^2 u = f(x)$
$u(a) = c_1 \quad u(b) = c_2$

where $b > a > 0$ and the homogeneous problem has a nontrivial solution.

15.4. Determine the solvability condition for the problem

$$u'' + \frac{1}{x} u' + \left(\lambda^2 - \frac{n^2}{x^2}\right) u = f(x)$$

$$u(0) < \infty \quad u'(a) = g$$

where a and g are constants and λ is a root of $J_n'(\lambda a) = 0$.

15.5. Determine the solvability condition for

(a) $\phi'' + \dfrac{n^2 \pi^2}{d^2} \phi = f(x)$
$\phi'(0) = 0 \quad \phi'(d) = \beta$

(b) $\phi'' + \gamma_n^2 \phi = f(x)$
$\phi'(0) = 0 \quad \phi'(1) - \alpha\phi(1) = \beta$

where $\gamma_n \tan \gamma_n = -\alpha$.

15.6. Determine the solvability conditions for

$$\frac{\partial^2 \phi}{\partial y^2} + \frac{\partial^2 \phi}{\partial z^2} + \left(\frac{n^2 \pi^2}{d^2} + \frac{m^2 \pi^2}{b^2}\right) \phi = f(y, z)$$

$$\phi_y(0, z) = 0 \quad \phi_z(y, 0) = 0$$

$$\phi_y(d, z) = \beta_1 \quad \phi_z(y, b) = \beta_2$$

15.7. Determine the solvability conditions for

$$\frac{d^2 \phi}{dr^2} + \frac{1}{r}\frac{d\phi}{dr} + \left(\gamma_{nm}^2 - \frac{n^2}{r^2}\right) \phi = f(r)$$

$$\phi(0) < \infty \quad \phi'(1) - \beta\phi(1) = \alpha$$

where γ_{nm} is a root of $\gamma J_n'(\gamma) - \beta J_n(\gamma) = 0$.

15.8. Determine the solvability conditions for

$$\frac{\partial^2 \phi}{\partial r^2} + \frac{1}{r}\frac{\partial \phi}{\partial r} + \frac{1}{r^2}\frac{\partial^2 \phi}{\partial \theta^2} + \gamma_{nm}^2 \phi = f(r) e^{in\theta}$$

464 SOLVABILITY CONDITIONS

$$\phi(0,\theta) < \infty \qquad \phi_r(1,\theta) = \alpha e^{in\theta}$$

where γ_{nm} is a root of $J_n'(\gamma) = 0$.

15.9. Determine the solvability conditions for

$$\frac{d^2u}{dr^2} + \frac{2}{r}\frac{du}{dr} + \lambda u = F(r)$$

$$u(a) = u_a$$

$$u(b) = u_b$$

when λ is an eigenvalue of the homogeneous problem.

15.10. Determine the solvability conditions for

$$\nabla^4 w - \lambda w = F(r)$$

$$w(0) < \infty \qquad w(1) = \beta_1 \qquad w'(1) = \beta_2$$

when λ is an eigenvalue of the homogeneous problem and

$$\nabla^2 = \frac{d^2}{dr^2} + \frac{1}{r}\frac{d}{dr}$$

15.11. Determine the solvability conditions for

$$\frac{d^2\phi_1}{dr^2} + \frac{1}{r}\frac{d\phi_1}{dr} + \left(\alpha_n^2 - \frac{1}{r^2}\right)\phi_1 = f_1(r)$$

$$\frac{d^2\phi_2}{dr^2} + \frac{1}{r}\frac{d\phi_2}{dr} - \left(\gamma_n^2 + \frac{1}{r^2}\right)\phi_2 = f_2(r)$$

$$\phi_1' = \mu_1\phi_1 + \mu_2\phi_2, \; \phi_2' = \mu_3\phi_1 + \mu_4\phi_2 \qquad \text{at} \qquad r = 1$$

$$\phi_1(0) < \infty \qquad a\phi_2'(a) - \phi_2(a) = 0 \qquad a > 1$$

where the homogeneous problem has a nontrivial solution.

15.12. Determine the solvability condition for

$$\frac{d^2\phi}{dr^2} + \left[\frac{1}{r} + \frac{T_0'}{T_0} + \frac{2ku_0'}{\omega - ku_0}\right]\frac{d\phi}{dr} + \left[\frac{(\omega - ku_0)^2}{T_0} - k^2 - \frac{m^2}{r^2}\right]\phi = f(r)$$

$$\phi(0) < \infty \qquad \phi' - \beta\phi = \alpha \qquad \text{at} \qquad r = 1$$

when T_0, u_0, and $f(r)$ are known functions of r; ω, k, β, and α are constants; m is an integer; and the homogeneous problem has a nontrivial solution.

15.13. In analyzing waves propagating in a duct, one might encounter the problem

$$\frac{\partial^2\phi}{\partial x^2} + \frac{\partial^2\phi}{\partial y^2} + 5\pi^2\phi = f(y)\sin 2\pi x$$

$$\frac{\partial\phi}{\partial y} = 0 \qquad \text{at} \qquad y = 0$$

$$\frac{\partial \phi}{\partial y} = \alpha \sin 2\pi x \quad \text{at} \quad y = 1$$

Show that the solvability condition is

$$\int_0^1 \cos \pi y \, f(y) \, dy = -\alpha$$

15.14. Determine the solvability conditions for

$$\frac{\partial}{\partial x}\left[p(x, y)\frac{\partial \phi}{\partial x}\right] + \frac{\partial}{\partial y}\left[q(x, y)\frac{\partial \phi}{\partial y}\right] + \lambda r(x, y)\phi = f(x, y)$$

$$\phi_x(0, y) = 0 \quad \phi_y(x, 0) = 0$$

$$\frac{\partial \phi}{\partial x}(a, y) - \alpha_1(y)\phi(a, y) = \beta_1(y)$$

$$\frac{\partial \phi}{\partial y}(x, b) - \alpha_2(x)\phi(x, b) = \beta_2(x)$$

15.15. Consider

$$\ddot{x} - \dot{y} + 2x + 3\epsilon x^2 + 2\epsilon y^2 = 0$$

$$\ddot{y} + \dot{x} + 2\delta y + 4\epsilon xy = 0$$

when $\delta = 1 + \epsilon\sigma$ and $\epsilon \ll 1$. Use the method of multiple scales to show that

$$x = A_1(T_1)e^{iT_0} + A_2 e^{2iT_0} + \text{c.c.} + \cdots$$

where

$$A_2' = \tfrac{1}{3}i\sigma A_2 - \tfrac{1}{2}iA_1^2$$

$$A_1' = \tfrac{1}{3}i\sigma A_1 - i\bar{A}_1 A_2$$

15.16. The free response of a two-degree of freedom system is governed by

$$\ddot{u}_1 + \tfrac{1}{2}\dot{u}_2 + \delta u_1 = \epsilon u_1 u_2$$

$$\ddot{u}_2 - \tfrac{1}{2}\dot{u}_1 + \tfrac{1}{2}u_2 = \epsilon u_1^2$$

where $\epsilon \ll 1$ and $\delta = \tfrac{1}{2} + \epsilon\sigma$. Show that the equations describing the complex amplitudes are

$$A_2' = \tfrac{1}{3}i\sigma A_2 + \tfrac{2}{3}A_1^2$$

$$A_1' = \tfrac{1}{3}i\sigma A_1 - \tfrac{4}{3}A_2 \bar{A}_1$$

15.17. Use the method of multiple scales to determine the equations describing the amplitudes and the phases of the system

$$\ddot{u}_1 + \omega_1^2 u_1 = \alpha_1 u_2 u_3$$

$$\ddot{u}_2 + \omega_2^2 u_2 = \alpha_2 u_1 u_3$$

$$\ddot{u}_3 + \omega_3^2 u_3 = \alpha_3 u_1 u_2$$

466 SOLVABILITY CONDITIONS

when $\omega_3 \approx \omega_1 + \omega_2$.

15.18. Consider the parametrically excited system of Section 15.3. Determine the equations describing the complex amplitudes when (a) $\Omega \approx 1$, (b) $\Omega \approx 2$, and (c) $\Omega \approx 4$.

15.19. Determine a first-order expansion for the eigenvalues of

$$u'' + \lambda u = \epsilon f(x) u$$
$$u(0) = 0 \quad u(1) = 0$$

when $\epsilon \ll 1$.

15.20. Consider the eigenvalue problem

$$u'' + [\lambda + \epsilon f(x)] u = 0$$
$$u(0) = 0 \quad u'(1) = 0$$

Show that

$$\lambda = (n + \tfrac{1}{2})^2 \pi^2 - 2\epsilon \int_0^1 f(x) \sin^2 (n + \tfrac{1}{2})\pi x \, dx + \cdots$$

15.21. Consider the problem

$$u^{iv} + 5u'' + [\lambda + \epsilon f(x)] u = 0 \quad \epsilon \ll 1$$
$$u(0) = u''(0) = u(\pi) = u''(\pi) = 0$$

Show that when $\lambda \approx 4$

$$\lambda = 4 + \epsilon \lambda_1 + \cdots$$

where

$$\lambda_1^2 + (f_{11} + f_{22})\lambda_1 + f_{11} f_{22} - f_{12}^2 = 0$$

$$f_{11} = \frac{2}{\pi} \int_0^\pi f(x) \sin^2 x \, dx \quad f_{22} = \frac{2}{\pi} \int_0^\pi f(x) \sin^2 2x \, dx$$

$$f_{12} = \frac{2}{\pi} \int_0^\pi f(x) \sin x \sin 2x \, dx$$

15.22. Determine the solvability conditions for

$$p_2(x) y'' + p_1(x) y' + p_0(x) y = f(x)$$

subject to the boundary conditions

(a) $y(0) = \beta_1 \quad y(1) = \beta_2$
(b) $y'(0) = \alpha_1 y(0) + \beta_1 \quad y'(1) = \alpha_2 y(1) + \beta_2$
(c) $y(0) = \alpha_{11} y'(0) + \alpha_{12} y'(1)$
$\quad y(1) = \alpha_{21} y'(0) + \alpha_{22} y'(1)$

15.23. Determine the conditions under which

$$p_3 y''' + p_2 y'' + p_1 y' + p_0 y = 0$$

is self-adjoint. Ans. $p_2 = \frac{3}{2} p_3'$, $p_0 = \frac{1}{2}(p_3''' - p_2'' + p_1')$

15.24. Determine the solvability conditions for

$$p_3(x) y''' + p_2(x) y'' + p_1(x) y' + p_0(x) y = f(x)$$

subject to the boundary conditions

(a) $y(0) = \beta_1 \quad y'(0) = \beta_2 \quad y(1) = \beta_3$

(b) $y''(0) = \alpha_{11} y(0) + \alpha_{12} y'(0) + \alpha_{13} y(1) + \alpha_{14} y'(1)$
$y''(1) = \alpha_{21} y(0) + \alpha_{22} y'(0) + \alpha_{23} y(1) + \alpha_{24} y'(1)$
$0 = \alpha_{31} y(0) + \alpha_{32} y'(0) + \alpha_{33} y(1) + \alpha_{34} y'(1)$

15.25. Prove that any homogeneous self-adjoint sixth-order differential equation can be written in the form

$$\frac{d^3}{dx^3}\left(A_3 \frac{d^3 y}{dx^3}\right) + \frac{d^2}{dx^2}\left(A_2 \frac{d^2 y}{dx^2}\right) + \frac{d}{dx}\left(A_1 \frac{dy}{dx}\right) + A_0 y = 0$$

15.26. Prove that any homogeneous self-adjoint differential equation of order $2m$ can be written in the form

$$\frac{d^m}{dx^m}\left(A_m \frac{d^m y}{dx^m}\right) + \frac{d^{m-1}}{dx^{m-1}}\left(A_{m-1} \frac{d^{m-1} y}{dx^{m-1}}\right) + \cdots + \frac{d}{dx}\left(A_1 \frac{dy}{dx}\right) + A_0 y = 0$$

15.27. Determine the solvability conditions for

$$p_5(x) y^v + p_4(x) y^{iv} + p_3(x) y''' + p_2(x) y'' + p_1(x) y' + p_0 y = f(x)$$

$$y(0) = \beta_1 \quad y'(0) = \beta_2 \quad y''(0) = \beta_3 \quad y(1) = \beta_4 \quad y'(1) = \beta_5$$

15.28. Consider the eigenvalue problem

$$\phi(s) = \lambda \int_0^\pi [\cos(s+t) + \epsilon K_1(s, t)] \phi(t) \, dt \qquad \epsilon \ll 1$$

Show that

$$\phi^{(1)} = \cos s + \cdots \qquad \phi^{(2)} = \sin s + \cdots$$

$$\lambda^{(1)} = 2\pi^{-1} + \epsilon \lambda_1^{(1)} + \cdots \qquad \lambda^{(2)} = -2\pi^{-1} + \epsilon \lambda_1^{(2)} + \cdots$$

Determine $\lambda_1^{(n)}$.

15.29. Determine a first-order uniform expansion for

$$\phi(s) = \lambda \int_{-1}^{1} [st + s^2 t^2 + \epsilon K_1(s, t)] \phi(t) \, dt \qquad \epsilon \ll 1$$

15.30. Determine a first-order uniform expansion for

$$\phi(s) = \lambda \int_0^1 [s - t + \epsilon K_1(s, t)] \phi(t) \qquad \epsilon \ll 1$$

15.31. Show that

$$L(y) = \frac{d}{v_3\,dx} \cdot \frac{d}{v_2\,dx} \cdot \frac{y}{v_1} = 0$$

is self-adjoint if and only if $v_3 = \pm v_1$.

15.32. Show that

$$L(y) = \frac{d}{v_4\,dx} \cdot \frac{d}{v_3\,dx} \cdot \frac{d}{v_2\,dx} \cdot \frac{y}{v_1} = 0$$

is self-adjoint if and only if

$$v_4 = \pm v_1 \qquad v_3 = \pm v_2$$

15.33. Show that

$$L(y) = \frac{d}{v_{n+1}\,dx} \cdot \frac{d}{v_n\,dx} \cdots \cdot \frac{d}{v_3\,dx} \cdot \frac{d}{v_2\,dx} \cdot \frac{y}{v_1} = 0$$

is self-adjoint if and only if

$$v_{n+1} = \pm v_1 \qquad v_n = \pm v_2 \qquad v_{n-1} = \pm v_3 \cdots$$

15.34. Consider the problem

$$\frac{\partial^2 \phi}{\partial r^2} + \frac{1}{r}\frac{\partial \phi}{\partial r} + \frac{1}{r^2}\frac{\partial^2 \phi}{\partial \theta^2} + \omega^2 \phi = 0$$

$$\frac{\partial \phi}{\partial r} - \alpha\phi = 0 \quad \text{at} \quad r = a + \epsilon f(\theta)$$

$$\frac{\partial \phi}{\partial r} = 0 \quad \text{at} \quad r = b + \epsilon g(\theta)$$

Determine a first-order expansion for ω.

15.35. The incompressible flow past a wavy wall is governed by the mathematical problem

$$\nabla^2 \phi = 0$$

$$\frac{\partial \phi}{\partial y} = -\epsilon k \frac{\partial \phi}{\partial x} \sin kx \quad \text{at} \quad y = \epsilon \cos kx$$

$$\phi \to Ux \quad \text{as} \quad y \to \infty$$

Show that

$$\phi = U[x + \epsilon \sin kx\, e^{-ky} + \tfrac{1}{2}\epsilon^2 k \sin 2kx\, e^{-2ky} + \cdots]$$

Discuss the uniformity of this expansion.

15.36. The inviscid incompressible flow past a slightly distorted circular body is governed by

$$\frac{\partial^2 \psi}{\partial r^2} + \frac{1}{r}\frac{\partial \psi}{\partial r} + \frac{1}{r^2}\frac{\partial^2 \psi}{\partial \theta^2} = 0$$

$$\psi \to U r \sin \theta \quad \text{as} \quad r \to \infty$$

$$\psi = 0 \quad \text{at} \quad r = a(1 - \epsilon \sin^2 \theta)$$

Show that

$$\psi = U\left(r - \frac{a^2}{r}\right)\sin\theta + \tfrac{1}{2}\epsilon U\left(\frac{3a^2}{r}\sin\theta - \frac{a^4}{r^3}\sin 3\theta\right) + \cdots$$

Discuss the uniformity of this expansion.

15.37. Determine the solvability conditions for

$$\nabla^4 \phi - \lambda \phi = F(r, \theta)$$

$$\phi(a, \theta) = 0 \qquad \phi(b, \theta) = 0$$

$$\frac{\partial \phi}{\partial r}(a, \theta) = f(\theta) \qquad \frac{\partial \phi}{\partial r}(b, \theta) = g(\theta)$$

15.38. In analyzing the nonparallel two-dimensional stability of incompressible flows past a sinusoidal wall, one encounters the inhomogeneous problem

$$i\alpha u + v' = f_1(y)$$

$$-i(\omega - \alpha U)u + U'v + i\alpha p - \frac{1}{R}(u'' - \alpha^2 u) = f_2(y)$$

$$-i(\omega - \alpha U)v + p' - \frac{1}{R}(v'' - \alpha^2 v) = f_3(y)$$

$$u(0) = c_1 \qquad v(0) = c_2$$

$$u, v \to 0 \quad \text{as} \quad y \to \infty$$

where U and f_n are known functions of y, where ω, α, R, and c_n are constants, and where the prime indicates the derivative with respect to y. When the homogeneous problem has a nontrivial solution, show that the solvability condition is

$$-\sum_{n=1}^{3}\int_0^\infty \phi_n f_n\, dy = -\frac{1}{R}c_1\phi_2'(0) - \frac{1}{R}c_2\phi_3'(0) - c_2\phi_1(0)$$

where the ϕ_n satisfy the adjoint problem

$$i\alpha\phi_2 - \phi_3' = 0$$

$$i\alpha\phi_1 - i(\omega - \alpha U)\phi_2 - \frac{1}{R}(\phi_2'' - \alpha^2 \phi_2) = 0$$

$$-\phi_1' + U'\phi_2 - i(\omega - \alpha U)\phi_3 - \frac{1}{R}(\phi_3'' - \alpha^2 \phi_3) = 0$$

470 SOLVABILITY CONDITIONS

$$\phi_2(0) = \phi_3(0) = 0$$

$$\phi_n \to 0 \quad \text{as} \quad y \to \infty \quad n = 2, 3$$

Hint: Multiply the governing equations by ϕ_1, ϕ_2, and ϕ_3, respectively; integrate each by parts from $y = 0$ to $y = \infty$ to transfer the derivatives from u, v, and p to the ϕ_n; add the resulting equations; collect coefficients of u, v, and p; follow the steps in Section 15.13.

15.39. In analyzing the stability of growing three-dimensional incompressible boundary layers over flat surfaces, one encounters the inhomogeneous problem

$$i\alpha u + i\beta w + v' = f_1(y)$$

$$-i\hat{\omega} u + U'v + i\alpha p - \frac{1}{R}(u'' - k^2 u) = f_2(y)$$

$$-i\hat{\omega} v + p' - \frac{1}{R}(v'' - k^2 v) = f_3(y)$$

$$-i\hat{\omega} w + i\beta p + W'v - \frac{1}{R}(w'' - k^2 w) = f_4(y)$$

$$u = v = w = 0 \quad \text{at} \quad y = 0$$

$$u, v, w \to 0 \quad \text{as} \quad y \to \infty$$

where $\hat{\omega} = \omega - \alpha U - \beta W$ and $k^2 = \alpha^2 + \beta^2$. Here U, W, and f_n are known functions of y, and α, β, R, and ω are constants. When the homogeneous problem has a nontrivial solution, show that the solvability condition is

$$\sum_{n=1}^{4} \int_0^\infty f_n \phi_n \, dy = 0$$

where the ϕ_n satisfy the adjoint problem

$$i\alpha \phi_2 + i\beta \phi_4 - \phi_3' = 0$$

$$i\alpha \phi_1 - i\hat{\omega}\phi_2 - \frac{1}{R}(\phi_2'' - k^2 \phi_2) = 0$$

$$-\phi_1' + U'\phi_2 - i\hat{\omega}\phi_3 - \frac{1}{R}(\phi_3'' - k^2 \phi_3) + W'\phi_4 = 0$$

$$i\beta \phi_1 - i\hat{\omega}\phi_4 - \frac{1}{R}(\phi_4'' - k^2 \phi_4) = 0$$

$$\phi_2 = \phi_3 = \phi_4 = 0 \quad \text{at} \quad y = 0$$

$$\phi_n \to 0 \quad \text{as} \quad y \to \infty \quad n = 2, 3, 4$$

15.40. Consider the eigenvalue problem

$$\phi_{xx} + \phi_{yy} + \lambda \phi = \epsilon x^2 \phi$$

$$\phi(x, 0) = \phi(x, \pi) = \phi(0, y) = \phi(\pi, y) = 0$$

Determine first-order expansions when λ is near 2 and 5.

15.41. Consider the problem

$$\nabla^2 \phi + \lambda \phi = \epsilon f(x, y, z)\phi$$

with ϕ vanishing on the surfaces of a cube of length π. Determine first-order expansions when $\lambda \approx 3$ and 6 if (a) $f = x^2$ and (b) $f = x^2 y$.

APPENDIX A

Trigonometric Identities

A.1. Basic Indentities

$$\sin^2 \alpha + \cos^2 \alpha = 1 \tag{A1}$$

$$\tan \alpha = \frac{\sin \alpha}{\cos \alpha} \quad \cot \alpha = \frac{\cos \alpha}{\sin \alpha} \quad \tan \alpha = \frac{1}{\cot \alpha} \tag{A2}$$

$$\sin (\alpha \pm \beta) = \sin \alpha \cos \beta \pm \cos \alpha \sin \beta \tag{A3}$$

$$\cos (\alpha \pm \beta) = \cos \alpha \cos \beta \mp \sin \alpha \sin \beta \tag{A4}$$

Adding the equations in (A3), we have

$$\sin \alpha \cos \beta = \tfrac{1}{2} [\sin (\alpha + \beta) + \sin (\alpha - \beta)] \tag{A5}$$

whereas subtracting the equations in (A3), we have

$$\cos \alpha \sin \beta = \tfrac{1}{2} [\sin (\alpha + \beta) - \sin (\alpha - \beta)] \tag{A6}$$

Similarly, adding and subtracting the equations in (A4), we have

$$\begin{aligned} \cos \alpha \cos \beta &= \tfrac{1}{2} [\cos (\alpha + \beta) + \cos (\alpha - \beta)] \\ \sin \alpha \sin \beta &= \tfrac{1}{2} [\cos (\alpha - \beta) - \cos (\alpha + \beta)] \end{aligned} \tag{A7}$$

It follows from (A2) through (A4) that

$$\tan (\alpha \pm \beta) = \frac{\sin (\alpha \pm \beta)}{\cos (\alpha \pm \beta)} = \frac{\sin \alpha \cos \beta \pm \cos \alpha \sin \beta}{\cos \alpha \cos \beta \mp \sin \alpha \sin \beta}$$

Dividing both numerator and denominator by $\cos \alpha \cos \beta$, we have

$$\tan (\alpha \pm \beta) = \frac{\tan \alpha \pm \tan \beta}{1 \mp \tan \alpha \tan \beta} \tag{A8}$$

Similarly,

$$\cot(\alpha \pm \beta) = \frac{\cot \alpha \cot \beta \mp 1}{\cot \beta \pm \cot \alpha} \tag{A9}$$

We let
$$\alpha + \beta = x \quad \text{and} \quad \alpha - \beta = y$$
so that
$$\alpha = \tfrac{1}{2}(x+y) \quad \text{and} \quad \beta = \tfrac{1}{2}(x-y)$$

Then, (A5) through (A7) can be rewritten as

$$\sin x + \sin y = 2 \sin \frac{x+y}{2} \cos \frac{x-y}{2}$$

$$\sin x - \sin y = 2 \cos \frac{x+y}{2} \sin \frac{x-y}{2} \tag{A10}$$

$$\cos x + \cos y = 2 \cos \frac{x+y}{2} \cos \frac{x-y}{2}$$

$$\cos y - \cos x = 2 \sin \frac{x+y}{2} \sin \frac{x-y}{2} \tag{A11}$$

Putting $\beta = \alpha$ in (A5) gives
$$\sin 2\alpha = 2 \sin \alpha \cos \alpha \tag{A12}$$

whereas putting $\beta = \alpha$ in (A7) gives

$$\cos 2\alpha = 2 \cos^2 \alpha - 1$$
$$\cos 2\alpha = 1 - 2 \sin^2 \alpha \tag{A13}$$

which, upon adding, yield
$$\cos 2\alpha = \cos^2 \alpha - \sin^2 \alpha \tag{A14}$$

To express $\sin 3\alpha$ in terms of $\sin \alpha$, we put $\beta = 2\alpha$ in (A3) and obtain
$$\sin 3\alpha = \sin \alpha \cos 2\alpha + \cos \alpha \sin 2\alpha$$

Using (A12) and (A13), we have
$$\sin 3\alpha = \sin \alpha (1 - 2 \sin^2 \alpha) + 2 \cos \alpha \sin \alpha \cos \alpha$$

But $\cos^2 \alpha = 1 - \sin^2 \alpha$; hence
$$\sin 3\alpha = \sin \alpha - 2 \sin^3 \alpha + 2 \sin \alpha (1 - \sin^2 \alpha)$$

or
$$\sin 3\alpha = 3 \sin \alpha - 4 \sin^3 \alpha \tag{A15}$$

which, upon solving for $\sin^3 \alpha$, gives

$$\sin^3 \alpha = \tfrac{1}{4} (3 \sin \alpha - \sin 3\alpha) \qquad (A16)$$

an identity that is frequently used. Similarly, we can express $\cos 3\alpha$ in terms of $\cos \alpha$ as follows. We put $\beta = 2\alpha$ in (A4) and obtain

$$\cos 3\alpha = \cos \alpha \cos 2\alpha - \sin \alpha \sin 2\alpha$$

which, upon using (A12) and (A13), becomes

$$\cos 3\alpha = \cos \alpha (2 \cos^2 \alpha - 1) - 2 \sin \alpha \sin \alpha \cos \alpha$$
$$= 2 \cos^3 \alpha - \cos \alpha - 2 \sin^2 \alpha \cos \alpha$$
$$= 2 \cos^3 \alpha - \cos \alpha - 2(1 - \cos^2 \alpha) \cos \alpha$$

according to (A12). Hence,

$$\cos 3\alpha = 4 \cos^3 \alpha - 3 \cos \alpha \qquad (A17)$$

or

$$\cos^3 \alpha = \tfrac{1}{4} (3 \cos \alpha + \cos 3\alpha) \qquad (A18)$$

which is frequently used.

A.2. Complex Quantities

Letting $x = i\theta$, where $i = \sqrt{-1}$, in (1.49), we have

$$e^{i\theta} = 1 + \frac{i\theta}{1!} + \frac{(i\theta)^2}{2!} + \frac{(i\theta)^3}{3!} + \frac{(i\theta)^4}{4!} + \frac{(i\theta)^5}{5!} + \frac{(i\theta)^6}{6!} + \cdots$$

But $i^2 = -1$, hence

$$e^{i\theta} = 1 + \frac{i\theta}{1!} - \frac{\theta^2}{2!} - \frac{i\theta^3}{3!} + \frac{\theta^4}{4!} + \frac{i\theta^5}{5!} - \frac{\theta^6}{6!} + \cdots$$

which can be rearranged into

$$e^{i\theta} = \left(1 - \frac{\theta^2}{2!} + \frac{\theta^4}{4!} - \frac{\theta^6}{6!} + \cdots\right) + i\left(\theta - \frac{\theta^3}{3!} + \frac{\theta^5}{5!} + \cdots\right) \qquad (A19)$$

Comparing the series in the parentheses with (1.47) and (1.48), we conclude that

$$e^{i\theta} = \cos \theta + i \sin \theta \qquad (A20)$$

Taking the complex conjugate of (A20) gives

$$e^{-i\theta} = \cos \theta - i \sin \theta \qquad (A21)$$

Adding (A20) and (A21), we have
$$\cos\theta = \tfrac{1}{2}(e^{i\theta} + e^{-i\theta}) \tag{A22}$$
Subtracting (A21) from (A20), we have
$$\sin\theta = \frac{1}{2i}(e^{i\theta} - e^{-i\theta}) \tag{A23}$$

Next, we show how one can use (A20) through (A23) to express $\cos^n\theta$ and $\sin^n\theta$ in Fourier series. It follows from (A22) that
$$\cos^3\theta = [\tfrac{1}{2}(e^{i\theta} + e^{-i\theta})]^3 = \tfrac{1}{8}(e^{i\theta} + e^{-i\theta})^3$$
$$= \tfrac{1}{8}(e^{3i\theta} + 3e^{i\theta} + 3e^{-i\theta} + e^{-3i\theta})$$
according to the binomial theorem. Rearranging, we have
$$\cos^3\theta = \tfrac{1}{8}(e^{3i\theta} + e^{-3i\theta}) + \tfrac{3}{8}(e^{i\theta} + e^{-i\theta})$$
$$= \tfrac{1}{8} \cdot 2\cos 3\theta + \tfrac{3}{8} \cdot 2\cos\theta$$
according to (A22). Hence,
$$\cos^3\theta = \tfrac{1}{4}(\cos 3\theta + 3\cos\theta)$$
in agreement with (A18). It follows from (A23) that
$$\sin^3\theta = \left[\frac{1}{2i}(e^{i\theta} - e^{-i\theta})\right]^3 = \frac{1}{8i^3}(e^{i\theta} - e^{-i\theta})^3$$
$$= -\frac{1}{8i}(e^{3i\theta} - 3e^{i\theta} + 3e^{-i\theta} - e^{-3i\theta})$$
$$= \tfrac{3}{4}\frac{e^{i\theta} - e^{-i\theta}}{2i} - \tfrac{1}{4}\frac{e^{3i\theta} - e^{-3i\theta}}{2i}$$
$$= \tfrac{3}{4}\sin\theta - \tfrac{1}{4}\sin 3\theta$$
in agreement with (A16).

To expand $\cos^n\theta$, for a general positive integer n, in a Fourier series, we note that
$$\cos^n\theta = [\tfrac{1}{2}(e^{i\theta} + e^{-i\theta})]^n = \frac{1}{2^n}(e^{i\theta} + e^{-i\theta})^n$$
Letting $a = \exp(i\theta)$ and $b = \exp(-i\theta)$ in (1.39b), we have
$$(e^{i\theta} + e^{-i\theta})^n = \sum_{m=0}^{n}\frac{n!}{m!(n-m)!}e^{i(n-m)\theta}e^{-im\theta} = \sum_{m=0}^{n}\frac{n!}{m!(n-m)!}e^{i(n-2m)\theta}$$

$$= e^{in\theta} + ne^{i(n-2)\theta} + \frac{n(n-1)}{2!} e^{i(n-4)\theta} + \cdots + \frac{n(n-1)}{2!}$$

$$\times e^{-i(n-4)\theta} + ne^{-i(n-2)\theta} + e^{-in\theta}$$

$$= e^{in\theta} + e^{-in\theta} + n[e^{i(n-2)\theta} + e^{-i(n-2)\theta}]$$

$$+ \frac{n(n-1)}{2} [e^{i(n-4)\theta} + e^{-i(n-4)\theta}]$$

$$+ \frac{n(n-1)(n-2)}{3!} [e^{i(n-6)\theta} + e^{-i(n-6)\theta}] + \cdots$$

Hence,

$$\cos^n \theta = \frac{1}{2^{n-1}} \left[\cos n\theta + n \cos(n-2)\theta + \frac{n(n-1)}{2!} \cos(n-4)\theta \right.$$

$$\left. + \frac{n(n-1)(n-2)}{3!} \cos(n-6)\theta + \cdots \right] \quad \text{for odd } n \quad \text{(A24)}$$

and

$$\cos^n \theta = \frac{1}{2^{n-1}} \left[\cos n\theta + n \cos(n-2)\theta + \frac{n(n-1)}{2!} \cos(n-4)\theta + \cdots \right.$$

$$\left. + \frac{n!}{2(\tfrac{1}{2}n)!(\tfrac{1}{2}n)!} \right] \quad \text{for even } n \quad \text{(A25)}$$

Similarly, to expand $\sin^n \theta$, we note that

$$\sin^n \theta = \left[\frac{1}{2i} (e^{i\theta} - e^{-i\theta}) \right]^n = \frac{1}{2^n i^n} (e^{i\theta} - e^{-i\theta})^n$$

Letting $a = \exp(i\theta)$ and $b = -\exp(-i\theta)$ in (1.39b), we have

$$(e^{i\theta} - e^{-i\theta})^n = \sum_{m=0}^{n} \frac{(-1)^m n!}{m!(n-m)!} e^{i(n-m)\theta} e^{-im\theta}$$

$$= \sum_{m=0}^{n} \frac{(-1)^m n!}{m!(n-m)!} e^{i(n-2m)\theta}$$

Then, when n is even,

$$(e^{i\theta} - e^{-i\theta})^n = e^{in\theta} - ne^{i(n-2)\theta} + \frac{n(n-1)}{2!} e^{i(n-4)\theta} + \cdots$$

$$+ \frac{n(n-1)}{2!} e^{-i(n-4)\theta} - ne^{-i(n-2)\theta} + e^{-in\theta}$$

$$= e^{in\theta} + e^{-in\theta} - n[e^{i(n-2)\theta} + e^{-i(n-2)\theta}]$$
$$+ \frac{n(n-1)}{2!}[e^{i(n-4)\theta} + e^{-i(n-4)\theta}] + \cdots$$

Hence,
$$\sin^n \theta = \frac{2}{2^n i^n}\left[\cos n\theta - n\cos(n-2)\theta + \frac{n(n-1)}{2!}\cos(n-4)\theta\right.$$
$$- \frac{n(n-1)(n-2)}{3!}\cos(n-6)\theta + \cdots$$
$$\left. + (-1)^{(1/2)n}\frac{n!}{2(\frac{1}{2}n)!(\frac{1}{2}n!)}\right], \quad \text{for even } n \qquad \text{(A26)}$$

When n is odd,
$$(e^{i\theta} - e^{-i\theta})^n = e^{in\theta} - ne^{i(n-2)\theta} + \frac{n(n-1)}{2!}e^{i(n-4)\theta} + \cdots$$
$$- \frac{n(n-1)}{2!}e^{-i(n-4)\theta} + ne^{-i(n-2)\theta} - e^{-in\theta}$$
$$= e^{in\theta} - e^{-in\theta} - n[e^{i(n-2)\theta} - e^{i(n-2)\theta}]$$
$$+ \frac{n(n-1)}{2!}[e^{i(n-4)\theta} - e^{-i(n-4)\theta}] + \cdots$$

Hence,
$$\sin^n \theta = \frac{1}{(2i)^{n-1}}\left[\sin n\theta - n\sin(n-2)\theta + \frac{n(n-1)}{2!}\sin(n-4)\theta\right.$$
$$\left. - \frac{n(n-1)(n-2)}{3!}\sin(n-6)\theta + \cdots \right] \quad \text{for odd } n \qquad \text{(A27)}$$

Putting $n = 4$ in (A25), we have
$$\cos^4 \theta = \tfrac{1}{8}(\cos 4\theta + 4\cos 2\theta + 3)$$

Putting $n = 5$ in (A24), we have
$$\cos^5 \theta = \tfrac{1}{16}(\cos 5\theta + 5\cos 3\theta + 10\cos \theta)$$

Putting $n = 4$ in (A26), we have
$$\sin^4 \theta = \tfrac{1}{8}(\cos 4\theta - 4\cos 2\theta + 3)$$

Putting $n = 5$ in (A27), we have
$$\sin^5 \theta = \tfrac{1}{16}(\sin 5\theta - 5\sin 3\theta + 10\sin \theta)$$

478 TRIGONOMETRIC IDENTITIES

To express $\cos^n \theta \sin^m \theta$ in a Fourier series, one can either use (A22) and (A23) directly or first express it in powers of $\cos \theta$ or $\sin \theta$ and then use (A22) and (A23).

A.3. Integrals

$$\int \cos \alpha x \cos \beta x \, dx = \frac{1}{2} \int \cos(\alpha + \beta)x \, dx + \frac{1}{2} \int \cos(\alpha - \beta)x \, dx$$

$$= \begin{cases} \dfrac{\sin(\alpha + \beta)x}{2(\alpha + \beta)} + \dfrac{\sin(\alpha - \beta)x}{2(\alpha - \beta)} & \beta \neq \alpha \\ \dfrac{\sin 2\alpha x}{4\alpha} + \dfrac{1}{2}x & \beta = \alpha \end{cases} \quad (A28)$$

$$\int \sin \alpha x \sin \beta x \, dx = \frac{1}{2} \int \cos(\alpha - \beta)x \, dx - \frac{1}{2} \int \cos(\alpha + \beta)x \, dx$$

$$= \begin{cases} \dfrac{\sin(\alpha - \beta)x}{2(\alpha - \beta)} - \dfrac{\sin(\alpha + \beta)x}{2(\alpha + \beta)} & \beta \neq \alpha \\ \dfrac{1}{2}x - \dfrac{\sin 2\alpha x}{4\alpha} & \beta = \alpha \end{cases} \quad (A29)$$

$$\int \sin \alpha x \cos \beta x \, dx = \frac{1}{2} \int \sin(\alpha + \beta)x \, dx + \frac{1}{2} \int \sin(\alpha - \beta)x \, dx$$

$$= \begin{cases} -\dfrac{\cos(\alpha + \beta)x}{2(\alpha + \beta)} - \dfrac{\cos(\alpha - \beta)x}{2(\alpha - \beta)} & \beta \neq \alpha \\ -\dfrac{\cos 2\alpha x}{4\alpha} & \beta = \alpha \end{cases} \quad (A30)$$

$$\int \cos^n \theta \sin \theta \, d\theta = -\frac{\cos^{n+1} \theta}{n+1} \quad (A31)$$

$$\int \sin^n \theta \cos \theta \, d\theta = \frac{\sin^{n+1} \theta}{n+1} \quad (A32)$$

Integrating both sides of each of (A24) through (A27), we obtain

$$\int \cos^n \theta \, d\theta = \frac{1}{2^{n-1}} \left[\frac{1}{n} \sin n\theta + \frac{n}{n-2} \sin(n-2)\theta + \frac{n(n-1)}{2!(n-4)} \sin(n-4)\theta \right.$$
$$\left. + \frac{n(n-1)(n-2)}{3!(n-6)} \sin(n-6)\theta + \cdots \right] \quad \text{for odd } n \quad (A33)$$

$$\int \cos^n \theta \, d\theta = \frac{1}{2^{n-1}} \left[\frac{1}{n} \sin n\theta + \frac{n}{n-2} \sin(n-2)\theta + \frac{n(n-1)}{2!(n-4)} \sin(n-4)\theta \right.$$

$$+ \cdots + \frac{n!\theta}{2(\tfrac{1}{2}n)!(\tfrac{1}{2}n)!} \Bigg] \quad \text{for even } n \tag{A34}$$

$$\int \sin^n \theta \, d\theta = \frac{2}{2^n i^n} \Bigg[\frac{1}{n} \sin n\theta - \frac{n}{n-2} \sin(n-2)\theta + \frac{n(n-1)}{2!(n-4)} \sin(n-4)\theta$$

$$+ \cdots + \frac{(-1)^{(1/2)n} n!\theta}{2(\tfrac{1}{2}n)!(\tfrac{1}{2}n)!} \Bigg] \quad \text{for even } n \tag{A35}$$

$$\int \sin^n \theta \, d\theta = \frac{1}{(2i)^{n-1}} \Bigg[-\frac{1}{n} \cos n\theta + \frac{n}{n-2} \cos(n-2)\theta$$

$$- \frac{n(n-1)}{2!(n-4)} \cos(n-4)\theta + \cdots \Bigg] \quad \text{for odd } n \tag{A36}$$

It follows from (A33) through (A36) that

$$\int_0^{2\pi} \cos^n \theta \, d\theta = \begin{cases} 0 & \text{if } n \text{ is odd} \\ \dfrac{n!\pi}{2^{n-1}(\tfrac{1}{2}n)!(\tfrac{1}{2}n)!} & \text{if } n \text{ is even} \end{cases} \tag{A37}$$

$$\int_0^{2\pi} \sin^n \theta \, d\theta = \begin{cases} 0 & \text{if } n \text{ is odd} \\ \dfrac{(-1)^{(1/2)n} n!\pi}{2^{n-1} i^n (\tfrac{1}{2}n)!(\tfrac{1}{2}n)!} & \text{if } n \text{ is even} \end{cases} \tag{A38}$$

APPENDIX B

Linear Ordinary-Differential Equations

A differential equation is an equation connecting the values of a function (called the dependent variable), the derivatives of this function, and certain known quantities. If the dependent variable is a function of a single variable (called the independent variable), the differential equation is called an ordinary-differential equation. If the dependent variable is a function of two or more independent variables, the differential equation is called a partial-differential equation. Thus, a general ordinary-differential equation is an equation of the form

$$F\left(\frac{d^n u}{dx^n}, \frac{d^{n-1} u}{dx^{n-1}}, \frac{du^{n-2}}{dx^{n-2}}, \ldots, \frac{d^2 u}{dx^2}, \frac{du}{dx}, u, x\right) = 0 \qquad (B1)$$

where the order of the highest derivative is called the order of the differential equation. Thus, (B1) is of order n.

An ordinary-differential equation of order n is said to be linear, if it is linear in the dependent variable u and its derivatives u', u'', ..., $u^{(n-2)}$, $u^{(n-1)}$, $u^{(n)}$. Thus, the most general linear ordinary-differential equation of order n has the form

$$p_n(x)\frac{d^n u}{dx^n} + p_{n-1}(x)\frac{d^{n-1} u}{dx^{n-1}} + \cdots + p_1(x)\frac{du}{dx} + p_0(x)u = f(x) \qquad (B2)$$

where the p_m, for $m = 0, 1, 2, \ldots, n$, and f are known functions of x. If $f(x) \equiv 0$, (B2) is called a homogeneous equation; otherwise, it is called an inhomogeneous equation.

Suppose that the $p_m(x)$, for $m = 0, 1, 2, \ldots, n$, are continuous functions on an interval $I = \{x | a \leq x \leq b\}$ that includes the point x_0 and suppose that $p_n(x)$ does not vanish at any point in I, then for any real numbers $\alpha_0, \alpha_1, \ldots, \alpha_{n-1}$, there exists a unique solution $u(x)$ that satisfies (B2) everywhere in I and satisfies the initial conditions

$$u(x_0) = \alpha_0 \qquad u'(x_0) = \alpha_1, \ldots, u^{(n-1)}(x_0) = \alpha_{n-1}$$

This is a statement of the fundamental existence theorem of linear ordinary-differential equations.

It is convenient to express (B2) in operator form. To this end, we let D denote the differential operator d/dx. Thus,

$$Du = \frac{du}{dx} = u'$$

$$D^2 u = D(Du) = D(u') = u''$$

$$D^n u = D(D^{n-1} u) = D[u^{(n-1)}] = u^{(n)}$$

Moreover, we define $D^0 u$ to be u. Hence, (B2) can be rewritten as

$$p_n D^n u + p_{n-1} D^{n-1} u + \cdots + p_1 Du + p_0 D^0 u = f(x) \tag{B3}$$

or in the convenient abbreviated form

$$L(u) = f(x) \tag{B4}$$

where the operator L is defined by means of the relation

$$L = p_n D^n + p_{n-1} D^{n-1} + \cdots + p_1 D + p_0 D^0 \tag{B5}$$

The operator L is called a linear operator and it has some interesting properties. If c is a constant and u is any function that possesses at least n derivatives, then

$$L(cu) = cL(u) \tag{B6}$$

because

$$L(cu) = p_n(cu)^{(n)} + p_{n-1}(cu)^{(n-1)} + \cdots + p_1(cu)' + p_0(cu)$$

$$= p_n cu^{(n)} + p_{n-1} cu^{(n-1)} + \cdots + cp_1 u' + p_0 cu$$

$$= c[p_n u^{(n)} + p_{n-1} u^{(n-1)} + \cdots + p_1 u' + p_0 u] = cL(u)$$

If u_1 and u_2 are any two functions that possess at least n derivatives, then

$$L(u_1 + u_2) = L(u_1) + L(u_2) \tag{B7}$$

because

$$L(u_1 + u_2) = p_n(u_1 + u_2)^{(n)} + p_{n-1}(u_1 + u_2)^{(n-1)} + \cdots + p_1(u_1 + u_2)'$$
$$+ p_0(u_1 + u_2) = p_n[u_1^{(n)} + u_2^{(n)}] + p_{n-1}[u_1^{(n-1)} + u_2^{(n-1)}] + \cdots$$
$$+ p_1(u_1' + u_2') + p_0(u_1 + u_2) = [p_n u_1^{(n)} + p_{n-1} u_1^{(n-1)} + \cdots$$
$$+ p_1 u_1' + p_0 u_1] + [p_n u_2^{(n)} + p_{n-1} u_2^{(n-1)} + \cdots + p_1 u_2' + p_0 u]$$

$$= L(u_1) + L(u_2)$$

It follows from (B6) and (B7) that

$$L(c_1 u_1 + c_2 u_2) = c_1 L(u_1) + c_2 L(u_2) \tag{B8}$$

and in general that

$$L(c_1 u_1 + c_2 u_2 + \cdots + c_m u_m) = c_1 L(u_1) + c_2 L(u_2) + \cdots + c_m L(u_m) \tag{B9}$$

where the c_i are constant.

B.1. Homogeneous Equations

Putting $f(x) \equiv 0$ in (B4) yields the homogeneous nth order ordinary-differential equation

$$L(u) = 0 \tag{B10}$$

If u_1 is a solution of (B10) and c is any constant, then cu_1 is also a solution of (B10) because

$$L(cu_1) = cL(u_1) = 0$$

PRINCIPLE OF SUPERPOSITION

If u_1 and u_2 are two solutions of (B10) and c_1 and c_2 are any two constants, then $c_1 u_1 + c_2 u_2$ is also a solution of (B10) because

$$L(c_1 u_1 + c_2 u_2) = c_1 L(u_1) + c_2 L(u_2) = c_1 \cdot 0 + c_2 \cdot 0 = 0$$

In general, if $u_1, u_2, \ldots, u_m$ are solutions of (B10) and $c_1, c_2, \ldots, c_m$ are any constants, then

$$c_1 u_1 + c_2 u_2 + \cdots + c_m u_m$$

is also a solution of (B10) because

$$L(c_1 u_1 + c_2 u_2 + \cdots + c_m u_m) = c_1 L(u_1) + c_2 L(u_2) + \cdots + c_m L(u_m)$$
$$= c_1 \cdot 0 + c_2 \cdot 0 + \cdots + c_m \cdot 0 = 0$$

This property is usually referred to as the principle of superposition. The function $c_1 u_1 + c_2 u_2 + \cdots + c_m u_m$ is usually referred to as a linear combination of $u_1, u_2, \ldots, u_m$.

LINEAR INDEPENDENCE

Let $u_1(x), u_2(x), \ldots, u_m(x)$ be a set of functions defined on the interval $I = [a, b]$. This set is said to be linearly dependent, if there exist constants $c_1, c_2, \ldots, c_m$, not all zero, such that

$$c_1 u_1(x) + c_2 u_2(x) + \cdots + c_m u_m(x) = 0 \tag{B11}$$

for all x in I; otherwise, this set is said to be linearly independent. The restriction that not all the c_i are zero is essential because (B11) holds for any set if all the c_i are zero. For example, the functions $2x - 1$ and $6x - 3$ are linearly dependent because

$$-3 \cdot (2x - 1) + 1 \cdot (6x - 3) \equiv 0$$

However, the functions $2x - 1$ and $5x - 3$ are linearly independent because it is impossible to find two constants c_1 and c_2, not both zero, such that

$$c_1(2x - 1) + c_2(5x - 3) = 0$$

An alternative but important method of testing for the linear dependence or independence of a set of functions $u_1(x), u_2(x), \ldots, u_m(x)$ involves their Wronskian determinant, which is defined by

$$W = W(u_1, u_2, \ldots, u_m) = \begin{vmatrix} u_1 & u_2 & \cdots & u_m \\ u_1' & u_2' & \cdots & u_m' \\ \vdots & \vdots & \cdots & \vdots \\ u_1^{(m-1)} & u_2^{(m-1)} & \cdots & u_m^{(m-1)} \end{vmatrix} \quad (B12)$$

If the functions $u_1(x), u_2(x), \ldots, u_m(x)$ are linearly dependent on the interval I, then there exist constants $c_1, c_2, \ldots, c_m$, not all zero, such that

$$c_1 u_1 + c_2 u_2 + \cdots + c_m u_m = 0 \quad (B13)$$

which, upon differentiation $(m - 1)$ times, yields

$$c_1 u_1' + c_2 u_2' + \cdots + c_m u_m' = 0 \quad (B14)$$

$$c_1 u_1'' + c_2 u_2'' + \cdots + c_m u_m'' = 0 \quad (B15)$$

$$\cdots \cdots \cdots \cdots \cdots \cdots \cdots$$

$$c_1 u_1^{(m-1)} + c_2 u_2^{(m-1)} + \cdots + c_m u_m^{(m-1)} = 0 \quad (B16)$$

At each point x in the interval I, (B13) through (B16) constitute a system of homogeneous linear algebraic equations for $c_1, c_2, \ldots, c_m$. Since the c_i are not all zero, the determinant of their coefficient matrix must be zero. But this determinant is the Wronskian determinant. Hence, $W(x) = 0$ at every point in I. If the Wronskian determinant is not zero at any point in I, then (B14) through (B16) have only the trivial solution $c_1 = c_2 = \cdots = c_m = 0$, and the functions $u_1(x), u_2(x), \ldots, u_m(x)$ are linearly independent. Hence, a set of functions is linearly dependent in I if and only if their Wronskian determinant vanishes at every point in I. For example, the Wronskian determinant of the functions $2x - 1$ and $6x - 3$ is

$$W = \begin{vmatrix} 2x - 1 & 6x - 3 \\ 2 & 6 \end{vmatrix} = 0$$

and hence, they are linearly dependent. On the other hand, the Wronskian determinant of the functions $2x - 1$ and $5x - 3$ is

$$W = \begin{vmatrix} 2x - 1 & 5x - 3 \\ 2 & 5 \end{vmatrix} = 1$$

and hence, they are linearly independent.

GENERAL SOLUTION

Equation (B10) has no more than n linearly independent solutions. For suppose that $u_1(x), u_2(x), \ldots, u_N(x)$, where $N > n$, are solutions of (B10), then

$$W(u_1, u_2, \ldots, u_n, \ldots, u_N)$$

$$= \begin{vmatrix} u_1 & u_2 & \cdots & u_n & \cdots & u_N \\ u_1' & u_2' & \cdots & u_n' & \cdots & u_N' \\ & & \cdots & & \cdots & \\ u_1^{(n)} & u_2^{(n)} & \cdots & u_n^{(n)} & \cdots & u_N^{(n)} \\ & & \cdots & & \cdots & \\ u_1^{(N-1)} & u_2^{(N-1)} & \cdots & u_n^{(N-1)} & \cdots & u_N^{(N-1)} \end{vmatrix} \quad (B17)$$

is zero because (B10) can be used to express the $(n+1)$th row as a linear combination of the first n rows. Hence, the solutions $u_1(x), u_2(x), \ldots, u_N(x)$, for $N > n$, are linearly dependent.

Next, we show that (B10) has exactly n linearly independent solutions on the interval I. To this end, we note that the fundamental existence theorem shows that (B10) has n unique solutions $u_1(x), u_2(x), \ldots, u_n(x)$ that satisfy the initial conditions

$$u_1(x_0) = 1 \quad u_1'(x_0) = u_1''(x_0) = \cdots = u_1^{(n-1)}(x_0) = 0$$

$$u_2(x_0) = 0 \quad u_2'(x_0) = 1 \quad u_2''(x_0) = \cdots = u_2^{(n-1)}(x_0) = 0$$

$$u_3(x_0) = u_3'(x_0) = 0 \quad u_3''(x_0) = 1 \quad u_3'''(x_0) = \cdots = u_3^{(n-1)}(x_0) = 0$$

$$\cdots$$

$$u_m(x_0) = \cdots = u_m^{(m-2)}(x_0) = 0 \quad u_m^{(m-1)}(x_0) = 1$$

$$u_m^{(m)}(x_0) = \cdots = u_m^{(n-1)}(x_0) = 0$$

$$\cdots$$

$$u_n(x_0) = u_n'(x_0) = \cdots = u_n^{(n-2)}(x_0) = 0 \quad u_n^{(n-1)}(x_0) = 1$$

It can be easily shown that the Wronskian determinant of these n solutions is unity at x_0. Thus, the Wronskian determinant of these solutions is not zero everywhere in I, and hence, they are linearly independent.

Any set of n linearly independent solutions of (B10) is called a fundamental set and any linear combination of these solutions is a solution of (B10). Thus, if $u_1(x), u_2(x), \ldots, u_n(x)$ is a fundamental set of (B10), then

$$u(x) = c_1 u_1(x) + c_2 u_2(x) + \cdots + c_n u_n(x) \quad (B18)$$

for any constants $c_1, c_2, \ldots, c_n$, is a solution of (B10) according to the principle of superposition. Since any solution of (B10) can be obtained from (B18) by a proper choice of the c_m, (B18) is called the general solution of (B10).

Before closing this section, we derive an expression that relates the value of the Wronskian determinant of a fundamental set at any point x in I to its value at any other point x_0. To this end, we write

$$W = \begin{vmatrix} u_1 & u_2 & \cdots & u_n \\ u_1' & u_2' & \cdots & u_n' \\ \vdots & \vdots & & \vdots \\ u_1^{(n-1)} & u_2^{(n-1)} & \cdots & u_n^{(n-1)} \end{vmatrix} \quad (B19)$$

Differentiating (B19) with respect to x and using the property of determinants, we have

$$W' = \begin{vmatrix} u_1' & u_2' & \cdots & u_n' \\ u_1' & u_2' & \cdots & u_n' \\ \vdots & \vdots & & \vdots \\ u_1^{(n-1)} & u_2^{(n-1)} & \cdots & u_n^{(n-1)} \end{vmatrix} + \begin{vmatrix} u_1 & u_2 & \cdots & u_n \\ u_1'' & u_2'' & \cdots & u_n'' \\ \vdots & \vdots & & \vdots \\ u_1^{(n-1)} & u_2^{(n-1)} & \cdots & u_n^{(n-1)} \end{vmatrix}$$

$$+ \cdots + \begin{vmatrix} u_1 & u_2 & \cdots & u_n \\ u_1' & u_2' & \cdots & u_n' \\ \vdots & \vdots & & \vdots \\ u_1^{(n)} & u_2^{(n)} & \cdots & u_n^{(n)} \end{vmatrix} \quad (B20)$$

All determinants in (B20) except the last one vanish because of the presence of repeated rows. Hence,

$$W' = \begin{vmatrix} u_1 & u_2 & \cdots & u_n \\ u_1' & u_2' & \cdots & u_n' \\ \vdots & \vdots & & \vdots \\ u_1^{(n)} & u_2^{(n)} & \cdots & u_n^{(n)} \end{vmatrix} \quad (B21)$$

Using (B10) to express the $u_m^{(n)}$ in terms of $u_m, u_m', \ldots, u_m^{(n-1)}$ and using properties of determinants, we rewrite (B21) as

$$W' = -\frac{p_{n-1}}{p_n} \begin{vmatrix} u_1 & u_2 & \cdots & u_n \\ u_1' & u_2' & \cdots & u_n' \\ \vdots & \vdots & & \vdots \\ u_1^{(n-1)} & u_2^{(n-1)} & \cdots & u_n^{(n-1)} \end{vmatrix}$$

$$- \frac{p_{n-2}}{p_n} \begin{vmatrix} u_1 & u_2 & \cdots & u_n \\ u_1' & u_2' & \cdots & u_n' \\ \vdots & \vdots & & \vdots \\ u_1^{(n-2)} & u_2^{(n-2)} & \cdots & u_n^{(n-2)} \\ u_1^{(n-2)} & u_2^{(n-2)} & \cdots & u_n^{(n-2)} \end{vmatrix}$$

$$-\cdots-\frac{p_0}{p_n}\begin{vmatrix} u_1 & u_2 & \cdots & u_n \\ u_1' & u_2' & \cdots & u_n' \\ \vdots & \vdots & \cdots & \vdots \\ u_1^{(n-2)} & u_2^{(n-2)} & \cdots & u_n^{(n-2)} \\ u_1 & u_2 & \cdots & u_n \end{vmatrix} \qquad (B22)$$

All the determinants in (B22) except the first one vanish owing to the presence of repeated rows. Moreover, comparing the first determinant in (B22) with that in (B19), we conclude that they are the same. Hence,

$$W' = -\frac{p_{n-1}}{p_n} W \qquad (B23)$$

whose solution is

$$W(x) = W(x_0) \exp\left[-\int_{x_0}^{x} \frac{p_{n-1}(\tau)}{p_n(\tau)} d\tau\right] \qquad (B24)$$

which relates $W(x)$ to $W(x_0)$.

B.2. Inhomogeneous Equations

Equation (B10) is called the associated homogeneous equation of (B4). It turns out that we can solve the inhomogeneous equation (B4) if we can find the general solution of the associated homogeneous equation and if we can also find just one particular solution of (B4). Specifically, if $u_1(x), u_2(x), \ldots, u_n(x)$ are n linearly independent solutions of the homogeneous equation (B10) and if $u_p(x)$ is a particular solution of (B4), then the general solution of (B4) is of the form

$$u = c_1 u_1 + c_2 u_2 + \cdots + c_n u_n + u_p \qquad (B25)$$

where the c_n are constants. The function

$$u_c = c_1 u_1 + c_2 u_2 + \cdots + c_n u_n \qquad (B26)$$

is called the *complementary function*. Thus, the general solution of (B4) is the sum of a particular solution and the complementary function. First, we verify that every solution of the form (B25) is a solution of (B4). Since $L(u_m) = 0$, for $m = 1, 2, \ldots, n$, and $L(u_p) = f$, we have

$$L(c_1 u_1 + c_2 u_2 + \cdots + c_m u_m + u_p) = c_1 L(u_1) + c_2 L(u_2) + \cdots + c_m L(u_m)$$
$$+ L(u_p) = f$$

verifying that (B25) is a solution of (B4). Next, we show that every solution of (B4) is of the form (B25). Let $u(x)$ be any solution of (B4), then

$$L(u) = f$$

and since
$$L(u_p) = f$$
then,
$$L(u - u_p) = L(u) - L(u_p) = f - f = 0$$

Hence, $u - u_p$ is a solution of the associated homogeneous equation $L(v) = 0$. Consequently,
$$v = u - u_p = c_1 u_1 + c_2 u_2 + \cdots + c_m u_m$$
which yields (B25).

If
$$L(u) = f^{(1)}(x) + f^{(2)}(x) + \cdots + f^{(k)}(x) \tag{B27}$$
and if
$$L[u_p^{(l)}] = f^{(l)}(x) \tag{B28}$$
then,
$$u_p = u_p^{(1)} + u_p^{(2)} + \cdots + u_p^{(k)} \tag{B29}$$
because
$$L(u_p) = L[u_p^{(1)} + u_p^{(2)} + \cdots + u_p^{(k)}] = L[u_p^{(1)}] + L[u_p^{(2)}] + \cdots + L[u_p^{(k)}]$$
$$= f^{(1)}(x) + f^{(2)}(x) + \cdots + f^{(k)}(x)$$

B.3. Solutions of Homogeneous Equations with Constant Coefficients

In this section, we determine the general solutions of homogeneous equations with constant coefficients, that is,
$$L(u) = 0 \tag{B30}$$
where
$$L = D^n + p_{n-1} D^{n-1} + p_{n-2} D^{n-2} + \cdots + p_2 D^2 + p_1 D + p_0 \tag{B31}$$
and the p_n are constant. We associate with the operator L a polynomial P, where
$$P(s) = s^n + p_{n-1} s^{n-1} + p_{n-2} s^{n-2} + \cdots + p_2 s^2 + p_1 s + p_0 \tag{B32}$$
and write
$$P(D) = D^n + p_{n-1} D^{n-1} + p_{n-2} D^{n-2} + \cdots + p_2 D^2 + p_1 D + p_0 \tag{B33}$$
We call $P(D)$ a polynomial operator and rewrite (B30) as
$$P(D)u = 0 \tag{B34}$$

Since
$$De^{sx} = se^{sx} \quad D^2 e^{sx} = s^2 e^{sx} \quad \cdots \quad D^m e^{sx} = s^m e^{sx}$$
it follows that
$$P(D)e^{sx} = P(s)e^{sx} \tag{B35}$$

The polynomial $P(s)$ is called the auxiliary polynomial associated with the operator polynomial $P(D)$. It follows from (B35) that if s_m is a root of $P(s)$, then $\exp(s_m x)$ is a solution of (B34). Hence, if $P(s)$ has the n distinct real roots $s_1, s_2, \ldots, s_n$, each of the functions
$$e^{s_1 x}, e^{s_2 x}, \ldots, e^{s_n x}$$
is a solution of (B34). Since they are linearly independent, the general solution of (B34) is
$$u = c_1 e^{s_1 x} + c_2 e^{s_2 x} + \cdots + c_n e^{s_n x} \tag{B36}$$
where the c_n are constants.

For example, we consider
$$u'' - 3u' + 2u = 0 \tag{B37}$$
which may be rewritten as
$$(D^2 - 3D + 2)u = 0$$
The associated auxiliary polynomial is
$$s^2 - 3s + 2 = (s - 2)(s - 1)$$
which has the roots $s = 1$ and 2. Hence, each of the functions $\exp(x)$ and $\exp \cdot (2x)$ is a solution of (B37). Since they are linearly independent, the general solution of (B37) is
$$u = c_1 e^x + c_2 e^{2x}$$
As a second example, we consider
$$u''' + 2u'' - u' - 2u = 0 \tag{B38}$$
whose associated auxiliary polynomial is
$$s^3 + 2s^2 - s - 2 = (s - 1)(s + 1)(s + 2)$$
It has the roots $s = 1, -1$, and -2. Hence, each of the functions
$$e^x \quad e^{-x} \quad e^{-2x}$$
is a solution of (B38). Since they are linearly independent, the general solution

$$u = c_1 e^x + c_2 e^{-x} + c_3 e^{-2x}$$

We should note that, in the general case, the roots of the associated auxiliary polynomial are neither all real nor all distinct. Next, we discuss these other cases starting with the case of complex roots.

THE CASE OF COMPLEX ROOTS

We consider a second-order equation whose associated auxiliary polynomial has the complex conjugate roots $\mu + i\omega$ and $\mu - i\omega$, where μ and ω are real. We note that if one of the roots is complex, the other root must be its complex conjugate because the auxiliary polynomial is real. Thus, each of the functions

$$e^{(\mu + i\omega)x} \quad \text{and} \quad e^{(\mu - i\omega)x}$$

is a solution of the equation. Since these solutions are linearly independent, the general solution is

$$u = c_1 e^{(\mu + i\omega)x} + c_2 e^{(\mu - i\omega)x} \tag{B39}$$

Using the identities

$$e^{i\omega x} = \cos \omega x + i \sin \omega x$$
$$e^{-i\omega x} = \cos \omega x - i \sin \omega x$$

we rewrite (B39) in real form as follows:

$$u = c_1 e^{\mu x}(\cos \omega x + i \sin \omega x) + c_2 e^{\mu x}(\cos \omega x - i \sin \omega x)$$
$$= e^{\mu x}[(c_1 + c_2) \cos \omega x + i(c_1 - c_2) \sin \omega x]$$

or

$$u = e^{\mu x}(A \cos \omega x + B \sin \omega x) \tag{B40}$$

where $A = c_1 + c_2$ and $B = i(c_1 - c_2)$ are arbitrary constants that can be considered real. Equation (B40) shows that each of the real and imaginary parts of either $\exp[(\mu + i\omega)x]$ or $\exp[(\mu - i\omega)x]$ is a solution.

Equation (B40) can be rewritten in one of the following convenient forms:

$$u = a e^{\mu x} \cos(\omega x - \beta) \tag{B41}$$

or

$$u = a e^{\mu x} \sin(\omega x + \theta) \tag{B42}$$

where

$$a = (A^2 + B^2)^{1/2} \quad \beta = \tan^{-1} \frac{B}{A} \quad \theta = \tfrac{1}{2}\pi - \beta$$

For example, we consider

$$u'' + \omega_0^2 u = 0 \tag{B43}$$

whose associated auxiliary polynomial is

$$s^2 + \omega_0^2$$

It has the roots $s = \pm i\omega_0$. Hence, each of the functions $\exp(i\omega_0 x)$ and $\exp(-i\omega_0 x)$ is a solution of (B43); equivalently, the real part $\cos \omega_0 x$ and the imaginary part $\sin \omega_0 x$ of $\exp(i\omega_0 x)$ are solutions of (B43). Hence, the general solution of (B43) can be expressed in one of the following forms:

$$u = A \cos \omega_0 x + B \sin \omega_0 x$$

or

$$u = a \cos(\omega_0 x - \beta)$$

or

$$u = a \sin(\omega_0 x + \theta)$$

As a second example, we consider

$$u'' + 2u' + 5u = 0 \tag{B44}$$

whose associated auxiliary polynomial is

$$s^2 + 2s + 5 = (s + 1 + 2i)(s + 1 - 2i)$$

Its roots are $s = -1 - 2i$ and $s = -1 + 2i$ so that each of the functions $\exp(-x - 2ix)$ and $\exp(-x + 2ix)$ is a solution of (B44). Equivalently, the real part $\exp(-x) \cos 2x$ and the imaginary part $-\exp(-x) \sin 2x$ of $\exp(-x - 2ix)$ are solutions of (B44). Hence, the general solution of (B44) can be expressed as

$$u = e^{-x}(A \cos 2x + B \sin 2x)$$

As a third example, we consider

$$u^{iv} - \omega_0^2 u = 0 \tag{B45}$$

whose associated auxiliary polynomial is

$$s^4 - \omega_0^2$$

Its roots are $s = \pm\sqrt{\omega_0}$ and $s = \pm i\sqrt{\omega_0}$. Hence, each of the functions

$$e^{\sqrt{\omega_0}\, x} \quad e^{-\sqrt{\omega_0}\, x} \quad e^{i\sqrt{\omega_0}\, x} \quad e^{-i\sqrt{\omega_0}\, x}$$

or

$$e^{\sqrt{\omega_0}\, x} \quad e^{-\sqrt{\omega_0}\, x} \quad \cos\sqrt{\omega_0}\, x \quad \sin\sqrt{\omega_0}\, x$$

is a solution of (B45). Therefore, the general solution of (B45) can be expressed as

HOMOGENEOUS EQUATIONS WITH CONSTANT COEFFICIENTS

$$u = A \cos \sqrt{\omega_0}\, x + B \sin \sqrt{\omega_0}\, x + C e^{\sqrt{\omega_0}\, x} + D e^{-\sqrt{\omega_0}\, x}$$

As a final example, we consider

$$u^{iv} + 2u''' + 6u'' + 2u' + 5u = 0 \tag{B46}$$

whose associated auxiliary polynomial is

$$s^4 + 2s^3 + 6s^2 + 2s + 5 = (s^2 + 1)(s^2 + 2s + 5)$$

Its roots are $s = \pm i$, $-1 - 2i$, and $-1 + 2i$. Hence, each of the functions

$$e^{ix} \quad e^{-ix} \quad e^{-x-2ix} \quad e^{-x+2ix}$$

or

$$\cos x \quad \sin x \quad e^{-x} \cos 2x \quad e^{-x} \sin 2x$$

is a solution of (B46). Therefore, the general solution of (B46) can be expressed as

$$u = A \cos x + B \sin x + e^{-x}(C \cos 2x + D \sin 2x)$$

THE CASE OF EQUAL ROOTS

We consider

$$u'' - 2\alpha u' + \alpha^2 u = 0 \tag{B47}$$

or

$$(D^2 - 2\alpha D + \alpha^2) u = (D - \alpha)^2 u = 0 \tag{B48}$$

The associated auxiliary polynomial is

$$(s - \alpha)^2$$

which has the repeated root α. Consequently, there is only one solution, namely $\exp(\alpha x)$, of the exponential form. Since (B47) is a second-order equation, we need to determine a second solution that is linearly independent from $\exp(\alpha x)$. To this end, we put

$$u = e^{\alpha x} v(x) \tag{B49}$$

Since

$$D(e^{\alpha x} v) = v D e^{\alpha x} + e^{\alpha x} D v = \alpha v e^{\alpha x} + e^{\alpha x} D v = e^{\alpha x}(D v + \alpha v)$$

$$= e^{\alpha x}(D + \alpha) v$$

$$D^2(e^{\alpha x} v) = D[D(e^{\alpha x} v)] = D[e^{\alpha x}(D + \alpha) v] = e^{\alpha x}(D + \alpha)^2 v$$

it follows that

$$P(D)(e^{\alpha x} v) = e^{\alpha x} P(D + \alpha) v \tag{B50}$$

Hence, substituting (B49) into (B48) yields

$$e^{\alpha x} D^2 v = 0 \quad \text{or} \quad D^2 v = 0$$

whose solution is $v = c_1 + c_2 x$. Therefore, the general solution of (B47) is

$$u = (c_1 + c_2 x) e^{\alpha x} \tag{B51}$$

In other words, $\exp(\alpha x)$ and $x \exp(\alpha x)$ are two linearly independent solutions of (B47).

As a second example, we consider

$$(D - \alpha)^3 u = 0 \tag{B52}$$

whose associated auxiliary polynomial is

$$(s - \alpha)^3$$

which has the root $s = \alpha$ with a multiplicity of three. Consequently, there is only one linearly independent solution of (B52) in the form of an exponential, namely, $\exp(\alpha x)$. To determine two other linearly independent solutions, we substitute (B49) into (B52), use (B50), and obtain

$$(D - \alpha)^3 u = (D - \alpha)^3 e^{\alpha x} v(x) = e^{\alpha x} D^3 v = 0$$

Hence,

$$D^3 v = 0$$

which has the general solution

$$v = c_1 + c_2 x + c_3 x^2$$

Consequently, the general solution of (B52) is

$$u = (c_1 + c_2 x + c_3 x^2) e^{\alpha x} \tag{B53}$$

It follows from (B53) that

$$e^{\alpha x} \quad x e^{\alpha x} \quad x^2 e^{\alpha x}$$

are linearly independent solutions of (B52).

As a third example, we consider

$$(D - \alpha)^m u = 0 \tag{B54}$$

which, upon using (B49) and (B50), becomes

$$D^m v = 0$$

whose solution is

$$v = c_1 + c_2 x + c_3 x^2 + \cdots + c_m x^{m-1}$$

Hence,

$$u = e^{\alpha x}(c_1 + c_2 x + c_3 x^2 + \cdots + c_m x^{m-1})$$

As a fourth example, we consider the following general equation with two repeated roots:

$$(D - \alpha_1)^m (D - \alpha_2)^n u = 0 \tag{B55}$$

Equation (B55) can be rewritten as

$$(D - \alpha_1)^m [(D - \alpha_2)^n u] = 0$$

Hence, (B55) is satisfied by any solution of the simpler problem

$$(D - \alpha_2)^n u = 0 \tag{B56}$$

Since the operators $(D - \alpha_1)^m$ and $(D - \alpha_2)^n$ are commutative, (B55) can be rewritten as

$$(D - \alpha_2)^n [(D - \alpha_1)^m u] = 0$$

Hence, (B55) is satisfied by any solution of the simpler problem

$$(D - \alpha_1)^m u = 0 \tag{B57}$$

The general solution of (B55) is the sum of the general solutions of (B56) and (B57) containing $(m + n)$ arbitrary constants, that is,

$$u = e^{\alpha_1 x}(c_1 + c_2 x + \cdots + c_m x^{m-1}) + e^{\alpha_2 x}(b_1 + b_2 x + \cdots + b_n x^{n-1})$$

$$\tag{B58}$$

As an application of (B58), we consider

$$(D^4 - 8D^2 + 16)u = 0 \quad \text{or} \quad (D^2 - 4)^2 u = 0 \tag{B59}$$

Its associated auxiliary polynomial $(s^2 - 4)^2$ has the roots

$$s = 2\text{(twice)} \quad s = -2\text{(twice)}$$

Thus, according to (B48), the general solution of (B59) is

$$u = (c_1 + c_2 x)e^{2x} + (b_1 + b_2 x)e^{-2x}$$

As a second application of (B58), we consider the following equation with complex roots:

$$(D^2 + \alpha_1^2)^2 (D^2 + \alpha_2^2)^3 u = 0 \tag{B60}$$

The associated auxiliary polynomial

$$(s^2 + \alpha_1^2)^2 (s^2 + \alpha_2^2)^3$$

has the roots

$$s = i\alpha_1 \text{(twice)} \quad -i\alpha_1 \text{(twice)} \quad i\alpha_2 \text{(three times)} \quad -i\alpha_2 \text{(three times)}$$

Hence, the general solution of (B60) can be written as

$$u = (a_1 + a_2 x)e^{i\alpha_1 x} + (b_1 + b_2 x)e^{-i\alpha_1 x}$$
$$+ (c_1 + c_2 x + c_3 x^2)e^{i\alpha_2 x} + (d_1 + d_2 x + d_3 x^2)e^{-i\alpha_2 x}$$

or in real form as

$$u = (A_1 + A_2 x)\cos\alpha_1 x + (B_1 + B_2 x)\sin\alpha_1 x$$
$$+ (C_1 + C_2 x + C_3 x^2)\cos\alpha_2 x + (D_1 + D_2 x + D_3 x^2)\sin\alpha_2 x$$

B.4. Particular Solutions of Inhomogeneous Equations with Constant Coefficients

We determine particular solutions of equations of the form

$$P(D)u = f(x) \qquad (B61)$$

where $f(x)$ consists of exponentials, circular functions, positive powers of x, and their products. Particular integrals corresponding to general functions $f(x)$ can be determined by using methods other than the symbolic method used here, such as the method of variation of parameters, examples of which are discussed in Chapters 4 through 11. We use the notation

$$u = \frac{1}{P(D)} f(x) \qquad (B62)$$

to denote a particular solution of (B61).

THE CASE $f(x) = \exp(\alpha x)$
Since

$$P(D)e^{\alpha x} = e^{\alpha x} P(\alpha)$$

then provided that $P(\alpha) \neq 0$, the particular integral (B62) can be rewritten as

$$u = \frac{1}{P(D)} e^{\alpha x} = \frac{e^{\alpha x}}{P(\alpha)} \qquad (B63)$$

One can easily verify that (B63) is a particular integral of (B61) because

$$P(D)u = P(D)\left[\frac{e^{\alpha x}}{P(\alpha)}\right] = \frac{e^{\alpha x} P(\alpha)}{P(\alpha)} = e^{\alpha x}$$

If $P(\alpha) = 0$, then $D - \alpha$ must be a factor of $P(D)$. We assume that α is a root of $P(\alpha)$ with multiplicity m so that $(D - \alpha)^m$ is a factor of $P(D)$. Hence, $P(D) = (D - \alpha)^m Q(D)$ where $Q(\alpha) \neq 0$. Then, using the notation (B62), we write a particular integral of (B61) as

$$u = \frac{1}{(D-\alpha)^m Q(D)} e^{\alpha x} = \frac{1}{(D-\alpha)^m} \left[\frac{1}{Q(D)} e^{\alpha x} \right]$$

Since $Q(\alpha) \neq 0$,

$$\frac{1}{Q(D)} e^{\alpha x} = \frac{e^{\alpha x}}{Q(\alpha)}$$

Hence,

$$u = \frac{1}{(D-\alpha)^m} \frac{e^{\alpha x}}{Q(\alpha)} = \frac{1}{Q(\alpha)} \cdot \frac{1}{(D-\alpha)^m} e^{\alpha x} \qquad (B64)$$

Using (B50) in (B64) with $v = 1$, we have

$$u = \frac{e^{\alpha x}}{Q(\alpha)} \cdot \frac{1}{D^m} \cdot (1) \qquad (B65)$$

where $1/D$ stands for the inverse of D, that is the operator that integrates once with respect to x, whereas $1/D^m$ integrates m times with respect to x. Thus,

$$\frac{1}{D^m} \cdot (1) = \frac{1}{D^{m-1}}(x) = \frac{x^m}{m!}$$

Hence, a particular integral of (B61) when $P(D) = (D-\alpha)^m Q(D)$, where $Q(\alpha) \neq 0$, is

$$u = \frac{e^{\alpha x}}{Q(\alpha)} \frac{x^m}{m!}$$

because

$$(D-\alpha)^m Q(D) u = (D-\alpha)^m Q(D) \left[\frac{e^{\alpha x} x^m}{Q(\alpha) m!} \right]$$

$$= Q(D) \left\{ (D-\alpha)^m \left[\frac{e^{\alpha x} x^m}{Q(\alpha) m!} \right] \right\}$$

$$= Q(D) \left\{ \frac{e^{\alpha x}}{Q(\alpha)} D^m \left[\frac{x^m}{m!} \right] \right\}$$

$$= Q(D) \left\{ \frac{e^{\alpha x}}{Q(\alpha)} \right\} = e^{\alpha x}$$

THE CASE $f(x) = \cos \alpha x$
Since

$$D^2(\cos \alpha x) = -\alpha^2 \cos \alpha x \qquad D^4(\cos \alpha x) = (-\alpha^2)^2 \cos \alpha x$$

then,

$$Q(D^2) \cos \alpha x = Q(-\alpha^2) \cos \alpha x$$

This suggests that we may determine a particular integral by replacing D^2 with $-\alpha^2$ whenever it occurs. Thus, a particular integral of

$$Q(D^2) u = \cos \alpha x \tag{B66}$$

is given by

$$u = \frac{1}{Q(D^2)} \cos \alpha x = \frac{\cos \alpha x}{Q(-\alpha^2)} \tag{B67}$$

provided that $Q(-\alpha^2) \neq 0$. This is so because

$$Q(D^2) u = Q(D^2) \left\{ \frac{\cos \alpha x}{Q(-\alpha^2)} \right\} = \frac{1}{Q(-\alpha^2)} Q(D^2) \{\cos \alpha x\}$$

$$= \frac{Q(-\alpha^2) \cos \alpha x}{Q(-\alpha^2)} = \cos \alpha x$$

An important special case of (B66) and (B67) is

$$\ddot{u} + \omega_0^2 u = \cos \omega t \tag{B68}$$

$$u = \frac{\cos \omega t}{\omega_0^2 - \omega^2} \qquad \omega \neq \omega_0 \tag{B69}$$

When $Q(-\alpha^2) = 0$, the preceding procedure needs to be modified. To this end, we note that

$$\cos \alpha x = \text{Real } (e^{i\alpha x})$$

so that (B66) can be rewritten as

$$Q(D^2) (\text{Real } u) = \text{Real } (e^{i\alpha x})$$

Hence, a particular solution of (B66) can be determined by determining a particular solution of

$$Q(D^2) u = e^{i\alpha x} \tag{B70}$$

and then taking its real part. A particular solution of (B70) can be expressed as

$$u = \frac{1}{Q(D^2)} e^{i\alpha x} \tag{B71}$$

Since $Q(-\alpha^2) = 0$,

$$Q(D) = (D + i\alpha)^m (D - i\alpha)^m F(D^2)$$

where $F(-\alpha^2) \neq 0$. Thus, (B71) can be rewritten as

INHOMOGENEOUS EQUATIONS WITH CONSTANT COEFFICIENTS 497

$$u = \frac{1}{(D+i\alpha)^m (D-i\alpha)^m F(D^2)} e^{i\alpha x}$$

$$= \frac{1}{(D-i\alpha)^m} \left\{ \frac{1}{(D+i\alpha)^m F(D^2)} e^{i\alpha x} \right\} \tag{B72}$$

where we separated the part of the operator that makes the denominator vanish. Performing the operation within the brackets in (B72), we have

$$u = \frac{1}{(D-i\alpha)^m} \left\{ \frac{e^{i\alpha x}}{(2i\alpha)^m F(-\alpha^2)} \right\} = \frac{1}{(2i\alpha)^m F(-\alpha^2)} \cdot \frac{1}{(D-i\alpha)^m} \{e^{i\alpha x}\}$$

which, upon using (B50) with $v = 1$, yields

$$u = \frac{e^{i\alpha x}}{(2i\alpha)^m F(-\alpha^2)} \cdot \frac{1}{D^m} \{1\} = \frac{x^m e^{i(\alpha x - \frac{1}{2}m\pi)}}{(2\alpha)^m F(-\alpha^2) m!} \tag{B73}$$

Taking the real part of (B73) yields the following particular solution of (B66):

$$u = \frac{x^m}{(2\alpha)^m F(-\alpha^2) m!} \cos(\alpha x - \tfrac{1}{2}m\pi) \tag{B74}$$

An important special case of (B66) and (B74) is

$$\ddot{u} + \omega_0^2 u = (D^2 + \omega_0^2)u = \cos \omega_0 t \tag{B75}$$

$$u = \frac{t}{2\omega_0} \cos(\omega_0 t - \tfrac{1}{2}\pi) = \frac{1}{2\omega_0} t \sin \omega_0 t \tag{B76}$$

The case of $f(x) = \sin \alpha x$ can be treated in a similar fashion. Thus, a particular solution of

$$Q(D^2)u = \sin \alpha x \tag{B77}$$

is given by

$$u = \frac{\sin \alpha x}{Q(-\alpha^2)} \quad Q(-\alpha^2) \neq 0 \tag{B78}$$

When $Q(-\alpha^2) = 0$, we can determine a particular solution as in the case $f(x) = \cos \alpha x$, except that we take the imaginary rather than the real part of (B73). Thus, a particular solution of

$$(D^2 + \alpha^2)^m F(D^2) u = \sin \alpha x \tag{B79}$$

is given be

$$u = \frac{x^m}{(2\alpha)^m F(-\alpha^2) m!} \sin(\alpha x - \tfrac{1}{2}m\pi) \tag{B80}$$

An important special case of (B79) and (B80) is

$$\ddot{u} + \omega_0^2 u = (D^2 + \omega_0^2)u = \sin \omega_0 t \tag{B81}$$

$$u = \frac{t}{2\omega_0} \sin(\omega_0 t - \tfrac{1}{2}\pi) = -\frac{1}{2\omega_0} t \cos \omega_0 t \tag{B82}$$

When $P(D)$ is a polynomial in D and D^2, the preceding procedures need to be modified. Instead of the general case, we consider the special case

$$(D^2 + 2D + 3)u = \cos \alpha x \tag{B83}$$

A particular solution of (B83) can be written as

$$u = \frac{1}{D^2 + 2D + 3} \cos \alpha x \tag{B84}$$

We now replace D^2 with $-\alpha^2$ in (B84) and obtain

$$u = \frac{1}{2D + 3 - \alpha^2} \cos \alpha x \tag{B85}$$

Next, we multiply both the numerator and denominator of the operator in (B85) with an operator that makes the denominator the difference between two squares; one of them involves D^2. Thus, we multiply the numerator and denominator in (B85) with the operator $2D - (3 - \alpha^2)$, that is,

$$u = \frac{2D - (3 - \alpha^2)}{(2D + 3 - \alpha^2)(2D - 3 + \alpha^2)} \cos \alpha x = \frac{2D - 3 + \alpha^2}{4D^2 - (3 - \alpha^2)^2} \cos \alpha x$$

$$= (2D - 3 + \alpha^2) \frac{1}{4D^2 - (3 - \alpha^2)^2} \cos \alpha x$$

$$= (2D - 3 + \alpha^2) \frac{\cos \alpha x}{-4\alpha^2 - (3 - \alpha^2)^2}$$

$$= \frac{2\alpha \sin \alpha x + (3 - \alpha^2) \cos \alpha x}{\alpha^4 - 2\alpha^2 + 9}$$

THE CASE $f(x) = x^m$, WHERE m IS POSITIVE

We consider

$$P(D)u = x^m \tag{B86}$$

a particular solution of which can be expressed as

$$u = \frac{1}{P(D)} x^m \tag{B87}$$

Next, we expand $1/P(D)$ in a Laurent series in D as

$$u = \frac{1}{P(D)} x^m = \frac{1}{D^k}(a_0 + a_1 D + a_2 D^2 + \cdots + a_{k+m} D^{k+m} + \cdots) x^m$$

$$= \frac{a_0}{D^k} + \frac{a_1}{D^{k-1}} + \cdots + a_{m+k} D^m + \cdots) x^m$$

$$= \frac{a_0 x^{m+k}}{(m+k)(m+k-1)\cdots(m+1)}$$

$$+ \frac{a_1 x^{m+k-1}}{(m+k-1)(m+k-2)\cdots(m+1)}$$

$$+ \cdots + m! a_{m+k}$$

As an example, we consider

$$D(D^2 + 3D + 2)u = x^2$$

Thus,

$$u = \frac{1}{D(D^2 + 3D + 2)} x^2 = \frac{1}{2D(1 + \frac{3}{2}D + \frac{1}{2}D^2)} x^2$$

$$= \frac{1}{2D}[1 - (\tfrac{3}{2}D + \tfrac{1}{2}D^2) + (\tfrac{3}{2}D + \tfrac{1}{2}D^2)^2$$

$$- (\tfrac{3}{2}D + \tfrac{1}{2}D^2)^3 + \cdots] x^2$$

$$= \frac{1}{2D}(1 - \tfrac{3}{2}D + \tfrac{7}{4}D^2 - \tfrac{15}{8}D^3 + \cdots) x^2$$

$$= \left(\frac{1}{2D} - \tfrac{3}{4} + \tfrac{7}{8}D - \tfrac{15}{16}D^2 + \cdots\right) x^2$$

$$= \tfrac{1}{6}x^3 - \tfrac{3}{4}x^2 + \tfrac{7}{4}x - \tfrac{15}{8}$$

We note that all the terms in the series beyond D^2 produce zero when they operate on x^2. Hence, they were not included in the expansion. Moreover, the constant term $-15/8$ is not needed because it can be absorbed into the homogeneous solution.

GENERAL CASE

We consider

$$(D^2 + 4\alpha^2)u = x^2 \cos 2\alpha x$$

Since $P(-\alpha^2) = 0$, we determine one of its particular solutions as the real part of a particular solution of

500 LINEAR ORDINARY-DIFFERENTIAL EQUATIONS

$$(D^2 + 4\alpha^2)u = x^2 e^{2i\alpha x}$$

Thus,

$$u = \frac{1}{D^2 + 4\alpha^2} x^2 e^{2i\alpha x} = \frac{1}{(D-2i\alpha)(D+2i\alpha)} x^2 e^{2i\alpha x}$$

$$= e^{2i\alpha x} \frac{1}{D(D+4i\alpha)} x^2$$

$$= \frac{e^{2i\alpha x}}{4i\alpha} \left(\frac{1}{D} - \frac{1}{4i\alpha} - \frac{D}{16\alpha^2} + \frac{D^2}{64i\alpha^3} + \cdots \right) x^2$$

$$= \frac{e^{2i\alpha x}}{4i\alpha} \left(\tfrac{1}{3}x^3 - \frac{x^2}{4i\alpha} - \frac{x}{8\alpha^2} + \frac{1}{32i\alpha^3} \right)$$

Hence,

$$u = \frac{1}{4\alpha} \operatorname{Real} \left[e^{2i\alpha x} \left(-\tfrac{1}{3}ix^3 + \frac{x^2}{4\alpha} + \frac{ix}{8\alpha^2} - \frac{1}{32\alpha^3} \right) \right]$$

$$= \frac{1}{16\alpha^2} \left(x^2 - \frac{1}{8\alpha^2} \right) \cos 2\alpha x + \frac{1}{12\alpha} \left(x^3 - \frac{3x}{8\alpha^2} \right) \sin 2\alpha x$$

Bibliography

Abramowitz, M. and I. A. Stegun (1964). *Handbook of Mathematical Functions.* Dover, New York.

Ahlfors, L. V. (1953). *Complex Analysis.* McGraw-Hill, New York.

Andronov, A., A. Vitt, and S. Khaikin (1966). *Theory of Oscillators.* Addison-Wesley, Reading, MA.

Arscott, F. (1964). *Periodic Differential Equations.* Pergamon, New York, p. 259.

Babich, V. M. (1970). *Mathematical Problems in Wave Propagation Theory.* Part I. Plenum, New York.

Babich, V. M. (1971). *Mathematical Problems in Wave Propagation Theory.* Part II. Plenum, New York.

Bateman, H. (1959). *Partial Differential Equations of Mathematical Physics.* Cambridge University Press, Cambridge.

Bellman, R. (1964). *Perturbation Techniques in Mathematics, Physics, and Engineering.* Holt, New York.

Bellman, R. (1953). *Stability Theory of Differential Equations.* McGraw-Hill, New York.

Bender, C. M. and S. A. Orszag (1978). *Advanced Mathematical Methods for Scientists and Engineers.* McGraw-Hill, New York.

Beyer, R. T. (1974). *Nonlinear Acoustics.* Naval Ship Systems Command, Washington, D.C.

Bhattacharya, R. N. and R. Ranga Rao (1976). *Normal Approximation and Asymptotic Expansions.* Wiley, New York.

Birkoff, G. and G. C. Rota (1962). *Ordinary Differential Equations.* Ginn and Company, Boston.

Blaquiére, A. (1966). *Nonlinear System Analysis.* Academic, New York.

Bleistein, N. and R. A. Handelsman (1975). *Asymptotic Expansions of Integrals.* Holt, Rinehart and Winston, New York.

Bloom, C. O. and N. D. Kazarinoff (1976). *Short Wave Radiation Problems in Inhomogeneous Media: Asymptotic Solutions.* Springer-Verlag, New York.

Bogoliubov, N. N. and Y. A. Mitropolsky (1961). *Asymptotic Methods in the Theory of Nonlinear Oscillations.* Gordon and Breach, New York.

Bolotin, V. V. (1963). *Nonconservative Problems of the Theory of Elastic Stability.* Pergamon, New York.

Bowman, F. (1958). *Introduction to Bessel Functions.* Dover, New York.

BIBLIOGRAPHY

Boyce, W. E. and R. C. DiPrima (1977). *Elementary Differential Equations and Boundary Value Problems.* Wiley, New York.

Brauer, F. and J. A. Nohel (1969). *The Qualitative Theory of Ordinary Differential Equations.* Benjamin, New York.

Brillouin, N. (1956). *Wave Propagation in Periodic Structures.* Dover, New York.

Brockett, R. W. (1970). *Finite Dimensional Linear Systems.* Wiley, New York.

Burgers, T. M. (1974). *The Nonlinear Diffusion Equation: Asymptotic Solutions and Statistical Problems.* Dordrecht-Holland, Amsterdam.

Butenin, N. (1965). *Elements of the Theory of Nonlinear Oscillations.* Blaisdell, New York.

Carrier, G. F., M. Krook, and C. E. Pearson (1966). *Functions of a Complex Variable.* McGraw-Hill, New York.

Carrier, G. and C. E. Pearson (1968). *Ordinary Differential Equations.* Blaisdell, Waltham, Mass.

Cesari, L. (1971). *Asymptotic Behavior and Stability Problems in Ordinary Differential Equations,* 3rd ed. Springer-Verlag, New York.

Chelomei, V. N. (1939). *The Dynamic Stability of Elements of Aircraft Structures.* Aeroflot, Moscow.

Chernov, L. A. (1960). *Wave Propagation in a Random Medium.* Dover, New York.

Coddington, E. A. and N. Levinson, (1955). *Theory of Ordinary Differential Equations.* McGraw-Hill, New York.

Cole, J. D. (1968). *Perturbation Methods in Applied Mathematics.* Blaisdell, Waltham.

Cole, R. H. (1968). *Theory of Ordinary Differential Equations.* Appleton-Century-Crofts, New York.

Copson, E. T. (1935). *An Introduction to the Theory of Functions of a Complex Variable.* Oxford University Press, Oxford.

Copson, E. T. (1965). *Asymptotic Expansions.* Cambridge University Press, Cambridge.

Courant, R. and D. Hilbert (1953). *Methods of Mathematical Physics.* Wiley, New York.

Cunningham, W. J. (1958). *Introduction to Nonlinear Analysis.* McGraw-Hill, New York.

Davis, H. (1962). *Introduction to Nonlinear Differential and Integral Equations.* Dover, New York.

De Bruijn, N. G. (1958). *Asymptotic Methods in Analysis.* North-Holland, Amsterdam.

Den Hartog, J. P. (1947). *Mechanical Vibrations.* McGraw-Hill, New York.

Dettman, J. W. (1974). *Introduction to Linear Algebra and Differential Equations.* McGraw-Hill, New York.

Dewar, M. J. S. and R. C. Dougherty (1975). *The PMO Theory of Organic Chemistry.* Plenum, New York.

Dimentberg, F. M. (1961). *The Flexural Vibration of Rotating Shafts.* Butterworth, London.

Dingle, R. B. (1973). *Asymptotic Expansions: Their Derivation and Interpretation.* Academic, New York.

Eckhaus, W. (1965). *Studies in Nonlinear Stability Theory.* Springer-Verlag, New York.

Eckhaus, W. (1973). *Matched Asymptotic Expansions and Singular Perturbations.* North-Holland, Amsterdam.

Erdelyi, A. (1956). *Asymptotic Expansions.* Dover, New York.

Erdelyi, A., W. Magnus, F. Gberhettinger, and F. G. Tricomi (1953). *Higher Transcendental Functions,* 3 Vols. McGraw-Hill, New York.

Erugin, N. P. (1966). *Linear Systems of Ordinary Differential Equations with Periodic and Quasi-Periodic Coefficients.* Academic, New York.

Evan-Iwanowski, R. M. (1969). *Resonance Oscillations in Mechanical Systems.* Elsevier, New York.

Fans, W. G. (1975). *Self-Adjoint Operators.* Springer-Verlag, New York.

Feshchenko, S. F., N. I. Shkil', and L. D. Nikolenko, (1967). *Asymptotic Methods in the Theory of Linear Differential Equations.* Elsevier, New York.

Friedrichs, K. O. (1965). *Advanced Ordinary Differential Equations.* Gordon and Breach, New York.

Fröman, N. (1965). *JWKB Approximation: Contributions to the Theory.* North-Holland, Amsterdam.

Garabedian, P. R. (1964). *Partial Differential Equations.* Wiley, New York.

Giacaghia, G. E. O. (1972). *Perturbation Methods in Nonlinear Systems.* Springer-Verlag, New York.

Gradshteyn, I. S. and I. W. Ryzhik (1965). *Tables of Integrals, Series, and Products.* Academic, New York.

Grabmüller, H. (1978). *Singular Perturbation Techniques Applied to Integro-Differential Equations.* Pitman, San Francisco.

Greenberg, M. D. (1978). *Foundations of Applied Mathematics.* Prentice-Hall, Englewood Cliffs.

Hale, J. K. (1963). *Oscillations in Nonlinear Systems.* McGraw-Hill, New York.

Hale, J. K. (1969). *Ordinary Differential Equations.* Wiley, New York.

Hagedorn, P. (1978). *Nichtlineare Schwingungen.* Akademische Verlagsgesellschft, Wiesbaden. W. Germany.

Hahn, W. (1967). *Stability of Motion.* Springer-Verlag, New York.

Hayashi, C. (1953a). *Forced Oscillations in Nonlinear Systems.* Nippon, Osaka, Japan.

Hayashi, C. (1964). *Nonlinear Oscillations in Physical Systems.* McGraw-Hill, New York.

Heading, J. (1962). *An Introduction to Phase-Integral Methods.* Methuen and Co., London.

Hildebrand, F. B. (1952). *Methods of Applied Mathematics.* Prentice-Hall, New York.

Hochstadt, H. (1964). *Differential Equations.* Holt, Rinehart and Winston, New York.

Hoffman, K. (1961). *Linear Algebra.* Prentice-Hall, Englewood Cliffs.

Ince, E. L. (1926). *Ordinary Differential Equations.* Longmans, Green, London.

Jeffreys, H. (1962). *Asymptotic Approximations.* Oxford University Press, Oxford.

Jeffreys, H. and B. S. Jeffreys (1966). *Methods of Mathematical Physics.* Cambridge University Press, Cambridge.

Kamke, E. (1947). *Differential-glechungen reeller Funktionen.* Chelsea, New York.

Kaplun, S. (1967). *Fluid Mechanics and Singular Perturbations*, P.A. Lagerstrom, L. N. Howard, and C. S. Liu, Eds. Academic, New York.

Kato, T. (1976). *Perturbation Theory for Linear Operators.* Springer-Verlag, New York.

Kauderer, H. (1958). *Nichtlineare Mechanik.* Springer-Verlag, Berlin.

Karpman, V. I. (1975). *Nonlinear Waves in Dispersive Media.* Pergamon, New York.

Kononenko, V. O. (1969). *Vibrating Systems with a Limiting Power Supply.* Iliffe, London.

Krylov, N. N. and N. N. Bogoliubov (1947). *Introduction to Nonlinear Mechanics.* Princeton University, Princeton.

BIBLIOGRAPHY

Kumar, K. (1962). *Perturbation Theory and the Nuclear Many Body Problem*. North-Holland, Amsterdam.

Landau, L. D. and E. M. Lifshitz (1965). *Quantum Mechanics, Non-Relativistic Theory*. Pergamon, London.

Lefschetz, S. (1959). *Differential Equations: Geometric Theory*. Wiley, New York.

Leibovich, S. and A. R. Seebass (1974). *Nonlinear Waves*. Cornell University Press, Ithaca.

Leipholz, H. E. (1970). *Stability Theory*. Academic, New York.

Liapunov, A. M. (1966). *Stability of Motion*. Academic, New York.

Lighthill, M. J. (1960). *Introduction to Fourier Analysis and Generalized Functions*. Cambridge University Press, Cambridge.

Lu, Y. -C. (1976). *Singularity Theory and an Introduction to Catastrophe Theory*. Springer-Verlag, New York.

McLachlan, N. W. (1947). *Theory and Application of Mathieu Functions*. Oxford University Press, New York.

McLachlan, N. W. (1950). *Ordinary Nonlinear Differential Equations in Engineering and Physical Sciences*. Clarendon Press, Oxford.

Magnus, W. and F. Oberhettinger (1947). *Formulas and Theorems for the Functions of Mathematical Physics*. Chelsea, New York.

Magnus, W. and S. Winkler (1966). *Hill's Equation*. Wiley, New York.

Malkin, I. G. (1956). *Some Problems in the Theory of Nonlinear Oscillations*. GITTL, Moscow.

Mathews, J. and R. L. Walker (1970). *Mathematical Methods of Physics*, 2nd ed., Benjamin, New York.

Meirovitch, L. (1970). *Methods of Analytical Dynamics*. McGraw-Hill, New York.

Merzbacher, E. (1970). *Quantum Mechanics*. Wiley, New York.

Miller, K. S. (1963). *Linear Differential Equations in the Real Domain*. W. W. Norton, New York.

Minorsky, N. (1947). *Non-Linear Mechanics*. J. W. Edwards, Ann Arbor, MI.

Minorsky, N. (1962). *Nonlinear Oscillations*. Van Nostrand, Princeton.

Mitropolsky, Y. A. (1965). *Problems of the Asymptotic Theory of Nonstationary Vibrations*. Daniel Davey, New York.

Morse, P. M. and H. Feshbach (1953). *Methods of Theoretical Physics*. McGraw-Hill, New York.

Murray, J. D. (1974). *Asymptotic Analysis*. Clarendon, Oxford.

Nayfeh, A. H. (1973). *Perturbation Methods*. Wiley, New York.

Nayfeh, A. H. and D. T. Mook (1979). *Nonlinear Oscillations*. Wiley, New York.

Oliver, F. W. J. (1974). *Asymptotics and Special Functions*. Academic, New York.

O'Malley, R. E. (1974). *Introduction to Singular Perturbations*. Academic, New York.

Piaggio, H. T. H. (1954). *An Elementary Treatise on Differential Equations and their Applications*. G. Bell and Sons, London.

Poincaré, H. (1892). *New Methods of Celestial Mechanics*, Vol. I–III (English transl.), NASA TTF-450, 1967.

Rabenstein, A. L. (1972). *Introduction to Ordinary Differential Equations*. Academic, New York.

BIBLIOGRAPHY

Reed, M. and B. Simon (1972). *Methods of Modern Mathematical Physics*. Academic, New York.

Rellich, F. (1969). *Perturbation Theory of Eigenvalue Problems*. Gordon and Breach, New York.

Rockland, C. (1975). *Hypoellipticity and Eigenvalue Asymptotics*. Springer-Verlag, New York.

Roseau, M. (1976). *Asymptotic Wave Theory*. North-Holland, Amsterdam.

Sagdeev, R. Z. and A. A. Galeev (1969). *Nonlinear Plasma Theory*. Benjamin, New York.

Schmidt, G. (1964). *Parametererregte Schwingungen*. VEB Deutcher Verlag der Wissenschaften, Berlin.

Scott, E. J. (1955). *Transform Calculus*. Harper and Brothers, New York.

Shtokalo, I. Z. (1961). *Linear Differential Equations with Variable Coefficients*. Gordon and Breach, New York.

Sibuya, Y. (1975). *Global Theory of a Second Order Linear Ordinary Differential Equation*. North-Holland, Amsterdam.

Siljak, D. D. (1969). *Nonlinear Systems*. Wiley, New York.

Sirovich, L. (1971). *Techniques of Asymptotic Analysis*. Springer-Verlag, New York.

Smirnov, V. I. (1964). *A Course of Higher Mathematics*, Vols. I–IV, Pergamon Press, Oxford.

Steinmann, O. (1971). *Perturbation Expansions in Axiomatric Field Theory*. Springer-Verlag, New York.

Struble, R. A. (1962). *Nonlinear Differential Equations*. McGraw-Hill, New York.

Tatarski, V. I. (1961). *Wave Propagation in a Turbulent Medium*. McGraw-Hill, New York.

Tondl, A. (1965). *Some Problems of Rotor Dynamics*. Chapman and Hall, London.

Van Dyke, M. (1975). *Perturbation Methods in Fluid Mechanics*. Parabolic Press, Stanford.

Wasow, W. A. (1965). *Asymptotic Expansions for Ordinary Differential Equations*. Wiley, New York.

Watson, G. N. (1944). *A Treatise on the Theory of Bessel Functions*. Macmillan, New York.

Whitham, G. B. (1974). *Linear and Nonlinear Waves*. Wiley, New York.

Whittaker, E. T. (1937). *Analytical Dynamics of Particles and Rigid Bodies*. 4th ed. Cambridge University Press, Cambridge.

Whittaker, E. T. (1961). *A Treatise on the Analytical Dynamics of Particles and Rigid Bodies*. Cambridge University Press, Cambridge.

Whittaker, E. T. and G. N. Watson (1962). *A Course of Modern Analysis*. Cambridge University Press, Cambridge.

Wilcox, C. H. (1964). *Asymptotic Solutions of Differential Equations and Their Applications*. Wiley, New York.

Wilcox, C. H. (1966). *Perturbation Theory and its Applications in Quantum Mechanics*. Wiley, New York.

Wylie, C. R. (1965). *Advanced Engineering Mathematics*. McGraw-Hill, New York.

Yakubovich, V. A. and V. M. Starzhinskii (1975). *Linear Differential Equations with Periodic Coefficients*. Wiley, New York.

Index

Abramowitz, M., 54, 102, 501
Accuracy of asymptotic series, 21-22
Accuracy of perturbation solution, 368, 369, 381, 383
Acoustic waves:
  in ducts carrying compressible flow, 445
  in ducts with sinusoidal walls, 388, 416-426
  in lined ducts with varying cross sections, 458-460
Adjoint:
  boundary conditions, 440-441, 449
  column vector, 448
  for acoustic equations, 446-447
  of an algebraic system of equations, 392
  of Fredholm's integral equation, 455
  of fourth-order equation, 433, 442-443
  matrix, 392
  operator, 406-412, 439
  of partial-differential equation, 459, 461
  of second-order differential equation, 403
  of second-order differential system, 402-404
  self, operator, 407
  of system of first-order equations, 448
Adjustment of frequency, 118, 197, 202, 207
Airplane, 198
Airy equation, 372, 376, 378, 384
Airy functions, 372, 375, 381
  first kind, 96-99, 105
  integral representation of, 96, 373
  relation to Bessel functions, 381, 384
  second kind, 106
Algebra, 238
Algebraic equations, 28-50
  cubic, 39-43
  higher-order, 43-45
  quadratic, 28-39
  solvability conditions for, 389-394
  transcendental, 45-50
Alternative theorem, see Fredholm's alternative theorem
Analytic function, 62, 63, 81, 90, 99, 107, 326, 329, 343
  definition of, 88
  not, 177
Annular:
  duct, 445
  plate, 388, 436
Aperiodic solutions, 242
Artificial parameter, 161
Ascent contour, 90, 92
Associated, homogeneous equation, 486, 487
  see also Auxiliary polynomial
Astronomers, 129
Astronomy, 113
Asymmetries, 415, 428
Asymptotic expansion, 22-23
  of Airy function, 96-99
  of Bessel functions, 23, 94-96, 105, 350-355, 358
  definition of, 22
  divergent, 21, 23
  elementary operations on, 24, 34
  for equations with large parameter, 362
  of integrals, 51-106
  in terms of a parameter, 28
  uniform, 24
  uniqueness of, 22
  see also Asymptotic series
Asymptotic matching principle, see Matching
Asymptotic method, 173
Asymptotic sequence, 22

507

508 INDEX

in fractional powers, 31, 35, 41
in inverse powers, 37, 38, 44
logarithms in, 22
Asymptotic series, 18-22
   accuracy of, 23-24
   vs. convergent series, 23, 24, 355
   definition of, 20, 21
   error in, 21
   failure of, 362
   see also Asymptotic expansion
Asymptotic sign, 21, 67, 346, 350
Apparent singularity, 357
Attenuation, 445
Augmented matrix, 391
Autonomous system, 196, 197, 201, 206
Auxiliary polynomial, 488-494
Averaging, method of, 108, 131, 132, 134, 148, 157, 158, 160, 191, 213-215
   for Duffing equation, 129-131, 209-213
   exercises involving, 254-256
   generalized, 146, 169-173, 175, 176
   for general nonlinear equation, 182-184, 187
   Krylov-Bogoliubov-Mitropolsky, 173-175
   for linear oscillator, 144-146
   for Mathieu equation, 253-254
   for multifrequency excitations, 226-230
   for Rayleigh equation, 155-157
   shortcomings of, 168-169
   vs. strained parameters, 243
   for systems with quadratic nonlinearities, 168-169

Bending of response curve, 202, 208
Bessel equation:
   modified, 358
   of order n, 427, 359
   of order one, 337, 358
   of order zero, 340
   in standard form 361
Bessel functions:
   asymptotic expansions of, 105, 355, 359, 382
   integral representation of, 94, 353, 358, 359
   modified, 358
   of order n, 428
   of order 1/2, 383
   of order 1/3, 381
   of order unity, 337-340, 358
   of order zero, 23, 94-96, 340-344, 350-355
   zeros of, 45-50
Bilinear, 408, 409, 440
Binomial theorem, 10, 11, 18, 29, 31, 34, 35, 36, 46, 53, 65, 109, 137, 148, 161, 164, 167, 475
Blended, *see* Matching
Bogoliubov, 501
   see also Krylov-Bogoliubov method; Krylov-Bogoliubov-Mitropolsky method technique
Bookkeeping device, 161
Boundary conditions:
   adjoint, 409-410
   disjoint, 448
   general, 401, 406-412, 450-451
   interfacial, 388, 452-454
   loss of, 259, 271
   mixed or nonseparable, 401
   transfer of, 418, 426
Boundary layer, 261
   equations, 9
   higher approximations of, 279-284
   interior, 292-296, 314, 318-320
   location, 271-277, 313-320
   nested, 304-307
   nonlinear, 307-320
   problem, 257-324
   problem, with two, s, 296-303
   by WKB method, 387
Boundary operator, 408-411
Boundary-value problem, *see* Boundary layer; Eigenvalue problem; Solvability conditions
Bounded solutions of Mathieu equation, 242
Branch:
   for outer expansion, 313
   singularity, 95
Brillouin, 364

Calculus, operational, 136
Canonical:
   boundary conditions, 315
   form, 238
   representation, 409, 440
Cauchy-Riemann equations, 89-91
Cauchy's theorem, 81, 90, 94
Celestial mechanicians, 129

Center, 114
  mass, 4, 5
Century, 113
Characteristic exponent, 240, 242, 243, 249
Circular:
  flow past, body, 468
  membrane, 426, 460
Clad rod, 452
Clockwise, 114
Closed trajectory, 114-117
Coefficients:
  method of undetermined, 136, 224
  slowly varying, 369-370
Col, 91
Cole, 313, 315
Combination, linear, 482
Combination resonance, 231, 232
  for acoustic waves in ducts, 422
  for multifrequency excitations, 219
  for parametrically excited gyroscopic systems, 399
  treated by averaging, 226-230
  treated by multiple scales, 219-226
Commutative operator, 493
Compatibility, condition, 388
  see Solvability conditions
Complementary function, 486
Complex notation, 126, 143
Composite expansion, 227-279
  for equation with constant coefficients, 284
  for equation with variable coefficients, 286, 289
  for multiple-deck problem, 307
  for nonlinear problem, 313
  for problem with two boundary layers, 303
  for turning-point problem, 292, 295, 296
Composite material, 452
Computer, 22, 29
  finite word length, 24, 355
Concomitant, 408, 440
Conditions, see Boundary conditions; Secular term; Solvability conditions
Connect, see Matching
Connection formula, 373-375
Conservative systems, 107, 134
Consistency, condition, 388

see also Solvability conditions
Consistent equations, 389, 390
Constant-coefficient equations:
  homogeneous, 487-494
  inhomogeneous, 494-500
Constant-level, 92
Constant-phase:
  contour, 94, 97
  path, 92
Continuous, piecewise, 177
Contour, 83-101, 354
  steepest descent, 92, 93, 99
  tracing of, 98
Contraction transformation, 307, 309
Convergence, 20
  vs. asymptotic, 22
  of asymptotic series, 18-22
  nonuniform, 262
  radius of, 327, 329
  see also Ratio test
Coordinates, natural, 22
Correction, 113
Coupling, nonlinear, 233, 394
  of torsional modes, 452
Cramer's rule, 224, 391
Crutching device, 161
Cumulative, 257
Curve:
  closed, 81
  integral, 114-116
  level, 91-93
  steepest, 92, 93
  see also Frequency-response
Cycle, see Limit

Damper, 2, 135
Damping, 134-191
  negative, 147
  ordering of, 205
Dashpot, 134
Debye, 93
Deck, multiple, 304-307
  triple, 304, 323
Deform, of contour of integration, 82, 93, 94, 97, 100
Degeneracy, 415, 428, 444
  removal of, 417, 432, 445
Degenerate:
  definition of, 415

fourth-order eigenvalue problem, 388, 442, 444-445
kernel, 456
least, 277, 372
second-order eigenvalue problem, 414-418
in vibration of near circular membrane, 428-431, 460-462
Denominator, 193
Dependent, linearly, 482-484
Derivative:
directional, 90
expansion method, *see* Multiple scales
first, missing, 360-361, 365
Descent, *see* Steepest descent
Determinant, Wronskian, 483-486
Detuning parameter, 35, 200, 206, 220, 222, 250, 399, 422
for acoustic waves in duct, 422
for combination resonance, 220, 222
for Mathieu equation, 250
for parametric excitation of gyroscopic systems, 399
for primary resonance, 206
for subharmonic resonance, 195
for superharmoic resonance, 200
Differentiable, 89
Differential equation:
definition of, 480
general solution of linear, 484
Dimensional analysis, 1-10
for Duffing equation, 108
for forced Duffing equation, 190
for linear oscillator, 134
for Mathieu equation, 234
for self-excited oscillator, 147
for system with quadratic nonlinearities, 160
Dirac delta function, 158
Discontinuous, 36, 261
Distinguished limit, 277, 285, 288, 290, 294, 297, 300, 310, 314, 315, 320, 323, 372
more than one, 304-307
Divisor, *see* Small
Divergence, 346, 352
illustration of, 21, 22
improvement of, 22
see also Ratio test
Domain, of validity, 265, 277

finite, 268
infinite, 307
inner, 277-279
outer, 277-279
see also Overlapping domains
Duct, heat transfer in, 381, 386, 388, 406, 458
Duffing equation, 160, 180, 184
exact solution, 113-118
forced oscillation of, 190-215
jump phenomenon for, 202-203, 208-209
phase plane for, 115
straightforward expansion for, 109-113
treated by averaging, 129-131
treated by Lindstedt-Poincaré method, 118-121
treated by multiple scales, 122-129
treated by renormalization, 121-122
variation of parameters for, 127-129
Dummy:
summation index, 330
variable of integration, 367

Earth, 1, 6
Edge layer, 257
see also Boundary layer
Eigenvalue:
of matrix, 238, 239
problems using WKB approximation, 366-369
problems with turning points, 379-386
Eigenvalue problem:
for boundary-layer stability, 435
degenerate second-order, 414-418
for problem with a regular singular point, 426-431
fourth-order, 441-447
simple second-order, 412-414
Eigenvector, of matrix, 238
Elastic waves, 418
Electrical applications, 257
Electromagnetic waves, 418
Electron, 1
Elementary function, 376
Ellipses, 5, 28
Elliptic integral:
first kind, 53, 117
second kind, 101

Energy, 114
Equations with large parameter, 360-387
  eigenvalue problems, 366-369
  eigenvalue problems with turning points, 379-383
  Langer transformation, 375-379
  Liouville-Green transformation, 364-366
  slowly varying coefficients, 369-370
  turning-point problems, 370-375
  WKB approximation, 361-364
  see also Derivative, first; Standard form
Equilibrium position, 107, 159, 175, 176
Error integral, 55
EST, 260
Euler:
  equation, 331-333, 335, 339, 340
  transformation, 22, 133
Exact solution:
  for algebraic equations, 31, 34
  for Duffing equation, 113-118
  for linear oscillator, 136, 137
  for simple boundary-value problem, 259-260
Excitation:
  external, 190, 191, 197, 202
  ordering of, 205
  parametric, 234
  see also Multifrequency excitations
Expansions, 10-12
  breakdown, 113
  error in form, 41
  nonuniform, 24, 113
  of integrands, 52-56
  overlap, 265
  pedestrian, 113
  steps in determining, 28-31
  see also Asymptotic expansion; Inner expansion; Inner inner expansion; Intermediate expansion, Outer expansion
Existence theorem, 481, 484
Exponential, form of expansion, 362

Factor, integrating, 51, 310
Factorial function, 56
Floquet:
  form, *see* Normal
  theory, 234, 236-243, 247-249
Flow:
  Compressible, 445

  laminar, 381
  past plate, 8
  past wavy wall, 468
  turbulent, 386
  see also Acoustic waves
Flow, viscous, past a plate, 8-10, 283
Fluid mechanics, 257
Forced oscillations:
  multifrequency, 216-233
  single frequency, 190-215
Fourier, series, 110, 178, 181, 183, 428, 429, 460
  transform, 51, 63
  of trigonometric functions, 475-478
Fourier integral, 88, 90
  generalized, 79
  by integration by parts, 63
  leading contribution to, 79-80
  transformation to Laplace integral, 81, 83
Fractional powers, 31
Fractions, partial, 153
Fredholm's alternative theorem:
  for integral equation, 456
  for Sturm-Liouville problem, 406
Fredholm's integral equation, 454
French, 113
Frequency:
  natural, 2, 160
  nonlinear, 108, 118, 162, 180
  of two mass centers, 5
Frequency-response equation, 198
  for primary resonance, 208-209
  for subharmonic resonance, 198-200
  for superharmonic resonance, 202-203
Frobenius form, 333-334, 341, 347
Frobenius solution, 333-344
Fundamental:
  existence theorem, 481, 483
  set of solutions, 484

Gamma function, 74, 75, 86, 87
  asymptotic expansion of, 105
  definition of, 70-72
Gauge functions, 12-18, 29
Geometric series, 19
Gradient, 90
Green:

512 INDEX

Liouville-, transformation, 361, 364-366, 376-377
 see also Identity
Gyroscopic systems, 394-400
 nonlinear, 394-397
 parametrically excited, 397-400

Heat transfer, 381, 386
Hermitian matrix, 392
Hills, 92
Homogeneous equation:
 definition of, 480
 properties of, 482-486
 solutions of, with constant coefficients, 487-494
Hour arm, 122
Hyperbola, 97

Identity:
 Green's, 408, 412, 416, 439, 441, 459, 460
 Lagrange's, 408
 trigonometric, 472-479
Illegitimate, 261
Incompatible, 424
 equations, 396
Inconsistencies, 388, 402, 424
Independence, linear, 482-484
Index, 331
Indices:
 differing by integer, 336-340
 differing by noninteger, 333-335
 equal, 333, 340-342
Indicial equation, 332, 334
Inertia force, 3
Inflection point, 74
Inhomogeneous equations, 388-471
 associated homogeneous equation, 486, 487
 definition of, 480
 properties of, 486-487
 solutions of, with constant coefficients, 494-500
Initial conditions, satisfaction of, 353
 treatment of, 111-113
Inner expansion, 266, 268, 277
 for equation with variable coefficients, 285-286, 290-291, 294
 for nonlinear problem, 310-311, 314-316
 for problem with two boundary layers, 297-302
 for simple boundary-value problem, 263, 280, 281
 for turning-point problem, 371-372
 two, s, 304-307
Inner inner expansion, 264, 266, 267
Inner product, 393
Inner variable, 268
 nonlinear, 277
 selection of, 271-276, 285, 288, 290
Instability, *see* Transition curves
Integrability:
 condition, 388
 see also Solvability conditions
Integral:
 asymptotic expansion of, 51-106, 354, 359
 curve, 114
 equation, 454-457
 of motion, 114
 powers, 31
 representation of Airy functions, 373
 representation of solutions of differential equations, 51-52, 353, 358
 see also Fourier integral; Laplace integral
Integrands, 118
 expansion of, 52-56
Integration, numerical, 315, 368, 381, 383
Integration by parts, 56-65, 80, 312
 applied to Airy function, 106, 373
 failure of, 64, 65, 68, 79, 94, 96
Interacting modes, 423, 426
Interaction, 169
 viscous-inviscid, 305
Intermediate expansion, 264, 267, 268
 for simple boundary-value problem, 282
 variable, 268
Intermolecular distance, 1
Irregular singular point, 328, 329
 at infinity, 344, 355
 solutions near, 344, 355
Iteration, 368, 369, 381, 383

Jordan form, 238
Jump phenomenon, 202, 203, 208, 209

Kernel, 456
Kramers, 364
Krylov-Bogoliubov method, 130
Krylov-Bogoliubov-Mitropolsky
　technique, 160, 169, 173-175

Lagrange, *see* Identity
Langer, 257, 284
Langer transformation, 257, 361, 371,
　375-379
Laplace integral, 81, 83, 84, 86-88, 90,
　95, 354
　generalized, 64, 79, 98
　by integration by parts, 61-63
Laplace's method, 18, 65-78
　applied to Airy function, 106, 373
　comparison of, with stationary phase, 90
Laplace transform, 51, 62, 373
Laurent series, 327, 498
Legendre polynomial, 106
Level curve, 91-93
l'Hospital's rule, 14-16, 58
Limit:
　cycle, 149-151
　definite, 277
　distinguished, 277
　not interchangeable, 262
Lindstedt-Poincaré technique, 108, 122,
　125, 131, 132, 134, 148, 160,
　162, 167, 172, 176
　applied to Duffing equation, 118-121
　applied to linear oscillator, 139-142
　applied to system with quadratic
　　nonlinearities, 164-165
　failure of, 141, 142, 152, 155
Linear, algebra, 238
　combination, 482
　equations, 480-500
　equations with large parameter, 360-
　　387
　equations with variable coefficients,
　　325-359
　independence, 482-484
　operator, 481
Linear oscillator, 134-146, 184
　exact solution, 136-139
　straightforward expansion, 135-136
　treaded by averaging, 144-146
　treated by Lindstedt-Poincaré technique,
　　139-142
　treated by multiple scales, 142-144
Lined ducts, 458
Liouville:
　-Green transformation, 361, 364-366,
　　376-377
　problem, 361
　Sutrm-, 406
Load, radial, 6, 7

Magnified scale, 257, 262
　moderately, 263
　more, 263
Main, resonance, *see* Primary
Matched asymptotic expansions,
　method of, 257, 279
　applied to equations with variable
　　coefficients, 284-296
　applied to simple boundary-value
　　problem, 270-279
　basic idea underlying, 265, 270
　comparison with multiple scales, 278,
　　279
　objective of, 267
Matching:
　basic idea, 265-268
　comparison of intermediate and
　　straightforward, 282
　intermediate, 268, 282
　multiple decks, 306, 307
　principle, 266, 267
　see also Van Dyke's matching principle
Matching of inner and outer expansions,
　equations with variable coefficients,
　　286, 288-289, 291-292, 294-295
　multiple deck problem, 306-307
　nonlinear problem, 311-312, 316-319
　problem with two boundary layers,
　　298-302
　simple boundary-value problem, 281-283
　turning-point problems, 372-375
Mathieu equation, 234-256
　straightforward expansion for, 235-236
　treated by averaging, 253-254
　treated by multiple scales, 249-253
　treated by strained parameters, 243-247
　treated by Whittaker's method, 247-249
Matrix:
　determinant of coefficient, 415, 417
　nonsingular, 408, 440
　sub, 410, 440

Membrane, 388, 406, 460
  vibration of, 426-431
Meromorphic function, 89
Minute arm, 122
Mitropolsky; see Krylov-Bogoliubov-method; Krylov-Bogoliubov-Mitropolsky technique
Mixed boundary conditions, 401
Mixed-secular term; see Secular term
Mode:
  acoustic, 420, 422
  interacting s, 423, 426
  torsional, 452
  in vibration of membrane, 428
Modeling, mathematical, 1
Modulation, of amplitude and phase, 196, 397, 400
Mook, 397, 400
Moon, 6
Mountain, 91
Multifrequency excitations, 216-233
  straightforward expansion for, 216-219
  treated by averaging, 226-230
  treated by multiple scales, 219-226
Multiple deck, 304-307
Multiple scales, method of, 108, 131, 132, 134, 146, 148, 156-158, 160, 169, 172, 173, 175, 176, 184, 191, 211, 212, 213-215, 388, 434
  application to nonlinear partial-differential equations, 279
  applied to acoustic waves, 422-426
  applied to boundary-layer problems, 257, 268-270
  applied to Duffing equation, 122-127
  applied to general nonlinear systems, 181-182, 187
  applied to linear oscillator, 142-144
  applied to Mathieu equation, 249-253
  applied to nonlinear gyroscopic system, 394-397
  applied to parametrically excited system, 397-400
  applied to primary resonances, 205-208
  applied to Rayleigh equation, 152-155
  applied to secondary resonances, 193-205
  applied to system with quadratic nonlinearities, 166-168
  comparion with matched asymptotic expansions, 278, 279
  excercises, 254-256
  vs. strained parameters, 243
  suited for boundary-layer problems, 265

Natural coordinates, 22
Navier, 279
Navier-Stokes equations, 8, 9, 279
Nayfeh, 397, 400
Newton-Raphson technique, 368, 369, 381, 383
Nonanalytic functions, 177, 185-189
Nonautonomous sytem, 196
Nondimensionalization, 1-10
Nonelementary functions, 376
Nonlinear, general weakly, systems, straightforward expansion for, 177-179
  treated by averaging, 182-184
  treated by multiple scales, 181, 182
  treated by renormalization, 179-180
Nonlinear oscillations, 269
Nonseparable boundary conditions, 401
Nonsingular transformation, 409
Nonuniform:
  convergence, 262
  expansion, 33, 108, 113
Nonuniformity 34, 120, 134
  for Duffing equation, 113, 118
  in expansion, 24
  for linear oscillator, 135-139
  manifestation of, 388
  for Rayleigh equation, 149
  region of, 26, 33
  for simple eigenvalue-problem, 258-262
Normal:
  form, 240-243, 247
  solution, 344, 348
Numerical integration, 315, 368, 369, 381, 383

Operation, illegitimate, 261
  not justified, 35
Operational calculus, 136
Operator, 481
  adjoint, 439
  commutative, 493
  fourth-order, 438
  polynomial, 487, 488
Orbit, 1, 129

Order:
  of differential equation, 480
  of magnitude, 1
  symbols, 17-18
  of term, 29, 205
Ordinary point, 326, 328, 329
  infinity, 342-343
  solutions near, 329-331
Orr-Sommerfeld equation, 435
Oscillations, rapid, 79, 80
Outer expansion, 266, 268, 277
  for equations with variable coefficients, 284-285, 287-288, 290, 293
  for nonlinear problem, 309, 313
  for problem with two boundary layers, 297
  for simple boundary-value probem, 261, 279, 280
  for triple-deck problem, 304
Outer variable, 268
  as guide for determining inner expansion, 310
Overlapping domains, 265-268, 271, 277

Parabola, 132
Parameter, 13, 127
  dimensionless, 3, 4, 6, 7, 9
  in Duffing equation, 108
  multiplies highest derivative, 257-324
  perturbation, 28
  see also Equations with large parameter; Strained parameter; Variation of parameter
Parametric excitation, 234
  of gyroscopic system, 397-400
  of two-degree-of-freedom system, 256
  see also Mathieu equation
Pass, mountain, 91
Path, steepest, 97
Pedestrian expansion, 113, 136
Period:
  of Duffing equation, 117, 118, 120, 121
  nonlinear, 118
  of two mass centers, 5
Periodic coefficients, *see* Mathieu equation; Parametric excitation
Periodic motion, 114-116, 141, 149
  for general nonlinear systems, 180

  for Mathieu equation, 240
  for primary resonance, 207
  for subharmonic resonance, 197
  for superharmonic resonance, 202
  see also Limit, cyle
Perturbation:
  parameter, 3, 4, 6, 7, 9, 10, 28
  singular, 36
  special method of, 108, 129
Perturbed equation, 28
Phase, plane:
  constant-, 90-92, 93, 97
  for Duffing equation, 114-116
  for Rayleigh equation, 149, 150
  stationary, 79, 80
  method of, 79-88
Plate:
  flow past, 8-10
  vibrations of, 6, 7, 388, 436-438
Poincaré, 20
  see also Lindstedt-Poincaré technique
Pole, 327-329
Polynomial:
  auxiliary, 488-494
  operator, 488-489
Primary resonance, 218
  definition of, 193
  treated by averaging, 212-213
  treated by multiple scales, 205-209
Product, inner, 393
Programmer, skill of, 24, 355
Propellers, 200

Quantum mechanics, 257, 379

Raphson, 368, 369, 381, 383
Rational fractions, 22
Ratio test, 10, 19, 20, 53, 55, 60, 61, 65, 66, 138, 139, 327, 331, 342, 346, 353
Rayleigh equation, 147, 150, 180, 184
  forced, 213-214
  see also Self-excited oscillators
Recurrence relations, 330, 334, 341
Reduced equation, 28, 30, 33, 109, 271
Reduced mass, 6
Redundant, 389, 390

Regular singular point, 328, 329, 406, 430
  infinity is, 343
  solutions near, 331-344
Renormalization, method of, 108, 122, 125, 131, 132, 148, 157, 158, 160, 165, 167, 172
  applied to Duffing equation, 121-122
  applied to general nonlinear systems, 179-180
  applied to Rayleigh equation, 151-152
  applied to system with quadratic nonlinearities, 162-164
  failure of, 141, 152, 155
Repeated root, 491-494
Representative point, motion of, 114, 117
Resonance, phenomena, 123, 215
  simultaneous, 222, 231, 232
  see also Combination resonance; Primary resonance; Subharmonic resonance; Superharmonic resonance
Resonant, values, 191, 215, 216
  frequencies, 193
Restoring force, 3, 107, 114, 134, 159
Reynolds number, 9
Ridges, 91
Riemann, 89-91, 93
Rigidity, of plate, 6
Roots:
  complex conjugate, 489-491
  equal, 491-494
Rudder, 200

Saddle, 91
Saddle point, 91, 92, 96, 97, 99, 100, 114
  method, *see* Steepest descent
Satellite, 1
Scales, 122, 123, 173
  combination, 262, 268
  effect of, on expansion, 262-265
  fast, 257
  magnified, 257, 262
  stretched, 257, 262
  strong dependence on, 264
  see also Stretching transformation
Scaling of dependent variable, 314
Second arm, 122
Secondary, resonance, 193-205, 209-212
  definition of, 193

  see also Combination resonance; Subharmonic resonance; Superharmonic resonance
Secular term, 113
  compounded at higher order, 136
  not leading to nonuniformity, 268
  elimination of, 126, 140, 141, 143, 144, 163-165, 175, 182, 388
  in forced Duffing equation, 193
  in solution of Duffing equation, 113, 120-122, 124, 215
  in solution of general nonlinear systems, 179
  in solution of Mathieu equation, 236
  in solution of Rayleigh equation, 149, 153
  in solution of systems with quadratic nonlinearities, 162
Self-adjoint, 414, 416, 424, 462
  algebraic system, 457
  equation with regular singular point, 429, 430
  fourth-order differential equation, 433-434, 443
  integral equation, 456
  making a second-order equation, 405
  matrix, 392
  operator, 407, 439
  partial-differential problem, 460, 461
  problem with interfacial boundary conditions, 454
  second-order differential system, 404, 410
  system of first-order equations, 449
  for vibration of plate, 436-438
Self-excited oscillators, 147-158
  straightforward expansion for, 148-150
  treated by averaging, 155-157
  treated by multiple scales, 152-155
  treated by renormalization, 151, 152
Separation of variables, 114, 116, 153, 171, 186, 284, 287, 306, 315, 360, 365, 378
  for acoustic waves in duct, 419-421, 423
  for vibration of membranes, 427
Separatrices, 114
Series, *see* Asymptotic series; Geometric series; Taylor series

Shock layer, 257, 314, 319
Siécle, 113
Similar matrices, 238
Singular, perturbation problem, 36, 262, 266
  transformation at turning point, 376
Singularity, 89, 91
  apparent, 351
  branch, 95
  classification of, 328, 329
  essential, 13
  in expansion, 24
  at infinity, 342-344
  isolated, 327
Singular point, 326
  infinity is, 343
  irregular, 328
  regular, 328
Sinusoidal:
  interface, 452
  walls, 418-426
Skin layer, 257
  see also Boundary layer
Small-divisor, 388
  for acoustic waves, 422
  conversion to secular term, 195, 424
  for forced oscillations of Duffing equation, 193
  leading to combination resonance, 218, 219, 399, 422
  leading to primary resonance, 205, 211
  leading to subharmonic resonance, 195
  leading to superharmonic resonance, 200
  for Mathieu equation, 236
  for multifrequency excitations, 216, 218, 219
  for parametrically excited gyroscopic systems, 399
Solid mechnaics, 257
Solvability conditions, 269, 388-471
  for algebraic equations, 389-394
  for differential system of equations, 445-447
  for differential systems with interfacial boundary conditions, 452-454
  example motivating, 401-402
  for fourth-order differential system, 432-438
  for general boundary conditions, 406-412
  for general differential systems of first-order equations, 447-452
  for general fourth-order differential system, 438-441
  for integral equations, 454-457
  for partial-differential equations, 458-462
  for second-order differential systems, 401-412
  sufficient, 456
Sound waves, 406
  see also Acoutic waves
Spacecraft, 6
Spaceship, 4, 5
Special method of perturbation, 108, 129
Spot, flat, 91
Spring, 2, 3, 134, 135, 159, 160
Stability:
  of boundary layer, 388, 469-470
  of nonparallel flows, 434
  see also Transition, curves
Standard form, equations in, 361
Stationary phase, method of, 79-88
  applied to Airy integral, 106, 373
  applied to Bessel function, 354
  comparison of, with Laplace's method, 90
Stationary point, 64, 78, 79, 80, 83, 84, 85, 87, 90, 114, 354
  contribution to leading term, 87
  value, 197, 202, 207
Steady-state solution, 141, 197, 198, 202, 221
Steepest, contour, 92, 98, 99
  descent, method of, 88-101, 373
  path, 92
Stegun, 54, 102
Stokes, 80, 84, 275, 279
  see also Stationary phase
Stokes lines and surfaces, 257
Stokes-Oseen flow, 10
Straightforward expansion, 134, 137
  for acoustic waves in duct, 418-422
  for circular function, 137-139
  for Duffing equation, 109-113
  for exponential function, 137, 138
  for forced Duffing equation, 191-193
  for general nonlinear systems, 177-179
  for linear oscillator, 135-136, 138

518 INDEX

for Mathieu equation, 235-236, 243
for multifrequency excitations, 216-219
for Rayleigh equation, 148-150
reason for breakdown, 118, 138, 139, 149
for simple boundary-value problem, 258
for systems with quadratic nonlinearities, 160-162
Strained parameters, method of, 234, 242, 436
  applied to degenerate second-order eigenvalue problem, 414-418
  applied to fourth-order eigenvalue problem, 442-445
  applied to Mathieu equation, 243-247
  applied to simple second-order eigenvalue problem, 412-414
  applied to vibrations of membrane, 426-431
Straining, mild, 257
Streamfunction, 434
Stretched variable, 257, 262, 271
  see also Inner variable
Stretching, transformation, 257, 262, 268, 274, 290, 294, 297, 300, 304, 305, 314, 371
  proper, 277
  see also Inner variable
Sturm-Liouville problem, 406
Subharmonic resonance, 195-200, 218, 219
  definition of, 198
Subnormal solution, 344, 348-350
Superharmonic resonance, 200-203, 211, 218
  definition of, 202
Superposition, principle of, 110, 111, 166, 178, 402, 482, 485
Switching from one expansion to another, 277
Symbolic method, 494-500
Symbols, order, 17-18
Symmetric:
  kernel, 456
  matrix, 392, 457
Symmetry, 415, 428

Table:
  3-1, 54
  4-1, 117
  4-2, 118
  14-1, 368
  14-2, 369
  14-3, 381
  14-4, 383
Taylor series, 11, 12, 52, 54, 55, 74, 75, 76, 78, 81, 84, 85, 98, 107, 121, 151, 163, 326-327, 329, 331, 343
  for cosine function, 137-139
  for exponential function, 137, 138
  to transfer boundary conditions, 418-419, 426
Tensor notation, 457
Torsional modes, 452
Trajectories, 114-116
Transcendental equation, 45-50
Transfer of boundary conditions, 418-419, 426
Transformation:
  of integral equation to algebraic equations, 457
  invert, 409
  nonsingular, 377, 409, 440, 450
  regularity of, 379
  see also Liouville; Stretching
Transforms, 51
Transient response, 141, 144, 146, 148, 157
Transition, curves, 243-254
  periodic shapes, 242, 243
  points, 257, 364, 370
  values, 240, 251
Transpose, 392, 448, 450
Trigonometric identities, 472-479
Triple deck, 304, 323
Turning-point problems, 361
  definition of, 364, 370
  eigenvalue problems, 379-386
  of order n, 377
  simple, 378-379
  treated by matched asymptotic expansions, 289-296, 371-375
  two, 381
Systems with quadratic nonlinearities, 159-176, 216, 231
  forced, 213
  straightforward expansion for, 160-162
  treated by averaging 168-169
  treated by generalized method of averaging 169-172

treated by Krylov-Bogoliubov-
Mitropolsky technique, 173-175
treated by Lindstedt-Poincaré technique,
164-165
treated by multiple scales, 166-168
treated by renormalization, 162-164
see also Multifrequency excitations

Unbounded motions, 226, 249, 251
of Mathieu equation, 240, 241
Uniformazion, 122
Unique, 480, 484
of Mathieu equation, 240, 241
Unstable motions, 203, 208, 231, 240, 243

Valleys, 91-93
van der Pol equation, 148, 157
see also Self-excited oscillators
van der Pol method, 130
van Dyke, 22, 282, 283
van Dyke's matching principle, 282-283
failure of, 283
Variation of parameters, 108, 136, 325-326, 494
for Duffing equation, 127-129
for general nonlinear equation, 182-183
for hard excitation of Duffing equations, 209-211
for inhomogeneous linear equations, 325-326
for linear oscillator, 144-145

for Mathieu equation, 253-254
for multifrequency excitations, 226-229
for particular solutions, 136
for Rayleigh equation, 155-156
for soft excitation of Duffing equation, 212-213
for systems with quadratic nonlinearities, 168
Vector solution, 325

Watch, 122, 123
Watson's lemma, 77, 95, 98, 101
statement of, 67, 68
Waveguide, 418
Wavenumber, 420, 422
Wave propagation, 452
see also Acoustic waves
Wentzel, 364
Whittaker's method, 243, 247-249, 253
exercises involving, 254, 255
limitations of, 249
Wings, 200
WKB, 257, 284, 370, 386, 387
approximation, 361-364, 372
breakdown of approximation, 364, 376-377
using, to solve eigenvalue problem, 366-369
WKBJ, method, 257
Wronskian, 236, 239, 483-486